T0249775

Middle Atmosphere Dynamics

This is Volume 40 in the
INTERNATIONAL GEOPHYSICS SERIES
A series of monographs and textbooks
Edited by RENATA DMOWSKA and JAMES R. HOLTON.

A complete list of the books in this series is available from the publisher.

Middle
Atmosphere
Dynamics

David G. Andrews
DEPARTMENT OF PHYSICS
OXFORD UNIVERSITY
OXFORD, ENGLAND

James R. Holton
DEPARTMENT OF ATMOSPHERIC SCIENCES
UNIVERSITY OF WASHINGTON
SEATTLE, WASHINGTON

Conway B. Leovy
DEPARTMENT OF ATMOSPHERIC SCIENCES
UNIVERSITY OF WASHINGTON
SEATTLE, WASHINGTON

Academic Press
San Diego New York Boston
London Sydney Tokyo Toronto

Find Us on the Web!   http://www.apnet.com

COPYRIGHT © 1987 BY ACADEMIC PRESS
ALL RIGHTS RESERVED.
NO PART OF THIS PUBLICATION MAY BE REPRODUCED OR
TRANSMITTED IN ANY FORM OR BY ANY MEANS, ELECTRONIC
OR MECHANICAL, INCLUDING PHOTOCOPY, RECORDING, OR
ANY INFORMATION STORAGE AND RETRIEVAL SYSTEM, WITHOUT
PERMISSION IN WRITING FROM THE PUBLISHER.

Academic Press
A Division of Harcourt Brace & Company
525 B Street, Suite 1900, San Diego, California 92101-4495

United Kingdom Edition published by
ACADEMIC PRESS INC. (LONDON) LTD.
24–28 Oval Road, London NW1 7DX

Library of Congress Cataloging in Publication Data

Andrews, David G.
    Middle atmosphere dynamics.

    (International geophysics series)
    Bibliography: p.
    Includes index.
    1. Middle atmosphere.   2. Dynamic meteorology.
3. Fluid dynamics.   4. Atmospheric circulation.
I. Holton, James R.   II. Leovy, Conway B.   III. Title.
IV. Series.
QC881.2.M53A53   1987   551.5'153   86-32256
ISBN 0–12–058575–8 (hardcover) (alk. paper)
ISBN 0–12–058576–6 (paperback) (alk. paper)

Contents

Chapter 4   Linear Wave Theory

Chapter 5   Extratropical Planetary-Scale Circulations

Chapter 6   Stratospheric Sudden Warmings

Chapter 7   The Extratropical Zonal-Mean Circulation

Chapter 8   Equatorial Circulations

Chapter 9   Tracer Transport in the Middle Atmosphere

Chapter 10   The Ozone Layer

Chapter 11   General Circulation Modeling

Chapter 12   Interaction between the Middle Atmosphere
and the Lower Atmosphere

Preface

During the past decade remarkable advances have been made in our understanding of the dynamics, physics, and chemistry of the middle atmosphere. Much of the observational and theoretical research that has led to this progress was stimulated by fears that human activities might adversely affect the ozone layer. The advances in our understanding of the chemistry of the stratosphere that have occurred during this period are quite well known. The parallel advances in our understanding of the dynamics of the middle atmosphere have been no less significant but are less widely appreciated. It is these that provide the primary motivation for the present book. The middle atmosphere cannot be properly understood without considering the complex interactions among dynamics, chemistry, and radiation. Because the chemical aspects of middle atmosphere science have been treated in depth elsewhere, the discussion of chemistry in the present work is limited to those aspects essential for a basic understanding of the ozone layer. A careful treatment of radiative heating and cooling processes is essential for an understanding of many of the features of the structure of the middle atmosphere. Since standard texts on atmospheric radiation devote very little space to aspects of the subject essential for middle atmosphere studies, we have included a much more complete treatment of radiative processes than is usually found in dynamics texts.

Middle Atmosphere Dynamics is intended for use in graduate courses on middle atmosphere dynamics for students with some background in dynamic meteorology or fluid dynamics. It will be useful to all research workers in meteorology, aeronomy, and atmospheric chemistry who are involved in any aspect of the study of the middle atmosphere. Furthermore, many of the basic dynamical and physical processes discussed also have broad applicability in other branches of atmospheric dynamics and will be of interest to those studying such areas as climate dynamics and planetary atmospheres.

We have not attempted to provide an exhaustive bibliography of original literature. References have been kept to a minimum within the text. Annotated lists of useful original papers, reviews, and monographs (organized by section headings) are given at the ends of the chapters. To avoid duplication, works cited within the text are not generally included in these lists. References are identified by author and year; complete citations are given in the bibliography at the end of the book.

Acknowledgments

We wish to thank the many colleagues who reviewed various portions of the manuscript and provided assistance with the diagrams. These include John Barnett, Steve Fels, David Fritts, Marvin Geller, John Gille, Rod Jones, Jeff Kiehl, Julius London, Michael McIntyre, Jerry Mahlman, Tim Palmer, Alan Plumb, Murry Salby, Mark Schoeberl, Keith Shine, and Susan Solomon. Many other colleagues have granted permission to use their published diagrams and provided us with prints. To all of them we express our appreciation.

We also wish to thank Marlene Anderson for her expert and patient typing of the manuscript.

JRH and CBL wish to acknowledge the support of the National Aeronautics and Space Administration (NASA) through its Upper Atmosphere Research Program. DGA acknowledges the support of a Royal Society Meteorological Office Research Fellowship.

Chapter 1 | Introduction

The atmosphere is conventionally divided into layers based on the vertical structure of the temperature field. These layers, the troposphere, stratosphere, mesosphere, and thermosphere, are separated by the tropopause, the stratopause, and the mesopause (Fig. 1.1). In the past, meteorologists often designated the entire region above the tropopause as the "upper atmosphere." Only fairly recently has the term "middle atmosphere" become popular in referring to the region from the tropopause (10–16 km) to the homopause (at approximately 110 km). In this part of the atmosphere eddy processes keep the constituents well mixed and ionization plays only a minor role. It is this region of the atmosphere that is the concern of this volume. The upper atmosphere will here be defined as the region above the homopause, where molecular diffusion begins to dominate over eddy mixing so that constituents become separated vertically according to their molecular masses, and increased ionization makes electromagnetic forces significant in the dynamics. This distinction between the middle and upper atmospheres is now widely accepted, although the term "upper atmosphere" still appears fairly frequently in reference to the stratosphere and mesosphere. Thus, the *Upper Atmosphere Research Satellite (UARS)* is actually designed primarily for observation of the middle atmosphere. However, there can be little doubt that the name "middle atmosphere" will eventually become the standard term for describing the layers of the atmosphere between about 10 and 100 km.

1.1 The Static Structure of the Middle Atmosphere

Atmospheric statics is the study of the relationship among the thermodynamic variables pressure, density, and temperature (p, ρ, T). Given the

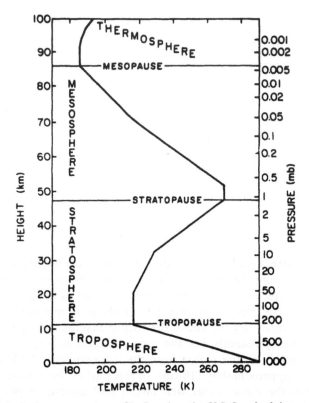

Fig. 1.1. Midlatitude temperature profile. Based on the *U.S. Standard Atmosphere* (1976).

vertical profile of temperature, the vertical profiles of pressure and density can be determined by use of the ideal gas law and the hydrostatic equation:

$$p = \rho RT, \tag{1.1.1}$$

$$\partial p / \partial z^* = -\rho g, \tag{1.1.2}$$

where R and g are the gas constant for dry air ($= 287 \text{ J K}^{-1} \text{ kg}^{-1}$) and the magnitude of the gravity acceleration, respectively, and z^* is the geometric height.

1.1.1   The Hypsometric Equation

The geopotential is defined as the work required to raise a unit mass to height z^* from mean sea level:

$$\Phi = \int_0^{z^*} g \, dz^*. \tag{1.1.3}$$

Using Eqs. (1.1.1)–(1.1.3) we can express the hydrostatic equation as

$$\partial\Phi/\partial\ln p = -RT, \qquad (1.1.4)$$

which upon integration in p (with horizontal coordinates held constant) yields the *hypsometric* equation

$$Z_2 - Z_1 = \frac{R}{g_0}\int_{p_2}^{p_1} T\, d\ln p. \qquad (1.1.5)$$

Here we have introduced the geopotential height Z defined by $Z \equiv \Phi/g_0$, where $g_0 = 9.80665$ m s^{-2} is the global average of gravity at mean sea level. As shown in Table 1.1, the difference between Z and z^* is negligible in the troposphere, but becomes increasingly significant in the upper mesosphere due to the decrease of g with height.

Table 1.1

The Relationship among Geometric Height z^*, Geopotential Height Z, Log-Pressure Height z, and Pressure p^a

z^*	Z(km)	z(km)	p(mb)
0	0.00	−0.09	1.01325 + 3
10	9.98	9.30	2.6499 + 2
20	19.94	20.27	5.5293 + 1
30	29.86	30.98	1.1970 + 1
40	39.75	40.98	2.8714 + 0
50	49.61	49.94	7.9779 − 1
60	59.44	58.97	2.1958 − 1
70	69.24	69.02	5.2209 − 2
80	79.01	80.25	1.0524 − 2
90	88.74	92.46	1.8359 − 3
100	98.45	104.68	3.2011 − 4

[a]From *U.S. Standard Atmosphere* (1976). Note that 1 mb = 100 Pa. The integers in the far-right column (preceded by plus or minus signs) indicate the power of ten by which the particular entry should be multiplied.

The geopotential height difference $\Delta Z = Z_2 - Z_1$ between the pressure levels p_2 and p_1 is referred to as the *thickness* of the layer. If a layer mean temperature is defined as

$$\langle T(p_2, p_1)\rangle \equiv \left(\int_{p_2}^{p_1} T\, d\ln p\right)\bigg/\left(\int_{p_2}^{p_1} d\ln p\right). \qquad (1.1.6)$$

then from Eq. (1.1.5)

$$Z = -[R\langle T(p, p_0)\rangle/g_0]\ln(p/p_0), \qquad (1.1.7)$$

where p_0 is the pressure at $Z = 0$. Thus, in an isothermal atmosphere of temperature $\langle T \rangle$, pressure decreases exponentially with height by a factor of $1/e$ per scale height ($H \equiv R\langle T \rangle / g_0$).

If temperature varies with height, then Eq. (1.1.7) will only be valid if $\langle T \rangle$ is computed from Eq. (1.1.6) for each value of p. For this reason it is convenient to replace geopotential height by a log-pressure vertical coordinate defined by

$$z \equiv -H \ln(p/p_s), \tag{1.1.8}$$

where p_s is a standard reference pressure (usually taken as 1000 mb or 10^5 Pa)[1] and H is a mean scale height ($\equiv RT_s/g_0$, where T_s is a constant reference temperature). In middle atmosphere studies it is common to let $H = 7$ km, corresponding to $T_s \approx 240$ K. In that case, as shown in Table 1.1, for the U.S. Standard Atmosphere the difference between the geometric height and the log-pressure vertical coordinate defined in Eq. (1.1.8) is quite similar in magnitude to the difference between z^* and Z throughout the middle atmosphere, although for extreme temperature profiles (e.g., the polar night) the difference between z^* and Z can be substantially greater than shown in the table. Note also from Table 1.1 that z and Z both increase monotonically with z^*; this follows generally from their definitions and the hydrostatic relation of Eq. (1.1.2). In theoretical developments throughout this book we shall nearly always use z as the vertical coordinate and generally shall refer to z simply as "height"; we shall also use the symbol g in place of g_0, for convenience. Only for detailed comparisons with observations should it be necessary to distinguish among z^*, Z, and z, or to make allowance for the variation of g with height.

1.1.2 The Vertical Temperature Profile

A standard model for the mean midlatitude temperature profile is shown in Fig. 1.1. The mean vertical distribution of T in the middle atmosphere can be approximately explained in terms of absorption and emission of radiation. Infrared emission by water vapor and clouds is primarily responsible for the temperature minimum at the tropopause. The temperature peak at the stratopause is due to the absorption of solar ultraviolet radiation by ozone, and the minimum at the mesopause is primarily due to the large decrease in ozone concentration at that level, which greatly reduces the absorption of solar radiation.

---

[1] Throughout most of this book we shall use the millibar as the unit of pressure, in accordance with standard meteorological practice: 1 mb = 100 Pa.

1.1.3 *Potential Temperature*

The potential temperature θ is by definition the temperature that a parcel of dry air at pressure p and temperature T would acquire if it were expanded or compressed adiabatically to the reference pressure $p_s = 1000$ mb:

$$\theta \equiv T(p_s/p)^\kappa \qquad (1.1.9a)$$

where $\kappa \equiv R/c_p \approx 2/7$ and c_p is the specific heat at constant pressure. Thus potential temperature, unlike temperature, is conserved for adiabatic flow. Because of this conservation property, θ is for some purposes superior to T as a field variable for characterizing the thermodynamic state of the atmosphere. Note that Eqs. (1.1.8) and (1.1.9a) imply

$$\theta = T \exp(\kappa z/H). \qquad (1.1.9b)$$

From Eqs. (1.1.1), (1.1.2), and (1.1.9) it is easily verified that

$$T \, \partial \ln \theta/\partial z^* = g/c_p + \partial T/\partial z^*, \qquad (1.1.10)$$

so that when the actual temperature *lapse rate*, $-\partial T/\partial z^*$, is less than the *adiabatic lapse rate*, g/c_p, potential temperature increases with height. The atmosphere is then said to be stably stratified or *statically stable*. The fact that potential temperature is a monotonically increasing function of height in a statically stable atmosphere means that θ can be used as an independent vertical coordinate. This "isentropic" coordinate system is discussed in Section 3.8.

1.1.4 *Static Stability and the Buoyancy Frequency*

The temperature profile of Fig. 1.1 indicates that the static stability in the stratosphere should be much greater than that of either the troposphere or the mesosphere. The most convenient measure of stability for dynamical studies is the square of the *buoyancy frequency*, which is the frequency of adiabatic oscillation for a fluid parcel displaced vertically from its equilibrium level in a stably stratified atmosphere (see, e.g., Holton, 1979, p. 50).

The buoyancy frequency squared, N_*^2, can be expressed in terms of potential temperature as

$$N_*^2 \equiv g \, \partial \ln \theta/\partial z^*. \qquad (1.1.11)$$

In log-pressure coordinates a slightly modified "buoyancy frequency," defined as

$$N \equiv N_*(T/T_s), \qquad (1.1.12)$$

proves to be the natural measure of stability (Gill, 1982, p. 184). From Eq. (1.1.8) with the aid of Eqs. (1.1.1) and (1.1.2) we find that $dz = (T_s/T) \, dz^*$, so that

$$N^2 = g\left(\frac{T}{T_s}\right)\frac{\partial \ln \theta}{\partial z} = \frac{R}{H}\left[\frac{\partial T}{\partial z} + \frac{\kappa T}{H}\right]; \qquad (1.1.13)$$

when $N^2 > 0$ the atmosphere is statically stable. In the stratosphere $N^2 \approx 5 \times 10^{-4}\,\mathrm{s}^{-2}$, while in the mesosphere $N^2 \approx 3 \times 10^{-4}\,\mathrm{s}^{-2}$. For analytic modeling it is often assumed that N^2 is constant throughout the middle atmosphere. However, it is important to realize that N^2 in the real atmosphere varies not only in the vertical but also with latitude, longitude, and season.

1.2   Zonal Mean Temperature and Wind Distributions

In the absence of eddy motions (i.e., departures from zonal symmetry) the middle atmosphere would be close to radiative equilibrium at all latitudes with a solstice temperature distribution similar to that shown in Fig. 1.2. Although the globally averaged temperature field at each altitude in the stratosphere and mesosphere is in approximate radiative equilibrium, eddy motions induce substantial local departures from equilibrium, especially in the winter stratosphere and near the mesopause in both winter and summer. The overall latitudinally dependent temperature distribution in the middle atmosphere arises from a balance between the net radiative drive (i.e., the sum of the solar heating and infrared heating or cooling) and the heat transport plus local temperature change (often called the dynamical heating or cooling) produced by these motions.

The net radiative heating distribution has a strong seasonal dependence, with maximum heating at the summer pole and maximum cooling at the winter pole. At the equinoxes the maximum heating is at the equator and there is cooling at both poles. The circulation in the meridional plane that dynamically balances this differential heating is often called the *diabatic circulation*. However, as discussed in Chapter 7, this circulation is primarily driven by eddy forcing, not by radiative heating directly. At the solstices the diabatic circulation consists of rising motion near the summer pole, a meridional drift into the winter hemisphere, and sinking near the winter pole. The Coriolis torque exerted by this meridional drift tends to generate mean zonal westerlies in the winter hemisphere and easterlies in the summer hemisphere that are in approximate geostrophic balance with the meridional pressure gradient. At the equinoxes the differential radiative drive is associated with a fairly weak diabatic circulation, with rising in the equatorial region and a poleward meridional drift in both the spring and autumn

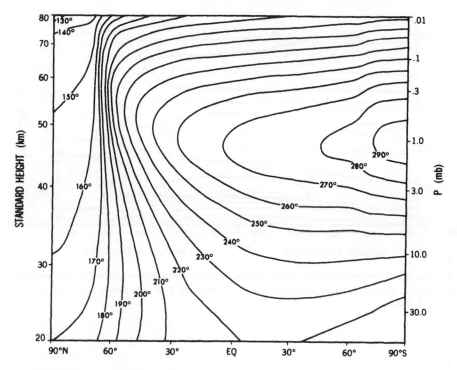

Fig. 1.2. Time-dependent "radiatively determined" temperature T_r for January 15 from a radiative–convective–photochemical model that is time-marched through an annual cycle. The surface temperatures are prescribed at their seasonally varying observed zonal mean values, and the tropospheric lapse rate is specified at 6.5 K km^{-1}. Cloudiness and the ozone below 25 km are prescribed at annual mean values; ozone above 25 km is computed by a detailed photochemical model. [From Fels (1985).]

hemispheres. The Coriolis torque thus generates weak mean zonal westerlies in both hemispheres.

Schematic cross sections of the longitudinally averaged solstice mean temperature and zonal wind fields for the middle atmosphere are shown in Figs. 1.3 and 1.4, respectively. The zonal mean wind is in approximate thermal wind balance with the temperature field, so that the vertical wind shear is proportional to the meridional temperature gradient, as can be qualitatively confirmed from the figures. Although the rather uniform increase of temperature from the winter pole to the summer pole in the region of 30–60 km is consistent with the radiative equilibrium distribution of Fig. 1.2, there are a number of features of the climatological temperature distribution that are not even qualitatively in accord with the distribution of radiative sources and sinks. These include the temperature increase in

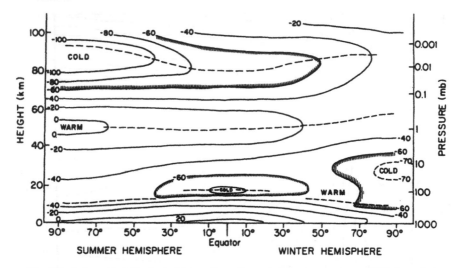

Fig. 1.3. Schematic latitude–height section of zonal mean temperatures (°C) for solstice conditions. Dashed lines indicate tropopause, stratopause, and mesopause levels. (Courtesy of R. J. Reed.)

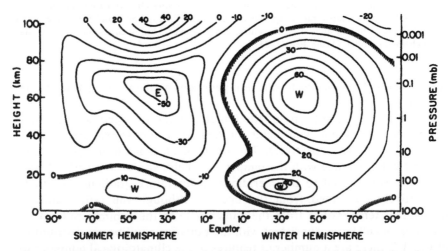

Fig. 1.4. Schematic latitude–height section of zonal mean zonal wind (m s^{-1}) for solstice conditions; W and E designate centers of westerly (from the west) and easterly (from the east) winds, respectively. (Courtesy of R. J. Reed.)

the lower stratosphere between the cold tropical tropopause and midlatitudes in the winter hemisphere, and the reversed temperature gradient above 60 km where temperatures increase uniformly from the summer pole to the winter pole. The summer polar mesopause is much colder than radiative equilibrium, while the winter polar temperatures are above radiative equilibrium throughout the entire middle atmosphere. Thus, dynamical processes must play an essential role in establishing the observed temperature and zonal wind distributions.

The cross sections of Figs. 1.3 and 1.4 should not be regarded as definitive climatologies. In reality there are substantial differences between the Northern and Southern Hemisphere solstice circulations (see Figs. 5.1 and 5.2). There is also a remarkable interannual variability in the middle atmosphere, especially in the winter hemisphere (see Chapter 12), so that many years of data are required to determine stable climatologies of the mean zonal wind and temperature distributions. Only in the past decade have adequate observations been available for the stratosphere, and for the mesosphere the data base is still unsatisfactory. However, the general features of the overall solstice circulation should be roughly as depicted in Figs. 1.3 and 1.4.

1.3 Composition of the Middle Atmosphere

In the lower and middle atmospheres, mixing by fluid motions on all scales tends to produce uniform mixing ratios for all gaseous constituents of the atmosphere. Only constituents with significant sources or sinks have spatially and temporally varying mixing ratios. For most such species the vertical variability is much greater than horizontal and temporal variability, so only vertical profiles are considered here. Horizontal and temporal variability are intimately related to dynamical transport processes. These will be discussed in Chapter 9.

The primary constituents in the lower and middle atmosphere are diatomic nitrogen and oxygen, which together account for 98.65% of the total mass of the dry atmosphere. The noble gas argon accounts for another 1.28%, so that the myriad of other species (often known as trace species) that have been detected in the atmosphere together account for less than 0.1% of the total mass. The primary constituents have no significant sources or sinks in the stratosphere or mesosphere, so that their mass fractions are nearly constant in height. In the thermosphere above 90 km the increasing molecular mean free paths lead to a gradual change from dominance of mixing by macroscopic fluid motions below 100 km to control by molecular diffusion above about 120 km. At altitudes where molecular diffusion dominates, each species (in the absence of sources or sinks) has an exponential decay of density with height with a scale height determined by its molecular mass. Thus the less massive species increasingly dominate with height. The region of the atmosphere where eddy mixing dominates is referred to as the *homosphere*, while the molecular diffusion region is the *heterosphere*. These are separated by the *homopause*, near 110 km, which is often considered to be the level where the two processes are of equal importance.

1.3.1 Measures of Trace Constituent Concentration

Several different variables are used to describe the concentration of trace gases in the atmosphere. For example, the absolute concentration can be expressed in terms of the number of molecules per unit volume, n_T; the mass per unit volume, ρ_T; or the partial pressure, p_T. Note that

$$n_T M_T = \rho_T N_a, \quad \text{and} \quad n_A M_A = \rho_A N_a, \tag{1.3.1}$$

where N_a is Avogadro's number ($= 6.022 \times 10^{26}$ molecules kmol^{-1}), M stands for the molecular weight ($\mathrm{kg\,kmol}^{-1}$), and the subscripts T and A stand for the trace constituent and "air" excluding the trace constituent, respectively. Since trace constituents in the middle atmosphere are present in such small quantities,

$$\rho_T \ll \rho_A \approx \rho, \quad \text{and} \quad p_T \ll p_A \approx p.$$

The partial pressure is related to the other concentration measures through the ideal gas law:

$$p_T = \rho_T R_* T / M_T, \quad p = \rho R_* T / M_A, \tag{1.3.2a,b}$$

where R_* is the universal gas constant ($= 8.314 \times 10^3 \mathrm{\,J\,K^{-1}\,kmol^{-1}}$) and where we have replaced p_A by p and ρ_A by ρ in Eq. (1.3.2b) with negligible error.

The number density, n_T, and the partial pressure, p_T, are often used for presentation of observational data. However, due to the compressibility of the atmosphere these are not conserved following the motion. For dynamical studies, *mixing ratio* is the preferred measure of concentration, since it is conserved following the motion in the absence of sources or sinks. Both *mass* mixing ratio (m) and *volume* mixing ratio (v; sometimes called the "mole fraction") are in common usage, although observational data are nearly always reported as volume mixing ratios. These are related to partial pressure as follows:

$$m = \rho_T / \rho = (p_T / p)(M_T / M_A), \tag{1.3.3a}$$

$$v = n_T / n_A = m(M_A / M_T) = p_T / p. \tag{1.3.3b}$$

Thus, $m \propto v = p_T/p$. Mass mixing ratios are usually used in thermodynamic and radiative transfer computations. Volume mixing ratios (or number densities) are essential in photochemical calculations involving the partitioning of molecules within groups. (For example, reactions involving conversions between O and O_3 conserve v but not m for the sum $O + O_3$.)

1.3.2 Major Trace Species

The radiatively active trace species water vapor, carbon dioxide, and ozone are the species of major importance for middle atmosphere dynamics. For convenience we refer to these collectively as "major" trace species. Of these three species only CO_2 is well mixed in most of the middle atmosphere, and even it has a detectable few percent decrease in fractional concentration with height due to the time lag in upward mixing of the CO_2 currently being produced by fossil fuel burning at the surface. In contrast, H_2O and O_3 are highly variable in space and time.

The variability of water vapor in the lower atmosphere is due entirely to the processes of evaporation, condensation, and sublimation that occur as part of the hydrological cycle. Although water vapor in the troposphere may have mixing ratios as high as 0.03 by volume, the stratosphere is observed to be extremely arid with mixing ratios in the range of 2-6 ppmv (parts per million by volume). The dryness of the stratosphere can be qualitatively accounted for if it is assumed that all air entering from the troposphere into the stratosphere passes through the extremely cold tropical tropopause (Fig. 1.3), where most of the remaining water content is frozen out. This "freeze-dry" model for the aridity of the stratosphere places important constraints on the nature of the mass exchange between the lower and middle atmosphere, as will be further discussed in Chapter 9. The extremely low water concentrations in the middle atmosphere make accurate measurement of the water vapor distribution very difficult. However, there is general agreement that the minimum concentration occurs in the lower stratosphere, with a gradual increase in the mixing ratio occurring above about 20 km due to the source provided by oxidation of methane. This type of profile is clearly shown in the water vapor measurements by the Limb Infrared Monitor of the Stratosphere (LIMS) satellite experiment (Fig. 1.5). Throughout the middle atmosphere the concentration of water vapor is too small for it to play much direct role in the local radiative heating or cooling. Radiatively, water vapor is important for the middle atmosphere primarily because the infrared emission by water vapor together with the vertical heat flux associated with convection are crucial for establishing the temperature

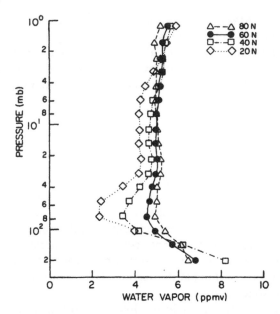

Fig. 1.5. Vertical profiles of water vapor mixing ratio at several latitudes measured by the LIMS instrument on the *Nimbus* 7 satellite for May 1–26, 1979. [From Remsberg *et al.* (1984b). American Meteorological Society.]

structure of the troposphere, and hence the temperature at the lower boundary of the middle atmosphere.

Ozone is, of course, the most significant trace species in the middle atmosphere. The absorption of solar ultraviolet insolation by ozone is the major radiative heat input for the middle atmosphere. By depleting the solar ultraviolet flux this absorption protects the biosphere from the damaging effects of ultraviolet radiation. The budget of atmospheric ozone, which involves very complex photochemical cycles depending on many trace species, both natural and manmade, is thus a major environmental concern. The observed spatial and temporal variability of many of the species involved in ozone photochemistry implies that transport and mixing by atmospheric motions are crucial aspects of the ozone problem. To a large extent it is the threat posed to the ozone layer by anthropogenic perturbations in atmospheric composition that has accounted for the rapid pace of middle atmospheric research in the past decade. The complex coupling among photochemical, radiative, and dynamical processes that ultimately controls the distribution of ozone still provides a major challenge for middle atmosphere research that will be discussed in Chapter 10.

The vertical profile of ozone for a midlatitude standard atmosphere is plotted in Figs. 1.6 and 1.7 in units of molecular concentration and mass

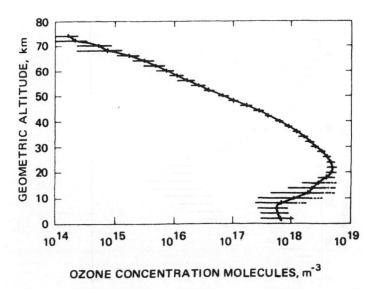

Fig. 1.6. Midlatitude standard ozone concentration profile (molecules m^{-3}). Horizontal bars show the standard deviation about the mean for observed profiles. [From the *U.S. Standard Atmosphere* (1976).]

mixing ratio, respectively. As Fig. 1.6 indicates, most of the mass of the ozone in the atmosphere is contained in the lower stratosphere with a maximum concentration near 22 km. However, the mixing ratio is a maximum at a much higher altitude ($\sim$37 km). Since most of the production of ozone molecules occurs above 30 km, the large molecular concentration below 30 km must result from downward transport by atmospheric motions. Thus, the observed ozone profile itself provides evidence of dynamical-chemical coupling. Despite its central role in the meteorology of the middle atmosphere, ozone is indeed a *trace* constituent. At the peak of the ozone layer the mixing ratio is only about 8–10 ppmv. If the entire column of ozone (i.e., the mass per unit area in a column of air extending from the surface to the top of the atmosphere) were brought to standard temperature and pressure (0°C, 1013.25 mb), the thickness of the column would only be about 3 mm!

1.3.3   Minor Trace Species

In addition to the three major radiatively active trace species, there are a large number of species present in sufficient concentrations to play significant roles in the chemistry of the middle atmosphere. We will refer

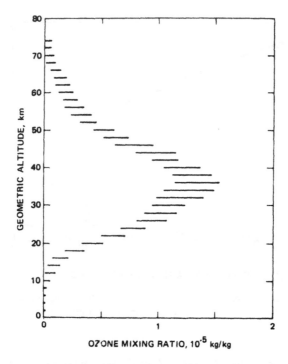

Fig. 1.7. The standard ozone profile of Fig. 1.6 plotted in terms of the mass mixing ratio. [From the *U.S. Standard Atmosphere* (1976).]

to these collectively as "minor" trace species. Of special significance for meteorological studies are the so-called "long-lived" trace species such as nitrous oxide (N_2O), methane (CH_4), and the chlorofluoromethanes (CF_2Cl_2 and $CFCl_3$). These species all have sources at the ground that are primarily natural for N_2O and CH_4 and entirely manmade for $CFCl_3$ and CF_2Cl_2. They are well mixed in the vertical in the troposphere and are destroyed in the stratosphere by oxidation and/or photodissociation. Thus, they all have vertical profiles in which mixing ratios decay with altitude in the middle atmosphere. Typical examples of midlatitude profiles for these tracers are shown in Fig. 1.8. The vertical gradient of mixing ratio varies enormously from the species with the slowest destruction rate in the stratosphere (CH_4) to the species with the most rapid destruction rate ($CFCl_3$). Such vertically stratified species provide excellent tracers for calibrating theoretical models of mass transport and diffusion in the middle atmosphere. Their observed meridional and temporal variability provide important clues as to the nature of transport and mixing by motion systems, as will be discussed in Chapter 9.

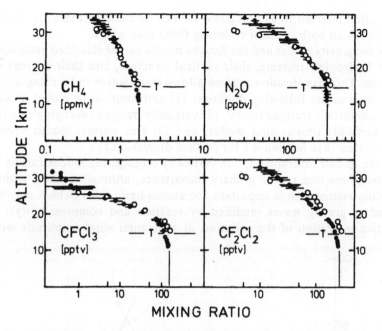

Fig. 1.8. Vertical volume mixing ratio profiles for several minor trace gases that have tropospheric sources and are photochemically destroyed in the stratosphere. All profiles are from balloon flights over Southern France (44°N). Full dots give average profiles for several summer flights. Open dots refer to a flight on October 21, 1982. [From Schmidt *et al.* (1984).]

1.4   The Vertical Distribution of Eddy Amplitudes

The observed longitudinally averaged zonal wind and temperature distributions in the middle atmosphere (Figs. 1.3 and 1.4) are maintained by the competing effects of differential radiative heating and the mechanical and thermal forcing by motions of all scales. Motions in the middle atmosphere occur on scales ranging from global-scale tides to microscale turbulent patches. Although turbulent mixing is thought to play a significant role in the momentum budget of the middle atmosphere, much of the momentum, heat, and tracer transport in the middle atmosphere is due to coherent wave motions of various classes. Discussion of such waves and their interactions with the mean flow will be a major theme of this book.

Wave motions in the atmosphere result from the competition between inertia and restoring forces acting on fluid parcels displaced from their equilibrium latitudes and/or elevations. For the waves of concern here, the restoring force is supplied either by gravity or by the poleward gradient of the planetary vorticity. The former is responsible for internal gravity waves (sometimes called buoyancy waves); the latter is responsible for Rossby waves (often referred to as planetary waves). In addition there are mixed modes in which both types of restoring force play a role.

Both the gravity modes and the Rossby modes can be classified according to their horizontal structure, their vertical structure, and their sources of excitation. This classification scheme allows these waves to be categorized on the basis of the following dualities: (1) extratropical or global modes versus equatorially trapped modes, (2) vertically trapped (decaying) modes versus vertically propagating modes, and (3) free normal modes versus forced modes. (See Section 4.1 for further discussion.)

For the middle atmosphere, it is vertically propagating modes forced in the troposphere that are of primary importance, although forcing within the middle atmosphere is important for atmospheric tides (which may be regarded as gravity waves modified by rotation and compressibility). A qualitative indication of the variation of horizontal wind amplitude with

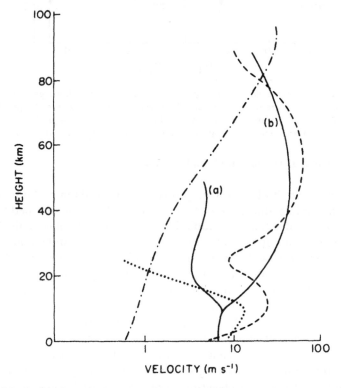

Fig. 1.9. Schematic vertical profiles of horizontal wind amplitudes corresponding to various types of atmospheric motions. Solid line: planetary waves (a) summer, (b) winter; dashed: zonal mean; dotted: synoptic scale; dotted–dashed: gravity waves. (Courtesy of Professor T. Matsuno.)

height for various types of motion is given in Fig. 1.9. Observations and theory (Section 4.5 and Chapter 5) indicate that the vertical structure of forced Rossby modes is critically dependent on their horizontal scale and on the mean zonal wind distribution. Synoptic-scale Rossby waves are strongly trapped in the troposphere and decay rapidly with height in the middle atmosphere, as indicated in Fig. 1.9. Planetary-scale Rossby waves, on the other hand, propagate vertically provided that their phase speeds are westward (but not too strongly westward) relative to the mean flow. Since the largest-amplitude planetary waves are stationary with respect to the ground, only in the winter hemisphere where the mean winds are westerly throughout the middle atmosphere are planetary wave amplitudes comparable to the magnitude of the mean zonal wind. In the summer hemisphere such waves cannot propagate past the level where the mean zonal flow vanishes, and hence are unable to penetrate beyond the lower stratosphere.

Nontidal gravity waves generated in the troposphere appear to have a rather broad two-dimensional spectrum of phase speeds that includes meridionally propagating modes as well as zonally propagating modes that are both westerly and easterly relative to the mean flow. Thus, during all seasons at least some gravity waves can propagate through the middle atmosphere without encountering critical levels at which the mean wind equals the phase speed. Observations indicate that in extratropical latitudes the vertical flux of gravity-wave activity is greater in the winter than in the summer. This seasonal dependence is probably due primarily to variations in the strength of tropospheric sources. In addition, there is an important seasonal dependence of the phase speed spectrum due to selective transmission caused by the presence of mean wind critical levels. During the summer (winter), waves that are easterly (westerly) relative to the mean flow are selectively filtered out so that there is a net upward transfer of westerly (easterly) momentum (see Sections 4.6.2 and 7.3).

Gravity-wave modes with sufficiently high Doppler-shifted frequencies will tend to propagate through the stratosphere without significant damping, so that energy density remains approximately constant in height and the horizontal velocity amplitude increases as the inverse square root of density, $\rho_0^{-1/2}$. Thus, although gravity waves (and atmospheric tides) contribute only a small part of the eddy horizontal wind variance in the troposphere and stratosphere, in the upper mesosphere they often appear to be the dominant modes. At some level in the mesosphere or lower thermosphere the amplitudes may become so great that such waves break down due to convective overturning or shear instability. Above 100 km, molecular diffusion becomes increasingly important in limiting wave amplitudes.

The vertical distribution of turbulence is very poorly known. However, since most turbulence in the middle atmosphere appears to result from gravity-wave breakdown, it is likely that the vertical profile of the wind variance due to turbulent motions is similar to that due to gravity waves.

1.5 Observational Techniques

Although observations of the stratosphere date from as early as 1902, when the French scientist Teisserenc de Bort discovered the stratosphere by means of balloon temperature soundings (see Hartmann, 1985, for a discussion), systematic observation of the middle atmosphere dates only from the International Geophysical Year (IGY) of 1957–1958. Knowledge concerning the temperatures, winds, and composition of the middle atmosphere has increased dramatically in the years following the IGY.

During the 1960s the operational meteorological radiosonde network provided sufficiently frequent soundings of the lower stratosphere up to about the 10-mb level so that the climatology of the lower stratosphere gradually was established at least in the Northern Hemisphere. In the same period a meteorological rocket network was established that provided wind and temperature measurements in the upper stratosphere and lower mesosphere. A network of balloon-borne ozonesondes was also operated for a relatively brief time. These *in situ* measurement techniques were only deployed at a limited number of locations (primarily the North American region) and could not provide global climatologies.

Observation of the global stratosphere really only began in 1969 with the launching of the *Nimbus* 3 satellite. This research satellite and subsequent research and operational satellites have provided more than a decade of global remote sounding of stratospheric temperatures and total ozone as well as limited measurements of profiles of ozone and a few other trace constituents.

Satellite temperature soundings generally are based on measurements of the infrared emission in the 15-μm band of carbon dioxide. Such measurements are done on an operational basis using nadir-viewing instruments. However, the vertical resolution of these devices is quite low, since they measure characteristic temperatures in layers 10–15 km deep (see Fig. 2.36). Limb-sounding radiometers [such as the LIMS and Stratospheric and Mesospheric Sounder (SAMS) instruments on *Nimbus* 7) are able to provide much better vertical resolution (approximately 3–5 km) in the middle atmosphere, since most of the signal is returned from the altitude at which the instrument scans the limb of the atmosphere (see Fig. 2.38).

Infrared emission instruments are not easily used for detecting minor trace species because of the low ratio of signal to noise. An alternative technique useful in some cases is the solar occultation method, in which the absorption of gas (or aerosol extinction) is measured by viewing the sun on the limb near sunrise or sunset (see Fig. 2.41). Good vertical resolution and high signal-to-noise ratios are possible, but the technique obviously can provide only limited latitudinal and diurnal coverage. Ozone has been measured using solar and stellar occultation, but the primary satellite observations of ozone are based either on infrared emission in the 9.6-μm band or on the backscatter of ultraviolet sunlight. The latter method is, of course, limited to sunlit regions.

Satellite remote observations to date have provided only temperature and constituent measurements. Horizontal winds have been deduced from temperature fields, using the thermal wind relationship to build upward from operational height analyses for some base level (e.g., 100 mb). The accuracy of such geostrophic winds is thus dependent on the base-level analyses, which may have substantial errors in data sparse regions.

Direct velocity measurements in the middle atmosphere have been made *in situ* by rocket soundings and remotely from the ground using several radar methods, operating in a variety of frequency ranges. These include the partial reflection drift method, meteor radars, and so-called MST (mesosphere–stratosphere–troposphere) radars. In the partial reflection drift technique, a triangular array of receivers is used to determine the drift velocities of ionized irregularities that partially reflect the radar signal. Only bulk horizontal motions can be sensed with this technique, and measurements are limited to the middle and upper mesosphere and lower thermosphere. Meteor radars are Doppler radars that measure line of sight velocities using returns from ionized meteor trials in the mesopause region. The MST technique utilizes very-high-frequency (VHF) radars (wavelengths of the order of a few meters) in Doppler mode to determine the drift velocities of back-scattering elements whose nature depends on the region being scanned. The echoes received by such radars from the troposphere and lower stratosphere are caused by refractive-index variations due to density fluctuations associated with neutral atmosphere (clear air) turbulence. In the mesosphere the echoes are produced by scattering due to fluctuations in free electron density associated with turbulence. With the MST technique it is possible to obtain three-dimensional velocity fields. MST radars have limited horizontal coverage but can produce data with high temporal and vertical resolution in a given locality. They are thus particularly appropriate for studying high-frequency components of the motion field, such as gravity waves and tides.

MST radars cannot provide information on the temperature fluctuations associated with atmospheric motions. Such information can be provided by ground-based lidar sounding. Lidars designed to detect the Rayleigh back-scatter from air molecules can yield density profiles in the altitude range of 30–90 km. From these the temperature profile can be computed using Eqs. (1.1.1) and (1.1.2). Other types of lidars can be used to provide profiles of ozone, aerosols, and other trace constituents in the middle atmosphere.

References

1.1. Atmospheric statics is discussed in more detail in standard meteorological texts such as Holton (1979a) and Wallace and Hobbs (1977).

1.3. *U.S. Standard Atmosphere, 1976* provides much useful information on the structure and composition of the atmosphere.

1.5. Remote sounding techniques are discussed in depth by Houghton *et al.* (1984). Gage and Balsley (1984) present a review of recent advances in remote sensing and provide an extensive bibliography of papers on both passive and active techniques.

Chapter 2 | Radiative Processes and Remote Sounding

2.1. Introduction

In any planetary atmosphere, the character of the circulation depends strongly on the magnitude and distribution of the net diabatic heating rate. In the earth's troposphere, the net diabatic heating rate is dominated by the imbalance between two large terms: transfer of heat from the surface, and thermal emission of radiation to space. Latent heat is a major component of the flux from the surface to the atmosphere, and clouds play a major role in the emission of radiation to space.

From the tropopause to the mesopause, the situation is much simpler. Net heating depends almost exclusively on the imbalance between local absorption of solar ultraviolet radiation and infrared radiative loss. In this region, ozone is the dominant absorber and carbon dioxide is the dominant emitter. Infrared emission by ozone and water vapor and solar absorption by water vapor, molecular oxygen, carbon dioxide, and nitrogen dioxide play secondary roles (Fig. 2.1). The distribution of the radiative sources and sinks due to these gases exerts a zero-order control on the large-scale seasonally varying mean temperature and zonal wind fields of the middle atmosphere. Infrared radiative emission also provides an important mechanism for damping dynamically forced temperature variations. These radiative processes will be described in this chapter.

Satellite observations of emitted, transmitted, and scattered radiation can be used to diagnose the global distributions of temperature and constituent concentrations throughout the middle atmosphere. Such observations have provided much of the basic data bearing on the dynamical state and transport processes in this region. Principles and applications of these

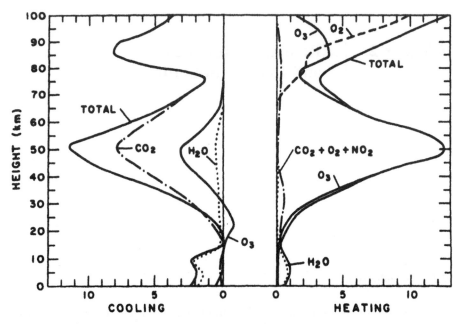

Fig. 2.1. Vertical distribution of heating due to absorption of solar radiation (right) and cooling due to emission of infrared radiation (left) in units of Kd^{-1}. [From London (1980), with permission.]

techniques to the middle atmosphere are considered in the last section of this chapter. In the future, satellite measurements of emitted radiance should also be useful for the direct determination of wind velocities.

2.2 Fundamentals

Calculation of the distribution of radiative sources and sinks requires the solution of the radiative transfer problem. This is essentially a problem of careful bookkeeping. The physics enters with the consideration of the actual processes of interaction between radiation and matter. In this section, the concepts and basic formalism of radiative transfer theory are introduced.

2.2.1 Radiative Transfer Quantities and the Equation of Transfer

Figure 2.2 illustrates the geometry of electromagnetic radiation crossing a plane surface S whose unit normal vector is $\hat{n}$. Radiant energy flux contained in a conical bundle of infinitesimal solid angle $d\Omega$ in the direction of unit vector $\hat{\Omega}$ is shown. The flux of energy per unit area of S in the

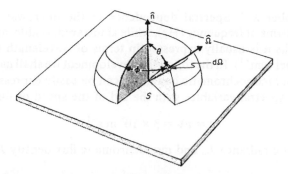

Fig. 2.2. Geometry of radiation crossing a plane surface S with unit normal $\hat{n}$. Radiance in the solid angle cone $d\Omega$ in the direction $\hat{\Omega}$ is described in terms of the spherical coordinates θ and ϕ.

bundle is $L(\hat{\Omega})\,d\Omega$, where $L(\hat{\Omega})$, the *radiance*, is the fundamental radiative transfer quantity. Its units are $\mathrm{W\,m^{-2}\,sr^{-1}}$.

The contribution of the radiant flux in this bundle to the total flux across S in the direction of $\hat{n}$ is $(\hat{n}\cdot\hat{\Omega})L(\hat{\Omega})\,d\Omega$, and the total flux across S into the half-space above S (2π steradians) is the *flux density F*, given by

$$F = \int_{2\pi} (\hat{n}\cdot\hat{\Omega})L(\hat{\Omega})\,d\Omega = \int_0^{2\pi}\int_0^{\pi/2} L(\phi,\theta)\cos\theta\sin\theta\,d\theta\,d\phi,$$

(2.2.1)

as expressed in terms of the azimuth and zenith angles ϕ and θ (see Fig. 2.2). Radiant energy impinging on S is called *irradiance*, while radiant energy emitted from S is called *emittance*. In this chapter this distinction will generally be avoided, and either quantity will be referred to as flux density. Units of F are $\mathrm{W\,m^{-2}}$.

The *net flux F_n* is the difference between flux density crossing S in the direction $\hat{n}$ and that crossing in the opposite direction, $-\hat{n}$. For example, if S is horizontal,

$$F_n = F_\uparrow - F_\downarrow$$

(2.2.2)

where $F_\uparrow$ and $F_\downarrow$ are flux densities in the upward and downward directions. For a Cartesian coordinate system, three values of F_n correspond to net flux values across the three coordinate planes of that system. These are the components in those coordinates of a vector, the *net flux vector* $\mathbf{F_n}$.

Because of the spectral dependences of radiation and its interaction with matter, it is necessary to consider *monochromatic* radiance and flux density. Spectral dependence is expressed in terms of frequency ν, wavelength λ, or wave number λ^{-1}. Spectral dependence in the microwave is usually expressed in terms of frequency (Hz), while at infrared, visible, and ultraviolet wavelengths it is usually expressed in terms of wavelength (cm or nm) or wave number (cm^{-1}). In the following development we shall use frequency exclusively, but monochromatic quantities can be easily expressed in terms of any of the spectral variables with the aid of the speed of light c, where

$$c = \nu\lambda \doteq 3 \times 10^8 \,\mathrm{m\,s^{-1}}. \tag{2.2.3}$$

Monochromatic radiance L_ν and monochromatic flux density F_ν are given by

$$L_\nu = \frac{dL}{d\nu}, \qquad F_\nu = \frac{dF}{d\nu}. \tag{2.2.4}$$

Units of L_ν are $\mathrm{W\,m^{-2}\,sr^{-1}\,(s^{-1})^{-1}}$ and units of F_ν are $\mathrm{W\,m^{-2}\,(s^{-1})^{-1}}$.

The rate of decrease (extinction) or increase of monochromatic radiance along a length element ds in the direction $\hat{\Omega}$ is linear in the amount of absorbing matter. Extinction is also linear in the monochromatic radiance. These facts are expressed in the *radiative transfer equation*

$$\frac{dL_\nu(\hat{\Omega})}{ds} = -k_\nu\rho_\mathrm{a}[L_\nu(\hat{\Omega}) - J_\nu(\hat{\Omega})]. \tag{2.2.5}$$

The density of the radiatively active gas ρ_a and two quantities k_ν and J_ν appear in this equation. These quantities characterize the interaction between radiation and matter. They depend on the local properties of the medium. The *extinction coefficient* k_ν, whose units are $\mathrm{m^2\,kg^{-1}}$, describes the extinction of L_ν along ds, while the *source function* J_ν describes the rate of increase.

Energy removed from L_ν by extinction can either increase the internal energy of matter at the point of extinction or can be immediately scattered at some angle to the incident beam, thereby providing an input to monochromatic radiance at the same frequency in a different direction. Thus k_ν can be expressed as the sum of an *absorption coefficient* a_ν and a *scattering coefficient* s_ν,

$$k_\nu = a_\nu + s_\nu. \tag{2.2.6}$$

The ratio of scattered energy to total energy lost by extinction at a point in the medium is the *single scattering albedo* $\bar{\omega}_\nu$,

$$\bar{\omega}_\nu \equiv s_\nu/k_\nu, \qquad (1 - \bar{\omega}_\nu) \equiv a_\nu/k_\nu. \tag{2.2.7}$$

Scattering contributes to emission of radiation in the direction of the scattered photons, and thereby contributes to the source function. The total scattering contribution to the source function from a volume element in the direction $\hat{\Omega}$, $J_{\nu,s}(\hat{\Omega})$, is an integral over all incident angles $\hat{\Omega}'$ filling 4π steradians,

$$J_{\nu,s}(\hat{\Omega}) = \frac{\bar{\omega}_\nu}{4\pi} \int_{4\pi} L_\nu(\hat{\Omega}') P_\nu(\hat{\Omega}, \hat{\Omega}') \, d\Omega'. \qquad (2.2.8)$$

The function $P_\nu(\hat{\Omega}, \hat{\Omega}')$ describes the angular distribution of the scattered radiation and also depends on the local properties of matter as well as the frequency; $P_\nu(\hat{\Omega}, \hat{\Omega}')$ is the *phase function*, and according to Eq. (2.2.8), it is normalized such that

$$\frac{1}{4\pi} \int_{4\pi} P_\nu(\hat{\Omega}, \hat{\Omega}') \, d\Omega' = 1. \qquad (2.2.9)$$

By itself, scattering does not change the internal energy of matter, but emission of radiation can also take place at the expense of the internal energy of the matter in a volume element. In order to describe this *thermal emission* it is necessary to distinguish between *local thermodynamic equilibrium (LTE)*, the condition under which *Kirchhoff's Law* applies, and the condition under which Kirchhoff's Law fails (*non-LTE*). For the most important radiatively active gases, LTE holds for pressures greater than about 0.1 mb but begins to fail at lower pressures.

Under LTE conditions with no scattering ($\bar{\omega}_\nu = 0$), Kirchhoff's Law states that the source function is equal to the *Planck function* B_ν,

$$J_\nu = B_\nu(T) = \frac{2h\nu^3}{c^2}(e^{h\nu/k_b T} - 1)^{-1}, \qquad (2.2.10)$$

where h and k_b are the Planck and Boltzmann constants. Note that B_ν is independent of direction; that is, it is *isotropic*. The integral of B_ν over all frequencies is the *blackbody radiance* $B(T)$,

$$B(T) = \int_0^\infty B_\nu(T) \, d\nu = \frac{\sigma}{\pi} T^4, \qquad (2.2.11)$$

where $\sigma = 5.67 \times 10^{-8} \, \mathrm{W \, m^{-2} \, K^{-4}}$, the Stefan–Boltzmann constant.

When LTE holds but scattering occurs ($\bar{\omega}_\nu \neq 0$), Kirchhoff's Law takes the more general form

$$J_\nu(\hat{\Omega}) = (1 - \bar{\omega}_\nu) B_\nu + \frac{\bar{\omega}_\nu}{4\pi} \int_{4\pi} L_\nu(\hat{\Omega}') P_\nu(\hat{\Omega}, \hat{\Omega}') \, d\Omega'. \qquad (2.2.12)$$

The source function for non-LTE conditions will be discussed in Section 2.6.

There is a useful integral form of the radiative transfer equation. As illustrated in Fig. 2.3, radiance at a position s along a path in the direction

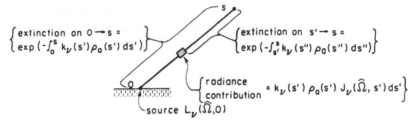

$$\left\{ \begin{array}{l} \text{extinction on } 0 \to s = \\ \exp\left(-\int_0^s k_\nu(s')\rho_0(s')\,ds'\right) \end{array} \right\} \qquad \left\{ \begin{array}{l} \text{extinction on } s' \to s = \\ \exp\left(-\int_{s'}^s k_\nu(s'')\rho_0(s'')\,ds''\right) \end{array} \right\}$$

$$\left\{ \begin{array}{l} \text{radiance} \\ \text{contribution} \end{array} \right\} = k_\nu(s')\,\rho_0(s')\,J_\nu(\hat{\Omega},s')\,ds'$$

source $L_\nu(\hat{\Omega},0)$

Fig. 2.3. Contributions to radiance at s from the path 0 to s.

of $\hat{\Omega}$ is given by

$$L_\nu(\hat{\Omega}, s) = L_\nu(\hat{\Omega}, 0) \exp\left[-\int_0^s k_\nu(s')\rho_a(s')\,ds'\right]$$

$$+ \int_0^s k_\nu(s')\rho_a(s')J_\nu(\hat{\Omega}, s') \exp\left[-\int_{s'}^s k_\nu(s'')\rho_a(s'')\,ds''\right]\,ds'.$$

$$(2.2.13)$$

The quantity $\int_{s'}^s k_\nu(s'')\rho_a(s'')\,ds''$ is called the *optical path* between s' and s. Equation (2.2.13) is the formal solution of the radiative transfer equation, Eq. (2.2.5). In accordance with the linear character of the problem, it expresses the fact that radiance $L_\nu(\hat{\Omega}, s)$ is composed of contributions from boundary radiance $L_\nu(\hat{\Omega}, 0)$ exponentially attenuated by the optical path length between 0 and s plus infinitesimal radiance contributions from volume elements at positions s' along the path, given by

$$k_\nu(s')\rho_a(s')J_\nu(\hat{\Omega}, s')\,ds'.$$

Each such contribution is attenuated exponentially by the matter contributing to the optical path between s' and s, and these contributions are integrated along the path.

2.2.2 Plane-Parallel Atmosphere Approximation

In atmospheric radiative transfer theory, two approximations are usually made. (1) The curvature of level surfaces due to the sphericity of the planet is negligible. (2) Properties of the medium and the radiation field depend only on the vertical coordinate. Together, these conditions comprise the *plane-parallel atmosphere approximation.*

When the plane-parallel atmosphere approximation holds, the net flux vector is vertical and it is appropriate to express the angular dependence of the radiation field in terms of polar coordinates (θ, ϕ) referenced to the local vertical so that θ is the *zenith angle* (measured from the vertical). Moreover, radiance is independent of azimuth ϕ. It follows from Eq. (2.2.1) that flux density in the upward direction is given by

$$F_\uparrow(z^*) = 2\pi \int_0^{\pi/2} L(\theta, z^*) \cos\theta \sin\theta \, d\theta = 2\pi \int_0^1 L(\mu, z^*)\mu \, d\mu,$$

(2.2.14a)

where $\mu \equiv \cos\theta$.

Flux density in the downward direction is given by

$$F_\downarrow(z^*) = 2\pi \int_\pi^{\pi/2} L(\theta, z^*) \cos\theta \sin\theta \, d\theta = 2\pi \int_0^1 L(-\mu, z^*)\mu \, d\mu.$$

(2.2.14b)

Note the use of the argument $-\mu$ to indicate downward radiance.

For plane-parallel atmosphere problems involving monochromatic radiation, the vertical coordinate z^* can be replaced by the *optical depth* $\tau_\nu(z^*)$. This is the optical path along the vertical between height z^* and the top of the atmosphere,

$$\tau_\nu(z^*) = \int_{z^*}^\infty k_\nu(z^{*\prime})\rho_a(z^{*\prime}) \, dz^{*\prime}.$$ (2.2.15a)

The notation τ_ν with subscript ν explicitly denotes the frequency dependence of the optical depth, which arises from the frequency dependence of k_ν. Optical depth can also be expressed in terms of the log-pressure vertical coordinate z:

$$\tau_\nu(z) = \int_z^\infty k_\nu(z')\rho_{a0}(z') \, dz'$$ (2.2.15b)

where $\rho_{a0}(z) = \rho_a(z)[T(z)/T_s] = \rho_a(z)[\rho_0(z)/\rho(z)]$ is the basic absorber density scaled to the reference temperature T_s used to define the reference scale height H [see Eq. (1.1.8)]. Note that $\rho_{a0}(z) = m_a(z)\rho_0(z)$, where m_a is the mass mixing ratio of the absorbing gas and $\rho_0(z) = \rho_s e^{-z/H}$ is the basic density in the log-pressure coordinate system, as defined in Section 3.1.1.

Solar radiation in a plane-parallel atmosphere is conveniently separated into radiation in the solar beam, *direct solar radiation,* and radiation scattered by the atmosphere, *diffuse solar radiation.* Direct solar radiation consists of a very large radiance in a very small solid angle. Outside the atmosphere on a surface normal to the direction of the solar radiation stream $\hat{\Omega}_0$, direct solar radiation contains a monochromatic flux density $S_{0\nu}$. The spectral integral of $S_{0\nu}$ is usually referred to as the *solar constant* S_0, although this term is inappropriate since S_0 is known to exhibit some variability in time. At level z within the atmosphere, the monochromatic flux of direct solar radiation across a level surface and incident at zenith angle $\theta_0 = \cos^{-1} \mu_0$ is

$$S_\nu(z) = \mu_0 S_{0\nu} e^{-\tau_\nu(z)/\mu_0}. \qquad (2.2.16)$$

The source function given by Eq. (2.2.12) can be rewritten to explicitly display the contributions to scattered radiance by the direct solar radiation and the diffuse radiation. The angular dependence of the direct solar radiation can be approximated by a delta function, and it follows from Eq. (2.2.16) that

$$J_\nu(\hat{\mathbf{\Omega}}) = (1 - \bar{\omega}_\nu) B_\nu + \frac{\bar{\omega}_\nu}{4\pi} \int_{4\pi} L_\nu(\hat{\mathbf{\Omega}}') P_\nu(\hat{\mathbf{\Omega}}, \hat{\mathbf{\Omega}}') \, d\Omega'$$

$$+ \frac{\bar{\omega}_\nu S_{0\nu}}{4\pi} P_\nu(\hat{\mathbf{\Omega}}, \hat{\mathbf{\Omega}}_0) e^{-\tau_\nu/\mu_0}. \qquad (2.2.17)$$

In Eq. (2.2.17), $L_\nu(\hat{\mathbf{\Omega}}')$ corresponds to diffuse radiance alone, and $J_\nu(\hat{\mathbf{\Omega}})$ contributes only to diffuse radiance. In most applications, spectral overlap between solar and thermal radiation can be neglected so the first term on the right side of Eq. (2.2.17) can be ignored at wavelengths shorter than about 4 μm while the last term can be neglected at longer wavelengths.

The formal solution to the problem of determining the diffuse monochromatic radiance in a plane parallel atmosphere can now be stated with the aid of the integral form of the radiative transfer equation [Eq. (2.2.13)] and suitable boundary conditions at the bottom and top of the atmosphere. In terms of optical depth, radiance in the upward direction is

$$L_\nu[\mu, \tau_\nu(z)] = L_\nu[\mu, \tau_\nu(0)] \exp\{-\mu^{-1}[\tau_\nu(0) - \tau_\nu(z)]\}$$

$$+ \int_{\tau_\nu(z)}^{\tau_\nu(0)} \mu^{-1} J_\nu[\mu, \tau_\nu(z')] \exp\{-\mu^{-1}[\tau_\nu(z') - \tau_\nu(z)]\} \, d\tau_\nu(z').$$

$$(2.2.18)$$

The radiance at the lower boundary $L_\nu[\mu, \tau_\nu(0)]$ depends on the monochromatic *emissivity* ε_ν of the surface, which is the ratio of the radiance emitted at the expense of the internal energy of the boundary to that of a blackbody at the same temperature. According to Kirchhoff's Law, the monochromatic reflectivity of an opaque surface is $(1 - \varepsilon_\nu)$. For an isotropically reflecting surface it follows that the boundary radiance $L_\nu[\mu, \tau_\nu(0)]$ required for Eq. (2.2.18) is the isotropic radiance

$$L_\nu[\tau_\nu(0)] = \varepsilon_\nu B_\nu[\tau_\nu(0)] + (1 - \varepsilon_\nu)\bar{L}_\nu[-\mu, \tau_\nu(0)]$$

$$+ (1 - \varepsilon_\nu)\mu_0 S_{0\nu} e^{-\mu_0^{-1}\tau_\nu(0)}, \qquad (2.2.19)$$

where $\bar{L}_\nu[-\mu, \tau_\nu(0)] \equiv 2 \int_0^1 L_\nu[-\mu, \tau_\nu(0)]\mu \, d\mu$ is the angular mean of L_ν. For thick clouds and most surface materials, $\varepsilon_\nu \approx 1$ at thermal infrared wavelengths ($\lambda \gtrsim 4 \, \mu m$), and the reflected contributions can often be neglected. Moreover, as in Eq. (2.2.17), B_ν can be neglected for $\lambda \lesssim 4 \, \mu m$ while $S_{0\nu}$ can be neglected for $\lambda \gtrsim 4 \, \mu m$. At visible as well as infrared wavelengths it is often assumed that ε_ν is isotropic, but this assumption can lead to serious errors for near-infrared and visible wavelengths over some surfaces, particularly at low sun angles.

The upper boundary condition for diffuse radiation is

$$L_\nu(-\mu, \tau_\nu = 0) = 0, \tag{2.2.20}$$

so that the radiative equation for downward radiance is simply

$$L_\nu[-\mu, \tau_\nu(z)] = \int_0^{\tau_\nu(z)} \mu^{-1} J_\nu[-\mu, \tau_\nu(z')]$$

$$\times \exp\{-\mu^{-1}[\tau_\nu(z) - \tau_\nu(z')]\} \, d\tau_\nu(z').$$

$$\tag{2.2.21}$$

2.2.3. Effect of Atmospheric Sphericity

In contrast to the troposphere, where horizontal inhomogeneities are often very large, the conditions for validity of the plane-parallel atmosphere approximation are generally well satisfied in the middle atmosphere except for direct solar radiation at large zenith angles ($\theta \gtrsim 80°$). At these large solar zenith angles, a correction must be applied to the expression for the direct solar radiation [Eq. (2.2.16)] to account for the sphericity of level surfaces.

Because the atmosphere is spherical, μ_0 varies along the path. The geometry is depicted in Fig. 2.4 for paths for which $\theta(z^*) < \pi/2$ and $\theta(z^*) > \pi/2$. In the former case, τ_ν in Eq. (2.2.15a) should be replaced by $\tau_{\nu,\text{cor}}$:

$$\tau_{\nu,\text{cor}}(z^*) = \mu_0(z^*) \int_{z^*}^\infty k_\nu(z^{*\prime})\rho_a(z^{*\prime}) \, dz^{*\prime}/\mu_0(z^{*\prime}), \tag{2.2.22}$$

where $\mu_0(z^{*\prime})$ varies along the path as indicated in Fig. 2.4. Because of the sphericity, solar radiation can reach the upper atmosphere when $\theta(z^*) > \pi/2$ provided that $\theta(z^*) < \theta_{\text{cutoff}}(z^*)$, and a straightforward modification of Eq. (2.2.22) is required in this case. Closed-form expressions employing tabulated functions known as *Chapman functions* can be used to evaluate the integrals in Eq. (2.2.22) in the special case of a constituent with a constant absorption coefficient whose density varies exponentially with height.

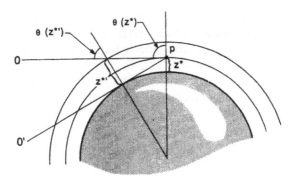

Fig. 2.4. Effect of spherical planetary shape on the geometry of direct solar radiation, shown with greatly exaggerated atmospheric thickness. Path $\overline{OP}$ corresponds to $\theta < \pi/2$; path O'P corresponds to $\theta > \pi/2$. In the latter case an observer at P sees the sun below the horizontal. The angle $\theta_{\text{cutoff}}(z^*)$ is the zenith angle at which O'P just intersects the horizon.

2.3 Gaseous Absorption Spectra

Solar photons are absorbed in the stratosphere and mesosphere primarily at ultraviolet wavelengths (0.1–0.4 μm) and to a lesser extent at visible (0.4–0.7 μm), near-infrared (0.7–4 μm), and X-ray wavelengths. The absorbed energy produces electronic, vibrational and rotational excitation, molecular dissociation, and ionization, but the energy used for ionization is of only minor importance for the total energy budget of the middle atmosphere. At altitudes below about 60 km, these processes are closely balanced by local recombination and collisional deexcitation so that most of the absorbed energy is *thermalized*—that is, it is realized locally as heat. In contrast, molecular collisions in the thermosphere and to some extent in the upper mesosphere are too infrequent to insure local thermalization of all absorbed energy (see Section 2.6). Dissociation and ionization occur primarily in continuous spectra, but excitation of electronic, vibrational, and rotational energy takes place in spectrally complex bands composed of large numbers of lines. In some cases, dissociation or ionization can also arise from absorption in complex spectral bands provided these are at wavelengths shorter than the threshold wavelength for dissociation or ionization.

Molecular absorption of thermal infrared radiation at wavelengths between 4 and 17 μm excites vibrational and rotational energy, producing *vibration–rotation* bands, which invariably have a complex structure. Positions of some important vibration–rotation bands are shown in Fig. 2.5. At wavelengths longer than 17 μm, absorption arises predominantly from transitions in molecular rotational energy, producing rotational lines and bands. In order to calculate diabatic heating rates, it is necessary to know

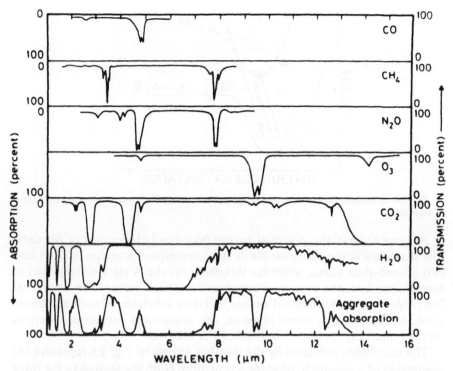

Fig. 2.5. Schematic spectra depicting the most important gaseous absorption features for infrared radiative transfer in the middle atmosphere. Absorptions correspond approximately to a normal incidence path through the atmosphere. [From Handbook of Geophysics and Space Environment, Air Force Cambridge Research Laboratory (1965), with permission.]

the strength, shape, and positions of the spectral lines responsible for absorption. The processes controlling these spectral characteristics are described here.

2.3.1 Molecular Energy Levels and Transitions: The Spectrum of O_2

Although many of the molecules of interest are polyatomic, the most important characteristics of molecular spectra can be illustrated by considering the simpler properties of diatomic molecules. Figure 2.6 is a schematic potential energy diagram of a diatomic molecule, AB. The potential energy versus internuclear distance is shown for the ground electronic state (X) and the first electronically excited state (A). For each of these states, potential energy first decreases and then increases with internuclear distance, corresponding to electrical forces that are repulsive at close range and attractive at longer range on either side of a stable equilibrium point.

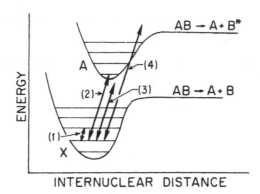

Fig. 2.6. Schematic potential energy diagram for a diatomic molecule showing four types of transitions.

At large separations, potential energy becomes independent of distance. The dissociation threshold for the X state corresponds to dissociation into two ground-state atoms, while the threshold for the A state corresponds to dissociation into one or more electronically excited atoms. The horizontal lines represent the vibrational energy levels or substates of each electronic state. As these energy levels increase, the vibrational oscillation energies increase until the dissociation threshold is reached.

The transitions indicated by the slanting arrows in Fig. 2.6 represent (1) absorption of a photon to produce a transition from the second to the third vibrational energy level of the X state,[1] (2) absorption producing a transition from the second vibrational energy level of the X state to the first level of the A state, (3) absorption producing photodissociation to two ground-state atoms, and (4) dissociation producing one excited state and one ground state atom. Note that the energy changes associated with the dissociations are continuous; in contrast, the discrete energy changes associated with transitions between vibrational energy levels produce discrete spectral bands.

The vibrational energy levels in turn contain further internal structure: each is divided into a number of rotational energy levels whose spacing is too small to be shown in Fig. 2.6. Hence, transitions such as (1) or (2) are multiple and are associated with a series of spectral lines.

These relationships between the energy levels and spectra are illustrated in Fig. 2.7, which shows the potential energy diagram and the ultraviolet absorption spectrum of O_2. These O_2 spectral features are important for the energetics of the mesosphere and for photochemistry in both stratosphere and mesosphere. The ordinate in Fig. 2.7b is the *cross section* σ_ν, which is

[1] For molecules with identical nuclei, like O_2 and N_2, photon absorption does not occur for this type of transition.

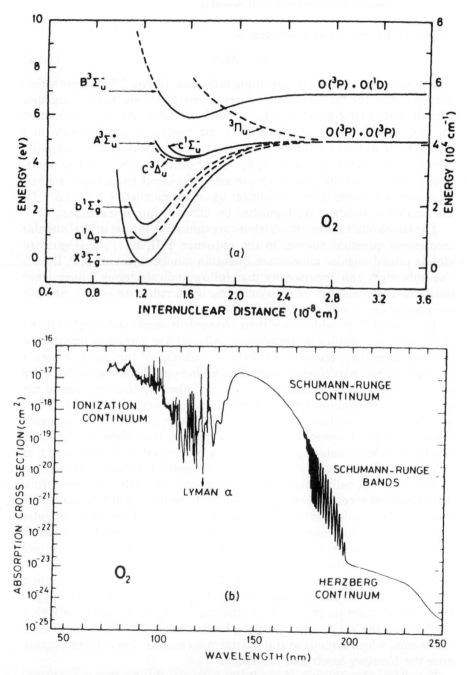

Fig. 2.7. (a) Simplified potential-energy diagram for O_2. [From Gilmore (1964), with permission.] (b) Ultraviolet spectrum of O_2. [From Brasseur and Solomon (1984). Copyright © 1984 by D. Reidel Company, Dordrecht, Holland.]

related to the absorption coefficient by

$$\sigma_\nu = M_a a_\nu \tag{2.3.1}$$

where M_a is the mass of the absorbing molecule. In Fig. 2.7a the potential-energy curves for the electronic energy levels of O_2 are shown, together with their *term symbols*. Vibrational as well as rotational levels are omitted for clarity. The term symbol specifies the electronic configuration and indicates the ordinal number of the level or *term* in a series of terms having the same *multiplicity* (series X, A, B, ..., or series X, a, b, ...). The multiplicity, indicated by the superscript preceding the Greek letter, is determined by the net electronic spin with integer quantum number S. It is $(2S + 1)$, the number of substates distinguished by different spin orientations.

The Greek letter in the term symbol corresponds to the net orbital angular momentum quantum number in the sequence $\Sigma, \Pi, \Delta, \ldots$, analogous to atomic orbital angular momentum quantum number symbols S, P, D, ... The subscripts and superscripts that follow indicate terms whose wave functions are symmetric or antisymmetric upon reflection $(+, -)$, or have even parity (g) or odd parity (u).

Term symbols provide more than convenient shorthand labels for the terms: they also display information on *allowed* and *forbidden* transitions. Allowed transitions can occur if there is a change of the dipole moment between the two participating terms. Such *electronic dipole transitions* have relatively large cross sections. In addition, other types of change in the electronic configuration of the molecule can be associated with absorption or emission of radiation, for example, *electronic quadrupole* or *magnetic dipole* changes. However, cross sections for such transitions are typically smaller than electronic dipole transitions by six orders of magnitude or more. Such transitions are included in the category of forbidden transitions.

Selection rules, derived from the quantum-mechanical theory of the terms, distinguish between allowed and forbidden transitions. Examples of selection rules are: spin change in a transition is forbidden; total angular momentum quantum number changes other than 0, ±1 are forbidden. A more detailed discussion of term symbols and selection rules can be found in Brasseur and Solomon (1984).

The forbidden transition $X\,^3\Sigma_g^- \rightarrow A\,^3\Sigma_u^+$ produces very weak absorption bands between 260 and 242 nm. Dissociation to two ground-state oxygen atoms (O ^{3}P) takes place at 242 nm (corresponding to an energy of 8.99×10^{-19} J or 5.58 eV). At this wavelength, the bands terminate in a weak continuum, which extends to still shorter wavelengths. These features comprise the *Herzberg bands* and *continuum*.

The allowed transition $X\,^3\Sigma_g^- \rightarrow B\,^3\Sigma_u^-$ is responsible for the strong bands between 200 and 175 nm, the *Schumann–Runge bands*. These bands terminate in the *Schumann–Runge continuum*, which extends to shorter wavelengths and corresponds to the dissociation

$$O_2 \rightarrow O(^3P) + O(^1D)$$

in which one oxygen atom emerges in the excited ^1D state. Molecules in the upper term of the Schumann–Runge bands, the $B\,^3\Sigma_u^-$ state, can undergo a spontaneous transition to the $^3\Pi_u$ state, but the latter has no potential energy minimum so that it is unstable and rapidly dissociates to two oxygen atoms in the ^3P ground state. As a result of this process, known as *predissociation*, the lifetime of O_2 molecules in the $B\,^3\Sigma_u^-$ state is exceptionally short.

Figure 2.8 shows the detailed structure of the Schumann–Runge bands and illustrates the rotational line structure of the vibrational bands. The vibrational levels involved in the transitions are indicated in the figure. For example, 1–0 corresponds to rotational lines in the transition between the first excited vibrational level of the B state and the ground vibrational level of the X state. This important band system shows a high degree of regularity as well as complexity. The regular appearance of the system at longer wavelengths arises from the regularity of the spacing of vibrational and rotational energy levels. At the shorter wavelengths, the number of lines from incommensurably overlapping bands becomes so great that complexity appears to be winning out. Between 56,000 and 57,000 cm^{-1}, the spectrum has a disordered, almost random appearance.

2.3.2 Line and Band Strength

The *strength S* of a spectral line or band is defined as

$$S = \int \sigma_\nu \, d\nu \tag{2.3.2}$$

where the integration is over the entire line or band. The cross section σ_ν is usually expressed in the units cm^2, so that the corresponding units for S are cm^2 s^{-1}. We turn now to a discussion of the factors controlling the strengths of lines and bands in a band system such as that shown in Fig. 2.8.

First, line or band strength for absorption is proportional to the ratio of the number density of molecules in the lower state of the transition, n_l, to the total number density of molecules of the absorbing gas, n_a. At thermodynamic equilibrium at temperature T, this ratio is given by the *Boltzmann factor*,

$$n_l/n_a = \frac{g_l\, e^{-E_l/k_b T}}{\sum_j g_j\, e^{-E_j/k_b T}}, \tag{2.3.3}$$

where E_l is the energy of level l and the summation is over all energy levels.

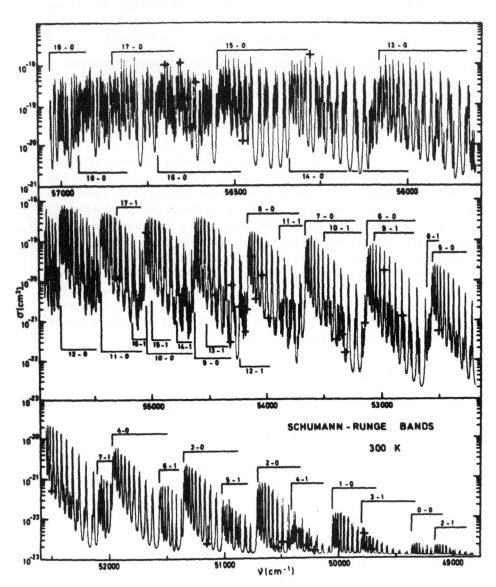

Fig. 2.8. Calculated spectrum of the Schumann–Runge band system shown with a few measured data points (+). [From Kockarts (1971). Copyright © 1971 by D. Reidel Publishing Company, Dordrecht, Holland.]

The integers g_l and g_j are *degeneracies*, the number of distinct states having energies E_l or E_j. For example, there are $2J + 1$ quantum-mechanically distinct orientations possible for a molecule whose angular momentum quantum number is J, and the corresponding degeneracy is $g_J = 2J + 1$. The denominator on the right side of Eq. (2.3.3) is the *partition function*, which gives the proper normalization, the summation over all states. It depends weakly on T, but not on the quantum numbers of energy level l.

The effect of the Boltzmann factor can be seen in the spectrum of Fig. 2.8 in two ways. First, within each vibrational band, the rotational quantum numbers of both the upper and lower states of each transition increase as the transitions progress from the strongest lines, near the short-wavelength edges of each band, toward weaker lines at longer wavelengths. Higher rotational quantum numbers for the lower states of each transition correspond to greater rotational energies of the lower state, and therefore, according to the Boltzmann distribution [Eq. (2.3.3)], to fewer molecules available for absorption. Second, vibrational bands corresponding to a transition whose lower state is an excited state have far fewer molecules available for absorption than bands corresponding to absorption from the ground vibrational level. Neglecting degeneracy , the ratio n_l/n_a is proportional to

$$e^{-E_l/k_bT} = e^{-h\nu_l/k_bT}$$

where $E_l = h\nu_l$ is the vibrational energy of the excited lower level. For typical vibrational transitions of diatomic molecules, the energy difference between adjacent vibrational levels gives $h\nu_l/k_bT \gtrsim 5$ at 273 K, so the number density ratios are typically ≤ 0.01. Comparing representative bands with lower-state quantum numbers 0 and 1 in Fig. 2.8, for example the 2–0 and 2–1 bands, it can be seen that the strengths of corresponding lines within the two bands typically have ratios $\sim 100:1$, and this is primarily due to the differences in populations of the energy levels as expressed by the Boltzmann factors.

The important consequence of these population effects is that absorption band structure is strongly temperature dependent through the factor $h\nu_l/k_bT$. Because this factor is large for vibrational transitions, the relative strengths of bands with different lower-state vibrational quantum numbers are extremely sensitive to temperature. On the other hand, differences in energy between rotational levels are much smaller, so the temperature sensitivity of the ratios of line strengths within a band is much less, although the overall envelope of the rotational lines within a single vibrational band is controlled by temperature.

The changes in the electronic configuration of the molecule in vibrational transitions or in combined electronic and vibrational transitions also influence the band strengths through a factor called the *oscillator strength*, which can be evaluated from calculations based on the quantum-mechanical description of the molecule. The relative oscillator strengths of vibrational bands within an electronic transition, such as the O_2 Schumann–Runge band system, are governed by the *Franck–Condon rule*, which states that the oscillator strength increases as the correspondence between the configurations of the upper and lower vibrational states increases. Together with the Boltzmann distribution, the oscillator strength controls the relative strengths of vibrational bands in a system such as the Schumann–Runge bands and is responsible for the general increase in strength toward shorter wavelengths shown in Fig. 2.8.

2.3.3 Line Shape and Line Width

Absorption lines have finite spectral width as a consequence of one or more of the following factors: finite lifetime Δt of the upper state, which leads to an uncertainty in the energy ΔE through the uncertainty principle

$$\Delta E \, \Delta t \gtrsim h/2\pi,$$

finite lifetime due to perturbation by molecular collisions, and Doppler frequency shifts due to relative thermal motions of the molecules. For a single line of strength S, the absorption cross section can be expressed in the form

$$\sigma_\nu = S f(\nu - \nu_0) \tag{2.3.4}$$

where $f(\nu - \nu_0)$ is a shape factor giving the relative cross section at a point displaced by $(\nu - \nu_0)$ frequency units from the line center ν_0. The function $f(\nu - \nu_0)$ is normalized so that $\int_{-\infty}^{\infty} f(\nu - \nu_0) \, d\nu = 1$.

The simplest model for the effect of finite lifetime is the *Lorentz line shape*,

$$f(\nu - \nu_0) = (\alpha_L/\pi)[(\nu - \nu_0)^2 + \alpha_L^2]^{-1} \tag{2.3.5}$$

where $\alpha_L = (2\pi \bar{t})^{-1}$ and $\bar{t}$ is the mean time between major perturbations of the excited state. The frequency at which $|\nu - \nu_0| = \alpha_L$ is the half-power point for the line, and α_L is called the *Lorentz half-width*.

Line broadening produced by finite lifetime of the upper state is called *natural broadening*. In this case, $\bar{t}$ is proportional to the mean lifetime of the upper state, the inverse of the molecular emission probability. For the O_2 Schumann–Runge bands, the mean lifetime in the upper state for some vibrational bands is $\sim 10^{-9}$ s and the corresponding Lorentz half-width (in wave number units) is ~ 1 cm^{-1}. Natural broadening is a significant broadening mechanism for these bands in the mesosphere and upper stratosphere. In contrast, typical lifetimes for upper states of vibration–rotation transitions in the infrared are ~0.1 s, and natural broadening is completely negligible compared with collisional broadening for these bands.

Broadening due to collisions, or *pressure broadening*, is a complex process, but it is usual to approximate each collision as an encounter that truncates the wave function corresponding to the state of the molecule by producing a large sudden phase shift. In this approximation, the "phase shift" approximation, the Lorentz line shape is a good approximation, with $\bar{t}$ the mean time between collisions. When typical kinetic theory values are used to represent molecular collisions,

$$\alpha_L \approx 0.07(p/p_s)(T_s/T)^{1/2} \quad \text{cm}^{-1}, \tag{2.3.6}$$

where $p_s = 1000$ mb and $T_s = 273$ K. At these reference pressure and temperature values, the Lorentz half-width usually falls in the range 0.05–0.11 cm^{-1}.

The Lorentz profile gives a good approximation to the shapes of pressure-broadened lines and is generally applicable at pressures typical of the middle stratosphere or greater, but there are some limitations. (1) There are significant variations in α_L from gas to gas, from band to band for the same gas, and in some cases even between rotational lines in the same band. Line width is also a function of the type of colliding molecule, or *broadening gas*. (2) Temperature dependence varies with the particular transition, but the actual temperature dependence is often stronger than that given by Eq. (2.3.6). The form $T^{-0.7}$ is more representative than $T^{-0.5}$. (3) Departures from the Lorentz shape occur at large distances from the line centers ($|\nu - \nu_0| \gg \alpha_L$). These extended wings can be important in the relatively transparent regions of the spectrum. Departures from the Lorentz shape in the extended wings are very difficult to measure and are a major source of uncertainty in atmospheric radiative transfer.

Even without pressure or natural broadening, finite line widths would arise because of molecular motion along the line of sight. This motion gives rise to the Doppler line shape

$$f(\nu - \nu_0) = (\pi \alpha_D)^{-1/2} \exp[-(\nu - \nu_0)^2/\alpha_D^2] \tag{2.3.7}$$

where

$$\alpha_D \equiv u_m \nu_0/c, \qquad u_m \equiv (2k_b T/M_a)^{1/2},$$

and M_a is the molecular mass. According to this definition, the spectral interval from line center to the half-power point (half-width) is $(\ln 2)^{1/2}\alpha_D$, not α_D.

When both Lorentz and Doppler broadening are important, the shape factor is given by the convolution integral of the Doppler and Lorentz shapes:

$$f(\nu - \nu_0, \alpha_D, y) = \alpha_D^{-1}\pi^{-3/2}y \int_{-\infty}^{\infty} \frac{e^{-x^2}\,dx}{y^2 + [x - (\nu - \nu_0)/\alpha_D]^2} \quad (2.3.8)$$

where $y = \alpha_L/\alpha_D$, and $x = u/u_m$. This is the *Voigt line shape*, applicable when $\alpha_L \lesssim \alpha_D$. In particular, it is applicable to the Schumann–Runge bands in the upper stratosphere and mesosphere, since $\alpha_D \cong 6 \times 10^{-2}\,\mathrm{cm}^{-1}$, comparable to the Lorentz half-width for natural broadening of some of the lines in these bands. For lines of vibration–rotation bands in the thermal infrared, $\alpha_D \approx 5 \times 10^{-4}\,\mathrm{cm}^{-1}$, and it follows from Eq. (2.3.6) that the Doppler and pressure-broadened half-widths for lines in these bands become comparable at altitudes of 30–40 km.

Figure 2.9 compares Doppler and Lorentz profiles when the half-widths are equal. The Doppler profile is concentrated near the center but falls off rapidly in the wings, while the Lorentz profile has very broad wings. The Voigt profile resembles the Doppler profile near the line center and the Lorentz profile in the line wings. Because line wings are often of great importance in atmospheric radiative transfer, the effects of the Lorentz profile are often important well above the altitude at which the Doppler and pressure-broadened half-widths are equal. At these altitudes, the Doppler profile underestimates the absorption coefficient in the line wings, but the Lorentz profile underestimates the spectral width of the regions near

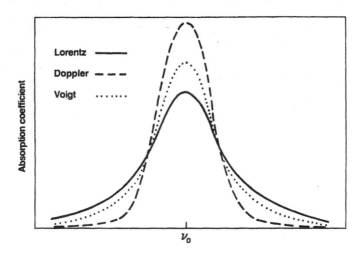

Fig. 2.9. Lorentz and Doppler line shapes for approximately equal half-widths and intensities. The corresponding Voigt profile is also shown.

the centers of strong absorption lines, since the Lorentz half-width continues to decrease with decreasing pressure.

Both the Doppler profile and the pressure-broadened Lorentz profile depend on temperature. However, the temperature dependence of line shape is usually of secondary importance compared with the temperature dependence of line and band strength.

2.3.4 Structure of Vibration–Rotation Bands

2.3.4.a Energy Levels

For a rigid rotating dipole, the energy levels E_J are given by

$$E_J \equiv BJ(J + 1), \qquad (2.3.9)$$

where B, the *rotational constant*, is inversely proportional to the moment of inertia of the dipole. The selection rule for a radiative transition is

$$\Delta J = \pm 1. \qquad (2.3.10)$$

The absorption spectrum corresponds to $\Delta J = +1$ and has the following characteristics. There is an absorption line corresponding to the transition $J = 0 \rightarrow J = 1$ at frequency $\nu = 2B/h$ or wave number $2B/hc$. Another line corresponding to $J = 1 \rightarrow J = 2$ occurs at frequency $4B/h$, etc. Thus, a series of lines occurs with uniform frequency spacing $2B/h$. The relative strengths of these lines are approximately proportional to the Boltzmann factor with degeneracy $(2J + 1)$,

$$S(J) \approx (2J + 1)e^{-BJ(J+1)/k_b T}. \qquad (2.3.11)$$

The most important modifications to this simple model of a rigidly rotating dipole arise from the following factors:

1. *Nonrigidity of the oscillator.*   At high rotation rates, the dipole separation and moment of inertia increase as a result of centrifugal stretching. The effect is to decrease line separations at large J values, and it may cause the progression of line positions with increasing J to reverse and produce a "band head," or sharp limit on the band.

2. *Complexity of the oscillator.*   Linear molecules have a single moment of inertia and behave much like this simple model. For nonlinear molecules there are moments of inertia about two or three axes, and rotational energy levels must be described by two or three quantum numbers. This leads to a far more complicated spectrum such that the spacing of individual rotational lines associated with changes of all of the rotational quantum numbers may appear quite random.

Despite these complexities, typical rotational line spacings are of order $2B/hc$ in wave numbers, or 0.1-1 cm^{-1}. There is considerable variation between bands, however, and line spacings are generally smaller for heavier molecules.

Molecules with permanent electric dipole moments, such as H_2O and O_3, have strong pure rotation spectra. Linear symmetric molecules such as CO_2, O_2, and N_2 have no electric dipole moment in the ground vibrational state and hence no electronic dipole rotational spectra. O_2 and N_2 lack dipole moments even when vibrating; consequently they have no vibrational or rotational features involving electronic dipole radiation.

To a first approximation, the lower vibrational energy levels of a molecule correspond to those of a linear harmonic oscillator and are uniformly spaced at levels $E_v = (v + \frac{1}{2})h\nu_0$ where $v = 0, 1, 2, \ldots$ is the *vibrational quantum number* and ν_0 is the fundamental frequency of the oscillator. The selection rule for dipole radiation for this simple oscillator is $\Delta v = \pm 1$ and the relative oscillator strengths for bands with upper-state quantum number v is proportional to v. In practice, anharmonicities of various kinds lead to departures from uniform spacing of the levels, to violations of the selection rule, and to weak "overtone" bands for which $|\Delta v| \geq 2$.

In a vibration–rotation transition, the selection rule for dipole radiation associated with a rotational energy change for a linear molecule such as CO_2 is either $\Delta J = 0, \pm 1$ or $\Delta J = \pm 1$. Such bands exhibit either two or three subbands corresponding to the two or three possibilities for ΔJ. The subband with $\Delta J = +1$ in absorption is called the *R branch* and lines on the short-wavelength end of the band. The band with $\Delta J = -1$ in absorption, called the *P branch*, falls on the long wavelength end. The band with $\Delta J = 0$, if it occurs, falls in the middle and is called the *Q branch*. A schematic spectrum of a linear molecule with an unresolved strong Q branch and with the ideal uniform line spacing $2B/h$ in the P and R branches is shown in Fig. 2.10.

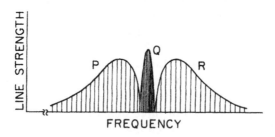

Fig. 2.10. Schematic vibration rotation band for a linear molecule showing the relationships between lines in the P, Q, and R branches.

2.3.4.b Infrared Spectrum of CO_2

Carbon dioxide is linear and symmetric in the ground state and has no pure rotational spectrum. The vibrational spectrum has three modes: symmetric stretching (ν_1), bending (ν_2), and asymmetric stretching (ν_3). Because of the symmetry of the ν_1 mode, transitions involving only ν_1 levels do not radiate, but the transition energy for ν_1 is very close to twice the transition energy for ν_2. As a result there is a strong interaction between ν_1 and ν_2, called *Fermi resonance*, and combined transitions involving changes in both ν_1 and ν_2 quantum numbers take place readily. Moreover, although there is no angular momentum about the molecular axis in the ground state, there is a contribution to this component of angular momentum in the excited ν_2 (bending) states. As a result of this, and of the Fermi resonant interaction with ν_1 for quantum numbers $v_2 \gtrsim 2$, these excited ν_2 energy levels split, and an additional angular momentum quantum number l is needed to represent these states. The vibrational state of CO_2 is described by the

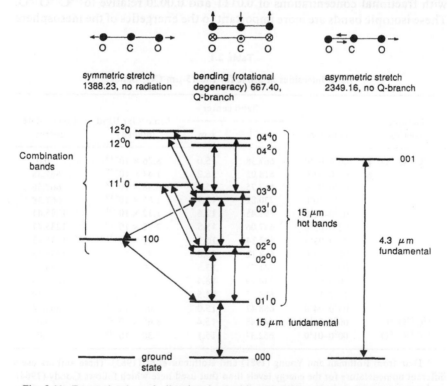

Fig. 2.11. Energy levels and vibrational transitions for CO_2. Note the large number of transitions corresponding to hot bands in the 15-μm region (667 cm^{-1}).

notation $(v_1\ v_2^l\ v_3)$; for example, 03^10 corresponds to $v_1 = 0$, $v_2 = 3$, $v_3 = 0$, and $l = 1$.

These energy levels are illustrated in Fig. 2.11, together with some of the allowed transitions producing bands in the 15-μm region. The transition 01^10–000 is the *fundamental* band. *Combination* bands such as 11^10–02^00 and 100–01^10 and *upper-state* bands such as 02^20–03^10 are together referred to as *hot bands*, since their strengths in absorption depend on molecular populations in excited states. As a consequence, intensities of these bands increase rapidly with temperature. The most important implication of Fig. 2.11 is that there are many CO_2 vibration–rotation bands in the 15-μm region, each with a slightly different central wave number, and each bearing a general resemblance to the schematic spectrum shown in Fig. 2.10. Strengths of all of these bands except the fundamental are very sensitive to temperature. This picture is further complicated by the occurrence of bands of minor isotopes of CO_2 in the same spectral region. The most important isotopic bands are the fundamentals of $^{13}C\,^{16}O\,^{16}O$ and $^{12}C\,^{18}O\,^{16}O$ with fractional concentrations of 0.0111 and 0.0020 relative to $^{12}C\,^{16}O\,^{16}O$. These isotopic bands are more important to the energetics of the mesosphere

Table 2.1

Band Intensities at 296 K for the 15-μm CO_2 Bands[a]

Isotopic species	Transition	Band center		Molecular band intensity[b]	Lower-state energy
		cm^{-1}	μm		
$^{12}C\,^{16}O\,^{16}O$	00^00–01^10	667.38	15.0	8.26×10^{-18}	0.0
	01^10–02^00	618.03	16.2	1.44×10^{-19}	667.38
	01^10–02^20	667.75	15.0	6.48×10^{-19}	667.38
	01^10–10^00	720.81	13.9	1.85×10^{-19}	667.38
	02^00–11^10	791.45	12.6	1.12×10^{-21}	1285.41
	02^00–03^10	647.06	15.5	2.22×10^{-20}	1285.41
	02^20–03^10	597.34	16.7	5.21×10^{-21}	1335.13
	02^20–03^30	668.12	15.0	3.82×10^{-20}	1335.13
	02^20–11^10	741.73	13.5	7.90×10^{-21}	1335.13
	10^00–03^10	544.29	18.4	2.72×10^{-22}	1388.19
	10^00–11^10	688.67	14.5	1.49×10^{-20}	1388.19
	03^30–04^40	668.47	15.0	2.00×10^{-21}	2003.24
$^{13}C\,^{16}O\,^{16}O$	00^00–01^10	648.48	15.4	8.60×10^{-20}	0.0
$^{12}C\,^{18}O\,^{16}O$	00^00–01^10	662.37	15.1	3.30×10^{-20}	0.0

[a] Data from Rothman and Young (1981) and Rothman *et al.* (1983). These authors use a different nomenclature for the energy levels than that used here which follows Goody (1964).

[b] Band intensities correspond to $S = \int \sigma_\nu\, d(\nu/c)$, in centimeters per molecule, and are reckoned with respect to the total number of CO_2 molecules, all isotopes.

than their concentrations would suggest. That is because their optical depths are sufficiently small even in the centers of the strongest lines that they can efficiently emit radiation to space (see Section 2.5). Table 2.1 lists properties of some of the important 15-μm CO_2 bands. Ramanathan *et al.* (1985) have found it necessary to include some 44 individual 15-μm bands in their investigations of climatic change effects of increasing CO_2 concentration.

Figure 2.12 shows spectra of CO_2 in the 15-μm region at low resolution. In addition to the P and R branches of the fundamental and some of the hot bands, Q branches of 02^20–03^10, 01^10–02^00, 02^00–03^10, 01^10–10^00, and 02^20–11^10 can be clearly identified with the help of the data in Table 2.1. Some of these are blended with the Q branches of the ν_2 fundamentals of $^{13}C\,^{16}O\,^{16}O$ and $^{12}C\,^{18}O\,^{16}O$.

A spectrum at much higher resolution is shown in Fig. 2.13. Note the unresolved Q branches, the apparently regular P branch in the bottom panel,

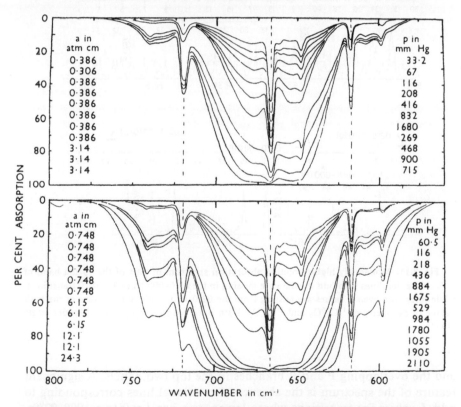

Fig. 2.12. Low-resolution laboratory absorption spectra of CO_2 at various combinations of optical path and pressure. Note the appearance of several Q branches as sharp absorption spikes. [From Burch *et al.* (1960), with permission.]

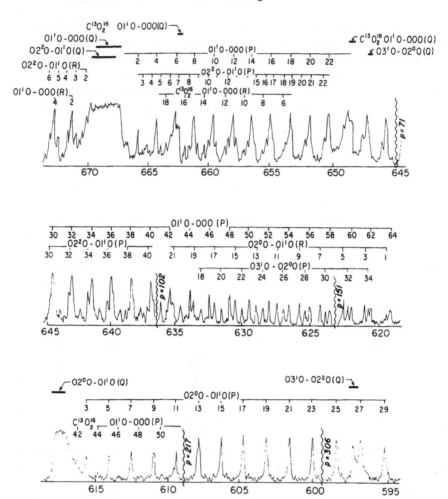

Fig. 2.13. Moderately high-resolution absorption spectrum of part of the 15-μm band of CO_2. Note that lines with odd J values are missing in the 01^10–000 and 03^10–02^00 transitions, while lines with even J values are missing from the 02^00–01^10 transition because of the high degree of symmetry of $^{12}C^{16}O_2$ and $^{13}C^{16}O_2$. Data have been replotted from Madden *et al.* (1957).

and the overlapping P and R branches in the top two panels. A significant feature of the spectrum is the absence of rotational lines corresponding to odd J values for transitions whose lower state has $l = 0$ (e.g., 000, 02^00). Because of the high degree of nuclear symmetry for $^{12}C^{16}O^{16}O$, antisymmetric rotational levels for these vibrational states are missing.

2.3.4.c   Infrared Spectrum of H_2O

The thermal infrared bands of H_2O are comparable in strength to those of CO_2, but the H_2O concentration is nearly two orders of magnitude smaller throughout most of the middle atmosphere ($\sim$5 ppmv compared with 340 ppmv). Consequently, H_2O has a relatively minor influence on thermal infrared exchange, but it is not totally negligible.

Water vapor is nonlinear and asymmetric. It has a dipole moment in the ground state and a strong rotational band extending from the 15-μm spectral region toward longer wavelengths. This is the most important water-vapor band for thermal radiative exchange in the atmosphere. Of the vibration–rotation bands, only the ν_2 fundamental at 6.3 μm influences thermal radiative exchange in the middle atmosphere, and then only at relatively high temperatures. A number of other bands, including higher-overtone bands, occur in the near-infrared and visible spectral regions. They are responsible for absorption of solar radiation and significant heating in the troposphere, but because of the low water-vapor concentration they make only a minor contribution to heating in the middle atmosphere.

Because water vapor is nonlinear, it has three angular momentum components, in contrast to one for a linear molecule in the ground state, and as a consequence its spectrum is relatively complex. The most important consequence of this structure is that the H_2O rotation levels are split into $2J + 1$ sublevels, where J is the quantum number for total angular momentum. These sublevels are irregularly spaced, and for the larger values of J, sublevels for different J values overlap. The result is a spectrum with a random appearance of line positions and intensities, as illustrated in Fig. 2.14. Although the spectrum of water vapor has been studied for many years, the band parameters are not as accurately known as those for CO_2.

2.3.4.d   Infrared Spectrum of Ozone

With mixing ratios ranging up to 15 ppmv and a strong thermal infrared spectrum, ozone is intermediate in importance to CO_2 and H_2O for infrared energy exchange in the middle atmosphere. It is also asymmetric, and it has a dipole moment in the ground state and a pure rotational spectrum. However, the important bands for thermal radiative exchange are the vibration–rotation bands. Strong ν_1 and ν_3 fundamentals at 1110 and 1045 cm^{-1} together with hot bands and minor isotopic bands in the same spectral region comprise the important 9.6-μm band system. The ν_2 fundamental at 701 cm^{-1} overlies the CO_2 15-μm band system and is at most of very minor significance for heating-rate calculations, but it has to be taken

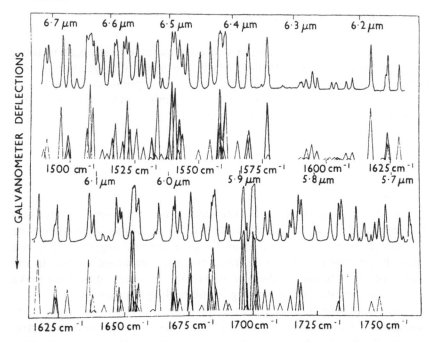

Fig. 2.14. Comparison between theoretical (triangles) and observed spectrum for H_2O in the 6.3-μm region. [From Nielsen (1941), with permission.]

into account in the calculation of accurate 15-μm band transmission functions for remote sensing purposes.

Because of the molecular asymmetry and the occurrence of hot and isotopic bands in the same spectral interval, the structure of the 9.6-μm band is very complex. Line positions and line strengths are rather randomly distributed, but average line spacings are smaller than those in the CO_2 or water-vapor bands. Spectroscopic parameters for O_3 are the least well known of the three gases, but because the contribution of ozone to the middle atmosphere cooling rate is generally less than one-third as large as that for CO_2, uncertainty in ozone spectroscopic parameters should make a relatively small contribution to net heating rate error.

2.3.4.e Other Gases

Minor contributions to the net heating of the atmosphere are made by infrared bands of methane (mixing ratio $\leq$1.6 ppmv), N_2O (mixing ratio $\leq$0.4 ppmv), HNO_3 (mixing ratio $\leq$0.01 ppmv), and the chlorofluoromethanes, $CFCl_3$ and CF_2Cl_2 (mixing ratios ~1 ppbv). In combination, plausible increases in these gases could enhance the "greenhouse effect" due to enhanced downward infrared radiation from the atmosphere by an amount comparable to that expected from doubling of the CO_2 concentration (see Section 12.3). At present, in the middle atmosphere, these gases contribute no more than a few hundredths of a degree per day to net heating. Since CH_4, N_2O, and $CFCl_3$ and CF_2Cl_2 have absorption bands in the relatively transparent atmospheric "window" region between 7 and 12 μm, so that these gases exchange radiation primarily with the warm underlying troposphere and surface, their main effect is to warm the extreme lower stratosphere and the tropopause region.

2.4   Transmission Functions

Exchange of radiation within the atmosphere and between the atmosphere and its surroundings is described by *transmission functions*. It is necessary to distinguish between *monochromatic* and spectrally averaged or *band* transmission functions and between *parallel beam* and *flux* transmission functions. The former describe the exchange of radiance and the latter describe the exchange of flux between isotropically emitting surfaces or layers. It is also useful to distinguish between transmission functions in the laboratory for which absorption and extinction coefficients are constant along optical paths and transmission functions in the atmosphere where absorption and extinction coefficients vary because of varying temperature and pressure along optical paths. In this and the following three sections, only gaseous absorption and emission (but no scattering) are considered, so that $\bar{\omega}_\nu = 0$ and $a_\nu = k_\nu$.

2.4.1   Definitions

The *monochromatic parallel beam transmission function* $T_\nu(s_1, s_2)$ appeared in Eq. (2.2.13) as the fraction of the monochromatic radiance leaving a point s_1 that reaches point s_2,

$$T_\nu(s_1, s_2) = \exp\left[-\int_{s_1}^{s_2} k_\nu(s)\rho_a(s) \, ds \right]. \qquad (2.4.1)$$

For the laboratory path at pressure p and temperature T, the extinction coefficient can be removed from the integration, and Eq. (2.4.1) can be expressed in the form

$$T_\nu(p, T, u) = \exp[-k_\nu(p, T)u(s_1, s_2)] \qquad (2.4.1')$$

where

$$u(s_1, s_2) = \int_{s_1}^{s_2} \rho_a \, ds \qquad (2.4.2)$$

is the optical mass between points s_1 and s_2. For a plane-parallel atmosphere, Eqs. (2.4.1) can be written in terms of the optical depth [Eq. (2.2.15b)],

$$T_\nu(s_1, s_2) = T_\nu(z_1, z_2, \mu) = \exp[-\mu^{-1} \Delta \tau_\nu(z_1, z_2)], \qquad (2.4.1'')$$

where $\Delta \tau_\nu(z_1, z_2) = |\tau_\nu(z_2) - \tau_\nu(z_1)|$. Note that $\mu^{-1} \Delta \tau_\nu(z_1, z_2)$ is positive for both upward and downward paths.

Calculation of the exchange of thermal radiation requires integration over the infrared spectrum. This integration involves thousands of absorption lines, each with a rapidly varying profile. On the other hand, the source function is generally a slowly varying function of frequency [e.g., the Planck function of Eq. (2.2.10)], so that it is convenient to work with a spectrally integrated transmission function. This quantity, the *band transmission function* $\bar{T}_r$, is defined by

$$\bar{T}_r = \Delta \nu_r^{-1} \int_{\Delta \nu_r} T_\nu \, d\nu, \qquad (2.4.3)$$

where $\Delta \nu_r$ is a spectral band width. This is usually taken to be wide enough to encompass many lines but narrow enough that spectral variations of the source function are small. Just as for the monochromatic transmission function, the functional dependences of the band transmission functions for laboratory and atmospheric cases are

$$\bar{T}_r(\text{lab}) = \bar{T}_r(p, T, u)$$

and

$$\bar{T}_r(\text{atmos}) = \bar{T}_r(z_1, z_2, \mu).$$

Two useful quantities that are closely related to $\bar{T}_r$ are the *band absorptivity*

$$\bar{A}_r = 1 - \bar{T}_r \qquad (2.4.4)$$

and the *equivalent width* (expressed in frequency units),

$$W_r = \Delta \nu_r(1 - \bar{T}_r) = \int_{\Delta \nu_r} [1 - T_\nu] \, d\nu. \qquad (2.4.5)$$

The latter is most often used in describing laboratory data, in which case

$$W_r = W_r(p, T, u).$$

Equivalent width is also known as the *integrated absorptance*. It can be applied either to a single isolated line or to a complete band such as the

9.6-μm band of O_3 or the 15-μm band of CO_2. The relationship between $W_r(p, T, u)$ and u for fixed p and T is called the *curve of growth*.

The *flux transmission function*, applicable to a plane-parallel atmosphere, is the μ-weighted angular average of Eq. (2.4.3),

$$\bar{T}_f(z_1, z_2) = \int_0^1 \mu \bar{T}_r(z_1, z_2, \mu) \, d\mu \cdot \left(\int_0^1 \mu \, d\mu \right)^{-1}$$

$$= 2 \int_0^1 \mu \bar{T}_r(z_1, z_2, \mu) \, d\mu. \tag{2.4.6}$$

For an isotropically radiating horizontal surface, this gives the fraction of the flux in band $\Delta \nu_r$ that reaches level z_2 after leaving level z_1 (or vice versa). The angular integration can be carried out explicitly,

$$\bar{T}_f(z_1, z_2) = \frac{2}{\Delta \nu_r} \int_{\Delta \nu_r} d\nu \int_0^1 \exp[-\Delta \tau_\nu(z_1, z_2)/\mu] \cdot \mu \, d\mu,$$

or, with the substitution $\eta \equiv \mu^{-1}$,

$$\bar{T}_f(z_1, z_2) = \frac{2}{\Delta \nu_r} \int_{\Delta \nu_r} E_3[\Delta \tau_\nu(z_1, z_2)] \, d\nu, \tag{2.4.7}$$

where $E_n(x)$ is the nth exponential integral,

$$E_n(x) = \int_1^\infty e^{-\eta x} \, d\eta / \eta^n. \tag{2.4.8}$$

We turn now to methods for evaluating transmission functions in the atmosphere from spectroscopic or laboratory data.

2.4.2 Line-by-Line Integration

The central problem is to carry out the spectral integration implied by Eq. (2.4.3), (2.4.5), or (2.4.7). Straightforward integration over the spectrum is the most accurate approach, but it is also extremely time consuming. For a bandwidth $\Delta \nu_r$ of 20 cm^{-1}, several hundred lines would have to be taken into account for the 15-μm CO_2 or 9.6-μm O_3 bands. This would imply several thousand spectral integration steps, which would have to be repeated 15–20 times to cover each relevant subband of width $\Delta \nu_r$ and for each pair of levels participating in the exchange. Since transmission functions depend on temperature and pressure, this process would have to be repeated for each atmospheric profile having a different temperature-pressure and/or absorbing gas-pressure relationship.

This is obviously an expensive process even for a very fast computer, but it is not prohibitive for limited calculations, and it can be used to provide benchmarks for approximate methods. It has the advantage that spectral variations are accurately evaluated and at the same time the effects of pressure and temperature on line strength and line width along the varying atmospheric paths can be accurately taken into account. Moreover, with the aid of a table or an efficient algorithm for obtaining the exponential integral, precise angular integration can also be incorporated in the calculation by using Eq. (2.4.7). The full Voigt line shape or accurate approximations to it can be readily used, the choice of approximation depending on the desired accuracy and the acceptable degree of complexity of the calculation. The accuracy of this approach is limited by the accuracy of the spectroscopic line parameters.

Fels and Schwarzkopf (1981) have carried out such a calculation for the 15-μm bands of carbon dioxide and have tabulated transmission functions between pairs of levels on a closely spaced grid between the surface and the upper mesosphere. For CO_2 the composition is uniform, but varying temperature profiles have to be taken into account. This was done by calculating transmission functions for a standard atmosphere temperature profile $T_0(p)$ and for "warm" and "cold" profiles, $T_0(p) + 25$ K and $T_0(p) - 25$ K. They were able to show that quadratic interpolation of transmission functions between these three sets of transmission functions using the mass-weighted mean temperature between the two endpoint levels of the transmission function provides a very accurate approximation to the exact flux transmission function for the entire 15-μm band system, provided that the actual temperature profile lies within the limits $T_0(p) + 25$ K and $T_0(p) - 25$ K. These tabulated values also incorporate a weighting factor to account for the spectral variation of the Planck function over the 15-μm band. Fels and Schwarzkopf have also carried out this calculation for higher CO_2 concentrations, since these may be relevant in the future.

2.4.3 Band Models

Because the concentrations of water vapor and ozone vary, no line-by-line method that is generalizeable to arbitrary temperature and composition profiles has been applied to these gases. This is one reason for considering a simplified transmission-function computation based on approximating the structure of vibration–rotation bands by specific models. Another advantage of this approach is that it makes clear the limiting behavior of transmission functions in important asymptotic regimes. Four specific issues will be dealt with in turn: treatment of the spectral integration along paths at constant pressure and temperature for Lorentz lines, treatment of varying pressure and temperature along the path, angular integration, and treatment of the transition to the Doppler line shape.

Consider first the absorptivity for a single line along a path at constant pressure and temperature $\bar{A}_1$ over an interval $\Delta\nu_r$ that is symmetric about the line center. This absorptivity is

$$\bar{A}_1 = \frac{1}{\Delta\nu_r} \int_{\Delta\nu_r} [1 - e^{-k_\nu u}] \, d\nu$$

$$= \frac{1}{\Delta\nu_r} \int_{\Delta\nu_r} \{1 - \exp[-\tilde{S}f(\nu)u]\} \, d\nu \qquad (2.4.9)$$

where $f(\nu)$ is the line-shape factor normalized to unity with line-center frequency set equal to zero and $\tilde{S}$ is the line strength referenced to absorber mass rather than to absorber number as in Eqs. (2.3.2) ($\tilde{S} \equiv S/M_a = \int k_\nu \, d\nu$). If the lines in the band are sufficiently well separated that their overlapping contributions to absorption are negligible, Eq. (2.4.9) can be approximated by

$$\bar{A}_1 \simeq \frac{1}{\delta} \int_{-\infty}^{\infty} \{1 - \exp[-\tilde{S}f(\nu)u]\} \, d\nu \qquad (2.4.10)$$

where δ is the mean spacing between lines. Above about 25 km, this isolated line approximation is sufficiently accurate for the P and R branches of CO_2 and for water vapor. It is also adequate for CO_2 Q branches and the O_3 9.6-μm band above about 40 km.

When $f(\nu)$ is the Lorentz line shape, the integral in Eq. (2.4.10) can be expressed in a closed form, known as the *Ladenberg–Reiche* function,

$$\bar{A}_1(y, \hat{u}) = \hat{u}e^{-\hat{u}}[I_0(\hat{u}) + I_1(\hat{u})] \qquad (2.4.11)$$

where

$$y = \alpha_L/\delta, \qquad \hat{u} = \frac{\tilde{S}u}{2\pi\alpha_L} \qquad (2.4.12)$$

and I_0 and I_1 are modified Bessel functions. The isolated line absorptivity, $\bar{A}_1$, has useful asymptotic limits that can be derived without reference to Eq. (2.4.11). For sufficiently small $\hat{u}$, the exponential in Eq. (2.4.10) can be replaced by the first two terms in its power series expansion, so that

$$\bar{A}_1 \to \frac{\tilde{S}u}{\delta} \int_{-\infty}^{\infty} f(\nu) \, d\nu = \tilde{S}u/\delta = 2\pi y\hat{u} \qquad \text{as} \quad \hat{u} \to 0. \qquad (2.4.13)$$

This *weak line limit* is valid for any line shape provided that absorption is sufficiently weak in the line center. On the other hand, if the line is fully absorbed, or *saturated*, in the line center and for some spectral distance on either side of the center, most of the contribution to $\bar{A}_l$ comes from the line wings for which the Lorentz profile can be approximated by $f(\nu) \approx \alpha_L/\pi\nu^2$. Substitution of this expression into Eq. (2.4.10) leads to the *strong line limit*,

$$\bar{A}_l \to 2y(2\pi\hat{u})^{1/2} = \frac{2}{\delta}(\tilde{S}u\alpha_L)^{1/2}. \tag{2.4.14}$$

The Ladenberg–Reiche formula interpolates between these two asymptotic regimes.

Next consider the mean absorptivity for a band containing an array of nonoverlapping lines whose strength distribution is described by the density function $\hat{Q}(\tilde{S})$ such that $\hat{Q}(\tilde{S})\, d\tilde{S}$ represents the probability that a line selected at random from the array has strength between $\tilde{S}$ and $\tilde{S} + d\tilde{S}$. It follows that the average absorptivity of the band is

$$\bar{A}_{sl} \approx \frac{1}{\delta} \int_{-\infty}^{\infty} \int_0^{\infty} \hat{Q}(\tilde{S})[1 - e^{-\tilde{S}f(\nu)u}]\, d\tilde{S}\, d\nu. \tag{2.4.15}$$

Equation (2.4.15) can be evaluated in closed form for the Lorentz line shape and some reasonably realistic line strength distributions. Absorptivities $\bar{A}_{sl}$ also have weak and strong limits, but mean absorptivities for the cases with distributed line strengths approach the strong line limit more slowly than the isolated single line absorptivity $\bar{A}_l$ because line arrays include very weak lines, which are slow to saturate.

When line overlap is important, the monochromatic transmission function for the overlapping lines is the product of the monochromatic transmission functions for the individual lines. Thus, for n overlapping lines.

$$T_\nu = \prod_{i=1}^{n} e^{-k_{\nu,i}u} = \exp\left[-\sum_{i=1}^{n} k_{\nu,i}u\right]. \tag{2.4.16}$$

A widely used model for incorporating the effects of line overlap into the evaluation of absorptivity is the *random model*. It is assumed that line centers are randomly distributed over the band width $\Delta\nu_r$, that the probability of any line being centered at a particular position in the interval is independent of the probability that any other line is at any other position in the interval, and that the line strength probability distribution for any line is independent of the strengths of all other lines. With these assumptions, Eq. (2.4.16) can be evaluated for any line shape, and the result is the general random model,

$$\bar{T}_n = \left(1 - \frac{W_{sl}}{n\delta}\right)^n, \tag{2.4.17}$$

where $W_{sl} \equiv \bar{A}_{sl}\delta$ is the average single-line equivalent width from Eqs.

(2.4.15). Equation (2.4.17) is usually applied in the limit $n \to \infty$, in which case

$$\bar{T}_n \to \bar{T}_r(\text{random}) = e^{-W_{sl}/\delta}. \tag{2.4.18}$$

Note that for strongly overlapping lines W_{sl}/δ may exceed unity.

Two distributions of line strength that are reasonable representations of actual line strength distributions for some bands are

$$\hat{Q}(\tilde{S}) = \frac{1}{\sigma} e^{-\tilde{S}/\sigma} \tag{2.4.19}$$

and

$$\tilde{Q}(\tilde{S}) = \tilde{S}^{-1} e^{-\tilde{S}/\sigma}, \tag{2.4.20}$$

where σ is a constant characterizing the distribution. Equation (2.4.20) is singular at the origin and so can be useful for representing a band with a relatively large number of very weak lines. With $\hat{u} \equiv \sigma u/2\pi\alpha_L$, the first of these leads to the *Goody model*,

$$\bar{T}_r(\hat{u}, y) = \exp[-2\pi y\hat{u}(1 + 2\hat{u})^{-1/2}], \tag{2.4.21}$$

while the second leads to the *Malkmus model*,

$$\bar{T}_r(\hat{u}, y) = \exp\{-2\pi y[(1 + 2\hat{u})^{1/2} - 1]\}. \tag{2.4.22}$$

These transmission functions have the following asymptotic limits. (1) If $\hat{u}$ is small, $\bar{T}_r \to e^{-2\pi y\hat{u}} = e^{-\sigma u/\delta}$ in both cases. In this limit, absorption is very weak in the centers of the lines of average strength and the transmission function has the same form as for spectrally uniform (gray) absorption with absorption coefficient σ/δ. If in addition $2\pi y\hat{u} \ll 1$, the lines are well separated and $\bar{T}_r \to 1 - 2\pi y\hat{u} = 1 - \sigma u/\delta$, the isolated weak line limit for a line of strength σ. (2) If $\hat{u}$ is large, $\bar{T}_r \to e^{-\pi y(2\hat{u})^{1/2}}$ for the Goody model and $\bar{T}_r \to e^{-2\pi y(2\hat{u})^{1/2}}$ for the Malkmus model. If in addition $y^2\hat{u} \ll (2\pi)^{-2}$, the lines are well separated and $\bar{T}_r(\text{Goody}) \to 1 - \pi y(2\hat{u})^{1/2}$ and $\bar{T}_r(\text{Malkmus}) \to 1 - 2\pi y(2\hat{u})^{1/2}$. These expressions correspond to the square-root regime of the isolated strong line.

Equations (2.4.21) and (2.4.22) are best thought of as physically motivated expressions for the band transmission function, which can be optimized in applications by fitting to characteristics of any particular spectral interval $\Delta\nu_r$ as determined from laboratory data or from calculations based on the fundamental spectroscopic parameters.

The Goody model applied to 10-cm^{-1} band intervals provides an excellent representation of transmission in the $6.3\text{-}\mu\text{m}$ and rotation bands of water vapor. Since the $9.6\text{-}\mu\text{m}$ band of ozone and particularly the $15\text{-}\mu\text{m}$ band system of CO_2 are less random, they are more difficult to represent with random models, but the Malkmus model has been shown by Kiehl and

Ramanathan (1983) to give an excellent representation of transmission in these bands, provided it is applied to narrow spectral intervals (~ 5 cm^{-1} in width).

The random models are most useful when applied to relatively narrow spectral divisions of a band. Whole-band absorptivity formulations are also available and can be used to provide transmission functions with accuracies of 10% or better for H_2O, O_3, CO_2, CH_4, and N_2O. One widely used form for such models that is particularly useful for weak bands is the exponential sum fit. In this approximation, the equivalent width is represented by

$$W_r(u) = \Delta \nu_r \left[1 - \sum_{i=1}^{n} f_i e^{-\tau_i} \right], \qquad (2.4.23a)$$

$$\tau_i = \left(\frac{k_i}{f_i} \right) \left(\frac{\tilde{S}u}{\Delta \nu_r} \right); \qquad (2.4.23b)$$

constrained by

$$\sum_{i=1}^{n} f_i = \sum_{i=1}^{n} k_i = 1, \qquad (2.4.23c)$$

where the model parameters k_i, f_i, $\tilde{S}$, and $\Delta \nu_r$ are obtained by fitting laboratory data. Note that in the weak absorption limit Eq. (2.4.23a) gives $W_r(u) \rightarrow \tilde{S}u$.

A whole band equivalent width formulation that is more appropriate to strong bands has been widely used in the stratospheric calculations by Cess and Ramanathan and their collaborators. The equivalent width W_r is represented by

$$W_r = 2A_0(T) \ln \left(1 + \sum_{i=1}^{n} \xi_i^{1/2} \right) \qquad (2.4.24a)$$

where

$$\xi_i = \frac{4 \alpha_{\nu i} \tilde{S}_i u}{A_0(T) \delta} \qquad (2.4.24b)$$

and the summation is over n overlapping bands of strengths $\tilde{S}_i$ with lines of Lorentz half-width $\alpha_{\nu i}$ and mean spacing δ in the spectral interval. The effective band width parameter $A_0(T)$ is determined by comparing W_r with laboratory data. Note that in the limit $\xi_i \rightarrow 0$, the dependence on half-width, band intensity, and optical path corresponds to the nonoverlapping strong line limit.

Water vapor is a special case because its absorption covers such broad spectral intervals. In this case, it is convenient to define an *emissivity* ε_l

appropriate for laboratory situations,

$$\varepsilon_1(p, T, u) = \int_0^\infty B_\nu(T)[1 - T_\nu(p, T, u)] \, d\nu/(\sigma T^4/\pi),$$

and the corresponding *flux emissivity*,

$$\varepsilon_f(p, T, z_1, z_2) = 2\pi \int_0^1 \mu \, d\mu \int_0^\infty B_\nu(T)[1 - T_\nu(z_1, z_2, \mu)] \, d\nu/\sigma T^4$$

$$= 2 \int_0^1 \varepsilon_1[\tilde{p}, \tilde{T}, \tilde{u}(\mu)]\mu \, d\mu, \tag{2.4.25}$$

where $\tilde{p}$, $\tilde{T}$, and $\tilde{u}$ are appropriate mean values for the slant path at angle $\cos^{-1}\mu$ between z_1 and z_2. Such expressions are particularly useful for water vapor because the temperature dependence arising from the Planck function through the factor $T^{-4}B_\nu(T)$ is quite small. Water vapor has strong absorptions on both extremes of the Planck function, the rotation and 6.3-μm bands, so that there is a high degree of cancellation between the opposing temperature dependences of the two bands. The resulting small temperature dependence of ε_f makes it possible to carry out surprisingly accurate flux calculations using simple expressions of the form

$$F_\uparrow(z) = \int_0^1 \sigma T^4 \, d\varepsilon_f(z', z), \tag{2.4.26}$$

with the convention that T corresponds to the surface temperature for values of ε_f such that $\varepsilon_f(z', z) > \varepsilon_f(0, z)$. However, when this formulation is used, it is necessary to take overlapping bands such as the 15-μm CO_2 band into account.

2.4.4 Treatment of Pressure and Temperature Variations along the Path

All of the models discussed above are applicable directly to parallel-beam transmission along a constant-pressure and constant-temperature path for the Lorentz line shape. Along vertical paths, the actual line shape is not Lorentz. It is a composite of high- and low-pressure Lorentz lines with broad high-pressure wings and a narrow low-pressure line center (Fig. 2.15). Approximate treatments of the variation of temperature and pressure along the atmospheric paths depend on the type of band model. For one-parameter models, such as Eq. (2.4.23), a scaling approximation can be employed.

In scaling approximations, pressure and temperature dependences are arbitrarily assumed to be separable from frequency dependence,

$$k_\nu(p, T) = \phi(p, T)\psi(\nu).$$

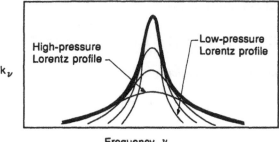

Fig. 2.15. Schematic composite showing how an actual line profile over a variable pressure path forms as a composite of the individual Lorentz profiles. The atmospheric line profile is not Lorentz in general; it is more sharply peaked because of low-pressure contributions, with broader wings due to high-pressure contributions.

Then

$$\Delta \tau_\nu = \int \psi(\nu)\phi(p, T)\, du = \psi(\nu) \int \phi(p, T)\, du$$

$$= k_\nu(p_r, T_r) \int [\phi(p, T)/\phi(p_r, T_r)]\, du = k_\nu(p_r, T_r)\tilde{u},$$

where

$$\tilde{u} \equiv \int [\phi(p, T)/\phi(p_r, T_r)]\, du$$

is a scaled absorber mass and p_r and T_r are reference pressure and temperature. Spectral integration can then be carried out at (p_r, T_r), or the laboratory transmission function measured at (p_r, T_r) can be used. Atmospheric transmission over path $\Delta \tau_\nu$ is then a function of the single parameter $\tilde{u}$, and we have the equivalence

$$\bar{T}_r(\tilde{u}, \text{atmosphere}) \leftrightarrow \bar{T}_r(u, p_r, T_r, \text{laboratory}). \qquad (2.4.27)$$

Pressure dependence is frequently represented by a power law, so that

$$\phi(p, T) = \bar{S}(T)(p/p_r)^n$$

where $\bar{S}$ is a mean line strength for the interval and the empirical exponent n is determined from laboratory measurements at varying pressures. With this approximation,

$$\tilde{u} = \int \frac{\bar{S}(T)}{\bar{S}(T_r)}(p/p_r)^n\, du. \qquad (2.4.28)$$

In the weak line limit, absorption is independent of pressure so that $n \approx 0$ is appropriate for weak absorption. In the strong line limit, absorption is proportional to the square root of the product of pressure and absorber mass so that $n \approx 1$ is appropriate for this limit, and in general $0 \lesssim n \lesssim 1$. The scaling approximation works fairly well for the troposphere, but it is a poor approximation if important contributions to radiative exchange are distributed over more than about two scale heights.

The random models are two-parameter models depending on an optical path parameter $\hat{u}$ and a line-width parameter y. For these models, there is a more satisfactory approximation, the *Curtis–Godson approximation*. In this approximation, the equivalence

$$\bar{T}_r(\hat{u}, \tilde{p}, \text{atmosphere}) \leftrightarrow \bar{T}_r(u, p, T, \text{laboratory}) \qquad (2.4.29)$$

is made. The equivalent width in the denominator of the Lorentz line shape [Eq. (2.3.5)] is replaced by a suitable mean value $\tilde{\alpha}$, so that, from Eq. (2.4.16),

$$\bar{T}_r \approx \frac{1}{\Delta \nu_r} \int_{\Delta \nu_r} d\nu \exp\left[-\int du \frac{\sum S_i \alpha_i}{\pi(\nu^2 + \tilde{\alpha}^2)} \right].$$

This expression, which varies from the exact expression only by the replacement of α by $\tilde{\alpha}$ in the denominator, is accurate in the strong line limit since $\tilde{\alpha}$ is neglected in the frequency integration in this case. The choice of $\tilde{\alpha}$ is then determined by requiring that this approximation provide an exact match to the weak limit,

$$\bar{T}_r(\text{weak}) \to 1 - \frac{1}{\Delta \nu_r} \int du \sum S_i.$$

Thus,

$$1 - \frac{1}{\Delta \nu_r} \int du \int_{\Delta \nu_r} \frac{\sum S_i \alpha_i}{\pi(\nu^2 + \tilde{\alpha}^2)} d\nu = 1 - \frac{1}{\Delta \nu_r} \int du \frac{\sum S_i \alpha_i}{\tilde{\alpha}}$$

$$= 1 - \frac{1}{\Delta \nu_r} \int du \sum S_i,$$

so that

$$\tilde{\alpha} = \int [\sum (S_i \alpha_i)] \, du \Big/ \int (\sum S_i) \, du.$$

Since $\alpha_i \approx (p/p_r)\alpha_r$ where α_r is the Lorentz width at standard pressure, and $\sum S_i = n\bar{S}(T)$, the path length and pressure in any two-parameter representation can be replaced by their Curtis–Godson means:

$$\tilde{p} = \frac{1}{\hat{u}} \int p[\bar{S}(T)/\bar{S}(T_r)] \, du \qquad (2.4.30a)$$

$$\tilde{u} = \int [\bar{S}(T)/\bar{S}(T_r)] \, du. \qquad (2.4.30b)$$

If the temperature dependence is neglected, the Curtis–Godson mean pressure is just the absorber mass weighted mean pressure. The Curtis–Godson approximation has been shown to be very accurate for absorbing gases whose concentration decreases monotonically with height. It is less accurate but still useful in the case of ozone, whose concentration increases with height in the troposphere and lower stratosphere.

2.4.5 Angular Integration

Next, consider the problem of converting the parallel-beam band transmission functions derived from band models into flux transmission functions. This can be done straightforwardly by evaluating the integrand of Eq. (2.4.6) at Gaussian quadrature points and carrying out the Gaussian integration. At each quadrature point, one simply replaces $\tilde{u}$ evaluated from Eq. (2.4.28) or (2.4.30b) with $\tilde{u}/\mu$. Convergence of the integration is rapid, and good accuracy can be achieved with only a small number (~ 2) of Gaussian points. However, a simpler approach is to take advantage of the fact that diffuse radiance for a plane-parallel atmosphere is at least approximately equivalent to direct beam radiance at some intermediate value of μ. In fact, heating rates in the troposphere and stratosphere can be calculated with an accuracy of $\sim 1\%$ with the approximation

$$\bar{T}_f(\tilde{p}, \tilde{u}) \simeq \bar{T}_r(\tilde{p}, \tilde{u}/\bar{\mu}), \qquad (2.4.31)$$

where

$$\bar{\mu} \approx \tfrac{3}{5}.$$

The reason for the surprising accuracy of this "diffuse flux factor approximation" for heating rate can be understood by returning to the exponential integral form of the flux transmission function of Eq. (2.4.7). It will be shown in Section 2.5 that the band heating rate at level z depends on the derivative of $\bar{T}_f$, $\partial \bar{T}_f(z, z')/\partial z$, or

$$\frac{\partial \bar{T}_f}{\partial z} = \pm \frac{2\rho_{a0}(z)}{\Delta \nu_r} \int_{\Delta \nu_r} k_\nu(z) \frac{dE_3(\Delta \tau_\nu)}{d(\Delta \tau_\nu)} \, d\nu$$

$$= \pm \frac{2\rho_{a0}(z)}{\Delta \nu_r} \int_{\Delta \nu_r} k_\nu(z) E_2(\Delta \tau_\nu) \, d\nu, \qquad (2.4.32)$$

where the sign depends on whether $\Delta \tau_\nu$ increases $(-)$ or decreases $(+)$ with z. The ν-dependence of the two factors in the integrand of Eq. (2.4.32) is

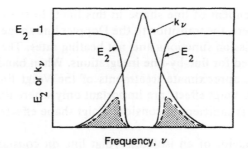

Fig. 2.16. Factors in the integrand of $\int \kappa_\nu E_2(\Delta\tau_\nu)\, d\nu$. Only the shaded areas contribute to the integral, and hence to $d\bar{T}_f/dz$.

depicted in Fig. 2.16 for a single line that is strong enough to be saturated in the center. Near the line center $E_2 \approx 0$, and in the far wings $k_\nu \approx 0$. Consequently, contributions to the integrand come only from the fairly narrow intervals in which $\Delta\tau_\nu \approx 1$. For any value of $\Delta\tau_\nu$, a quantity $\bar{\mu}^{-1}$ can be found such that

$$2E_3(\Delta\tau_\nu) = \exp(-\bar{\mu}^{-1}\Delta\tau_\nu),$$

where $1.5 \le \bar{\mu}^{-1} \le 2$. For $\Delta\tau_\nu = 1$, $\bar{\mu}^{-1} \approx \frac{5}{3}$. Thus, the accuracy and widespread applicability of the diffuse flux factor approximation stems from the fact that exchange in the strong bands of interest is dominated by spectral intervals in which $\Delta\tau_\nu \approx 1$. For very weak bands, the lines may be unsaturated near the centers and this may not be the case; instead, exchange may be dominated by smaller values of $\Delta\tau_\nu$, and a larger value of $\bar{\mu}^{-1}$ would be appropriate. For such bands, Gaussian quadrature may be used. Alternatively, where an exponential sum fit to whole band absorptivity is used, the diffuse flux factor can be used with great accuracy with $\bar{\mu}^{-1}$ given by

$$\bar{\mu}^{-1} = 1.5 + \frac{0.5}{1 + 4\tau_i + 10\tau_i^2} \tag{2.4.33a}$$

with τ_i obtained from Eq. (2.4.23b),

$$\tau_i = \sum_j \left(\frac{k_i}{f_i}\right)_j \left(\frac{\tilde{S}u}{\Delta\nu_r}\right)_j, \tag{2.4.33b}$$

where the summation is over all gases overlapping in the interval $\Delta\nu_r$ (Ramanathan et al., 1985).

2.4.6 Voigt Line-Shape Effects

The treatment up to this point is applicable to Lorentz lines. It is also applicable to any line shape in the weak line limit, since transmission functions are independent of line shape in this limit. In the upper stratosphere and mesosphere, the transition to the Doppler line shape has a strong influence on transmission functions and net heating rates. The exact Voigt line shape can be used for line-by-line integrations. When band models are employed, however, approximate treatments of the Voigt line shape are needed. Fortunately, Voigt effects are important only where line overlap is negligible, so that it is sufficient to consider Voigt shape effects for isolated lines.

Consider the behavior of an isolated Voigt line on constant-pressure-constant-temperature paths as a function of two parameters, $\hat{u} = \tilde{S}u/2\pi\alpha_L$ and $d = 2\alpha_L/\tilde{\alpha}_D$, where $\tilde{\alpha}_D = (\ln 2)^{1/2}\alpha_D$ is the Doppler half-width. Since the Voigt shape resembles the Doppler shape in the line center for $d < 1$, the maximum line-center strength is $\sim(\pi\tilde{\alpha}_D)^{-1}\tilde{S}u/(\ln 2)^{1/2}$. If this is sufficiently small, the weak line approximation applies and $\bar{A}_1 \approx \tilde{S}u/\delta$. On the other hand, if the line is saturated throughout the Doppler line core region, only the Lorentz wings influence the transmission and $\bar{A}_1 \approx 2(\tilde{S}u\alpha_L)^{1/2}/\delta$. Between these regimes, if d is sufficiently small that the line has a definite Doppler core, the "shoulders" and wings of the Doppler core dominate transmission. Since $f(\nu)$ falls off rapidly in the Doppler wings, the growth of $\bar{A}_1$ with $\hat{u}$ is relatively slow in this regime. Despite this relatively slow growth of absorption in the Doppler regime, the principal effect of the Voigt shape on atmospheric absorption and net heating rate is to increase the rate of change of absorption and the heating rate over values that would be obtained in the corresponding low pressure Lorentz regime. This is because the Lorentz line continues to narrow and strengthen in the line center as the pressure drops, so that absorption for a strong Lorentz line remains in the strong line regime even at very low pressure if the Voigt shape is neglected. In contrast, the transition to the weak line regime at low pressure does occur for Voigt lines because of the finite lower limit to line half-width and the corresponding maximum strength at line center, and $\bar{A}_1$ varies much more rapidly with absorber mass in the weak line regime than in the strong line regime.

Several approximate methods of incorporating the Lorentz-Doppler transition into band models have been suggested. One approach is to incorporate a smooth transition from Lorentz equivalent width W_L to Doppler equivalent width W_D. An interpolation formula suggested by Rodgers and Williams (1974) for this purpose is

$$W = \left[W_L^2 + W_D^2 - \frac{2W_L W_D}{(\tilde{S}u)^2} \right]^{1/2} \qquad (2.4.34)$$

Alternatively, Ramanathan (1976) has suggested that W_L be used when $dW_L/du > dW_D/du$ and that W_D be used otherwise. Fels (1979) has shown that the line shape

$$f(\nu) = C, \qquad |\nu| < |\nu_0| \qquad \text{(core)},$$

$$f(\nu) = \alpha_L / \pi\nu^2, \qquad |\nu| > |\nu_0| \qquad \text{(wings)},$$

$$\nu_0 = \varepsilon\alpha_L + \beta\alpha_D, \qquad (\varepsilon, \beta \text{ are constants } \sim 1) \qquad (2.4.35)$$

$$C = \frac{1}{2\nu_0} - \frac{\alpha_L}{\pi\nu_0^2},$$

might be used to approximate the behavior of isolated lines in the Voigt regime. This approximation reduces to the correct weak line limit and mimics the effect of strong line regimes that depend on the value of α_L / α_D. It can also be easily incorporated into random band models. Figure 2.17 shows the behavior of equivalent width for emission of radiation to space using the Goody random model and Eq. (2.4.35). The overlapping Lorentz, nonoverlapping strong line, weak line, and Doppler regimes can be clearly distinguished. Because actual band transmission functions involve the integrated effects of lines with a wide range of strengths, approximate treatments such as Eqs. (2.4.34) and (2.4.35), which are asymptotically

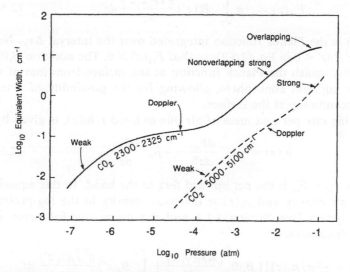

Fig. 2.17. Log–log plot of equivalent width versus pressure for the path between the indicated pressure level and space for a strongly absorbing spectral interval (CO_2, 2300–2325 cm^{-1}) and a weakly absorbing interval (CO_2, 5000–5100 cm^{-1}). The Goody random model with the Curtis–Godson approximation and the Fels approximation to the Voigt line shape have been used. Note the linear behavior in both weak and nonoverlapping strong line limits. Heating rate per unit mass due to solar absorption in these bands is proportional to W/p times the slope of these curves.

correct in weak and strong line limits, generally produce accuracies of
~20% or better in heating-rate calculations.

2.5   Infrared Radiative Exchange and Radiative Damping

In a plane-parallel atmosphere, consider the upward and downward flux
densities integrated over a band of width $\Delta\nu_r$ [$F_{r\uparrow}(z)$ and $F_{r\downarrow}(z)$] at a
log-pressure level z in an atmosphere bounded below by a surface radiating
as a blackbody at $z = 0$, and unbounded above. If the band is sufficiently
narrow that the Planck function is essentially constant over the band and,
moreover, if $\bar{\omega}_\nu = 0$ for the entire band, if surface emissivity is unity, and
if local thermodynamic equilibrium prevails, it follows from Eqs. (2.4.1″),
(2.4.3), (2.2.18), (2.2.21), and (2.4.6) that

$$F_{r\uparrow}(z) = \pi B_r(0_-)\bar{T}_r(0, z) + \pi \int_0^z B_r(z')\frac{\partial \bar{T}_r(z, z')}{\partial z'}\,dz' \qquad (2.5.1)$$

and

$$F_{r\downarrow}(z) = -\pi \int_z^\infty B_r(z')\frac{\partial \bar{T}_r(z, z')}{\partial z'}\,dz', \qquad (2.5.2)$$

where $B_r(z)$ is the Planck function integrated over the interval $\Delta\nu_r$. Note
that $\partial\bar{T}_r(z, z')/\partial z' < 0$ in Eq. (2.5.2), so that $F_{r\downarrow}(z) > 0$. The notation $B_r(0_-)$
is used to distinguish the Planck function at the surface from that of the
immediately adjacent atmosphere, allowing for the possibility of a temperature discontinuity at the surface.

The heating rate per unit mass of air due to band r, $h_r(z)$, is given by

$$h_r(z) = -\rho^{-1}\frac{dF_{n,r}}{dz^*} = -\rho_0^{-1}\frac{dF_{n,r}}{dz}, \qquad (2.5.3)$$

where $F_{n,r} = F_{r\uparrow} - F_{r\downarrow}$ is the net upward flux in the band. In this equation
$\rho(z)$ is the air density and $\rho_0(z)$ is the basic density in the log-pressure
coordinate system [see Section 3.1.1 and the discussion following Eq.
(2.2.15)]. It follows that

$$h_r(z) = -[\pi/\rho_0(z)]\left\{ B_r(0_-)\frac{d\bar{T}_r(0, z)}{dz} + \int_0^z B_r(z')\frac{\partial^2 \bar{T}_r(z, z')}{\partial z\,\partial z'}\,dz' \right.$$

$$\left. + \int_z^\infty B_r(z')\frac{\partial^2 \bar{T}_r(z, z')}{\partial z\,\partial z'}\,dz' - B_r(z)\left[\frac{\partial \bar{T}_r(z, z')}{\partial z'}\right]_{z'=z_-}^{z'=z_+} \right\},$$

$$(2.5.4)$$

the last term giving the contribution due to the jump in $\partial\bar{T}_r(z, z')/\partial z'$ at $z' = z$.

2.5.1 Exchange Integral Formulation

Equation (2.5.4) can be rearranged to give

$$
h_r(z) = [\pi/\rho_0(z)]\Bigg\{ -B_r(z)\frac{d\bar{T}_f(z,\infty)}{dz} - [B_r(0_-) - B_r(z)]\frac{d\bar{T}_f(z,0)}{dz}
$$

$$
- \int_0^z [B_r(z') - B_r(z)]\frac{\partial^2\bar{T}_f(z,z')}{\partial z\,\partial z'}\,dz'
$$

$$
- \int_z^\infty [B_r(z') - B_r(z)]\frac{\partial^2\bar{T}_f(z,z')}{\partial z\,\partial z'}\,dz' \Bigg\}.
\tag{2.5.5}
$$

The four terms in this "exchange integral formulation" of the heating-rate equation represent the contributions to infrared heating at z due to the possible exchanges of photons between levels. The first term represents "exchange" with space. Since downward flux from space can be neglected, this term always contributes to *cooling*. The second term corresponds to exchange with the underlying surface. Since $d\bar{T}_f(z,0)/dz < 0$, it contributes to heating if $B_r(0_-) > B_r(z)$. Similarly, the last two terms represent contributions due to exchanges with underlying and overlying layers and contribute to heating wherever $B_r(z') > B_r(z)$. Note that $\partial^2\bar{T}_f/\partial z\,\partial z'$ is always negative or zero.

The essence of the exchange problem is expressed by the factor $\partial\bar{T}_f(z,z')/\partial z$. It is often convenient to express the exchange integral formulation in terms of a corresponding function that has been normalized to lie in the range 0–1. This normalization yields the function

$$
\Gamma_r(z,z') = \frac{\Delta\nu_r}{2\rho_{a0}(z)\tilde{S}_r(z)}\left|\frac{\partial\bar{T}_f(z,z')}{\partial z}\right|
$$

$$
= \frac{1}{\tilde{S}_r(z)}\int_0^1 d\mu \int_{\Delta\nu_r} k_\nu(z)\exp[-\Delta\tau_\nu(z,z')/\mu]\,d\nu
$$

$$
= \frac{1}{\tilde{S}_r(z)}\int_{\Delta\nu_r} k_\nu(z)E_2[\Delta\tau_\nu(z,z')]\,d\nu.
\tag{2.5.6}
$$

The term $\Gamma_r(z,z')$ represents the probability that a photon emitted in band r between z and $z + dz$ will escape to level z' before being reabsorbed, and for this reason it is appropriately called the *escape function*. Note that, unlike $T_r(z,z')$, Γ_r is not symmetric in z and z', so that the order of its arguments is important. Examples of $\Gamma_r(z,z')$ are shown in Fig. 2.18. The term $\Gamma_r(z,z')$ decays monotonically away from z with an extremely sharp peak and a first-order discontinuity at z. For a gas whose concentration does not increase with height, $\Gamma_r(z,z')$ decreases more rapidly above than below z. As illustrated in Fig. 2.18, it generally has a small but finite value as $z \to \infty$,

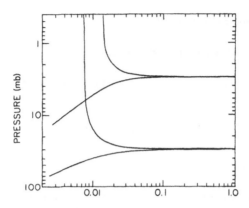

Fig. 2.18. Escape functions $\Gamma_r(z, z')$ for the 15-μm band of CO_2 centered at the 30 and 3 mb levels. [From Leovy (1984).]

corresponding to the probability of photon escape to space. In terms of Γ_r, Eq. (2.5.5) can be rewritten in the compact Stieltjes integral form,

$$h_r(z)\{[2\pi m_a(z)\tilde{S}_r(z)]^{-1} \Delta\nu_r\} = \int_0^1 {}_{z'<z} [B_r(z') - B_r(z)] \, d\Gamma_r(z, z')$$

$$+ \int_0^1 {}_{z'>z} [B_r(z') - B_r(z)] \, d\Gamma_r(z, z'), \qquad (2.5.7)$$

where m_a is the absorber mass mixing ratio, and it is understood that $B_r(z') = B_r(0_-)$ for $\Gamma_r(z, z') \leq \Gamma_r(z, 0)$ in the first integral and $B_r(z') = 0$ for $\Gamma_r(z, z') \leq \Gamma_r(z, \infty)$ in the second integral, so that the boundary terms are incorporated in the integrals.

2.5.2 Approximations for the Exchange Integrals

It is evident that, because of the sharply peaked character of the escape function and the absence of radiation returned from space, the dominant contribution to $h_r(z)$ will often be the radiation-to-space term,

$$h_r(z) \approx 2\pi m_a(z)S_r(z) \, \Delta\nu_r^{-1}B_r(z)\Gamma_r(z, \infty)$$

$$= [\pi/\rho_0(z)]B_r(z)\frac{d\bar{T}_f}{dz}(z, \infty). \qquad (2.5.8)$$

Rodgers and Walshaw (1966) have shown that, except close to the lower boundary, or in regions of large curvature of the vertical profile of $B_r(z)$,

Eq. (2.5.8) is a remarkably accurate approximation for both water vapor and CO_2. Thus, under a wide range of circumstances, the exchange terms can be neglected and $h_r(z)$ can be represented by Eq. (2.5.8). This approximation is called the *cool-to-space approximation*. Suppose that the necessary conditions for validity of this equation hold (sufficient distance from the lower boundary and sufficiently small temperature-profile curvature), and a dynamical disturbance produces a small local perturbation T' from an equilibrium temperature T_e. The temperature T_e may have been maintained by a balance between nonradiative as well as radiative terms. If the curvature of the vertical T' profile is sufficiently small, and if T' itself is small enough to permit linearization of the Planck function, then

$$\left(\frac{\partial T'}{\partial t}\right)_{\text{rad}} \approx -\frac{\pi}{\rho_0(z)c_p}\left\{\sum_r \left(\frac{dB_r}{dT}\right)_{T=T_e}\left[\frac{d\bar{T}_f(z,\infty)}{dz}\right]\right\}T' \equiv -K_{\text{rad}}(z,T_e)T',$$

(2.5.9)

where the summation over r includes all relevant spectral bands. This simple approximation is known as the *Newtonian cooling approximation*, and $K_{\text{rad}}(z,T_e)$ is the *Newtonian cooling coefficient*.

Unfortunately, above the stratopause, the radiation-to-space approximation breaks down for the 15-μm bands of CO_2, and net cooling or heating due to these bands is the small difference between cooling due to radiation to space and heating due to exchange with underlying layers (Fig. 2.19). At these altitudes, the CO_2 bands have become sufficiently transparent that mesospheric layers can "see" the distant warm stratopause.

Nevertheless, the applicability and simplicity of the cool-to-space approximation, which depends only on the *local* source function, suggest

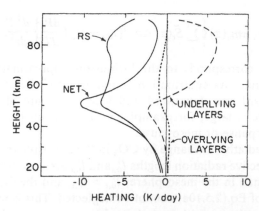

Fig. 2.19. Contributions of radiation to space (RS) and exchanges with underlying and overlying layers to the net cooling of the tropical middle atmosphere. [From Leovy (1984).]

the substitution of a Taylor-series expansion of B_r in the integrals of Eq. (2.5.7) in order to obtain more accurate local approximations in terms of B_r and its derivatives. The result is

$$h_r(z) \approx 2\pi m_a(z)\tilde{S}_r(z)\,\Delta\nu_r^{-1}\left\{-B_r(z)\Gamma_r(z,\infty) + [B_r(0_-) - B_r(z)]\Gamma_r(z,0)\right.$$

$$\left. + \sum_{m=1}^{\infty}\frac{1}{m!}[(-1)^m l_{d,r}^m + l_{u,r}^m]\frac{d^m B_r}{dz^m}\right\} \tag{2.5.10a}$$

where

$$l_{d,r}^m = \int_{\Gamma_r(z,0)}^1 |z'-z|^m\,d\Gamma_r(z,z'), \qquad l_{u,r}^m = \int_{\Gamma_r(z,0)}^1 (z'-z)^m\,d\Gamma_r(z,z') \tag{2.5.10b}$$

are the mth moments of effective radiation lengths downward and upward from level z in band r.

If the atmosphere is sufficiently opaque both upward and downward that both radiation to space and radiation to the underlying surface can be neglected, the summation in Eq. (2.5.10a) converges rapidly. Moreover, the first moments $l_{d,r}^1$ and $l_{u,r}^1$ may be approximately equal and tend to cancel each other. In this case, the most important term in the series is

$$\tfrac{1}{2}[l_{d,r}^2 + l_{u,r}^2]\frac{d^2 B_r}{dz^2} \approx \tfrac{1}{2}[l_{d,r}^2 + l_{u,r}^2]\frac{dB_r}{dT}\frac{d^2 T}{dz^2}$$

[neglecting a term containing $(d^2 B_r/dT^2)(dT/dz)^2$], and the total infrared net heating rate due to r radiatively active bands is approximately

$$h(z) \equiv \sum_r h_r(z)$$

$$\approx \pi m_a(z)\left\{\sum_r \tilde{S}_r(z)\,\Delta\nu_r^{-1}[l_{d,r}^2 + l_{u,r}^2]\frac{dB_r}{dT}\right\}\frac{d^2 T}{dz^2}. \tag{2.5.11}$$

Equation (2.5.11) corresponds to the "diffusion approximation" to the radiative heating rate. Its contribution is important if the atmosphere is very opaque upward and downward and local curvature of the vertical temperature profile is large, for example, near inversion bases and tops, and near the tropopause or mesopause.

In the case of radiative exchange by CO_2 in the mesosphere, the upward and downward effective radiation lengths $l_{d,r}^1$ and $l_{u,r}^1$ are very different. Since $[1 - \Gamma_r(z,\infty)]$ is small in the mesosphere, $l_{d,r}^1 > l_{u,r}^1$, and the first derivative term in the series of Eq.(2.5.10a) cannot be neglected. This is why the term corresponding to exchange with underlying layers is so large in the mesosphere in Fig. 2.19.

Another limiting regime occurs if the atmosphere is opaque upward, but transparent downward, and there is a substantial temperature difference between level z and the lower boundary. In this case, the contribution of exchange with the underlying surface can be of equal or greater magnitude than radiation to space, so that, approximately,

$$h(z) \approx -2\pi m_a(z) \sum_r (\tilde{S}_r(z)\Delta \nu_r^{-1}\{B_r(z)\Gamma_r(z, \infty)$$

$$- [B_r(0_-) - B_r(z)]\Gamma_r(z, 0)\}). \tag{2.5.12}$$

This approximation applies to radiative transfer by ozone in the 9.6-μm band in the lowest part of the stratosphere. Since there is very little ozone below the lower stratosphere, the second term in Eq. (2.5.12) is dominant there. It produces a net heating, since generally $B_r(0_-) > B_r(z)$ for z in the lower stratosphere. However, this heating rate is very sensitive to the effective radiative temperature of the lower boundary, and hence to the height, emissivity, and coverage of clouds in the troposphere. Above about 25 km, the atmosphere begins to become more transparent upward than downward in the 9.6-μm band, with the result that cooling to space [the first term in Eq. (2.5.12)] dominates and the band heating rate becomes negative.

In practical calculations of h_r, the integrals in Eq. (2.5.4) are represented by quadratures. The term $B_r(z')$ is represented by an interpolation formula with the general form

$$B_r(z') = \sum_j a_j(z', z_j)B_r(z_j),$$

where the quantities $a_j(z', z_j)$ are suitable interpolation functions for the discrete set of levels z_j whose values increase with j. Then

$$h_r(z_i) = \sum_j R_{ij}(z_i, z_j)B_r(z_j)$$

$$= \left\{ \sum_j R_{ij}B_r(z_i) + \sum_{j<i} R_{ij}[B_r(z_j) - B_r(z_i)] \right.$$

$$\left. + \sum_{j>i} R_{ij}[B_r(z_j) - B_r(z_i)] \right\}, \tag{2.5.13}$$

where

$$R_{ij}(z_i, z_j) = [\pi/\rho_0(z_j)] \int a_j(z', z_j) \frac{\partial^2 \bar{T}_f(z_i, z')}{\partial z \, \partial z'} \, dz'.$$

The quantities R_{ij} are elements of a square matrix called the *Curtis matrix*. The first form of Eq. (2.5.13) corresponds to Eq. (2.5.4), and the second to the exchange form of Eq. (2.5.5), with the first term corresponding to radiation to space, the second to exchange with the underlying layers and the ground, and the third to exchanges with overlying layers. Curtis matrices have been applied to the 15-μm CO_2 bands and used to calculate middle atmosphere cooling rates by several investigators. As a consequence of the sharply peaked character of the escape functions, particularly close attention must be paid to the quadrature treatment of the Planck-function profile in evaluating contributions to R_{ij} from layers adjacent to the one for which heating is being calculated. Quadrature errors are likely to be largest in the mesosphere, where exchange with underlying layers is important.

The Curtis matrix elements give directly the heating rate response to a unit Planck-function change in any other layer. By expressing B_r in the perturbation form,

$$B_r(T) = B_r(T_e) + \left(\frac{dB_r}{dT}\right)_{T_e} \delta T, \qquad (2.5.14)$$

one can also assess quite directly the radiative damping rate for small temperature perturbations of arbitrary shape (Fig. 2.20). In general, damping rate decreases with increasing vertical scale of temperature perturbations, reaching its smallest value when the entire atmospheric column is perturbed by the same amount. The damping rate per kelvin of temperature perturbation in this case is the Newtonian cooling coefficient.

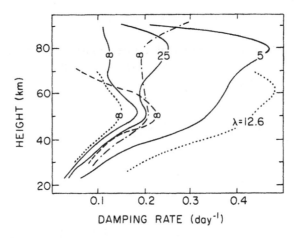

Fig. 2.20. Damping rates for the 15-μm band for temperature perturbations from the standard atmosphere: Wehrbein and Leovy (1982, thin solid), Dickinson (1973, dashed), Fels (1982, dotted), Apruzese *et al.* (1982, dash–dot). Curves are for radiation to space (∞) and for 5- and 25-km "boxcar" temperature perturbations. Also shown is the damping for a sinusoidal disturbance of vertical wavelength $\lambda = 12.6$ km. [From Leovy (1984), with permission.]

An empirical estimate of the Newtonian cooling coefficient has been obtained by Ghazi *et al.* (1985) by combining observed temperature and ozone fluctuations with corresponding detailed calculations of heating rate. They inferred a vertical profile of K_{rad} with a maximum of 0.15 $(day)^{-1}$ between 40 and 45 km, generally similar to the dotted curve marked (∞) in Fig. 2.20. They also found that temperature correlated ozone fluctuations arising from temperature-dependent photochemistry contribute an additional effective temperature damping rate of as much as 0.12 $(day)^{-1}$ in the lower mesosphere.

2.5.3 *Fourier Representation of Radiative Damping*

An alternative and more widely applicable approach to the problem of scale dependent radiative damping follows a suggestion of Spiegel (1957). Small temperature perturbations, linearized with respect to the Planck function as in Eq. (2.5.14), can be expanded in Fourier series in the spatial coordinates in order to derive a perturbation heating rate as a function of wave number. This approach has been applied specifically to a plane-parallel atmosphere by Sasamori and London (1966), and subsequently justified by Fels (1982), who demonstrated that conditions for its validity correspond to those of an extended form of the *WKBJ* approximation. Specifically, the basic state properties must vary slowly in the vertical in comparison with m^{-1}, the local inverse vertical wave number. Moreover, it is necessary that $m^{-1} < 2H$, where H is the scale height. These conditions are satisfied for most internal gravity waves and the diurnal tidal modes in the middle atmosphere.

The scale-dependent radiative damping time $\tau_R(m, z)$ is defined by

$$\frac{\partial \hat{T}(m, z)}{\partial t} \approx -\tau_R(m, z)^{-1} \hat{T}(m, z) \qquad (2.5.15)$$

where $\hat{T}$ is the complex Fourier amplitude of the component of the temperature perturbation with vertical wave number m. The dependence on height indicated in Eq. (2.5.15) is in the slowly varying sense of the *WKBJ* approximation. Inverse damping time τ_R^{-1} assumes a particularly simple form when the wavelength is short, so that the *WKBJ* approximation is well satisfied and the boundary terms can be neglected.

Fels (1984) has calculated τ_R^{-1} for the 15-μm CO_2 band using Eq. (2.5.15) for disturbances of short vertical wavelength in the mesosphere and using a slightly more general expression that accounts for boundary effects and line overlap for stratospheric disturbances. His results for the standard atmosphere temperature profile are shown in Fig. 2.21. For O_3, the maximum

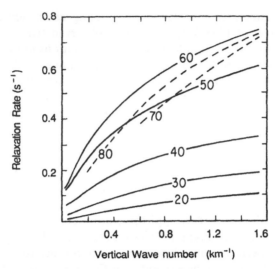

Fig. 2.21. Dependence of radiative damping rates at several altitudes (given in km) on vertical wavelength. The solid curves have been replotted from Fels (1984), and the dashed curves are from Fels (1982).

damping rate is reached near the stratopause. In the upper stratosphere and mesosphere, damping due to O_3 infrared exchange is independent of wavelength for moderate and short wavelengths above the stratopause. This behavior corresponds to the *optically thin limit* in which ozone is essentially transparent for distances comparable to these short wavelengths. For CO_2, which is much more opaque, damping rate increases with increasing m, rapidly at first, and then more slowly, and for fixed m it reaches a maximum in the lower mesosphere. In this limit,

$$\tau_R^{-1} \to \frac{4\pi\rho_a}{c_p\rho}(dB_r/dT)\sum_i S_i.$$

In the CO_2 calculations shown in Fig. 2.21, Fels has taken departures from local thermodynamic equilibrium into account at 70 and 80 km using the two-level model formulation that will be described in the next section. Non-LTE effects are largely responsible for the upward decrease in damping rate above a lower mesosphere maximum.

2.6. Departure from Local Thermodynamic Equilibrium

If local thermodynamic equilibrium (LTE) prevailed at all altitudes, unimpeded radiation to space in the centers of the strong Doppler lines of the ν_2 fundamental of CO_2 would cool the atmosphere very strongly.

In the optically thin limit, the heating rate (in $K\,s^{-1}$) is given approximately by the radiation-to-space term from Eq. (2.5.8),

$$h_r(z)/\rho \approx -\frac{2\pi m_a}{c_p} B_r(z)\tilde{S}_r\Gamma_r(z,\infty)\,\Delta\nu_r^{-1}$$

$$\approx -\frac{2\pi m_a}{c_p} B_r(z)\tilde{S}_r\,\Delta\nu_r^{-1},$$

since $\Gamma_r(z,\infty) \to 1$ as $z \to \infty$. Using the CO_2 mixing ratio and the ν_2 fundamental band strength from Table 2.1, this rate of cooling exceeds 100 K day^{-1} at typical lower-thermosphere temperatures. These large hypothetical cooling rates do not occur because below the altitude at which the optically thin limit is reached, the rate of collisional transfer from kinetic to vibrational energy limits energy loss. As a result, excited vibrational levels at these altitudes are not populated according to the Boltzmann distribution, Kirchhoff's Law breaks down, and the source function departs strongly from the Planck function. Cooling rates at and above the mesopause are smaller than this optically thin LTE limit by more than a factor of 10. In fact, the rapid increase in temperature above the mesopause is attributable largely to the rapid decrease in radiative cooling efficiency above that level. The non-LTE behavior of CO_2 influences mesopause temperature structure on Venus and Mars as well as on Earth. In this section, we develop the non-LTE formulation in terms of the Einstein coefficients and consider two specific non-LTE models.

2.6.1   The Einstein Coefficients

The interaction between radiation and matter can be described in terms of three fundamental processes: *spontaneous emission, absorption,* and *induced emission* (Fig. 2.22). The last, the process responsible for lasing, yields induced photons identical in phase, polarization, and direction of

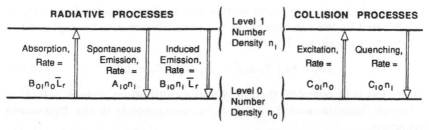

Fig. 2.22. Radiative and collisional transitions connecting the ground state and lowest vibrational level of a molecule.

travel to the incident photons. Consider a vibration–rotation transition with upper vibrational level 1, molecular population n_1, and ground-state level 0, population n_0, for an absorbing gas whose total number density is n_a. The rates of photon emission and absorption per unit volume are proportional to n_1 and n_0, and in the cases of absorption and induced emission, they are also proportional to the band and angular average radiance,

$$\bar{L}_r = \frac{1}{4\pi S_r} \int_{4\pi} \int_{\Delta\nu_r} \sigma_\nu L_\nu \, d\nu \, d\Omega \equiv \frac{1}{4\pi} \int_{4\pi} \int_{\Delta\nu_r} \tilde{f}(\nu) L_\nu \, d\nu \, d\Omega, \quad (2.6.1)$$

where $\tilde{f}(\nu)$ is a generalized "band-shape factor," which takes into account rotational structure as well as line shape. The proportionality constants are the *Einstein coefficients* A_{10}, B_{01}, and B_{10}, defined such that

$$A_{10}n_1 = \text{rate of spontaneous emission,}$$

$$B_{01}n_0\bar{L}_r = \text{rate of absorption,} \qquad (2.6.2)$$

$$B_{10}n_1\bar{L}_r = \text{rate of induced emission,}$$

with each of these terms expressed in units of photons $m^{-3}\,s^{-1}$. Thus, A_{10} is the spontaneous emission probability per molecule. Each of these coefficients depends on the quantum-mechanical details of the transition but is independent of the properties of the radiation field or the thermodynamic state of the gas.

The volume heating rate for the band is equal to the rate of decrease of monochromatic radiance with distance, $-dL_\nu/ds$, integrated over all frequencies in the band and all solid angles. This can readily be shown from the definition of h_r or from the heating-rate equation in the form of Eq. (2.5.3). Hence, applying the radiative transfer equation [Eq. (2.2.5)], the heating rate per unit volume is

$$\rho h_r = -\int_{4\pi} \int_{\Delta\nu_r} \frac{dL_\nu}{ds} \, d\nu \, d\Omega = 4\pi S_r n_a (\bar{L}_r - J_r). \qquad (2.6.3)$$

For the spectrally narrow bands of interest, frequency in the band can be approximated by the constant value ν_r. Then ρh_r can also be expressed in terms of the Einstein coefficients,

$$\rho h_r = h\nu_r(B_{01}n_0\bar{L}_r - A_{10}n_1 - B_{10}n_1\bar{L}_r). \qquad (2.6.4)$$

If the gas is in complete thermodynamic equilibrium, then $h_r = 0$, $\bar{L}_r$ is the Planck function, and the ratio n_1/n_0 corresponds to the Boltzmann distribution,

$$n_1/n_0 = g_{10}e^{-h\nu_r/k_b T} \qquad (2.6.5)$$

where $g_{10} \equiv g_1/g_0$, the degeneracy ratio. For the ν_2 fundamental of CO_2, $g_{10} = 2$. It follows from Eq. (2.6.4) that

$$B_{01}/B_{10} = g_{10}$$

and

$$A_{10} = \frac{2h\nu_r^3}{c^2} B_{10} = \frac{2h\nu_r^3}{c^2 g_{10}} B_{01}. \tag{2.6.6}$$

These relations hold under nonequilibrium as well as equilibrium conditions, since the Einstein coefficients are independent of the state of the gas or the radiation field. This is an application of the *principle of detailed balance*.

Comparison of Eqs. (2.6.3) and (2.6.4) shows that

$$S_r = \frac{h\nu_r}{4\pi} B_{01}(n_0/n_a)[1 - g_{10}^{-1}(n_1/n_0)] \simeq \frac{h\nu_r}{4\pi} B_{01}, \tag{2.6.7}$$

under nonequilibrium as well as equilibrium conditions, or, from Eq. (2.6.6),

$$A_{10} \approx \frac{8\pi\nu_r^2}{g_{10}c^2}\left(\frac{n_a}{n_0}\right)[1 - g_{10}^{-1}(n_1/n_0)]^{-1} S_r \approx \frac{8\pi\nu_r^2}{g_{10}c^2} S_r. \tag{2.6.8}$$

Moreover, with the aid of Eqs. (2.2.10) and (2.6.6),

$$J_r = \left(\frac{h\nu_r A_{10} n_1}{4\pi S_r n_a}\right) = \frac{[1 - e^{-h\nu_r/k_b T}]}{[1 - g_{10}^{-1}(n_1/n_0)]} B_r\left[\left(\frac{n_1}{n_0}\right)\left(\frac{n_{0eq}}{n_{1eq}}\right)\right] \simeq B_r\left(\frac{n_1}{n_{1eq}}\right), \tag{2.6.9}$$

where n_{1eq} and n_{0eq} are thermal equilibrium number densities. In Eqs. (2.6.7), (2.6.8) and (2.6.9), S_r, A_{10}, and J_r have been approximated by dropping terms of order $\exp(-h\nu_r/k_b T)$. For the ν_2 fundamental of CO_2 these terms are smaller than the retained terms by a factor of order 100 under normal atmospheric conditions. Neglect of these terms is equivalent to ignoring induced emission, and they will be neglected throughout the subsequent development. Note that Eq. (2.6.8) shows how the emission probability per molecule, A_{10}, is related to the measured band strength.

Equation (2.6.9) shows that the source function for the band is proportional to the ratio of the actual number density in the excited state to the equilibrium number density. Under LTE conditions, $n_1 = n_{1eq}$, and Kirchhoff's Law holds. In the more general case, the problem of determining the source function reduces to that of determining n_1.

2.6.2   The Two-Level Model

Under LTE conditions, the rate of population of level 1 by collisions that convert translational kinetic energy to vibrational energy (excitation) is exactly balanced by the reverse process (quenching), collisional conversion of level-1 vibrational energy to translational kinetic energy (see Fig. 2.22). These rates can be represented by $C_{01}n_0$ for the first process and $C_{10}n_1$ for the second. The kinetic coefficients C_{01} and C_{10} are both proportional to molecular number densities, and their ratio can be determined from application of the detailed balance principle by setting $(n_1/n_0) = (n_{1eq}/n_{0eq})$. At equilibrium,

$$C_{01} = C_{10}g_{10}e^{-h\nu_r/k_bT}, \qquad (2.6.10)$$

where the collision parameter C_{10} is

$$C_{10} = k_0\eta n, \qquad (2.6.11)$$

with k_0 the molecular collision frequency, η the fraction of collisions responsible for vibrational–translational (V-T) energy conversion, and n the total number density of all gas molecules. Equation (2.6.10) holds under nonequilibrium as well as equilibrium conditions, but C_{10} is a function of temperature.

The simplest non-LTE model is the *two-level model*, in which only collisional and radiative interactions between level 1 and the ground state are considered. Under LTE conditions, the collisional rates are much faster than the radiative rates, so that the collisional rates are very nearly in steady-state balance. Under non-LTE conditions this is no longer the case, but a very close balance exists between the sum of radiative and collisional rates tending to populate and depopulate level 1. As shown in Fig. 2.22, this balance is

$$(A_{10} + C_{10})n_1 \approx (B_{01}\bar{L}_r + C_{01})n_0. \qquad (2.6.12)$$

Solving for (n_1/n_0), making use of Eqs. (2.6.5), (2.6.6), (2.6.9), and (2.2.10), and neglecting terms of order $\exp(-h\nu_r/k_bT)$ gives

$$J_r = \frac{\bar{L}_r + C_{01}/B_{01}}{1 + C_{10}/A_{10}} = \frac{\bar{L}_r + \phi B_r}{1 + \phi}, \qquad \phi \equiv C_{10}/A_{10}. \qquad (2.6.13)$$

The behavior of the source function is controlled by the ratio of the time constant for radiative deexcitation of level 1, A_{10}^{-1}, to the time constant for collisional deexcitation, C_{10}^{-1}. For large ϕ, $J_r \to B_r$ and Kirchhoff's Law applies, but for $\phi \to 0$, $J_r \to \bar{L}_r$; the source function reduces to the source function for isotropic conservative scattering.

For the ν_2 fundamental of CO_2, $A_{10} = 1.51 \text{ s}^{-1}$. Because conversion between kinetic and vibrational energy is an improbable event, only about one collision in 10^5 is effective in deexciting level 1, that is, $\eta \approx 10^{-5}$. Since $k_0 \approx 2 \times 10^{-16} \text{ m}^3 \text{ s}^{-1}$, this gives

$$\phi \approx \frac{2 \times 10^{-21} n}{1.51}$$

(with n in molecules m^{-3}), so that the transition between LTE and non-LTE regimes centers around the level at which $n \approx 0.75 \times 10^{21} \text{ m}^{-3}$, or 75 km. This is the *vibrational relaxation level* in the earth's atmosphere.

The effect of vibrational relaxation on the heating rate can be made clear by substituting Eq. (2.6.13) into Eq. (2.6.3):

$$\rho h_r = 4\pi S_r n_a (\bar{L}_r - J_r) = \frac{4\pi S_r n_a \phi}{(1 + \phi)}(\bar{L}_r - B_r). \qquad (2.6.14)$$

For large ϕ (LTE), h_r is proportional to $(\bar{L}_r - B_r)$, but as $\phi \to 0$, $h_r \to 0$. The decoupling between the radiation field and the kinetic energy of the gas is responsible for the vanishing infrared cooling rate at altitudes far above the vibrational relaxation level.

To see this behavior more clearly, assume that the radiation-to-space approximation applies. Then

$$\rho h_r(z) = -2\pi S_r n_a(z) J_r(z) \Gamma_r(z, \infty).$$

But eliminating $\bar{L}_r$ between Eqs. (2.6.13) and (2.6.14) gives

$$J_r = B_r + \frac{\rho h_r}{4\pi S_r n_a \phi}, \qquad (2.6.15)$$

so that

$$J_r = B_r / [1 + \tfrac{1}{2}\Gamma_r(z, \infty)\phi^{-1}], \qquad (2.6.16)$$

and

$$h_r = -\frac{2\pi S_r \rho^{-1} n_a \Gamma_r(z, \infty) B_r}{1 + \tfrac{1}{2}\Gamma_r(z, \infty)\phi^{-1}}. \qquad (2.6.17)$$

Thus, in the radiation-to-space approximation, vibrational relaxation is controlled not by ϕ^{-1}, but by $\tfrac{1}{2}\Gamma_r(z, \infty)\phi^{-1}$. If the atmosphere is sufficiently opaque, $\Gamma_r(z, \infty)$ is small and excited-state populations can be maintained near their equilibrium values by radiation even if ϕ is small. For the ν_2 fundamental of CO_2, the level at which $\tfrac{1}{2}\Gamma_r(z, \infty)\phi^{-1} \approx 1$ occurs about 10 km above the level at which $\phi^{-1} \approx 1$, so that, for this band, the source function tends to equal the Planck function up to about 80–85 km (Fig. 2.23). At

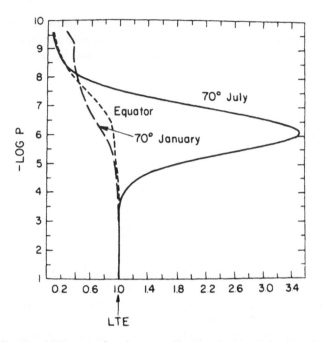

Fig. 2.23. Non-LTE source function normalized by the Planck function for the transition $01^1 0$–000 of $^{12}C^{16}O_2$ for three combinations of latitude and season. The ordinate is $-\log($pressure in mb$)$ and has the following approximate relationship to altitude: 0, 47 km; 2, 63 km; 4, 76 km; 6, 87 km; 8, 98 km. [From Dickinson (1984), with permission.]

very high altitudes, the gas becomes transparent, $\Gamma_r(z, \infty) \to 1$, and $\frac{1}{2}\Gamma_r(z, \infty)\phi^{-1}$ becomes large. Then

$$h_r \to -4\pi S_r \rho^{-1} n_a \phi B_r = -\tfrac{1}{2} g_{10} \frac{c^2}{\nu_r^2} \rho^{-1} n_a C_{10} B_r$$

$$\approx -h\nu_r \rho^{-1} C_{10} g_{10} n_0 e^{-h\nu_r / k_b T} = -h\nu_r C_{01} n_0. \qquad (2.6.18)$$

In this limit, the rate of cooling is just $h\nu_r$ times the rate of collisional excitation of level 1. Both the volume heating rate and the heating rate per unit mass vanish as $z \to \infty$.

The two-level non-LTE problem can be posed in the form of an integral equation. As described by Houghton (1977), this integral equation can be solved for the source function by expressing the heating rate integral in the Curtis matrix forms derived from Eq. (2.5.13),

$$h_r(z_i) = \sum_j R_{ij} J_r(z_i),$$

which can be expressed in the matrix form,

$$\mathbf{h}_r = \mathbf{RJ}. \qquad (2.6.19)$$

In matrix notation Eq. (2.6.15) is

$$\mathbf{J}_r = \mathbf{B}_r + \mathbf{Eh}_r, \tag{2.6.20}$$

where $\mathbf{E}$ is a diagonal matrix of dimension equal to the number of columns in $\mathbf{R}$. The elements of $\mathbf{E}$ are

$$E_{ij} = [4\pi S_r n_a(z_i)\phi(z_i)]^{-1}\rho\delta_{ij}. \tag{2.6.21}$$

Combining Eqs. (2.6.19) and (2.6.20) gives

$$\mathbf{J}_r = [\mathbf{I} - \mathbf{ER}]^{-1}\mathbf{B}_r, \tag{2.6.22}$$

or

$$\mathbf{h}_r = [\mathbf{I} - \mathbf{RE}]^{-1}\mathbf{RB}_r. \tag{2.6.23}$$

Equation (2.6.23) has been widely used to calculate heating rates under non-LTE conditions. In these applications, the cooling rate has been calculated for the entire 15-μm band system so that it is implicitly assumed that the hot and isotopic bands in this spectral region have the same source function as the fundamental. The source functions for these relatively weak bands are imperfectly coupled to the fundamental, however, so that contributions from the hot and isotopic bands are not accurately calculated by this method. In the region above the mesopause where CO_2 is transparent even in the strong line centers of the ν_2 fundamental, the cooling rate due to the fundamental dominates the total cooling, and the resulting error is not serious. Just below the mesopause, where non-LTE effects are important, the contributions of the hot and isotopic bands are large and the error can be significant. A model more complex than the two-level model is needed for this region. Such a model will be discussed in the next section.

Before leaving the two-level model, a comment on the role of rotational fine structure is needed. Throughout the preceding discussion, it has been implicitly assumed that the rotational-level populations are not affected by the interplay of radiative and kinetic processes that produce non-LTE conditions. As a consequence, Einstein coefficients were needed only for the vibrational transition as a whole, not for the individual rotational transitions comprising the fine structure. The fine structure of the band was assumed to be determined by the kinetic temperature of the gas alone. This major simplification was possible because collisional exchange between translational and rotational energy is very efficient, occurring at nearly every collision. As a consequence, the rotational energy levels remain in thermal equilibrium up to a level at which the density is $\sim 10^{-5}$ times the vibrational relaxation density, well above the level where cooling by CO_2 infrared emission ceases to play a significant role.

2.6.3 Dickinson's Multilevel Model

The processes considered in the last section can be summarized in the following energy-transfer steps:

$$CO_2 (v_2 = 1) \leftrightarrows CO_2 \text{ (ground state)} + h\nu \ (15 \ \mu m), \qquad (2.6.24)$$

the fundamental radiative transitions, and

$$\text{Air} + CO_2 (v_2 = 1) \leftrightarrows CO_2 \text{ (ground state)} + \text{air} + KE, \qquad (2.6.25)$$

the fundamental V-T transitions, where KE is kinetic energy.

Adequate treatment of the hot bands requires at a minimum that the radiative and kinetic transitions involving the 100, 02^00, and 02^20 levels be taken into account. These levels are so closely spaced that collisional equilibration between them must be very rapid, as it is for the rotational levels. Consequently, they can be treated as a single level, $v_2 = 2$, with degeneracy 4. The most important transitions affecting the population of this level are the vibration–vibration (V-V) transitions

$$CO_2 (v_2 = 2) + CO_2 \text{ (ground state)} \leftrightarrows 2CO_2 (v_2 = 1), \qquad (2.6.26)$$

the V-T transitions

$$\text{Air} + CO_2 (v_2 = 2) \leftrightarrows CO_2 (v_2 = 1) + \text{air} + KE, \qquad (2.6.27)$$

and the radiative transition

$$CO_2 (v_2 = 2) \leftrightarrows CO_2 (v_2 = 1) + h\nu \ (15 \ \mu m). \qquad (2.6.28)$$

Because 15-μm band Doppler lines are well separated, the spectrum of this radiative transition is well separated from that of the fundamental, even though both occupy the same broad spectral interval. Additional interactions involving higher vibrational levels occur but are usually of minor importance.[2] Because little energy exchange is involved, the V-V transitions are rapid and the forward and backward transitions have nearly the same bimolecular rate constant, K_{VV}. On the other hand, the V-T exchanges involve large energy transfers and are very slow. The rate for the forward reaction of Eq. (2.6.27), K_{VT2}, is $\sim 10^{-5} k_0$, comparable with the rate K_{VT1} of the forward reaction of Eq. (2.6.25).

The fundamental isotopic bands are coupled to the fundamental of the principal isotope by V-V reaction of the type

$$CO_2 \ (v_2 = 1, \text{ main isotope}) + CO_2 \text{ (ground state, minor isotope)}$$

$$\leftrightarrows CO_2 \text{ (ground state, main isotope)}$$

$$+ CO_2 \ (v_2 = 1, \text{ minor isotope}). \qquad (2.6.29)$$

Because the energy changes are again small, these reaction rates are also

---

[2] Excitation of higher vibrational levels by transformation of absorbed solar radiation may sometimes render these interactions important, however.

fast, and forward and backward reaction rates are nearly equal. In addition, the isotopic fundamental radiative transitions [Eq. (2.6.24)] and lowest-level V-T transitions [Eq. (2.6.25)] must be separately treated for each isotope. Isotopic hot bands are not important.

Dickinson (1984) has extended the two-level model to account for these additional processes. This involves the solution of a coupled set of integral equations that are generalizations of Eq. (2.6.12) and involve the number densities of the first two excited vibrational levels of $^{12}C\,^{16}O\,^{16}O$ and the first excited level of $^{13}C\,^{16}O\,^{16}O$ and $^{12}C\,^{16}O\,^{18}O$. For the collisional interaction between level 1 and the ground state, he considered the additional effect of atomic oxygen, which appears to exchange energy efficiently with vibrationally excited CO_2 and could be important above the mesopause.

Dickinson's calculated source functions for three latitudes for the ν_2 fundamental of $^{12}C\,^{16}O_2$ are shown in Fig. 2.23. In each case the source function is shown normalized by the local value of the Planck function. Throughout the winter hemisphere and in the tropics, these normalized source functions decrease with height above the level at which $\frac{1}{2}\Gamma_r(z, \infty)\phi^{-1} \approx 1$, in accordance with the radiation to space approximation [Eq. (2.6.16)]. At high latitudes in the summer hemisphere, the normalized source functions increase to a strong maximum near the mesopause before decreasing at still higher levels. This is a consequence of the extremely steep lapse rate in this region and the corresponding dominance of upwelling radiation from the warm stratopause region. This effect is more pronounced for the hot and isotopic bands than for the fundamental. The effect of this upwelling radiance on heating is illustrated in Fig. 2.24. The upwelling radiance is sufficient to produce net heating near the summer mesopause. Figure 2.25 shows the altitude–latitude distribution of CO_2 infrared cooling for the solstices as calculated by Dickinson. Cooling maxima occur near the stratopause and in the lower thermosphere, with the calculated magnitude of the latter dependent on the uncertain rate of V-T exchange with atomic oxygen, so these values are somewhat uncertain. Weak mesopause warming extends from the summer polar region into the tropics. This feature is sensitive to details of the vertical temperature profile.

Absorption of solar photons in near-infrared bands can result in a cascading of quanta into the $v_2 = 2$ level and from there into $v_2 = 1$ quanta or thermal energy via the processes represented by Eqs. (2.6.24)–(2.6.27). The efficiency with which these quanta can cascade to the $v_2 = 1$ level depends on rates of V-V transfer to N_2, O_2, and H_2O, as well as "leakage" via emission in the 4.3-μm band of CO_2 and the 6.3-μm band of H_2O. These processes were first decribed in detail by Houghton (1969). Williams (1971) has calculated that solar absorption by CO_2 produces heating of as much as 1 K day^{-1} near the mesopause.

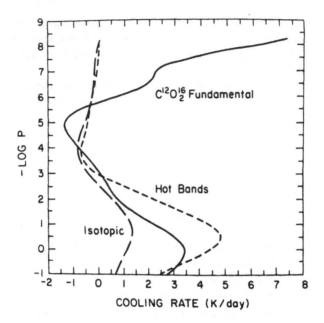

Fig. 2.24. Contributions of the $^{12}C^{16}O_2$ fundamental and the hot and isotopic bands to cooling rate for July at 70°N latitude. The ordinate scale is the same as in Fig. 2.23. [From Dickinson (1984), with permission.]

2.7 Absorption of Solar Radiation

So far, this chapter has emphasized emission, absorption and transfer of thermal infrared radiation, the process responsible for cooling the middle atmosphere. In the global mean, this is approximately balanced at each level by absorption of solar radiation.

From the direct solar beam, the rate of energy absorption per unit volume for a spectral interval $\Delta \nu_r$, $ph_r(z^*)$, follows from Eq. (2.2.16),

$$ph_r(z^*) = -\mu_0 \int_{\Delta \nu_r} S_{0\nu} \frac{d}{dz^*} e^{-\tau_\nu(z^*)/\mu_0} \, d\nu$$

$$= n_a(z^*) \int_{\Delta \nu_r} \sigma_\nu S_{0\nu} e^{-\tau_\nu(z^*)/\mu_0} \, d\nu. \qquad (2.7.1)$$

This is closely related to the direct beam photodissociation rate per molecule,

$$P(z^*) = \int_{\nu > \nu_{diss}} e_\nu \sigma_\nu \tilde{S}_{0\nu} e^{-\tau_\nu(z^*)/\mu_0} \, d\nu, \qquad (2.7.2)$$

where ν_{diss} is the dissociation threshold, e_ν is the quantum efficiency for photodissociation (frequently $e_\nu \approx 1$ for $\nu > \nu_{diss}$), and $\tilde{S}_{0\nu} \equiv S_{0\nu}/h\nu$ is the

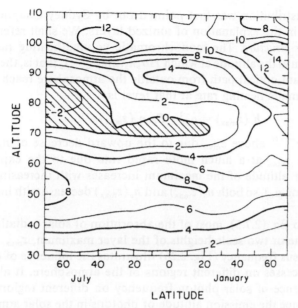

Fig. 2.25. Numbers on contours give total CO_2 15-μm band cooling rate in degrees Kelvin per day for the January and July 1972 COSPAR International Reference Atmospheres. [From Dickinson (1984), with permission.]

monochromatic solar flux density expressed in photons rather than energy units. If more than one gas absorbs in a spectral interval,

$$\rho h_r(z^*) = \sum_i n_{a,i} \int_{\Delta\nu_r} \sigma_{\nu,i} S_{0\nu} e^{-\tau_\nu(z^*)/\mu_0} \, d\nu, \qquad (2.7.3)$$

where, in both Eqs. (2.7.2) and (2.7.3), τ_ν includes contributions from all absorbing gases (number densities $n_{a,i}$),

$$\tau_\nu(z^*) = \sum_i \int_{z^*}^{\infty} \sigma_{\nu,i} n_{a,i} \, dz^{*\prime}. \qquad (2.7.4)$$

In order to see how heating per unit volume depends on frequency and height, consider the monochromatic version of Eq. (2.7.1) in an atmosphere whose scale height is constant and equal to H so that $z^* = z$ (see Section 1.1.1). The monochromatic volume heating rate ρh_ν is

$$\rho h_\nu(z) \equiv \frac{\rho \, dh_r}{d\nu} = n_a(z)\sigma_\nu S_{0\nu} e^{-\tau_\nu(z)/\mu_0}$$

$$= \sigma_\nu S_{0\nu} n_a(0) \exp[-z/H - H\sigma_\nu n_a(0)\mu_0^{-1} e^{-z/H}]. \qquad (2.7.5)$$

This vertical distribution was first pointed out by Sydney Chapman, who first applied it in the explanation of ionized layers. We shall refer to it as *Chapman layer structure.* The monochromatic volume heating rate has a single maximum at the altitude at which $H\sigma_\nu n_a(z)/\mu_0 = 1$, that is, the altitude at which the slant optical path from outside the atmosphere reaches unity. The monochromatic heating rate at that level is

$$h_\nu(z_{max}) = \rho^{-1}\sigma_\nu S_{0\nu} n_a(z_{max})e^{-1}. \qquad (2.7.6)$$

It falls off as $e^{-z/H}$ above z_{max} due to the upward decrease of n_a, and it falls off below z_{max} at a much more rapid rate due to the exponential absorption. The altitude of the maximum increases with increasing zenith angle (decreasing μ_0), so both $n_a(z_{max})$ and $h_\nu(z_{max})$ decrease with increasing zenith angle.

According to Eq. (2.7.5), most of the absorption of solar radiation takes place within one or two scale heights of the layer maximum, z_{max}. Thus, a plot of z_{max} versus frequency (Fig. 2.26) illustrates the influence of different absorption processes on different regions of the atmosphere. It also illustrates the influence of solar photon frequency on different regions of the atmosphere. Since the emission altitude of photons in the solar atmosphere and the processes responsible for photon emission depend on frequency, there is a direct link between altitudes and processes in the solar atmosphere and altitudes and processes in the earth's atmosphere.

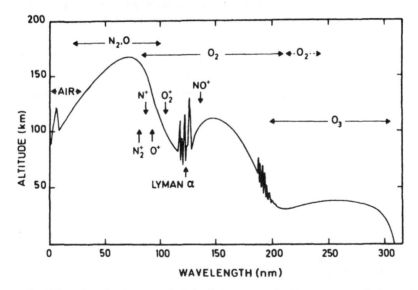

Fig. 2.26. Altitude of unit optical depth for normal incidence solar radiation. Principal absorbers and ionization thresholds are indicated. [From Herzberg (1965), with permission.]

At visible wavelengths ($\lambda > 310$ nm) solar radiation penetrates to the surface, but below about 310 nm in the ultraviolet, ozone absorption of solar radiation in a photodissociation continuum begins to shield the surface. This continuum absorbs most strongly at about 40 km and 250 nm, and it is responsible for shielding the surface from the biologically damaging radiation below 300 nm. Molecular oxygen absorbs in the Herzberg dissociation continuum below 240 nm, and oxygen becomes the dominant absorber below about 200 nm. The rate of absorption and consequent O_2 photodissociation near 200 nm is strongly dependent on the opacity due to ozone in the same spectral region. Between 175 and 200 nm, the Schumann–Runge bands are responsible for most absorption, and their complicated structure influences solar photon penetration in the mesosphere and upper stratosphere. The Schumann–Runge continuum controls solar photon penetration between 135 and 175 nm in the lower thermosphere and mesopause region.

Complex O_2 spectral features between 115 and 135 nm allow some solar radiation to penetrate to the upper mesosphere. A most important gap occurs at 121.6 nm, precisely the wavelength of hydrogen Lyman α, the strongest solar emission line. Because NO and alkali metals, which are present in trace amounts near the mesopause, ionize at wavelengths longer than 121.6 nm, this penetrating radiation is largely responsible for the formation of the D and lower E ionization layers in the upper mesosphere and lower thermosphere. X-rays and cosmic rays, which can also penetrate into the mesopause region, are responsible for the remainder of the ionization there. The bulk of the ionization in the earth's atmosphere is formed higher in the thermosphere, however, and is due to the ionization of the major gases at wavelengths between 10 and 100 nm.

2.7.1 Solar Radiation Flux

The information needed to calculate solar heating rates and photodissociation and ionization rates includes the spectral distribution of solar flux and its time variability, the molecular absorption cross sections, and the dissociation and ionization efficiencies. The relationship between the solar spectrum and the structure of the sun is well described by Brasseur and Solomon (1984), so only a brief account will be given here.

Figure 2.27 depicts the solar spectrum. Most solar radiant energy flux is emitted from the visible surface of the sun, the photosphere, at an effective blackbody temperature of about 6000 K. The solar spectrum at wavelengths longer than about 300 nm corresponds closely to a blackbody at this temperature. Temperature decreases upward through the photosphere to a minimum of about 4600 K at the base of the chromosphere, then rises to

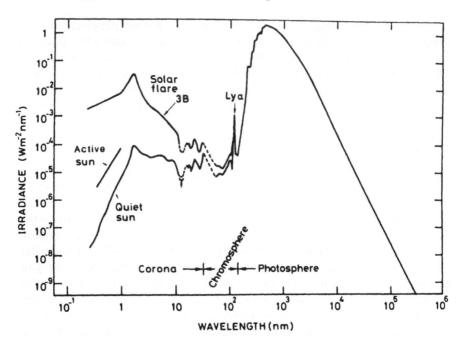

Fig. 2.27. Solar spectrum under quiet and active conditions and spectrum of an intense solar flare. [From Smith and Gottlieb (1974), with permission.]

10^6 K or more in the corona. Between 121.6 and 300 nm, the effective blackbody temperature of the solar spectrum reflects these lower temperatures between the photosphere and the base of the chromosphere. At 121.6 nm, the wavelength of Lyman α, and at shorter ultraviolet wavelengths, the solar spectrum is dominated by emission lines originating in the chromosphere and corona.

The variability of the solar atmosphere increases with increasing height, and as a result, the variability of the solar spectrum increases with decreasing wavelength. This is illustrated schematically in Fig. 2.27. Variability occurs on the time scales of the 26-day solar rotation, the 11-year sunspot cycle, and with a variety of short-lived solar disturbances, particularly flares. In the past, solar variability at ultraviolet wavelengths has been estimated from its correlation with solar flux at the radio wavelength 10.7 cm, which can be measured at the Earth's surface, but more recently ultraviolet variability has been monitored directly by satellites. At wavelengths affecting the thermal structure of the middle atmosphere ($\sim$160–700 nm), measured peak-to-peak variations in solar flux range from about 10% at 160 nm to about 5% at 200 nm. At wavelengths longer than 200 nm, measured solar variability drops rapidly and is less than 1% at 300 nm. However, the variability may be somewhat greater than this, since available measurements do not yet cover a full solar cycle.

2.7.2 Absorption of Solar Radiation by Gases

Molecular oxygen absorption was discussed in Section 2.3.1, and absorption cross sections are shown in Figs. 2.7 and 2.8. Because of the complex structure of the Schumann–Runge bands and their temperature dependence, constant broadband absorption cross sections are not suitable. The World Meteorological Organization has provided tables of broadband transmission functions for this spectral region for the standard atmosphere temperature distribution. These nonexponential transmission functions take into account the spectral complexity of the bands. Photodissociation rates of several minor constituents in the upper stratosphere are sensitive to transmission in the Schumann–Runge spectral region, so it is important to have an accurate broadband description of this transmission. The photodissociation efficiency of O_2 is nearly unity at wavelengths shorter than the dissociation threshold at 242 nm.

Ozone absorption takes place in three dissociation continua in which dissociation efficiency is close to unity: the Hartley band from 200 to 310 nm, the Huggins bands, which blend with the Hartley band near 310 nm and extend to 350 nm, and the much weaker Chappuis bands extending from 440 to 800 nm (Fig. 2.28). The Hartley and Chappuis bands are smooth continua whose cross sections are independent of temperature and pressure. The Huggins bands have a diffuse banded structure (not shown in Fig. 2.28)

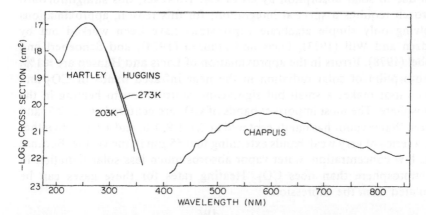

Fig. 2.28. Absorption spectrum of ozone. Note the temperature dependence in the Huggins bands. Weak band structure in this 310 to 350-nm spectral region has been suppressed in this presentation. Data from World Meteorological Organization (WMO, 1986).

with some temperature dependence. Band maxima are insensitive to temperature, but the minima show variations of up to 30% over the temperature range 200–290 K, with lower values at low temperatures. Despite the small absorption cross sections, the Chappuis bands are of great importance because they occur at the peak of the solar spectrum and absorb in the lower stratosphere and troposphere. Shorter wavelength radiation is nearly absent at these levels because of absorption higher up. Because of the Chappuis bands, ozone in sunlight is subject to rapid photodissociation right down to the surface.

Ultraviolet absorption by H_2O, NO_2, HNO_3, and many other gases is vital to the chemistry of the middle atmosphere but does not contribute significantly to heating, so these absorption cross sections will not be discussed in detail here. Readers are referred to Brasseur and Solomon and the report by the World Meteorological Organization (WMO, 1986). In general, gases with dissociation thresholds at wavelengths longer than about 300 nm are dissociated by the intense portion of the solar spectrum and are not shielded by overlying O_3 and O_2, so they dissociate rapidly in sunlight at all altitudes; lifetimes of individual molecules against dissociation are $\leq 10^3$ s. In addition to O_3, examples of species which dissociate rapidly in sunlight are NO_2, NO_3, and $HOCl$. On the other hand, molecules with dissociation thresholds at shorter wavelengths have relatively long dissociation lifetimes at all levels. Their dissociation rates decrease rapidly below the ozone layer and are also sensitive to zenith angle. Examples are H_2O, HNO_3, HCl, and O_2 (see Chapter 10).

If scattering is neglected, Eq. (2.7.1) can be used to calculate heating rates due to solar absorption by O_3 or O_2. However, this straightforward approach requires a spectral integration; for this reason, approximations involving only simple algebraic expressions have been worked out by Lindzen and Will (1973), Lacis and Hansen (1974), and Schoeberl and Strobel (1978). Errors in the approximation of Lacis and Hansen are $\leq 1\%$.

Absorption of solar radiation in the near infrared bands of CO_2 and water vapor makes a small but significant contribution to heating in the stratosphere. The most important bands of CO_2 are centered at 4.3, 2.7, and 2.0 μm. Water vapor has bands centered at 2.7, 1.9, 1.6, and 1.1 μm, together with a series of very weak bands extending to 0.55 μm in the visible. Because of its low concentration, water vapor absorbs much less solar radiation in the stratosphere than does CO_2. Heating rates for these gases can be calculated from the expression

$$h_r(z^*) = -\rho^{-1} S_{0,r} \frac{dW_r}{dz^*}(z^*, \mu_0), \qquad (2.7.7)$$

where $W_r(z^*, \mu_0)$ is the band equivalent width for the direct solar path from level z^* to space, and $S_{0,r}$ is the solar flux averaged over band r. The value of $W_r(z^*, \mu_0)$ can be calculated using band models and tabulated spectroscopic data as described in Section 2.4. Allowance must be made for overlap between CO_2 and water-vapor bands and for the transition to the Voigt line shape in the upper stratosphere.

Although daily mean values of solar heating rate can be approximated by using a suitable daily mean value of μ_0 together with a heating-rate multiplier corresponding to the illuminated fraction of the day, calculations of high accuracy are carried out by numerical integration of the solar heating rate over the illuminated period.

2.7.3 Effects of Scattering

For the middle atmosphere, three scattering processes play a role: reflection and scattering from the underlying surface and troposphere, including clouds and aerosols; *in situ* molecular scattering (Rayleigh scattering); and *in situ* scattering by aerosols, particularly volcanic aerosols. Of these, the first is by far the most important. It is relatively easy to incorporate in calculations, but because the heating rate between the tropopause and 30 km is very sensitive to this upwelling solar flux, it is also sensitive to the distribution of clouds and aerosols.

The method of Lacis and Hansen is illustrated in Fig. 2.29. The stratosphere is divided into layers, and the solar heating is calculated by taking the difference between the direct solar radiation incident at the top of each layer and that emergent at the bottom of the layer, and adding the difference between the upwelling reflected solar flux incident at the bottom and emergent at the top. The solar flux at any interface depends on the total ozone amount along the path, as specified by the algebraic approximations derived by Lacis and Hansen. For the direct solar flux, the total ozone amount is the product of the vertical ozone column above the interface and the effective zenith angle, with the latter corrected appropriately for sphericity. The total ozone amount for upwelling radiation is the sum of two components, that due to the path for direct solar radiation reaching the underlying surface (clouds or ground) and that due to the diffusely reflected path from the surface to the interface. For the latter, Lacis and Hansen use an effective "diffuse flux approximation" factor 1.9. The upwelling flux is proportional to the reflectivity of the underlying surface, whose prescription depends on cloud cover. For clear conditions, the reflectivity is assumed to be the average over all reflection angles of the reflectivity of a surface of known albedo overlain by a Rayleigh scattering atmosphere. For cloudy conditions, the reflectivity is that of a diffusely reflecting cloud alone, and is assumed to depend on the optical depth of the cloud in the visible.

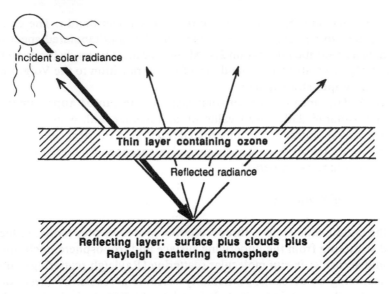

Fig. 2.29. Illustration of the Lacis and Hansen (1974) scheme for calculating ozone absorption. Solar energy deposited in the thin sublayer of the ozone layer is calculated by differencing the incident direct solar flux between the bottom and top of the layer, and differencing the upwelling diffuse flux between the top and bottom of the layer. The upwelling flux is assumed to come from an underlying reflecting layer whose albedo depends on surface properties, solar zenith angle, and the occurrence and properties of clouds.

This method provides an efficient and reasonably accurate calculation of ozone heating rate, accounting for upwelling solar radiation in the stratosphere. It does not account for scattering in the stratosphere or the interaction between scattering and absorption by ozone in the troposphere. As an alternative to the use of prescribed underlying surface and atmosphere reflectivities, Earth's reflectivity as measured by satellites can be used. The reflectivity seen by a satellite at wavelengths longer than about 340 nm is almost the same as the reflectivity seen looking downward from the stratosphere.

Background aerosol concentrations are not large enough to have a signicant effect on the energy balance of the stratosphere, but the aerosol effect can be important following major eruptions such as that of El Chichon in 1982 (see Section 12.5.2). The volcanic aerosols consist mainly of small sulfuric acid particles (radius ~1 μm) whose single scattering albedo is high, ~0.98–1.00. Because they absorb infrared radiation emitted by underlying ground or cloud, and because they increase the effective path of sunlight through the ozone layer, they tend to warm the lower stratospheric region in which they reside. If the volcanic cloud contains significant numbers of relatively absorptive soil particles, direct absorption of sunlight also contributes to heating.

Polar stratospheric water ice clouds and noctilucent clouds forming in the upper mesosphere may also have a small influence on heating in the middle atmosphere. Polar stratospheric clouds occur frequently in the southern hemisphere winter polar vortex and more rarely in the northern hemisphere winter polar vortex in the 15- to 25-km layer. The influence of these clouds on the radiation budget has been studied by Pollack and McKay (1985), who showed it to be generally unimportant except possibly in the southern winter polar vortex. Noctilucent clouds occur during summer near the polar mesopause. Though of great interest as an indicator of water-vapor condensation in the upper mesosphere, these clouds probably have only a minor effect on the radiation balance. Not only are the clouds extremely thin, but at these altitudes gas-particle heat exchange is inefficient in much the same way that translation-to-vibration energy exchange in CO_2 is inefficient.

2.8   Radiative Equilibrium Temperature and Heating-Rate Distributions

In the preceding sections the components of the radiative energy balance of the middle atmosphere have been considered. In this section, we consider the contributions of these components to the net heating rate and the radiative equilibrium temperature distribution that these contributions would produce in the absence of dynamical processes.

2.8.1   Net Heating and Its Components

The earliest detailed calculation of net heating rate in the middle atmosphere was carried out by Murgatroyd and Goody (1957) and included absorption of solar radiation by O_2 and O_3 and emission and exchange of longwave radiation by CO_2 and O_3. A number of calculations have been carried out since the pioneering work of Murgatroyd and Goody and have refined but not greatly altered the original conclusions. Most of the middle atmosphere is quite close to radiative equilibrium, such that there is near cancellation between large heating and cooling terms. Except in the polar regions, net heating-rate magnitudes rarely exceed 2 K day^{-1}, but there is a strong net cooling (up to 10–15 K day^{-1}) in the winter polar region between the stratopause and mesopause, and somewhat weaker net heating (up to ~5 K day^{-1}) 'in the summer polar lower mesosphere. These substantial imbalances in polar regions exert a strong control on the seasonally varying temperatures and zonal winds.

A recent calculation by Kiehl and Solomon (1986) illustrates these points (Fig. 2.30). This calculation made use of detailed global distributions of temperature and O_3 concentration measured by an instrument on board a

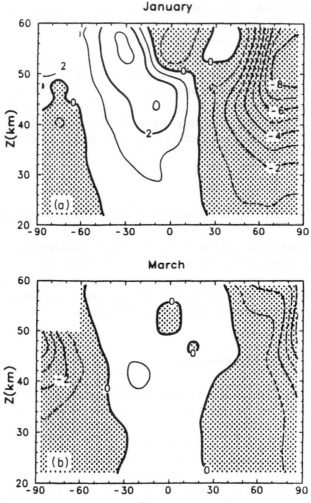

Fig. 2.30. Net heating rates (K day^{-1}) calculated for January and March. Temperatures and ozone concentrations used in the calculation were obtained from the Limb Infrared Monitor of the Stratosphere (LIMS) on the *Nimbus* 7 spacecraft during 1979. Due to lack of LIMS data, climatological temperatures and ozone concentrations were used south of 60°S, and there are no calculations above 50 km south of 60°S. [From Kiehl and Solomon (1986), with permission.]

polar orbiting satellite, the Limb Infrared Monitor of the Stratosphere (LIMS) instrument on board the *Nimbus* 7 satellite (see Section 2.9.3). Calculations were not made above 60 km or, in the south polar region, above 50 km due to lack of data. During January and May this calculation shows strong net cooling in the winter polar region, moderate net heating in the summer lower mesosphere, and very weak net heating elsewhere. During March there is strong cooling in both polar regions. There is considerable structure in the heating distribution above the stratopause in low latitudes, especially in January. This structure is real and is associated with the strong semiannual temperature cycle at low latitudes.

Figure 2.31 shows long-wavelength contributions to the balance by CO_2, O_3, and H_2O during January 1979. The patterns of the three contributions are similar, with maximum cooling near the stratopause, except that the 9.6-μm band of ozone contributes strong warming in the lower equatorial stratosphere. This feature is due to absorption of upwelling long-wavelength radiation, and it is sensitive to the distribution of tropospheric cloudiness. Water vapor and O_3 make negligible contributions to long-wavelength cooling above 60 km, and the upward extension of the CO_2 long-wavelength cooling into the upper mesosphere according to Dickinson is shown in Fig. 2.25. Note the good agreement between the calculations of Dickinson and Kiehl and Solomon in the region of overlap, despite the use of different temperature data sets.

Heating due to absorption of solar radiation by O_3, NO_2, and O_2 calculated by Kiehl and Solomon is shown in Fig. 2.32. In the region shown, O_2 and NO_2 heating rates are negligible but O_2 heating becomes important in the upper mesosphere due to strong heating in the Schumann–Runge bands. Figure 2.33 shows heating rates from London that extend the heating due to absorption of solar radiation by O_3 and O_2 above the mesopause. Absorption by O_2 is responsible for the strong high level heating maximum. Note, however, that London's heating rates are significantly greater than those calculated by Kiehl and Solomon in the region of overlap.

Experimental verification of the heating rates shown in Figs. 2.25 and 2.29–2.32 is not currently feasible because the flux divergences are so small. However, an indirect check of the validity of these calculations is possible. In the stratosphere and lower mesosphere, dynamical contributions to global mean heating at any level are small, so that global mean net radiative heating should be small throughout this region. In the middle and upper mesosphere, dynamical contributions to global mean net heating may be larger, but a small value of calculated global mean net radiative heating should still provide a rough check on calculations. Table 2.2 gives global mean net heating rates at various levels from the calculations of Kiehl and Solomon and Dickinson.

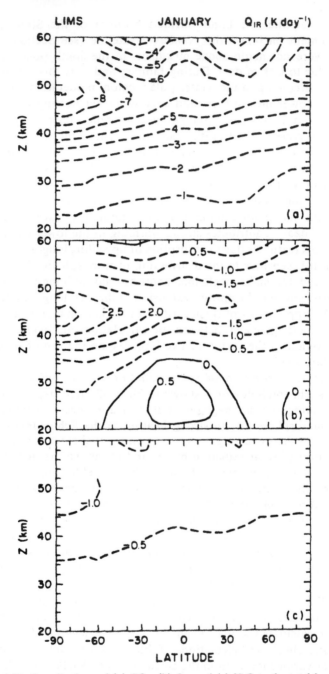

Fig. 2.31. Contributions of (a) CO_2, (b) O_3, and (c) H_2O to the total long-wave cooling during January 1979. [From Kiehl and Solomon (1986), with permission.]

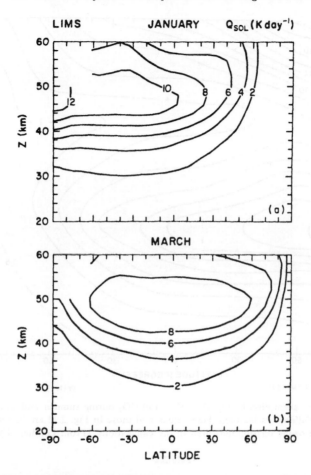

Fig. 2.32. Contributions of O_3, O_2, and NO_2 to heating rate during January and March 1979. [From Kiehl and Solomon (1986), with permission.]

The small values of calculated global mean net heating in Table 2.2 support the validity of the components of the calculations. The discrepancy in the lower mesosphere in the calculation of Kiehl and Solomon (1986) may be due largely to errors in the input ozone and temperature data. Solomon *et al.* (1986) have shown that the LIMS ozone measurements may be in error in this region for the following reason. Energy of solar ultraviolet radiation used to dissociate O_3 will be released when O_3 reforms during recombination. A fraction of this energy will appear as vibrational excitation of O_3, and some will cascade from high vibrational levels into the upper levels of transitions in the 9.6-μm spectral region. The resulting enhanced

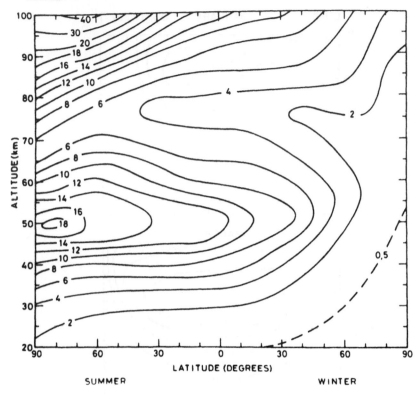

Fig. 2.33. Solar absorption by O_3, O_2, NO_2, and CO_2 during summer and winter, from London (1980). Differences between these values and those in Fig. 2.32 are largely due to differences in the ozone distributions used in the calculations. [From London (1980), with permission.]

Table 2.2

Calculated Globally Averaged Net Heating Rate

Altitude (km)	Net heating (K/day)[a]	Altitude (km)	Net heating (K/day)[b]
65	+0.2	25	−0.3
70	+0.8	30	−0.6
75	+0.6	35	−0.7
80	+0.7	40	−0.9
85	+1.7	45	−0.6
90	+4.7	50	−0.7
95	+2.4	55	+0.3
100	+1.5	60	+1.4

[a] Dickinson (1984).
[b] January mean, Kiehl and Solomon (1986).

9.6-μm band radiation would have been interpreted as a spurious O_3 excess in the analysis of the LIMS measurements.

2.8.2   Radiative Equilibrium

The radiative equilibrium temperature distribution is of interest as an indicator of the amount of temperature change that must be effected by the circulation. It can be calculated from

$$\frac{\partial T}{\partial t} = \frac{1}{\rho c_p}\left(\sum h_{r,ir} + \sum h_{r,uvv}\right),\tag{2.8.1}$$

where $\sum_r h_{r,ir}$ and $\sum_r h_{r,uvv}$ are the volume heating-rate contributions due to infrared and ultraviolet-plus-visible bands, respectively. Equation (2.8.1) is integrated from arbitrary initial conditions until a steady state is achieved.

Figure 2.34 shows the results of such a calculation. In the tropics and subtropics the radiative equilibrium temperature distribution is close to that observed. At high latitudes, particularly in the winter hemisphere, radiative equilibrium temperatures are far from observed temperatures. In the winter polar cap region, only infrared radiation transferred from the troposphere maintains nonzero temperatures. In radiative equilibrium, temperature decreases continuously from the summer to winter pole at all levels, in marked contrast to the observed temperature gradients in the upper mesosphere and lower stratosphere. Notice that this temperature distribution is very similar to that obtained from the time-marched radiative calculation shown in Fig. 1.2.

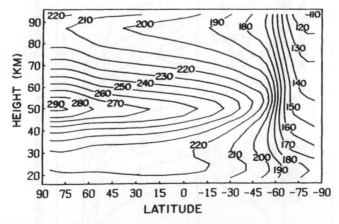

Fig. 2.34. Radiative equilibrium temperature distribution for northern (left) summer solstice. [From Wehrbein and Leovy (1982), with permission.]

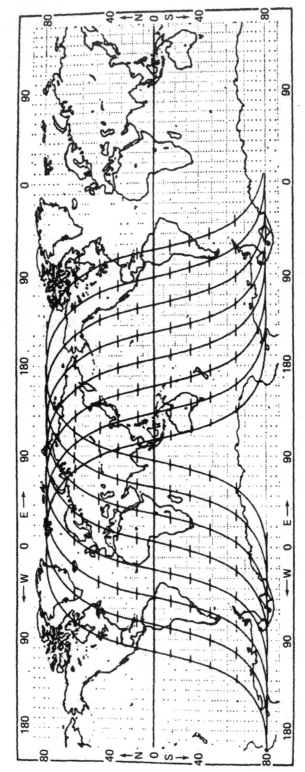

Fig. 2.35. Ground tracks for several orbits of a typical sun-synchronous polar orbiting satellite. [From Houghton *et al.* (1984), with permission.]

2.9   Remote Sounding

Until relatively recently, our knowledge of the distributions of temperature, winds, and such important trace gases as ozone and water vapor was limited to data from the worldwide radiosonde network, extending no higher than about 30 km, from a limited number of rocket profiles, and from some balloon and aircraft measurements obtained in connection with special experiments. None of these provided truly global coverage. This situation underwent a radical change beginning in the late 1960s with the advent of satellites carrying sensors capable of sounding the middle atmosphere. Examples of temperature and constituent distributions derived from satellite sounding are shown throughout this book.

The most useful satellite data sets for the middle atmosphere have so far been generated by polar orbiting satellites in the *Nimbus, NOAA,* and *Explorer* series. These satellites operate in sun-synchronous circular orbits near an altitude of 1000 km. The characteristic period of these orbits is slightly more than 100 min, so that there are typically 13–14 orbits per day. Because the earth is spinning beneath the satellite orbit, the subsatellite track slips westward by about 30° longitude from one orbit to the next. Sun-synchronous orbits maintain a constant angle between the earth–sun line and the line joining the earth's center and the satellite at the point where the satellite orbit crosses the ecliptic plane.

The satellite orbit slowly precesses to maintain constant-local-time satellite crossings of each latitude. The precession rate is accomplished by inclining the satellite orbit about 7° from the earth's polar axis. The lowest-order non-spherically symmetric component of the earth's gravity field provides the torque necessary to maintain the proper precession rate. The satellite orbit crosses each latitude circle twice each orbit, once ascending in latitude and once descending in latitude. Because the orbital period is not ordinarily commensurate with the earth's rotation period, orbits on successive days do not overlay. Instead, the spaces between adjacent orbital ground tracks on a given day are gradually filled in on succeeding days. A typical segment of the subspacecraft track of a sun-synchronous satellite is shown in Fig. 2.35.

A brief discussion of the use of radiance data acquired by satellites to infer properties of the middle atmosphere is given in this section.

2.9.1   Nadir Sounding of Temperature

The spectral distribution of upwelling radiance contains information about the vertical distribution of temperature and gaseous constituent concentrations. Since atmospheric opacity varies with wavelength and the depth from which emergent radiation is emitted varies with opacity, different depths in the atmosphere are sensed by measurements at different wavelengths.

The upwelling radiance L_r in spectral band r sensed on a nadir viewing satellite is given by

$$L_r = \int_{0_-}^{\infty} \int_{\Delta \nu_r} \hat{I}(\nu) B_\nu(z) \frac{dT_\nu(z, \infty)}{dz} \, d\nu \, dz$$

$$= \int_{0_-}^{\infty} B_r(z) w_r(z) \, dz \qquad (2.9.1)$$

where

$$w_r(z) \equiv \int_{\Delta \nu_r} \hat{I}(\nu) \frac{dT_\nu(z, \infty)}{dz} \, d\nu,$$

and $\hat{I}_\nu$ is the instrument spectral response function, normalized such that

$$\int_{\Delta \nu_r} \hat{I}(\nu) \, d\nu = 1,$$

and the transmission function is for the vertical viewing path. The notation 0_- is used to denote inclusion of the lower boundary in the integration. The quantity $w_r(z)$, called the "weighting function," determines the contribution of different atmospheric levels to L_r.

In order to examine the properties of w_r, consider a hypothetical instrument capable of sensing monochromatic radiance at frequency ν that is used to measure radiance emitted by a uniformly mixed absorbing gas with basic absorber density (scaled to reference temperature T_s) $\rho_{a0}(z) = \rho_{a0}(0)e^{-z/H} = \rho_{a0}(0)(p/p_0)$ and a constant absorption coefficient k_ν. The transmission function is

$$T_\nu(z, \infty) = \exp[-k_\nu H \rho_{a0}(0)(p/p_0)],$$

where p_0 is the pressure at the level at which $\rho_{a0} = \rho_{a0}(0)$ (e.g., the ground), and

$$w_r(z) = \frac{dT_\nu(z, \infty)}{dz} = -\frac{p}{H} \frac{dT_\nu}{dp}$$

$$= k_\nu \rho_{a0}(0)(p/p_0) \exp[-k_\nu H \rho_{a0}(0)(p/p_0)]. \qquad (2.9.2)$$

This function has the pressure dependence of the Chapman layer structure for normal incidence radiance. Like the Chapman layer, it has a single maximum where the optical depth, $\tau_\nu = k_\nu H \rho_{a0}(z)$, is equal to 1. The upwelling radiance is determined primarily by the Planck function near this level. Most importantly, this function has a characteristic width of about two scale heights, and this is the characteristic thickness of the region sensed at the instrument frequency ν.

It might seem that any real weighting function for a uniformly mixed gas would have a characteristic width of at least two scale heights, since in general w_r would be a superposition of monochromatic weighting functions, each corresponding to a different value of k_ν. This is not quite true, however, since k_ν is pressure-dependent. If k_ν corresponds to the wings of pressure broadened Lorentz lines,

$$k_\nu = S\alpha_L \pi^{-1}(\nu - \nu_0)^{-2} = k_{\nu 0}(p/p_0)$$

where

$$k_{\nu 0} \equiv S\alpha_{L0} \pi^{-1}(\nu - \nu_0)^{-2}$$

[using Eqs. (2.3.5) and (2.3.6) with p_s replaced by p_0], so that

$$w_r(z) = k_{\nu 0}\rho_{a0}(0)(p/p_0)^2 \exp[-\tfrac{1}{2}k_{\nu 0}H\rho_{a0}(0)(p/p_0)^2]. \qquad (2.9.3)$$

This weighting function has a shape similar to that of Eq. (2.9.2), but it is only half as broad. This is the narrowest physically realizable weighting function for nadir sounding of the Planck function.

An approximate representation of the vertical temperature profile can be obtained from multichannel satellite radiometer measurements in the 15-μm CO_2 band. The set of equations, one for each channel in the form of Eqs. (2.9.1), can be inverted to obtain a set of temperatures defining the approximate profile. The large width of the weighting functions severely limits the vertical resolution of these temperature profiles. In principle, resolution might be increased by using measurements from a large number of channels with closely spaced weighting functions. However, when there is strong overlap between the weighting functions, inversion of radiances to yield temperatures is unstable in the presence of any instrumental noise. Small radiance errors in one channel will be compensated by magnified errors in the temperatures retrieved for levels corresponding to the centers of weighting functions for adjacent overlapping channels. These errors in turn will propagate still further from the initial level. For realistic instrumental noise levels, the best resolution that can be achieved is comparable to the vertical width of the weighting functions.

Three nadir-viewing satellite instruments that have provided useful data on stratospheric temperature distributions are the Satellite Infrared Spectrometer (SIRS), a grating spectrometer flown on the *Nimbus* 3 and 4 satellites during the early 1970s, the Infrared Interferometer Spectrometer (IRIS), a Michelson interferometer flown on *Nimbus* 3 and on the *Mariner*

9 and *Voyager* planetary missions, and the High Resolution Infrared Radiation Sounder (HIRS), a multichannel filter radiometer flown on the *Nimbus* 6 and *TIROS-N* satellites in the late 1970s. Similar instrumentation has been flown by the Soviet Union on the *Meteor* satellite series during the same period. All of the weighting functions for these instruments peak at or below the 10-mb level, so they provide useful temperature data only for the middle and lower stratosphere.

In order to reach higher levels with nadir-viewing thermal infrared sensors, the technique of correlation spectrometry has been used. The first correlation spectrometers flown on a satellite were the Selective Chopper Radiometers (SCR) flown on *Nimbus* 4 and 5, and these have provided useful data from 1970 to 1978. In this technique, radiance from the underlying atmosphere first passes through filters to isolate broad regions in the 15-μm band, and the beam is rapidly cycled by means of a chopper between a direct path and path through an absorption cell containing a fixed amount of CO_2. By subtracting the two signals for the same location, a viewing path passing through the CO_2 cell and a path avoiding the cell, it is possible to isolate the component of terrestrial radiation emitted in the strong lines of the Q branch of the CO_2 fundamental. Since this radiation is emitted from relatively high altitude, the corresponding weighting function peaks at high altitude. By using a second cell with a larger CO_2 path, it is also possible to partially isolate the contribution to terrestrial radiance from the wings of the Lorentz lines in the P and R branches of the 15-μm fundamental. This produces a relatively narrow weighting function, approaching in width the theoretical limit discussed earlier in this section. Thus, the SCR instruments have produced a valuable data set covering the entire region from the ground to the stratopause.

A second type of gas-cell correlation spectrometer is the Pressure-Modulator Radiometer (PMR), first flown on *Nimbus* 6. A second-generation version of this instrument, the Stratospheric and Mesospheric Sounder (SAMS), was flown on *Nimbus* 7, providing data beginning in 1978. In the PMR technique, a CO_2 absorption cell is also used, but the pressure in the cell is varied by means of an oscillating piston. As the pressure in the cell varies, different 15-μm-band terrestrial radiance contributions are isolated. For example, at the lowest cell pressure (0.5 mb), the Doppler cores of the Q-branch lines of the fundamental are isolated. This gives a weighting function peaked in the upper mesosphere. Weighting functions for the SCR and PMR instruments are shown in Figs. 2.36 and 2.37. It can be seen that the entire region from tropopause to mesopause is covered by these instruments, but the coverage between 55 and 75 km in the mesosphere is sparse, so vertical resolution of temperature retrievals is poor in this region. In the case of SAMS, this situation was improved by operating the instrument in

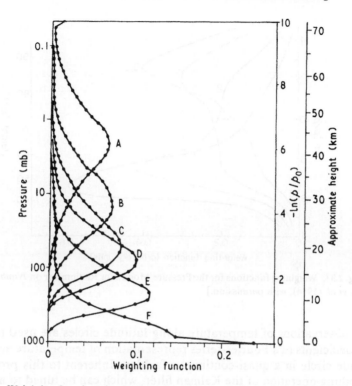

Fig. 2.36. Weighting functions for the Selective Chopper Radiometer on *Nimbus* 4. [From Abel *et al.* (1970), with permission.]

a limb-scanning mode (see Section 2.9.3). However, the main purpose of SAMS was to provide composition information. PMR cells are also used in the *TIROS-N* Operational Vertical Sounder (TOVS). Three cells, which have weighting functions that can be used to retrieve temperature from about 30 mb to 0.4 mb, comprise the Stratospheric Sounding Unit (SSU) of TOVS.

2.9.2 Geopotential and Wind

Temperatures retrieved from satellites are asynoptic: new observations are added continuously as the satellite moves in its orbit. However, a number of techniques are available to convert these evolving data sets into synoptic temperature distributions at specified times. A convenient practical method for doing this utilizes the Kalman filter as described by Rodgers (1976b).

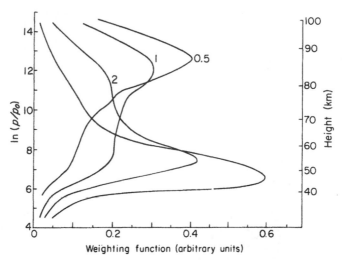

Fig. 2.37. Weighting functions for the Pressure Modulator Radiometer on *Nimbus* 5. [From Curtis *et al.* (1974), with permission.]

New observations of temperature along latitude circles are used to update the coefficients in a Fourier-series representation of temperature around the latitude circle in a quasi-continuous fashion. Inherent to this process is a smoothing operation of the Kalman filter, which can be tuned to match the precision of the observations, so that noise is suppressed.

Once the global distribution of temperature has been mapped, the distribution of geopotential on constant pressure surfaces can be obtained by integrating the hydrostatic equation from a base pressure level on which the geopotential is known from analysis of radiosonde observations. This base level is usually chosen to be 100 or 50 mb.

Instead of temperature, thickness of deep layers, typically spanning an order of magnitude range in pressure, can be retrieved. These retrievals can be carried out with good accuracy by linear regression, in which observed radiance is regressed against thickness derived from rocket measurements. Nadir satellite data with their broad weighting functions are well suited to this simple retrieval technique. The resulting geopotential surfaces have provided a sound basis for diagnoses of middle atmosphere circulation, although the vertical resolution is very limited.

Once the distribution of geopotential on middle atmosphere pressure surfaces has been deduced, geostrophic or gradient winds can be readily calculated, as can a variety of diagnostic quantities relevant to momentum and energy balances and wave propagation. Gradient or higher-order approximations to the rotational wind generally provide better estimates of diagnostic quantities than geostrophic winds in the middle atmosphere (Elson, 1986). Despite the numerous approximations and limitations of the nadir-viewing satellite data, the time variations and relationships between these derived quantities have given a remarkably consistent picture of dynamical processes in the stratosphere in a number of studies, the results of which are discussed elsewhere in this volume.

In the future, we may look for direct satellite determinations of winds in the middle atmosphere based on advanced optical techniques. These will resolve the Doppler shifts of spectral lines resulting from winds averaged over domains whose dimensions are several hundred kilometers in the horizontal and a few kilometers in the vertical. An instrument capable of this type of measurement is to be flown on the Upper Atmosphere Research Satellite (*UARS*).

2.9.3   Limb Sounding of Temperature

As an alternative to nadir sounding, temperature profiles can be derived from a radiometer with a narrow field of view scanning vertically through the limb of the planet and measuring radiance in the 15-μm band (Fig. 2.38). This limb-sounding technique has several advantages over nadir sounding. The most important of these is that the weighting functions are determined primarily by the instrument field of view, and their width is therefore limited more by signal-to-noise considerations than by funda-

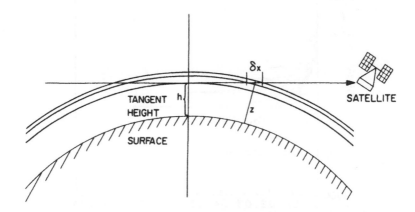

Fig. 2.38. Limb viewing geometry. The satellite instrument scans through the atmosphere on the limb at tangent height *h*. Radiance is received from elements of length δx at height *z* along the tangent path. [From Houghton *et al.* (1984), with permission.]

mental physical principles (Fig. 2.39). Since the instrument views a long atmospheric optical path (typically about 40 times the length of the nadir optical path to the altitude of the tangent point), and views it against the background of cold space, the sensitivity is high, and measurements can be easily extended well into the mesosphere. An additional advantage is that an "onion-peeling" technique, working from the top down, can be used to retrieve temperature, so that retrieval errors at one level do not propagate upward, although they do propagate downward.

The technique has some disadvantages. It is relatively sensitive to the presence of aerosols in the lower stratosphere, which can produce serious errors, and it cannot be consistently used to probe below the tropopause. It requires very precise knowledge of field of view and is sensitive to stray light. It also requires precise knowledge of spacecraft attitude and attitude change rates so that instrument pointing can be determined. If attitude change rates vary slowly compared with the limb-scan time, this rate can be nearly eliminated by combining data from adjacent upward and downward scans. If spacecraft attitude is not accurately known so that the absolute height scale on the limb is uncertain, a "two-color" technique, suggested by Gille and House (1971), can be used to locate a reference pressure level on each limb profile. Temperatures can then be associated with pressures at all other profile levels by working upward or downward from the reference

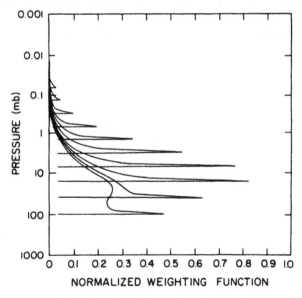

Fig. 2.39. Weighting functions for one of the 15-μm band channels of the Limb Infrared Monitor of the Stratosphere (LIMS). [From Bailey and Gille (1986), with permission.]

pressure level, using the hydrostatic equation and the retrieved temperatures. In this technique, simultaneous colocated observations in two 15-μm channels, one wide band and one narrow band, are used. Temperature–pressure profiles are retrieved using data from each channel starting from an assumed reference pressure at the same height on each profile. The two temperature–pressure retrievals will show systematic disagreement unless the assumed reference pressure corresponds to the actual pressure at that height. Thus, the true reference pressure can be deduced through an iterative correction procedure.

Thus far, limb-scanning instruments have viewed the horizon at a fixed angle with respect to the spacecraft track. As a result, sampling in longitude has been limited to the longitudinal sampling frequency of the spacecraft orbit. In contrast, nadir sampling has incorporated scanning across the spacecraft track with slightly off-nadir viewing, and this is done with the TOVS instrument to give complete longitudinal sampling at the resolution of the TOVS footprint.

The vertical resolution and sensitivity to high-altitude signals available in limb sounding are responsible for a considerable advance in quality of the global middle atmosphere data sets. In addition to SAMS, two limb-scanning radiometers have flown in *Nimbus* satellites, the Limb-Sounding Infrared Radiometer (LRIR) on *Nimbus* 6, and the Limb Infrared Monitor of the Stratosphere (LIMS) on *Nimbus* 7.

2.9.4 Sounding of Composition

A complete understanding of the chemistry of the middle atmosphere requires information on the distributions of nearly 50 minor gaseous species. The role played by some of these is discussed briefly in Chapter 10. Among the more important are O_3, O, H_2O, HNO_3, NO, NO_2, CH_4, CO, HCl, N_2O, $CFCl_3$, and CF_2Cl_2. Equation (2.9.1) can be inverted to retrieve $dT_\nu(z, \infty)/dz$, from which the vertical distribution of absorbing gas can be derived, provided that $B_\nu(z)$ is known independently and the atmosphere is in LTE. In practice this approach is very difficult from nadir-sounding observations, because of the low concentrations and resulting weak absorptions for most of the gases of interest. Even in principle, the information content of upwelling radiance for composition determination is strongly dependent on the vertical temperature profile. If the atmosphere is isothermal, no vertical structure information can be derived from low-resolution upwelling radiances.

A high degree of overlap between the spectral features of many of the gases of interest is another complication. As a result of these difficulties, useful retrievals of thermal radiance from nadir viewing have been restricted to total column water vapor, total column O_3, and some constraints on the vertical profile of water vapor.

In contrast, the long viewing path, cold space background, and dependence of weighting functions on viewing geometry that characterize limb emission measurements have made it possible to infer near-global distributions of several important gases in the middle atmosphere. These include O_3 and H_2O (LRIR), O_3, H_2O, HNO_3, and NO_2 (LIMS), and CH_4, N_2O, CO, and NO (SAMS). Additional global measurements of NO_2, H_2O, and O_3 have been made with limb-viewing radiation sensors on the *Solar Mesosphere Explorer (SME)* satellite. Global distributions of several additional species are expected to be obtained from *UARS* limb measurements of thermal radiance.

Ozone is the most important variable species in the middle atmosphere, and its prominent ultraviolet spectrum lends itself to distribution determinations from measurements of backscattered solar radiance. Since the radiance source is Rayleigh-scattered sunlight for which the scattering cross section is strongly wavelength-dependent, the pressure range of the atmospheric layer from which the radiance is scattered depends strongly on wavelength. On the other hand, attenuation of the scattered radiation, and hence the radiance received at the satellite, depends on the amount of overlying O_3. As a consequence, the ultraviolet spectrum of the earth shows a broad deep minimum centered near 250 nm. The signature of this feature in backscattered solar irradiance has been widely used to retrieve O_3 vertical profiles. The most recent instruments to be used for this purpose are the Backscatter Ultraviolet Spectrometer (BUV) used on *Nimbus* 4 and on *Atmospheric Explorer* 5, and the Solar Backscatter Ultraviolet Spectrometer (SBUV) used on *Nimbus* 7 and on current operational satellites. Weighting functions for the SBUV instrument are shown in Fig. 2.40.

The total column amount of ozone has been measured globally since 1978 with the Total Ozone Mapping Spectrometer (TOMS) flown on several satellites. This technique uses differential absorption in reflected near-ultraviolet radiance to infer the total mass of ozone along the viewing path. The TOMS is a scanning instrument, so that measurements are available globally at high spatial resolution. Total ozone variations are dominated by variations in ozone amount in the lower stratosphere and by variations in tropopause height, so these data are especially useful for assessing the behavior of the tropopause at high spatial resolution and in regions that are poorly sampled by radiosondes.

Ozone vertical distribution below the level of the concentration maximum, about 25 km, is difficult to retrieve by either the ultraviolet backscatter technique or the thermal limb emission technique. Measurements of direct absorption of radiance along the path from the rising or setting sun (i.e.,

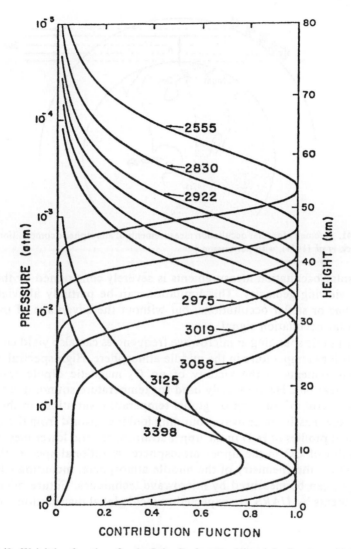

Fig. 2.40. Weighting functions for the Solar Backscatter Ultraviolet Spectrometer (SBUV) on the *Nimbus* 7 satellite. Note the absence of weighting functions peaked in the lower stratosphere. [From Yarger and Mateer (1976), with permission.]

measurements of radiance during *solar occultations*) can provide fairly precise estimates of the concentrations of many species over a broad altitude range, including O_3 in the difficult region between the tropopause and 25 km (Fig. 2.41). This *solar occultation* technique has also been used extensively to infer vertical distributions of stratospheric aerosols. The spatial distribu-

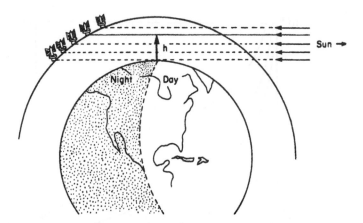

Fig. 2.41. Geometry for solar occultation measurements of atmospheric composition. [From McCormick *et al.* (1979), with permission.]

tion of solar occultation measurements is severely constrained by the sun-satellite viewing geometry. This limitation can be partially alleviated by using lunar or stellar occultations, but without the advantage of the very strong solar occultation signal.

Atmospheric sounding at microwave frequencies can also yield composition as well as temperature in the middle atmosphere. High-spectral-resolution measurements in the vicinity of an O_2 magnetic dipole transition centered near 60 GHz are widely used for temperature sounding and have provided useful information on global temperature variations in the lower stratosphere. Passive microwave sounding looking upward from the ground has yielded profiles of H_2O in the upper stratosphere and lower mesosphere and profiles of CO in the upper mesosphere. Additional species that are important for the chemistry of the middle atmosphere, including ClO, O_3, and H_2O_2, can be measured by microwave techniques. Future microwave measurements by *UARS* are expected to yield global distributions of these species.

References

2.1. An excellent overview of radiation in the middle atmosphere is given by Brasseur and Solomon (1984). Contributions of various gases to radiative heating and cooling are reviewed by London (1980). Applications of satellite remote sensing to the middle atmosphere are reviewed by Houghton *et al.* (1984).

2.2. Radiative transfer formalism is developed in the texts of Chandrasekhar (1950), Goody (1964), and Liou (1978). Chandrasekhar develops general methods for solving the radiative transfer equation in multiply scattering atmospheres; Goody emphasizes radiative transfer in atmospheres with absorption and emission by molecular gases; Liou provides a clear development of the theory of scattering by spherical particles and presents a variety of methods for solving the radiative transfer equation. Spherical atmosphere effects and the Chapman function are discussed by Swider and Gardner (1957).

2.3. Physical processes responsible for molecular absorption, the structure of vibration-rotation bands and line shape are given in Goody (1964), and Brasseur and Solomon (1984). A good basic reference for this topic is Levine (1974). The semi-classical theory of transition strength is given in Houghton and Smith (1974). Molecular oxygen absorption is discussed by Frederick *et al.* (1983), and the Schumann-Runge bands in particular are discussed by Kockarts (1976), and by Allen and Frederick (1982). A recent tabulation of molecular oxygen absorption coefficients is given in WMO (1986). Infrared spectral data for CO_2, H_2O, O_3, and several other relevant gases are discussed by Drayson *et al.* (1984). Tabulations of positions, strengths, and temperature dependent line widths in the infrared and microwave regions are available in the AFGL Absorption Line Compilation as described by Rothman and Young (1981), and Rothman *et al.* (1983), and detailed discussion of spectroscopic data sets is given in Appendix B of WMO (1986).

2.4. Transmission functions are discussed in Goody (1964), Kondratyev (1969), Paltridge and Platt (1976), and Houghton (1977). The random models presented here were given by Goody (1952a), and Malkmus (1967). Whole band models are given by Ramanathan (1976), and Lacis and Hansen (1974). The Curtis-Godson approximation was suggested by Curtis (1952) and Godson (1953). Angular integration and errors of approximate approaches are discussed by Rodgers and Walshaw (1966). Approximations to the Voigt lines shape are given by Goldman (1968).

2.5. The exchange integral formulation of radiative transfer presented in this section was first given by Curtis (1956) and applied by Curtis and Goody (1956). The concept of the escape function was introduced by Dickinson (1972), and a simple parameterization of radiative exchange suitable for the stratosphere was suggested by Dickinson (1973). The spectral approach to radiative damping is discussed by Fels (1985).

2.6. The effects of departures from local thermodynamic equilibrium are quantitatively discussed by Goody (1964), Murgatroyd and Goody (1957), Kuhn and London (1969), Dickinson (1972), Kutepov and Shved (1978), Apruzese, Strobel, and Schoeberl (1984), and Fomichev *et al.* (1986). The most complete non-LTE model and calculation for the mesopause region published to date is that of Dickinson (1984).

2.7. The description of the vertical absorption profile for incident radiation originates with Chapman (1931). Deposition of solar radiation and its effects are discussed by Whitten and Poppoff (1971) and Brasseur and Solomon (1984). WMO (1986) is an excellent source of data on solar flux and its variability and on the ultraviolet absorption spectra of oxygen and ozone. Effects of scattered radiation in the stratosphere have been analyzed by Meier *et al.* (1982), Nicolet *et al.* (1982), and Herman and Mentall (1982). Approximate treatments of solar absorption have been given by Cogley and Borucki (1976) and Strobel (1978). Pollack and Ackerman (1983) have investigated the radiative effects of the El Chichon aerosol on the stratosphere.

2.8. Plass (1956a,b) carried out early studies of middle atmosphere cooling rates due to the 9.6-μm ozone and 15-μm CO_2 bands. Calculations of net heating rate for the middle atmosphere have been carried out by Murgatroyd and Goody (1957), Kuhn and London (1969), Wehrbein and Leovy (1982), Rosenfield et al. (1986), and Gille and Lyjak (1986). Drayson (1967) and Fomichev et al. (1986) have carried out a valuable assessment of the long wave contribution. Akmaev and Shved (1982) and Fomichev and Shved (1985) have provided accurate simplified treatments of long wave cooling by the CO_2 15-μm band and the O_3 9.6-μm band. Dopplick (1972, 1979) has presented a valuable calculation of net heating in the troposphere and stratosphere. Radiative equilibrium is discussed by Goody (1952), Leovy (1964a), and Apruzese et al. (1984).

2.9. Houghton et al. (1984) provide a valuable review of the principles and applications of remote sounding from satellites. The discussion of weighting functions in this section follows Houghton and Smith (1970). Hanel et al. (1971) discuss the IRIS instrument and display informative infrared spectra of Earth. The PMR is discussed by Taylor et al. (1972). Limb scanning infrared radiometer measurements of temperature, ozone, and water vapor are described in Gille and Russell (1984), Gille et al. (1984), Remsberg et al. (1984a,b), and Russell et al. (1984); SAMS measurements of CH_4 and N_2O are described by Jones and Pyle (1984). SBUV ozone measurements are described by McPeters et al. (1984), and TOMS measurements of total ozone are described by Bhartia et al. (1984). McCormick et al. (1984) present solar occultation measurement results for ozone, and McCormick et al. (1979) describe solar occultation measurements of stratospheric aerosol. The use of microwave radiances to infer temperature variations in the lower stratosphere has been described by Newman and Stanford (1985); microwave measurements of water vapor have been presented by Bevilacqua et al. (1983), and Waters et al. (1981) have discussed microwave measurements of ClO, O_3, and H_2O_2.

Chapter 3 | Basic Dynamics

3.1 Introduction

The atmosphere is a shallow envelope of compressible gas surrounding an approximately spherical, rotating planet. The equations of motion in a rotating frame for such a gas are well known, but in their most general form they are far more complicated than necessary or desirable for application to the large- and medium-scale meteorological phenomena considered in this book. Scale analysis, involving an investigation of the relative orders of magnitude of the various terms in the relevant equations, shows that several simplifications to the equations can be made. In particular, the vertical momentum equation can be replaced by hydrostatic balance, the Coriolis force associated with the horizontal component of the earth's rotation vector can be neglected, and the distance r from any point in the atmosphere to the center of the earth can be replaced by a mean radius a. The resulting approximate set of equations is called the *primitive equations*.

3.1.1 The Primitive Equations in Log-Pressure Coordinates on the Sphere

Although the geometric height $z^* \equiv r - a$ is the most obvious choice of vertical coordinate, the primitive equations take slightly simpler forms when other vertical coordinates are used, and throughout most of this book we shall use the "log-pressure" coordinate

$$z \equiv -H \ln(p/p_s) \tag{3.1.1}$$

113

introduced in Eq. (1.1.8). (The inverse relation

$$p = p_s e^{-z/H},$$ (3.1.2)

should be noted.) Using z as the vertical coordinate and spherical coordinates in the horizontal, the primitive equations take the following form:

$$\frac{Du}{Dt} - \left(f + \frac{u \tan \phi}{a}\right) v + \frac{\Phi_\lambda}{a \cos \phi} = X,$$ (3.1.3a)

$$\frac{Dv}{Dt} + \left(f + \frac{u \tan \phi}{a}\right) u + \frac{\Phi_\phi}{a} = Y,$$ (3.1.3b)

$$\Phi_z = H^{-1} R\theta \, e^{-\kappa z/H},$$ (3.1.3c)

$$\frac{[u_\lambda + (v \cos \phi)_\phi]}{a \cos \phi} + \frac{(\rho_0 w)_z}{\rho_0} = 0,$$ (3.1.3d)

$$\frac{D\theta}{Dt} = Q,$$ (3.1.3e)

(e.g., Holton, 1975). These express, respectively, momentum balance in the zonal and meridional directions, hydrostatic balance in the vertical, continuity of mass, and the thermodynamic relation between diabatic heating and the material rate of change of potential temperature.

In Eq. (3.1.3) Φ denotes the geopotential [see Eq. (1.1.3)], θ the potential temperature [see Eq. (1.1.9)], and the following new notation has been used:

Horizontal coordinates: (λ, ϕ) = (longitude, latitude).
"Velocity" components:

$$(u, v, w) \equiv \left[(a \cos \phi) \frac{D\lambda}{Dt}, a \frac{D\phi}{Dt}, \frac{Dz}{Dt}\right],$$

where D/Dt is the material derivative, or time rate of change following the fluid motion, whose expression in the present coordinates is

$$\frac{D}{Dt} \equiv \frac{\partial}{\partial t} + \frac{u}{a \cos \phi} \frac{\partial}{\partial \lambda} + \frac{v}{a} \frac{\partial}{\partial \phi} + w \frac{\partial}{\partial z}.$$

Note that w is not in general equal to the geometric vertical velocity Dz^*/Dt; however, the difference is generally insignificant except near the ground.

Coriolis parameter (the vertical component of the earth's rotation vector): $f \equiv 2\Omega \sin \phi$, where $\Omega = 2\pi$ (sidereal day)$^{-1}$ = 7.292×10^{-5} s^{-1} is the earth's rotation rate.

Unspecified horizontal components of friction, or other nonconservative mechanical forcing: (X, Y).

Diabatic heating term: $Q \equiv (J/c_p)e^{\kappa z/H}$, where J is the diabatic heating rate per unit mass, which in the middle atmosphere equals the net *radiative* heating rate per unit mass, $-\rho_0^{-1}\partial F_n/\partial z$ (see Section 2.5), plus a small thermal conduction term; note that J/c_p is often expressed in units of Kelvins per day.

Basic density: $\rho_0(z) \equiv \rho_s e^{-z/H}$ where $\rho_s \equiv p_s/RT_s$. Thus $\rho_0 = p/RT_s$ by Eq. (3.1.2); some authors use p instead of ρ_0 in Eq. (3.1.3d).

Some partial derivatives with respect to λ, ϕ, and z are denoted by suffixes.

The primitive equations are frequently written using the temperature T instead of potential temperature θ, in which case Eqs. (3.1.3c, e) are replaced by

$$\Phi_z = H^{-1}RT \tag{3.1.3c'}$$

and

$$\frac{DT}{Dt} + \frac{\kappa Tw}{H} = \frac{J}{c_p}, \tag{3.1.3e'}$$

respectively.

A further quantity of considerable dynamical importance is Ertel's potential vorticity P (Rossby (1940), Ertel (1942)), defined in general as $\rho^{-1}\boldsymbol{\omega}_a \cdot \nabla\theta$ where $\boldsymbol{\omega}_a$ is the absolute vorticity. Under the approximations that lead to the primitive equations, it is given by

$$\rho_0 P \equiv \theta_z \left[f - \frac{(u\cos\phi)_\phi}{a\cos\phi} + \frac{v_\lambda}{a\cos\phi} \right] - \frac{\theta_\lambda v_z}{a\cos\phi} + \frac{\theta_\phi u_z}{a} \tag{3.1.4}$$

in log-pressure coordinates. A rather lengthy calculation, starting from Eq. (3.1.3), shows that

$$\frac{DP}{Dt} = (\rho_0 a \cos\phi)^{-1} \left[-\frac{\partial(X\cos\phi, \theta)}{\partial(\phi, z)} + \frac{\partial(Y, \theta)}{\partial(\lambda, z)} - \frac{\partial(Q, v)}{\partial(\lambda, z)} + \frac{\partial(Q, m)}{\partial(\phi, z)} \right], \tag{3.1.5}$$

where

$$\frac{\partial(A, B)}{\partial(x, y)} \equiv \frac{\partial A}{\partial x}\frac{\partial B}{\partial y} - \frac{\partial A}{\partial y}\frac{\partial B}{\partial x}$$

and

$$m \equiv a\Omega\cos^2\phi + u\cos\phi$$

is a^{-1} times the absolute zonal angular momentum per unit mass. Note that if the mechanical forcing (X, Y) and the diabatic heating Q both vanish, the right-hand side of Eq. (3.1.5) vanishes, and P is conserved following the motion. This "conservable" property of P is one reason why it is of such interest to dynamical meteorologists (see Sections 5.2.3 and 6.2.4, for example). A physical interpretation of the conservation of P is given in Section 3.8.2.

3.1.2   Boundary Conditions

To solve the primitive equations of Eqs. (3.1.3), or any approximate set of equations derived from them, it is of course necessary to apply suitable boundary conditions. These depend on the particular physical problem under consideration; some typical examples will be discussed here.

3.1.2.a   Conditions at the Lower Boundary

1. If the lower boundary is the *ground*, the shape of the topography should be specified in terms of the geometric height z^*, rather than z (this is a slight inconvenience of log-pressure coordinates): for example,

$$z^* = h(x, y, t) \qquad \text{at the ground.}$$

(The t-dependence is a mathematical device that is sometimes useful for idealized initial-value problems; for example, a mountain might be "grown" so as to set up a flow in an unambiguous manner.) Since the ground is a material surface, the kinematic boundary condition is

$$\frac{D}{Dt}(z^* - h) = 0 \qquad \text{at} \quad z^* = h.$$

If viscosity is important, further conditions are required, but we shall not need them in this book. In terms of Φ, we have, from Eq. (1.1.3),

$$\frac{D\Phi}{Dt} = g\frac{Dh}{Dt} \qquad \text{at} \quad \Phi = \int_0^h g \, dz^* \approx gh, \tag{3.1.6a}$$

where the latter approximation relies on the fact that g is essentially constant over the altitude range of the earth's topography.

2. Some models specify the geopotential or geometric height of a given log-pressure level, for example,

$$\Phi(x, y, z_0, t) = F(x, y, t), \tag{3.1.6b}$$

where $z_0 = -H \ln(p_0/p_s) = $ constant; p_0 might be near the tropopause (say

$p_0 = 100$ mb). This is usually the easiest form of lower boundary condition to implement in log-pressure coordinates.

3. It is sometimes convenient in simple mathematical models to specify the log-pressure at a lower material boundary. Thus $z = \zeta(x, y, t)$, say, at the boundary; by analogy with (1), the kinematic boundary condition is then

$$w \equiv \frac{Dz}{Dt} = \frac{D\zeta}{Dt} \quad \text{at} \quad z = \zeta(x, y, t). \tag{3.1.6c}$$

This condition is less suitable than Eq. (3.1.6a) or (3.1.6b) for use in detailed simulation of the atmosphere.

3.1.2.b Conditions at the Upper Boundary

Numerical general circulation models of the middle atmosphere, such as those to be discussed in Chapter 11, employ a finite number of levels in the vertical and usually have to include an effectively "rigid" upper boundary. For example, a model formulated in z coordinates might take $w = 0$ at some large but finite height z_1. A rigid lid of this type will tend to lead to unrealistic reflections of wave disturbances that reach it, and large dissipative terms are usually introduced near the upper boundary in an attempt to damp such waves and minimize spurious reflections. These dissipation terms are primarily a numerical expedient and normally have little physical basis.

Simpler linear models of wave disturbances in the middle atmosphere can often adopt more satisfactory dynamical upper boundary conditions, which fall into one of two categories:

1. Disturbances are *trapped* or *evanescent*: that is, they tend to zero with increasing height (where a suitable measure of a disturbance might be its energy per unit volume).

2. Vertically propagating disturbances obey a *radiation condition*: that is, they transfer "information" upward, and not downward, at great heights. This condition tacitly assumes that mean atmospheric conditions do not allow significant reflection of vertically travelling disturbances at great heights, so that a clear distinction can be made between upward and downward propagation. The radiation condition then states that only the upward-propagating disturbances—as identified perhaps by a "group velocity" argument—exist above some height z_2.

We shall examine cases (1) and (2) in detail in Chapter 4, when specific examples of wave motions are discussed.

3.1.2.c Conditions at Side Boundaries

The conditions here depend on the geometry of the atmospheric model under consideration. On the sphere, it is only necessary that all variables be bounded at the poles. In idealized cases where attention is fixed on a "channel" with vertical walls parallel to latitude circles $\phi = \phi_1, \phi_2$, say, $v = 0$ is taken on these walls.

3.2.   The Beta-Plane Approximation and Quasi-Geostrophic Theory

The primitive equations (3.1.3) are still a complicated set, despite the simplifications that have been used in deriving them. Moreover, they are capable of describing a very wide range of atmospheric flows, from slow motions of global scale to quite rapid, medium-scale disturbances. To focus on the larger-scale, slower motions, at least in the extratropical regions, we can introduce further approximations to obtain the *quasi-geostrophic equations*.

3.2.1   The Primitive Equations on a Beta-Plane

Before making the *dynamical* approximations that result in the quasi-geostrophic equations, it is first convenient to make a *geometrical* simplification, by replacing the spherical coordinates (λ, ϕ) by eastward and northward cartesian coordinates (x, y), and restricting the flow domain to some neighborhood of the latitude ϕ_0. This task can be carried out in a formally rigorous manner; however, since the resulting primitive equations are intuitively reasonable approximations to the full set of Eqs. (3.1.3), we shall only state them here. They are:

$$\frac{Du}{Dt} - fv + \Phi_x = X, \tag{3.2.1a}$$

$$\frac{Dv}{Dt} + fu + \Phi_y = Y, \tag{3.2.1b}$$

$$\Phi_z = H^{-1}R\theta e^{-\kappa z/H}, \tag{3.2.1c}$$

$$u_x + v_y + \rho_0^{-1}(\rho_0 w)_z = 0, \tag{3.2.1d}$$

$$\frac{D\theta}{Dt} = Q, \tag{3.2.1e}$$

with

$$\frac{D}{Dt} \equiv \frac{\partial}{\partial t} + u\frac{\partial}{\partial x} + v\frac{\partial}{\partial y} + w\frac{\partial}{\partial z}.$$

Here x is eastward distance and y northward distance from some origin (λ_0, ϕ_0), subscripts x and y denote partial derivatives, and other symbols are as before, with the exception that now

$$f = f_0 + \beta y, \qquad (3.2.1f)$$

where $f_0 \equiv 2\Omega \sin \phi_0$ and $\beta \equiv 2\Omega a^{-1} \cos \phi_0$. Note that Eq. (3.2.1f) is a formally valid approximation to $f = 2\Omega \sin \phi$ if $|y| \ll a \cot \phi_0$; the terms in $\tan \phi$ in Eqs. (3.1.3a, b) are negligible if u and v are comparable in magnitude and their horizontal length scales are much less than $a \cot \phi_0$. The linear variation of f with y captures the most important dynamical effect of the variation of $2\Omega \sin \phi$ with latitude; this "beta-effect" was first pointed out by Rossby (1939). Equations (3.2.1) are called the "beta-plane" versions of the primitive equations. To simulate the periodicity around latitude circles on the sphere, it is often convenient to consider the beta-plane to be periodic in x with period $2\pi a \cos \phi_0$.

3.2.2 Geostrophic Balance and the Thermal Wind Equations

Having simplified the geometry in this way, we can next use the fact that for large-scale, low-frequency, extratropical flows, approximate *geostrophic balance* holds: that is, the Coriolis terms $(-fv, fu)$ in Eqs. (3.2.1a, b) are roughly balanced by the horizontal gradients of geopotential. Thus the horizontal wind $(u, v, 0)$ satisfies

$$u \approx u_g, \qquad v \approx v_g, \qquad (3.2.2)$$

where the *geostrophic wind* $\mathbf{u}_g \equiv (u_g, v_g, 0)$ is defined in terms of the geopotential by

$$(u_g, v_g) \equiv (-\psi_y, \psi_x), \qquad (3.2.3)$$

where

$$\psi \equiv f_0^{-1}(\Phi - \Phi_0) \qquad (3.2.4)$$

is called the geostrophic stream function and $\Phi_0(z)$ is a suitable reference geopotential profile; note that the definition of ψ involves f_0 and not f. From the hydrostatic balance of Eq. (3.2.1c), we have

$$\theta_e \equiv \theta - \theta_0(z) = HR^{-1}f_0 e^{\kappa z/H}\psi_z, \qquad (3.2.5)$$

where $\theta_0(z) = HR^{-1}e^{\kappa z/H}\Phi_{0z}$ is a reference potential temperature. Likewise, using Eq. (3.1.3c'),

$$T - T_0(z) = HR^{-1}f_0\psi_z, \qquad (3.2.5')$$

where $T_0(z) = e^{-\kappa z/H}\theta_0(z)$ is a reference temperature. [Possible choices for $T_0(z)$ might be the midlatitude profile sketched in Fig. 1.1, or a global mean profile; $\Phi_0(z)$ would then be obtained by vertical integration, subject to the boundary condition $\Phi_0(0) = 0$.] Combining Eqs. (3.2.3) and (3.2.5) or (3.2.5′) to eliminate ψ by cross differentiation, we obtain the "thermal wind" equations

$$\frac{\partial u_g}{\partial z} = -\frac{R}{Hf_0} e^{-\kappa z/H} \frac{\partial \theta}{\partial y} = -\frac{R}{Hf_0} \frac{\partial T}{\partial y}, \tag{3.2.6a}$$

$$\frac{\partial v_g}{\partial z} = \frac{R}{Hf_0} e^{-\kappa z/H} \frac{\partial \theta}{\partial x} = \frac{R}{Hf_0} \frac{\partial T}{\partial x}, \tag{3.2.6b}$$

which relate the vertical shear of the geostrophic wind components to horizontal potential temperature (or temperature) gradients. Note also from Eq. (3.2.3) that $\partial u_g/\partial x + \partial v_g/\partial y \equiv 0$, so by the continuity equation [Eq. (3.2.1d)] the geostrophic wind is associated with a vertical "velocity" w_g that satisfies $(\rho_0 w_g)_z \equiv 0$. To ensure that w_g is bounded as $z \to \infty$, we must therefore take $w_g \equiv 0$.

3.2.3 Quasi-Geostrophic Flow

It will be observed that Eqs. (3.2.2)–(3.2.4) are first approximations to the horizontal momentum equations (3.2.1a, b), provided that the accelerations Du/Dt and Dv/Dt and the nonconservative terms X and Y are ignored, and $f_0 + \beta y$ is replaced by f_0. To examine these approximations more closely, and to investigate the time development of the geostrophic flow [which is not predicted by Eqs. (3.2.2)–(3.2.4)], we define ageostrophic velocities, denoted by a subscript a, thus:

$$u_a \equiv u - u_g, \qquad v_a \equiv v - v_g, \qquad w_a \equiv w. \tag{3.2.7}$$

We suppose that U is a typical order of magnitude of the geostrophic wind speed $|u_g|$, and that L is a typical horizontal length scale, so that $\partial/\partial x$ and $\partial/\partial y$ are $O(L^{-1})$. It can then be shown that Eqs. (3.2.2)–(3.2.4) are valid first approximations, with $|u_a| \ll |u_g| \sim U$, $|v_a| \ll |v_g| \sim U$, if the following conditions are satisfied:

(a) $\text{Ro} \equiv U/f_0 L \ll 1$.

(b) $\partial/\partial t \ll f_0$.

(c) $\beta L \ll f_0$,

(d) $|X|, |Y| \ll f_0 U$.

$$\tag{3.2.8}$$

Condition (a) states that the Rossby number Ro, which measures the ratio of the nonlinear terms $\mathbf{u} \cdot \nabla(u, v)$ to the Coriolis terms $(-f_0 v, f_0 u)$ in Eqs. (3.2.1a,b), should be small. Likewise, condition (b) states that the ratio of the time derivatives $(\partial u/\partial t, \partial v/\partial t)$ to the Coriolis terms should be small. Condition (c) allows the use of f_0 rather than f in Eq. (3.2.4), while condition (d) ensures that friction is small. These conditions make precise the restriction to "large-scale, low-frequency motions" mentioned above.

Given conditions (3.2.8), the next approximation to Eqs. (3.2.1a,b,d,e) beyond geostrophic balance is a set of equations describing "quasi-geostrophic flow":

$$D_g u_g - f_0 v_a - \beta y v_g = X, \tag{3.2.9a}$$

$$D_g v_g + f_0 u_a + \beta y u_g = Y, \tag{3.2.9b}$$

$$u_{ax} + v_{ay} + \rho_0^{-1}(\rho_0 w_a)_z = 0 \tag{3.2.9c}$$

$$D_g \theta_e + w_a \theta_{0z} = Q, \tag{3.2.9d}$$

where

$$D_g \equiv \frac{\partial}{\partial t} + u_g \frac{\partial}{\partial x} + v_g \frac{\partial}{\partial y}$$

is the time derivative following the geostrophic wind. It is assumed that the departure θ_e from the reference potential temperature $\theta_0(z)$ is always small, in the sense that $|\theta_{ez}| \ll \theta_{0z}$, so that $w_a \theta_z$ can be replaced by $w_a \theta_{0z}$, as in Eq. (3.2.9d). This is a fair approximation in the middle atmosphere.

The quasi-geostrophic set of Eqs. (3.2.9) still appears quite complicated; however, we now combine the members to yield a single useful and illuminating equation, Eq. (3.2.14). First, we construct the *vorticity equation*

$$D_g \zeta_g = f_0 \rho_0^{-1}(\rho_0 w_a)_z - X_y + Y_x \tag{3.2.10}$$

by taking $(\partial/\partial x)$ [Eq. (3.2.9b)] $-\partial/\partial y$ [Eq. (3.2.9a)] and using the identities $\mathbf{u}_{gx} \cdot \nabla v_g \equiv 0$, $\mathbf{u}_{gy} \cdot \nabla u_g \equiv 0$, which follow from Eq. (3.2.3), and $D_g(f_0 + \beta y) \equiv \beta v_g$, together with Eq. (3.2.9c). Here

$$\zeta_g \equiv f_0 + \beta y - u_{gy} + v_{gx} = f_0 + \beta y + \psi_{xx} + \psi_{yy}$$

is the geostrophic approximation to the beta-plane form of the vertical component of the absolute vorticity, $f - u_y + v_x$. The first term on the right of Eq. (3.2.10) is called a "stretching" term, since it can generate vorticity by differential vertical motion.

The next step is to eliminate w_a between the thermodynamic equation [Eq. (3.2.9d)] and the vorticity equation [Eq. (3.2.10)]. We therefore multiply Eq. (3.2.9d) by ρ_0/θ_{0z}; this is a function of z alone, and can be taken through the D_g operator, giving

$$D_g(\rho_0 \theta_e/\theta_{0z}) + \rho_0 w_a = \rho_0 Q/\theta_{0z}. \tag{3.2.11}$$

Using Eq. (3.2.5), we can write

$$\theta_e / \theta_{0z} = f_0 \psi_z / N^2, \tag{3.2.12}$$

where

$$N^2(z) \equiv H^{-1} R \theta_{0z}(z) e^{-\kappa z/H} \equiv \frac{R}{H} \left(T_{0z} + \frac{\kappa T_0}{H} \right). \tag{3.2.13}$$

The term N is thus the log-pressure buoyancy frequency corresponding to the reference temperature profile $T_0(z) \equiv \theta_0(z) e^{-\kappa z/H}$ [cf. Eq. (1.1.13)]. As indicated in Section 1.1.4, the atmosphere is statically stable if N^2 and θ_{0z} are positive.

On combining the z derivative of Eq. (3.2.11) with Eq. (3.2.10), using the identity $\mathbf{u}_{gz} \cdot \nabla (\rho_0 \theta_e / \theta_{0z}) \equiv 0$ [which follows from Eq. (3.2.6) and the fact that ρ_0 / θ_{0z} and θ_0 depend on z alone], and substituting Eq. (3.2.12), we obtain the *quasi-geostrophic potential vorticity equation*

$$D_g q_g = -X_y + Y_x + f_0 \rho_0^{-1} (\rho_0 Q / \theta_{0z})_z, \tag{3.2.14}$$

where

$$q_g \equiv \zeta_g + f_0 \rho_0^{-1} (\rho_0 \theta_e / \theta_{0z})_z \tag{3.2.15a}$$

$$\equiv f_0 + \beta y + \psi_{xx} + \psi_{yy} + \rho_0^{-1} (\rho_0 \varepsilon \psi_z)_z \tag{3.2.15b}$$

is the quasi-geostrophic potential vorticity and

$$\varepsilon(z) \equiv f_0^2 / N^2(z). \tag{3.2.16}$$

Equation (3.2.14) gives the time development of q_g; note that it does not involve ageostrophic velocities. Moreover, if the flow is frictionless ($X = Y = 0$) and adiabatic ($Q = 0$), then $D_g q_g = 0$ and q_g is conserved following the geostrophic wind. Given q_g at any instant, and appropriate boundary conditions, the elliptic operator on the right of Eq. (3.2.15b) can in principle be inverted to obtain ψ, and hence u_g, v_g, and θ or T, using Eqs. (3.2.3) and (3.2.5).

The quasi-geostrophic potential vorticity q_g, defined by Eq. (3.2.15), should be contrasted with Ertel's potential vorticity P, defined by Eq. (3.1.4). In particular, q_g is *not* generally the quasi-geostrophic approximation to P. Furthermore, if $X = Y = Q = 0$, q_g is conserved following the horizontal geostrophic flow under the quasi-geostrophic conditions (3.2.8), while P is conserved following the total flow, even when quasi-geostrophic scaling is not valid: see Eq. (3.1.5). For these reasons, some authors call q_g the "pseudo-potential vorticity," although when there is no danger of confusion, we shall simply call q_g the "potential vorticity" for short. (Note, however, that certain analogies exist between formulas involving q_g in log-pressure coordinates and formulas involving P in isentropic coordinates: see Section 3.8.3.)

Another useful equation that can be derived from Eq. (3.2.9) is the *omega equation*, obtained by eliminating the $\partial/\partial t$ terms in Eqs. (3.2.9a,d). It is a *diagnostic* equation (that is, it involves no time derivatives) for obtaining the ageostrophic velocity w_a from the geostrophic quantity ψ and its derivatives. Some special cases of the omega equation will be discussed in Sections 3.3 and 3.5.

We note finally that several versions of the quasi-geostrophic equations have been derived in spherical coordinates. Although none of these is entirely satisfactory in every respect, some examples are useful in modeling the middle atmosphere and are mentioned in Chapters 5 and 6.

3.3 The Eulerian-Mean Equations

Many of the middle atmosphere phenomena to be discussed in this book can be regarded as involving the interaction of a mean flow with disturbances ("waves" or "eddies") that are superimposed upon it. This interaction is generally a two-way process, for the mean-flow configuration can strongly modify the propagation of the disturbances, while the disturbances themselves can bring about significant mean-flow changes, through rectified nonlinear effects.

We shall mostly be concerned with cases where the mean is a *zonal mean*, to be denoted by an overbar: thus, for example,

$$\bar{u}(\phi, z, t) = (2\pi)^{-1} \int_0^{2\pi} u(\lambda, \phi, z, t)\, d\lambda. \qquad (3.3.1a)$$

The departure from the zonal mean will be denoted by a prime:

$$u'(\lambda, \phi, z, t) \equiv u - \bar{u}. \qquad (3.3.1b)$$

It should be emphasized that this separation into mean and disturbance quantities is primarily a mathematical device and may not be the most natural physical separation in all cases; for example, in many tropospheric applications a *time* mean may be more useful. However, the zonal average has proved a satisfactory tool for the investigation of most of the stratospheric and mesospheric phenomena to be discussed in this book. Moreover, the theory of the interaction of waves with the zonal-mean flow is more highly developed than that for the corresponding interaction of transient disturbances with the time-mean flow.

The average defined by Eq. (3.3.1a) is an example of an *Eulerian* mean, since it is taken over λ at fixed values of the coordinates ϕ, z, and t. Another type of average, to be discussed in Section 3.7, is the Lagrangian mean, which is taken over a specified set of moving fluid parcels.

Separating each variable into a zonal-mean part and a disturbance part, as in Eq. (3.3.1b), substituting into Eq. (3.1.3), taking the zonal average, and performing some straightforward manipulations, we obtain the following set of primitive equations for the Eulerian-mean flow in spherical coordinates:

$$\bar{u}_t + \bar{v}[(a \cos \phi)^{-1}(\bar{u} \cos \phi)_\phi - f] + \bar{w}\bar{u}_z - \bar{X}$$
$$= -(a \cos^2 \phi)^{-1}(\overline{v'u'} \cos^2 \phi)_\phi - \rho_0^{-1}(\rho_0 \overline{w'u'})_z, \qquad (3.3.2a)$$

$$\bar{v}_t + a^{-1}\bar{v}\bar{v}_\phi + \bar{w}\bar{v}_z + \bar{u}(f + \bar{u}a^{-1} \tan \phi) + a^{-1}\bar{\Phi}_\phi - \bar{Y}$$
$$= -(a \cos \phi)^{-1}(\overline{v'^2} \cos \phi)_\phi - \rho_0^{-1}(\rho_0 \overline{w'v'})_z - \overline{u'^2}a^{-1} \tan \phi, \qquad (3.3.2b)$$

$$\bar{\Phi}_z - H^{-1}R\bar{\theta}e^{-\kappa z/H} = 0, \qquad (3.3.2c)$$

$$(a \cos \phi)^{-1}(\bar{v} \cos \phi)_\phi + \rho_0^{-1}(\rho_0 \bar{w})_z = 0, \qquad (3.3.2d)$$

$$\bar{\theta}_t + a^{-1}\bar{v}\bar{\theta}_\phi + \bar{w}\bar{\theta}_z - \bar{Q}$$
$$= -(a \cos \phi)^{-1}(\overline{v'\theta'} \cos \phi)_\phi - \rho_0^{-1}(\rho_0 \overline{w'\theta'})_z. \qquad (3.3.2e)$$

(The subscript t denotes a time derivative.) In these equations the terms involving mean quadratic functions of disturbance variables have been written on the right. Given these "rectified eddy-forcing" terms, together with suitable expressions for $\bar{X}$, $\bar{Y}$, and $\bar{Q}$ and appropriate boundary and initial conditions, Eqs. (3.3.2) comprise a closed set of equations for predicting the time development of the zonal-mean circulation. A similar set of primitive equations can be written down for the Eulerian-mean flow on a beta-plane, starting from Eq. (3.2.1) and replacing Eq. (3.3.1a) by

$$\bar{u}(y, z, t) \equiv a_0^{-1} \int_0^{a_0} u(x, y, z, t) \, dx,$$

where $a_0 \equiv 2\pi a \cos \phi_0$ is the length of the latitude circle at $\phi = \phi_0$.

In the case of quasi-geostrophic flow on a beta-plane we first note that

$$\bar{v}_g \equiv a_0^{-1} \int_0^{a_0} \psi_x \, dx = 0,$$

by periodicity, so the zonal-mean geostrophic wind $(\bar{u}_g, \bar{v}_g, 0)$ is purely zonal. Dropping subscripts g on geostrophic quantities, but retaining subscripts a on ageostrophic variables, we then obtain the set

$$\bar{u}_t - f_0 \bar{v}_a - \bar{X} = -(\overline{v'u'})_y, \qquad (3.3.3a)$$

$$\bar{\theta}_t + \bar{w}_a \theta_{0z} - \bar{Q} = -(\overline{v'\theta'})_y, \qquad (3.3.3b)$$

$$\bar{v}_{ay} + \rho_0^{-1}(\rho_0 \bar{w}_a)_z = 0, \qquad (3.3.3c)$$

$$f_0 \bar{u}_z + H^{-1}Re^{-\kappa z/H}\bar{\theta}_y = 0, \qquad (3.3.3d)$$

from Eqs. (3.2.9a,c,d) and (3.2.6a), after a little manipulation. These again form a closed set for the mean-flow variables $(\bar{u}, \bar{\theta}, \bar{v}_a, \bar{w}_a)$, given the rectified eddy-forcing terms on the right, $\bar{X}$, $\bar{Q}$, and suitable boundary conditions. [The zonal mean of the y-momentum equation, Eq. (3.2.9b), then supplies $\bar{u}_a$, if $\bar{Y}$ and further rectified eddy terms are given.] From Eqs. (3.2.14) and (3.2.3) it is easy to obtain the zonal-mean quasi-geostrophic potential vorticity equation

$$\bar{q}_t + \overline{(v'q')}_y = -\bar{X}_y + f_0 \rho_0^{-1} (\rho_0 \bar{Q}/\theta_{0z})_z. \tag{3.3.4}$$

The y derivative of this equation can also be obtained by elimination of $(\bar{\theta}_t, \bar{v}_a, \bar{w}_a)$ from Eqs. (3.3.3), using Eqs. (3.2.3) and (3.2.15), provided that Eq. (3.5.10) is used to relate the mean northward eddy potential vorticity flux $\overline{v'q'}$ to $\overline{v'u'}$ and $\overline{v'\theta'}$; see the end of Section 3.5. One can also eliminate $\bar{u}_t$ and $\bar{\theta}_t$ from Eqs. (3.3.3) to obtain diagnostic equations for $\bar{v}_a$ and $\bar{w}_a$ analogous to the omega equation mentioned above. A similar equation, Eq. (3.5.8), will be discussed below.

3.4 Linearized Disturbances to Zonal-Mean Flows

In the preceding section we briefly discussed the separation of atmospheric flows into Eulerian zonal-mean and "wave" or "eddy" parts, and presented sets of equations governing the zonal-mean flow. Similar sets of equations also hold for the disturbances to the zonal mean: exact forms of such equations are given for example by Holton (1975), Eqs. (2.19)–(2.22). In practice, these disturbance equations are most useful in studies of *small-amplitude* departures from the zonal-mean state, when they can be linearized in the disturbance amplitude and perhaps solved numerically, or even analytically in simple idealized cases. A variety of solutions of this type will be described later in this book, and we shall present here the appropriate sets of linearized equations, for future reference.

We first consider a steady, zonally symmetric basic flow, which is purely zonal and unforced. Denoting the basic state by an overbar and suffix zero, we thus have $\bar{v}_0 = \bar{w}_0 = 0$, and, from Eq. (3.1.3b,c),

$$\left(f + \frac{\bar{u}_0 \tan \phi}{a} \right) \bar{u}_0 + a^{-1} \bar{\Phi}_{0\phi} = 0, \tag{3.4.1a}$$

$$\bar{\Phi}_{0z} = H^{-1} R \bar{\theta}_0 e^{-\kappa z/H}, \tag{3.4.1b}$$

in the spherical, primitive equation case.[1] [The presence of a basic diabatic

---

[1] Note that $\bar{\theta}_0(\phi, z)$, $\bar{T}_0(\phi, z)$, and $\bar{\Phi}_0(\phi, z)$ are not in general equal to the reference profiles $\theta_0(z)$, $T_0(z)$, and $\Phi_0(z)$ introduced in Section 3.2.2.

heating $\bar{Q}_0$ would be associated with nonzero $(\bar{v}_0, \bar{w}_0)$ by Eq. (3.1.3e), and would introduce extra complications that will not concern us here.] Note that elimination of $\bar{\Phi}_0$ from Eqs. (3.4.1a,b) gives the thermal wind equation for the basic state [cf. Eq. (3.2.6a)]:

$$\left(f + \frac{2\bar{u}_0 \tan \phi}{a}\right)\frac{\partial \bar{u}_0}{\partial z} = \frac{-R}{aH}\frac{\partial \bar{\theta}_0}{\partial \phi} e^{-\kappa z/H} = \frac{-R}{aH}\frac{\partial \bar{T}_0}{\partial \phi}. \tag{3.4.1c}$$

We now consider small disturbances to this basic state. Thus all primed quantities, as defined in Section 3.3, will be taken to be $O(\alpha)$, where α is a dimensionless amplitude parameter that is much less than 1. Furthermore, $\bar{u} - \bar{u}_0$, $\bar{\Phi} - \bar{\Phi}_0$, $\bar{\theta} - \bar{\theta}_0$, $\bar{v}$, $\bar{w}$, $\bar{X}$, $\bar{Y}$ and $\bar{Q}$ must all be $O(\alpha^2)$ or smaller. The self-consistency of these conditions on the disturbances and the mean flow follows from Eqs. (3.3.2) and (3.4.1); for example, the eddy-forcing terms on the right of Eqs. (3.3.2a,b,e) can lead to $O(\alpha^2)$ departures of $\bar{u}$ from $\bar{u}_0$, and $X = O(\alpha^2)$ would do likewise.

Substitution of

$$u = \bar{u}_0 + u' + O(\alpha^2), \qquad v = v' + O(\alpha^2), \qquad \text{etc.}$$

into the primitive equations of Eq. (3.1.3) and use of Eq. (3.4.1) then give the following set of linear equations for the disturbances:

$$\bar{D}u' + [(a \cos \phi)^{-1}(\bar{u} \cos \phi)_\phi - f]v'$$
$$+ \bar{u}_z w' + (a \cos \phi)^{-1}\Phi'_\lambda = X', \tag{3.4.2a}$$

$$\bar{D}v' + (f + 2\bar{u}a^{-1}\tan \phi)u' + a^{-1}\Phi'_\phi = Y', \tag{3.4.2b}$$

$$\Phi'_z = H^{-1}R\theta'e^{-\kappa z/H}, \tag{3.4.2c}$$

$$(a \cos \phi)^{-1}[u'_\lambda + (v' \cos \phi)_\phi] + \rho_0^{-1}(\rho_0 w')_z = 0, \tag{3.4.2d}$$

$$\bar{D}\theta' + a^{-1}\bar{\theta}_\phi v' + \bar{\theta}_z w' = Q'. \tag{3.4.2e}$$

Here terms of $O(\alpha^2)$ have been neglected and (consistent with this approximation) $\bar{u}_0$ and $\bar{\theta}_0$ have been replaced by $\bar{u}$ and $\bar{\theta}$, respectively, to simplify the notation. Moreover,

$$\bar{D} \equiv \frac{\partial}{\partial t} + \frac{\bar{u}}{a \cos \phi}\frac{\partial}{\partial \lambda} \tag{3.4.3}$$

is the time derivative following the basic flow. The corresponding beta-plane versions can be obtained in a similar manner from Eqs. (3.2.1).

An analogous linearization procedure can be performed for the full quasi-geostrophic set of Eq. (3.2.9). We just note here the equations for a steady basic geostrophic zonal flow, which follow from Eqs. (3.2.3)–(3.2.5):

$$\bar{u} = -\bar{\psi}_y = -f_0^{-1}\bar{\Phi}_y, \tag{3.4.4a}$$

$$\bar{\theta} - \theta_0(z) = HR^{-1}f_0 e^{\kappa z/H}\bar{\psi}_z, \tag{3.4.4b}$$

and the linearized version of the quasi-geostrophic potential vorticity equation, Eq. (3.2.14):

$$\bar{D}q' + v'\bar{q}_y = -X'_y + Y'_x + f_0\rho_0^{-1}(\rho_0 Q'/\theta_{0z})_z. \qquad (3.4.5)$$

Here

$$\bar{D} \equiv \frac{\partial}{\partial t} + \bar{u}\frac{\partial}{\partial x} \qquad (3.4.6)$$

is the time derivative following the basic flow,

$$q' \equiv \psi'_{xx} + \psi'_{yy} + \rho_0^{-1}(\rho_0 \varepsilon \psi'_z)_z \qquad (3.4.7)$$

is the disturbance potential vorticity, and

$$\bar{q}_y \equiv \beta - \bar{u}_{yy} - \rho_0^{-1}(\rho_0 \varepsilon \bar{u}_z)_z \qquad (3.4.8)$$

is the basic northward potential vorticity gradient (sometimes called "effective beta"). The subscript g on geostrophic quantities has again been omitted here, as has the subscript zero on $\bar{u}$, etc. Note that Eq. (3.4.8) follows from the y derivative of the zonal mean of Eq. (3.2.15), together with Eq. (3.4.4a). Versions of Eq. (3.4.5) in spherical coordinates can also be written down; an example is Eq. (5.3.1).

The technique of expansion in the small-amplitude parameter α, described here, can be carried to higher orders. For example, on using $O(\alpha)$ wave solutions of Eq. (3.4.2) to calculate the rectified eddy terms on the right of the Eulerian-mean equations, Eqs. (3.3.2) [or, more conveniently, of the transformed set of Eqs. (3.5.2)], one can in principle calculate the $O(\alpha^2)$ back-effect of the waves on the mean flow. Examples of this method will be discussed in Sections 6.3.1 and 8.3.2. Of course, such an asymptotic expansion in amplitude can only describe weakly nonlinear aspects of the interaction of waves and mean flows. Apart from some rather exceptional circumstances under which exact analytical solutions for finite-amplitude disturbances can be constructed, the behavior of large-amplitude waves must be investigated by numerical solution of the full nonlinear equations.

3.5 The Transformed Eulerian-Mean Equations

The Eulerian-mean sets of equations presented in Section 3.3 were obtained by a straightforward separation of the atmospheric variables into mean and disturbance parts, and averaging and manipulation of the basic equations. However, it is less easy to anticipate how the zonal-mean flow will respond, for example, to a specified "eddy momentum flux" $\overline{v'u'}$ or "eddy heat flux" $\overline{v'\theta'}$ in the quasi-geostrophic set of Eqs. (3.3.3), or in turn to anticipate what physical properties of the waves control these eddy fluxes. To investigate questions like these, it is convenient to transform the mean-flow equations to an alternative form. In the spherical case, the approach is first to define a *residual mean meridional circulation* $(0, \bar{v}^*, \bar{w}^*)$ by

$$\bar{v}^* \equiv \bar{v} - \rho_0^{-1}(\rho_0 \overline{v'\theta'}/\bar{\theta}_z)_z, \tag{3.5.1a}$$

$$\bar{w}^* \equiv \bar{w} + (a \cos \phi)^{-1}(\cos \phi \, \overline{v'\theta'}/\bar{\theta}_z)_\phi. \tag{3.5.1b}$$

(Other definitions of the residual circulation are also possible.) On substituting for $(\bar{v}, \bar{w})$ in Eqs. (3.3.2), using Eqs. (3.5.1) and rearranging, the following *transformed Eulerian-mean* (TEM) set is obtained:

$$\bar{u}_t + \bar{v}^*[(a \cos \phi)^{-1}(\bar{u} \cos \phi)_\phi - f] + \bar{w}^*\bar{u}_z - \bar{X}$$

$$= (\rho_0 a \cos \phi)^{-1}\nabla \cdot \mathbf{F}, \tag{3.5.2a}$$

$$\bar{u}(f + \bar{u}a^{-1}\tan \phi) + a^{-1}\bar{\Phi}_\phi = G, \tag{3.5.2b}$$

$$\bar{\Phi}_z - H^{-1}R\bar{\theta}e^{-\kappa z/H} = 0, \tag{3.5.2c}$$

$$(a \cos \phi)^{-1}(\bar{v}^* \cos \phi)_\phi + \rho_0^{-1}(\rho_0 \bar{w}^*)_z = 0, \tag{3.5.2d}$$

$$\bar{\theta}_t + a^{-1}\bar{v}^*\bar{\theta}_\phi + \bar{w}^*\bar{\theta}_z - \bar{Q} = -\rho_0^{-1}[\rho_0(\overline{v'\theta'}\,\bar{\theta}_\phi/a\bar{\theta}_z + \overline{w'\theta'})]_z. \tag{3.5.2e}$$

The vector $\mathbf{F} \equiv (0, F^{(\phi)}, F^{(z)})$ is known as the *Eliassen–Palm flux* (EP flux); its components are given by

$$F^{(\phi)} \equiv \rho_0 a \cos \phi(\bar{u}_z \overline{v'\theta'}/\bar{\theta}_z - \overline{v'u'}), \tag{3.5.3a}$$

$$F^{(z)} \equiv \rho_0 a \cos \phi\{[f - (a \cos \phi)^{-1}(\bar{u} \cos \phi)_\phi]\overline{v'\theta'}/\bar{\theta}_z - \overline{w'u'}\}; \tag{3.5.3b}$$

note that

$$\nabla \cdot \mathbf{F} \equiv (a \cos \phi)^{-1}\frac{\partial}{\partial \phi}(F^{(\phi)} \cos \phi) + \frac{\partial F^{(z)}}{\partial z}$$

in spherical, log-pressure coordinates. In Eq. (3.5.2b) G represents all the terms that lead to a departure from gradient-wind balance between $\bar{u}$ and $\bar{\Phi}$; it can readily be calculated from Eqs. (3.3.2b) and (3.5.1). In most meteorological applications G is small and only produces slight deviations from gradient wind balance; its dynamical effects are usually only of secondary importance.

At first sight, the transformed set of Eqs. (3.5.2) appears to have no particular advantage over the Eulerian-mean equations of Eqs. (3.3.2). However, a more detailed investigation, to be described in Section 3.6, shows that the rectified eddy-forcing terms on the right of Eqs. (3.5.2a,e) depend on certain basic physical properties of the wave or eddy disturbances. For example, Eq. (3.6.1) will show that $\nabla \cdot \mathbf{F} = 0$ if the disturbances are steady, linear, frictionless, and adiabatic and if the mean flow is conservative to $O(\alpha)$; a similar result holds for the expression on the right of Eq. (3.5.2e). By contrast, under the same linear, steady, conservative conditions, the forcings on the right of Eqs. (3.3.2a,e) are nonzero in general. Further discussion of these results, and some important consequences, will be presented in the next section.

Similar sets of transformed equations can be derived for beta-plane geometry. We shall just discuss the quasi-geostrophic case, for which a residual circulation can be defined by

$$\bar{v}^* \equiv \bar{v}_a - \rho_0^{-1}(\rho_0 \overline{v'\theta'}/\theta_{0z})_z, \qquad \bar{w}^* \equiv \bar{w}_a + (\overline{v'\theta'}/\theta_{0z})_y; \qquad (3.5.4)$$

from Eqs. (3.3.3) the following quasi-geostrophic TEM set can readily be obtained:

$$\bar{u}_t - f_0 \bar{v}^* - \bar{X} = \rho_0^{-1}\mathbf{\nabla} \cdot \mathbf{F}, \qquad (3.5.5a)$$

$$\bar{\theta}_t + \bar{w}^* \theta_{0z} - \bar{Q} = 0, \qquad (3.5.5b)$$

$$\bar{v}_y^* + \rho_0^{-1}(\rho_0 \bar{w}^*)_z = 0, \qquad (3.5.5c)$$

$$f_0 \bar{u}_z + H^{-1}Re^{-\kappa z/H}\bar{\theta}_y = 0. \qquad (3.5.5d)$$

In this quasi-geostrophic beta-plane case,

$$\mathbf{F} \equiv (0, -\rho_0 \overline{v'u'}, \rho_0 f_0 \overline{v'\theta'}/\theta_{0z}). \qquad (3.5.6)$$

Note that the only explicit appearance of eddy-forcing terms here is in $\rho_0^{-1}\mathbf{\nabla} \cdot \mathbf{F}$ in the transformed mean zonal momentum equation, Eq. (3.5.5a); in particular, the eddy forcing of the quasi-geostrophic transformed mean thermodynamic equation, Eq. (3.5.5b), is negligible. Thus, as far as their effects on the mean tendencies $\bar{u}_t$ and $\bar{\theta}_t$ and on the residual circulation $(\bar{v}^*, \bar{w}^*)$ are concerned, the eddy momentum flux $\overline{v'u'}$ and eddy heat flux $\overline{v'\theta'}$ do not act separately [as might have been expected from the untransformed Eqs. (3.3.3)] but in the combination

$$\mathbf{\nabla} \cdot \mathbf{F} \equiv -(\rho_0 \overline{v'u'})_y + (\rho_0 f_0 \overline{v'\theta'}/\theta_{0z})_z.$$

This latter point can be emphasized by solving Eqs. (3.5.5) [or Eqs. (3.3.3)] to find the mean tendencies and the residual circulation. For example, it can be shown that

$$\rho_0 \left[\frac{\partial^2}{\partial y^2} + \frac{1}{\rho_0}\frac{\partial}{\partial z}\left(\rho_0 \varepsilon \frac{\partial}{\partial z}\right) \right] \bar{u}_t = (\mathbf{\nabla} \cdot \mathbf{F} + \rho_0 \bar{X})_{yy} - (\rho_0 f_0 \bar{Q}/\theta_{0z})_{yz} \quad (3.5.7)$$

and

$$\rho_0 \left[\frac{\partial^2}{\partial y^2} + \frac{1}{\rho_0}\frac{\partial}{\partial z}\left(\rho_0 \varepsilon \frac{\partial}{\partial z}\right) \right] f_0 \bar{v}^*$$
$$= -[\rho_0 \varepsilon (\rho_0^{-1}\mathbf{\nabla} \cdot \mathbf{F} + \bar{X})_z]_z - (\rho_0 f_0 \bar{Q}/\theta_{0z})_{yz}. \qquad (3.5.8)$$

(The second of these is related to the omega equation.) Here the rectified effects of the eddies are expressed by the terms in $\nabla \cdot \mathbf{F}$ on the right, while friction and diabatic heating are given by $\bar{X}$ and $\bar{Q}$, respectively. Note that these forcing terms generally produce nonlocal responses in $\bar{u}_t$ and $\bar{v}^*$, since the operator in square brackets on the left of Eqs. (3.5.7) and (3.5.8) is elliptic. To solve for $\bar{u}_t$, $\bar{v}^*$, etc., given the right-hand sides, boundary conditions must be imposed. (See Appendix 3B.)

We conclude this section by mentioning a useful alternative form for $\rho_0^{-1} \nabla \cdot \mathbf{F}$, valid under quasi-geostrophic scaling. It can be derived by simple manipulations using the following identities, which stem from Eqs. (3.2.3), (3.2.12), and (3.4.7):

$$u' = -\psi_y', \qquad v' = \psi_x', \qquad \theta'/\theta_{0z} = f_0 \psi_z'/N^2,$$
$$q' = \psi_{xx}' + \psi_{yy}' + \rho_0^{-1}(\rho_0 \varepsilon \psi_z')_z. \tag{3.5.9}$$

Some integrations by parts, and use of the fact that the x derivatives of zonal-mean quantities vanish, then yield

$$\overline{v'q'} = -(\overline{v'u'})_y + \rho_0^{-1}(\rho_0 f_0 \overline{v'\theta'}/\theta_{0z})_z,$$
$$= \rho_0^{-1} \nabla \cdot \mathbf{F}. \tag{3.5.10}$$

This important quasi-geostrophic relationship between the northward eddy flux of potential vorticity and the divergence of the Eliassen–Palm flux can be used, together with Eq. (3.4.8), to show that Eq. (3.5.7) is equivalent to the y derivative of the quasi-geostrophic potential vorticity equation, Eq. (3.3.4): see the end of Section 3.3.

3.6 The Generalized Eliassen–Palm Theorem and the Charney–Drazin Nonacceleration Theorem

It was mentioned in the previous section that the divergence of the EP flux $\nabla \cdot \mathbf{F}$, unlike the convergence of the eddy momentum flux in Eq. (3.3.2a), depends on certain basic physical properties of the flow. This was given as the main reason for using the TEM set of Eqs. (3.5.2) in preference to the Eulerian-mean equations, Eqs. (3.3.2); we now discuss this point in more detail.

The foundations for the theory to be described were laid in a pioneering paper by Eliassen and Palm (1961). They considered steady, linear waves on a basic zonal flow $\bar{u}(\phi, z)$, with no frictional or diabatic effects ($X = Y = Q = 0$). Using a set of linear disturbance equations essentially equivalent to Eqs. (3.4.2) (but in pressure, rather than log-pressure, coordinates and beta-plane, rather than spherical, geometry), with $\partial/\partial t = 0$ and $X' = Y' = Q' = 0$, they proved the identity

$$\nabla \cdot \mathbf{F} \equiv 0. \tag{3.6.1}$$

Thus the divergence of the Eliassen–Palm flux (as it has come to be called) vanishes for linear, steady, conservative waves on a purely zonal basic flow. This result was extended to include nonzero X', Y', Q' and to allow for spherical geometry by Boyd (1976), and to include time-varying wave amplitudes (i.e., "wave transience") as well by Andrews and McIntyre (1976a, 1978a). The latter's *generalized Eliassen–Palm theorem* takes the form

$$\frac{\partial A}{\partial t} + \nabla \cdot \mathbf{F} = D + O(\alpha^3), \tag{3.6.2}$$

where A and D, like $\mathbf{F}$, are mean quadratic functions of disturbance quantities; however, unlike $\mathbf{F}$, their explicit primitive-equation forms generally involve parcel displacements and are quite complicated. [The simpler quasi-geostrophic versions are given in Eqs. (3.6.6), (3.6.7), and (3.6.10).] The "density" A appearing in Eq. (3.6.2) is called the "wave-activity density"; its time derivative represents wave-transience effects, vanishing for steady waves. The quantity D contains the frictional and diabatic effects X', Y', and Q', and thus vanishes for conservative waves. The $O(\alpha^3)$ term, where α is the wave amplitude as before, represents nonlinear wave effects, and vanishes for purely linear waves. [Note that Eq. (3.6.2), like the linearized Eqs. (3.4.2) from which it is derived, requires that $\bar{X}$, $\bar{Y}$, and $\bar{Q}$ are no larger than $O(\alpha^2)$, so that the basic flow is essentially zonal.]

The generalized EP theorem [Eq. (3.6.2)] makes explicit the dependence of $\nabla \cdot \mathbf{F}$ on the physical properties of wave transience and nonconservative wave effects. (Investigation of its detailed dependence on nonlinear wave processes generally involves going to higher orders in α.) More fundamentally, when the terms on the right of Eq. (3.6.2) are zero, it takes the form of a *conservation law* for wave properties: such laws are of considerable interest in many branches of physics. Note that it is simpler in structure than the *wave-energy equation*, which takes the form

$$\frac{\partial}{\partial t} \left[\tfrac{1}{2} \rho_0 (\overline{u'^2} + \overline{v'^2} + \overline{\Phi_z'^2}/N^2) \right] + \nabla \cdot (0, \rho_0 \overline{v'\Phi'}, \rho_0 \overline{w'\Phi'})$$

$$= -\rho_0 a^{-1} [\bar{u}_\phi + \bar{u} \tan \phi] \overline{v'u'} - \rho_0 \bar{u}_z \overline{w'u'} - \rho_0 a^{-1} \bar{\theta}_\phi \overline{v'\Phi_z'}/\bar{\theta}_z \tag{3.6.3}$$

for linear, conservative waves. This equation can be derived by taking $u' \times$ Eq. (3.4.2a) $+ v' \times$ Eq. (3.4.2b) $+ (\Phi_z'/\bar{\theta}_z) \times$ Eq. (3.4.2e), averaging, setting $X' = Y' = \cdot Q' = 0$, and using Eqs. (3.4.2c,d) and (3.2.13). The terms on the right of Eq. (3.6.3) are generally nonzero in the presence of a mean shear flow and represent an exchange of energy between the mean flow and the disturbances: no such exchange terms appear in the generalized EP theorem [Eq. (3.6.2)]. (In this respect the latter theorem is similar to the law of conservation of wave action: see Section 3.7.1 and Appendix 4A.) For these reasons Eq. (3.6.2) has certain advantages as a diagnostic of wave propagation in complicated mean flows, and will be used for such a purpose later in this book: see, for example, Sections 4.5.5, 5.2.2, 6.2.3, and 6.3.2.

We can now return to the TEM equations [Eqs. (3.5.2)], and use Eq. (3.6.2) to substitute for $\nabla \cdot \mathbf{F}$ on the right of Eq. (3.5.2a). By Eq. (3.6.1) that term vanishes if the disturbances are steady and linear, and the flow is conservative; a similar result can be shown to hold for the expression on the right of Eq. (3.5.2e). It then follows that under such hypotheses, and with appropriate boundary conditions (Appendix 3B), a possible mean flow satisfying Eq. (3.5.2) is given by

$$\bar{u}_t = \bar{\theta}_t = \bar{v}^* = \bar{w}^* = 0.$$

This is an example of a *nonacceleration theorem*, of which a first case was noted by Charney and Drazin (1961). It shows how the waves induce no mean-flow changes under the stated conditions; such a result is not at all obvious from the untransformed Eulerian-mean equations, for which the eddy-forcing terms on the right of Eqs. (3.3.2a,e) do not generally vanish when $\nabla \cdot \mathbf{F} = 0$, but induce a nonzero Eulerian-mean circulation $(\bar{v}, \bar{w})$ that precisely cancels their effect. [Note incidentally that zonally symmetric oscillations, involving a significant contribution $-\partial \bar{v}^*/\partial t$ to G in Eq. (3.5.2b), are possible in principle; however, these are not forced by the waves.]

As a result of the theory described above, there has been a recent emphasis on the physical processes that violate the nonacceleration theorem. Examples of such processes in wave, mean-flow interaction phenomena in the middle atmosphere will occur several times in this book.

We note finally the explicit quasi-geostrophic form of the generalized EP theorem. This can be derived most readily from the linearized potential vorticity equation, Eq. (3.4.5), which can be written

$$\bar{D}q' + v'\bar{q}_y = Z' + O(\alpha^2), \tag{3.6.4}$$

where $Z' \equiv -X'_y + Y'_x + f_0\rho_0^{-1}(\rho_0 Q'/\theta_{0z})_z$ and the $O(\alpha^2)$ term is the error incurred by linearization. On multiplying by $\rho_0 q'/\bar{q}_y$, taking $\rho_0(\bar{q}_y)^{-1}$ through the $\bar{D}$ operator with $O(\alpha^3)$ error [since $\bar{q}_t = O(\alpha^2)$ by Eq. (3.3.4) under the present hypotheses that $\bar{X} = \bar{Q} = O(\alpha^2)$], and averaging, we obtain

$$\frac{\partial}{\partial t}(\tfrac{1}{2}\rho_0 \overline{q'^2}/\bar{q}_y) + \nabla \cdot \mathbf{F} = \rho_0 \overline{Z'q'}/\bar{q}_y + O(\alpha^3), \tag{3.6.5}$$

using Eq. (3.5.10). This is of the form of Eq. (3.6.2) with quasi-geostrophic wave-activity density

$$A = \tfrac{1}{2}\rho_0 \overline{q'^2}/\bar{q}_y, \tag{3.6.6}$$

proportional to the "wave potential enstrophy," $\tfrac{1}{2}\overline{q'^2}$, and nonconservative term

$$D = \rho_0 \overline{Z'q'}/\bar{q}_y. \tag{3.6.7}$$

Note that A is positive definite if $\bar{q}_y > 0$ and is then a natural measure of wave amplitude. A useful alternative form for A is in terms of the northward parcel displacement η', defined by

$$\bar{D}\eta' = v' + O(\alpha^2). \tag{3.6.8}$$

From Eqs. (3.6.4) and (3.6.8) it follows that if $Z' = 0$,

$$q' = -\eta'\bar{q}_y + O(\alpha^2), \tag{3.6.9}$$

given suitable initial conditions, say $\eta' = q' = 0$ at $t = 0$. From Eqs. (3.6.6) and (3.6.9),

$$A = \tfrac{1}{2}\rho_0 \overline{\eta'^2}\bar{q}_y \tag{3.6.10}$$

if $Z' = 0$. When $Z' \neq 0$ we can retain Eq. (3.6.10) as the wave-activity density, but the corresponding nonconservative term D differs from that given in Eq. (3.6.7).

3.7 The Lagrangian Approach

3.7.1 Finite-Amplitude Theory

The generalized EP theorem and its corollary, the nonacceleration theorem, were derived in the previous section for disturbances of small amplitude α. However, many wave-like phenomena in the middle atmosphere are of large amplitude, and it is natural to inquire whether similar results apply to such waves. As yet, the only finite-amplitude results using the formalism adopted above have been rather restricted in character, although a promising approach has been developed by Killworth and McIntyre (1985).

Further progress along these lines has come from a rather different procedure, using a generalized Lagrangian-mean (GLM) theory, rather than the Eulerian-mean formalism discussed above. The GLM approach is quite technical in nature, and only a brief descriptive outline will be given here.

As its name implies, the GLM formalism involves taking averages following fluid parcels, rather than averaging over a set of coordinates fixed in

(λ, ϕ, z, t)-space, as with the Eulerian mean [Eq. (3.3.1a)]. The simplest Lagrangian mean to visualize is a time average following a single parcel; that this can differ significantly from the Eulerian mean is demonstrated by the trajectory in Fig. 3.1, which traces the motion of a single parcel in a hypothetical oscillatory flow whose Eulerian time mean is zero. By contrast, as shown by the mean drift of the parcel towards increasing x and y, the time-mean velocity following the parcel is nonzero for this flow.

For many meteorological purposes, however, a Lagrangian zonal average is required, and this can be described as follows. Consider an initially undisturbed, purely zonal basic flow (on a beta-plane, for simplicity), and fix attention on a thin, infinitely long tube of fluid, lying along the x axis (Fig. 3.2a). Suppose that some waves are excited: the tube will distort in a wavy manner (Fig. 3.2b) and its *mean* motion in the yz plane can be defined as the motion of a line R, which is parallel to the x axis and passes through the tube's center of mass as viewed in the yz plane. This construction gives the y and z components $(\bar{v}^{L}, \bar{w}^{L})$ of the Lagrangian-mean motion at the current position of R. A more general approach, associating each fluid parcel (P_{T}, say) in the tube with a suitably defined reference point P_{R} on R, allows the definition of a parcel displacement vector $\boldsymbol{\xi} = P_{R}P_{T}$, and also enables the Lagrangian means of other variables ($\bar{u}^{L}, \bar{\theta}^{L}$, etc.) to be defined.

A mathematical theory can be constructed using ideas like these; it provides in principle an exact finite-amplitude conservation law of the form

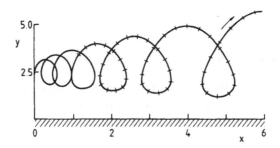

Fig. 3.1. Path of a single fluid parcel in the two-dimensional velocity field $u = 0.01t \cos(x - t)$, $v = 0.01yt \sin(x - t)$, satisfying the incompressibility condition $u_x + v_y = 0$. The path $[X(t), Y(t)]$ was calculated numerically by solving the equations $dX/dt = u(X, t)$, $dY/dt = v(X, Y, t)$, starting at the initial position $X = 0$, $Y = 2.5$. Part of the path is marked at equal time intervals $\Delta t = 0.5$. In addition to the clear mean drift to the right, the marks show that the parcel spends more time further from the wall as t increases so that its time-mean position (averaged over several cycles) also drifts away from the wall. On the other hand, a parcel starting on the wall $y = 0$ must remain on the wall since $v = 0$ there; the Lagrangian-mean motion is therefore divergent, even though (u, v) is nondivergent and the Eulerian-mean motion is zero. [After McIntyre (1980b).]

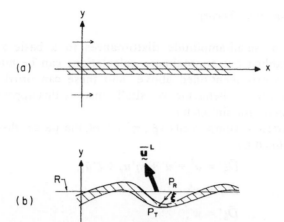

Fig. 3.2. Schematic illustration of the definitions of the Lagrangian-mean velocity $\bar{u}^L$ and the parcel displacement ξ. The material tube is shown hatched. See text for details.

of Eq. (3.6.2) (but with no error term), which generalizes the *wave-action* law of Bretherton and Garrett (1968). The theory also gives a set of equations for the Lagrangian-mean flow, which leads to a finite-amplitude nonacceleration theorem. These equations bear some formal similarity to the TEM set of Eqs. (3.5.2); however, it should be emphasized that the residual circulation $(0, \bar{v}^*, \bar{w}^*)$ is generally *not* the same as the Lagrangian mean meridional circulation $(0, \bar{v}^L, \bar{w}^L)$. These two velocities differ by terms representing transience, nonconservative effects, and nonlinearity in the waves. An important consequence of this difference is that, while the residual circulation has zero mass flux divergence, that is, $\nabla \cdot (0, \rho_0 \bar{v}^*, \rho_0 \bar{w}^*) = 0$ [see Eq. (3.5.2d)], the Lagrangian-mean flow is generally divergent, owing to the dispersion of parcels about their reference positions when waves are transient: see Eq. (9.4.16). (The same effect is also evident in the example shown in Fig. 3.1.)

Direct application of finite-amplitude GLM theory to atmospheric data, or even to numerical models of the atmosphere, encounters serious practical difficulties (see Section 6.3.2), although the theory has an obvious conceptual value for discussion of the transport of quasi-conservative tracers (see Section 9.4.2). A modified version of the theory, based on the use of the quasi-conservative tracers θ and P, may perhaps turn out to be of more practical meteorological benefit.

3.7.2   Small-Amplitude Theory

In the case of small-amplitude disturbances to a basic zonal flow
$[\bar{u}(y, z), 0, 0]$, explicit but approximate calculations can be made of the
Lagrangian quantities mentioned above, and these can provide helpful
insights into atmospheric behavior. We shall illustrate this approach, using
beta-plane geometry for simplicity.

First, the Cartesian components (ξ', η', ζ') of the parcel displacement
vector $\boldsymbol{\xi}'$ are defined by

$$\bar{D}\xi' = u^l \equiv u' + \eta'\bar{u}_y + \zeta'\bar{u}_z, \qquad (3.7.1a)$$

$$\bar{D}\eta' = v', \qquad (3.7.1b)$$

$$\bar{D}\zeta' = w', \qquad (3.7.1c)$$

with $O(\alpha^2)$ error [note that Eq. (3.7.1b) is the same as Eq. (3.6.8)]; furthermore,

$$\overline{\xi'} = \overline{\eta'} = \overline{\zeta'} = 0. \qquad (3.7.2)$$

Using the linearized continuity equation

$$u'_x + v'_y + \rho_0^{-1}(\rho_0 w')_z = 0 \qquad (3.7.3)$$

[cf. Eqs. (3.2.1d) and (3.4.2d)] together with Eqs. (3.7.1), it can be shown
that $\bar{D}[\xi'_x + \eta'_y + \rho_0^{-1}(\rho_0\zeta')_z] = 0$, and thus

$$\xi'_x + \eta'_y + \rho_0^{-1}(\rho_0\zeta')_z = 0, \qquad (3.7.4)$$

given suitable initial conditions.

An example of the calculation of $\boldsymbol{\xi}'$ and thus the approximate orbits of
fluid particles, given the disturbance velocity (u', v', w'), will be presented
in Section 4.5.3 for the case of planetary waves. A knowledge of particle
orbits for small-amplitude wave disturbances is useful, for example, in the
interpretation of tracer transport in the presence of such waves: see
Chapter 9.

The theory given here provides a useful physical interpretation of the
Eliassen–Palm flux (Section 3.5) for waves of small amplitude α. Consider
a material surface, initially at pressure p_1 and $z_1 \equiv -H \ln(p_1/p_s)$, that is
distorted by the waves. The zonal pressure force exerted by the fluid above
the surface on that below is

$$F_1 = g^{-1}\overline{p\frac{\partial\Phi}{\partial x}}\bigg|_{\text{surface}} \qquad \text{per unit horizontal area,}$$

since $z^* = g^{-1}\Phi$ is the geometric altitude of the surface. Linearizing about
z_1 we have

$$F_1 = g^{-1}\overline{p'\Phi'_x}|_{z_1} + O(\alpha^3), \qquad \text{because} \quad \bar{\Phi}_x \equiv 0.$$

Further, since $z_1 + \zeta' = -H \ln[(p_1 + p')/p_s]$ from Eq. (3.1.1), we have $p' = -p_1 \zeta'/H + O(\alpha^2)$. Then $F_1 = -\rho_0(z_1)\zeta'\Phi'_x$, since $\rho_0(z_1) = p_1/RT_s = p_1/gH$: see Section 3.1.1. For quasi-geostrophic flow, $\Phi'_x = f_0 v'$, from the linearized versions of Eqs. (3.2.3) and (3.2.4), and $\zeta' = -\theta'/\theta_{0z}$ from Eq. (3.7.1c) and the linearized form of Eq. (3.2.9d) if the flow is adiabatic ($Q' = 0$). Hence, $F_1 = \rho_0 f_0 \overline{v'\theta'}/\theta_{0z}$: this equals the quasi-geostrophic expression for $F^{(z)}$, by Eq. (3.5.6).

This result, that $F^{(z)}$ equals the "form drag" per unit horizontal area across an initially isobaric material surface disturbed by small-amplitude adiabatic waves, also holds for flow described by the primitive equations if the waves are steady and frictionless as well. A similar interpretation holds for $F^{(y)}$ in terms of the force across a distorted material surface initially given by $y = $ constant. It follows that $\nabla \cdot \mathbf{F}$ equals the net zonal pressure force, per unit volume in xyz space, on a small, initially zonal material tube of fluid that is distorted by the waves; $\rho_0^{-1}\nabla \cdot \mathbf{F}$ is the corresponding force per unit mass. A finite-amplitude analog of the result for $F^{(z)}$ holds in isentropic coordinates: see Section 3.9.

Small-amplitude theory can also be used to derive an approximation to the Lagrangian mean of any quantity χ. The general definition of the Lagrangian mean of χ is

$$\bar{\chi}^L(\mathbf{x}, t) \equiv \overline{\chi[\mathbf{x} + \boldsymbol{\xi}'(\mathbf{x}, t), t]}, \tag{3.7.5}$$

where $\mathbf{x} + \boldsymbol{\xi}'(\mathbf{x}, t)$ is the current position of the particle whose mean position is $\mathbf{x}$. Application of the identity $\chi \equiv \bar{\chi} + \chi'$ together with a Taylor expansion of Eq. (3.7.5) gives

$$\bar{\chi}^L(\mathbf{x}, t) = \bar{\chi}(\mathbf{x}, t) + \bar{\chi}^S(\mathbf{x}, t) \tag{3.7.6}$$

where the *Stokes correction* $\bar{\chi}^S$ is defined by

$$\bar{\chi}^S = \overline{\boldsymbol{\xi}' \cdot \nabla \chi'} + \tfrac{1}{2}\overline{\xi'_j \xi'_k} \frac{\partial^2 \bar{\chi}}{\partial x_j \, \partial x_k} + O(\alpha^3), \tag{3.7.7}$$

and summation over all values of the indices j, k is implied. If χ is a velocity component, say u, $\bar{u}^S$ is known as the Stokes drift, and represents the difference between the Lagrangian-mean velocity $\bar{u}^L$ and the Eulerian-mean velocity $\bar{u}$. The fact that this quantity can be nonzero was pointed out in 1847 by Stokes, who applied a time average to water waves (see also Fig. 3.1). A calculation of the Stokes drift and Lagrangian-mean flow for planetary waves will be given in Section 4.5.3.

3.8   Isentropic Coordinates

3.8.1   The Primitive Equations in Isentropic Coordinates

In this chapter we have up to now been using the log-pressure variable z as a vertical coordinate in the equations of motion. We conclude, however, with a brief discussion of the primitive equations in isentropic coordinates (also called θ coordinates), where the potential temperature (a function of the entropy per unit mass) is used as the vertical coordinate.

The first, and most obvious, reason for the use of isentropic coordinates is that the isentropic "vertical velocity," $D\theta/Dt$, equals the diabatic heating term Q [see Eq. (3.1.3e)]. Thus in adiabatic flow, when $Q = 0$, this "velocity" vanishes, and there is no motion across the isentropic surfaces $\theta =$ constant; isentropic coordinates are therefore partly Lagrangian in character.

In spherical geometry, the primitive equations in θ coordinates are

$$\tilde{D}u - \left(f + \frac{u \tan \phi}{a}\right)v + (a \cos \phi)^{-1}M_\lambda = X - Qu_\theta, \qquad (3.8.1a)$$

$$\tilde{D}v + \left(f + \frac{u \tan \phi}{a}\right)u + a^{-1}M_\phi = Y - Qv_\theta, \qquad (3.8.1b)$$

$$\sigma_t + (a \cos \phi)^{-1}\{(\sigma u)_\lambda + (\sigma v \cos \phi)_\phi\} = -(\sigma Q)_\theta \qquad (3.8.1c)$$

$$M_\theta = \Pi(p) \equiv c_p(p/p_s)^\kappa = c_p e^{-\kappa z/H} \qquad (3.8.1d)$$

$$\sigma \equiv -g^{-1}p_\theta. \qquad (3.8.1e)$$

Here

$$\tilde{D} \equiv \frac{\partial}{\partial t} + \frac{u}{a \cos \phi}\frac{\partial}{\partial \lambda} + \frac{v}{a}\frac{\partial}{\partial \phi} \qquad (3.8.2)$$

is the time derivative following the components of the flow on an isentrope, and has been used in preference to $D/Dt \equiv \tilde{D} + Q\,\partial/\partial\theta$, so that the nonconservative cross-isentropic advection terms involving $Q\,\partial/\partial\theta$ can be written with the other nonconservative terms on the right of Eqs. (3.8.1a,b). The quantity σ is the "density" in (λ, ϕ, θ) space, as can be shown by considering the mass contained within a volume element lying between isentropes at potential temperatures θ and $\theta + \delta\theta$: see Fig. 3.3. The quantity M is called the Montgomery stream function, and is defined by

$$M \equiv c_p T + \Phi \equiv \theta\Pi(p) + \Phi \qquad (3.8.3a,b)$$

where Eq. (3.8.3b) uses Eqs. (1.1.9a) and (3.8.1d); $\Pi(p)$ is known as the Exner function. Subscripts θ denote partial derivatives, and other derivatives here are of course taken at constant θ, not at constant z. The remaining variables are as defined in Section 3.1.1; a brief discussion of the derivation of Eqs. (3.8.1) is given in Appendix 3A. Note incidentally that difficulties arise in regions where $\partial\theta/\partial z^*$ is zero or negative corresponding to neutral or unstable stratification: in such regions θ does not increase monotonically with the geometric height z^*, and σ becomes infinite or negative.

3.8.2 Ertel's Potential Vorticity in Isentropic Coordinates

As mentioned at the end of Section 3.7.1, and elsewhere in this book (e.g., Sections 5.2.3, 6.2.4, and 9.1), there are advantages in using θ and Ertel's potential vorticity P as tracers of atmospheric motion. The most convenient way of doing this is by plotting contours of P on isentropic surfaces. In θ coordinates, Ertel's potential vorticity is given by

$$P = \tilde{\zeta}/\sigma \qquad (3.8.4a)$$

where

$$\tilde{\zeta} = f - \frac{(u \cos \phi)_\phi}{a \cos \phi} + \frac{v_\lambda}{a \cos \phi}, \qquad (3.8.4b)$$

is the vertical component of absolute isentropic vorticity. Using Eqs. (3.8.1) it can be shown that P satisfies

$$\tilde{D}P = (\sigma a \cos \phi)^{-1}[-(X \cos \phi)_\phi + Y_\lambda - Q_\lambda v_\theta$$
$$+ Q_\phi u_\theta \cos \phi] + PQ_\theta - QP_\theta. \qquad (3.8.5)$$

Details of the derivation of this equation are given in Appendix 3A, together with a method of verifying that Eq. (3.8.5) reduces to Eq. (3.1.5) on transformation to z coordinates.

Isentropic coordinates provide an enlightening physical explanation of potential vorticity conservation. Consider for simplicity a frictionless ($X = Y = 0$), adiabatic ($Q = 0$) flow, and focus attention on a small material circuit C lying initially on an isentrope (Fig. 3.3). This circuit moves with the fluid but always remains on the same isentrope, since the flow is adiabatic. Its horizontally projected area δA changes according to

$$\frac{D}{Dt} \delta A = (\delta A)\Delta, \qquad (3.8.6a)$$

where $\Delta = (a \cos \phi)^{-1}[u_\lambda + (v \cos \phi)_\phi]$ is the isentropic divergence [equal to $u_x + v_y$ in (x, y, θ) coordinates]. Equation (3.8.6a) can be proved most easily for a rectangular circuit $\delta A = \delta x\, \delta y$, using $D\, \delta x/Dt = \delta u \approx u_x\, \delta x$, etc. From Eq. (3.8.1c) with $Q = 0$ (so that $\tilde{D} = D/Dt$) we also have

$$\frac{D\sigma}{Dt} = -\sigma\Delta, \qquad (3.8.6b)$$

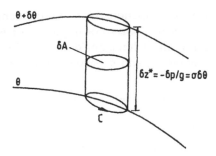

Fig. 3.3. An elementary volume in (x, y, θ) space or (λ, ϕ, θ) space. This has horizontally projected cross-sectional area δA and geometric height δz^*; its mass is therefore $\delta m = \rho \, \delta A \, \delta z^*$, where ρ is the physical density. By the hydrostatic relation of Eq. (1.1.2), $\rho \delta z^* = -g^{-1} \, \delta p$, where δp is the pressure difference between the top and bottom of the element. But

$$-g^{-1} \, \delta p = -g^{-1} \left(\frac{\partial p}{\partial \theta} \right) \delta\theta = \sigma \, \delta\theta \qquad \text{by} \quad \text{Eq. (3.8.1e)}$$

and so $\delta m = \sigma \, \delta A \, \delta\theta$. The quantity σ is therefore the "density" in (λ, ϕ, θ) space. The circuit C is discussed in Section 3.8.2.

while the vorticity equation, obtained by taking the horizontal curl of Eqs. (3.8.1a,b), is

$$\frac{D\tilde{\zeta}}{Dt} = -\tilde{\zeta}\Delta \tag{3.8.6c}$$

[see Eq. (3A.5)]. On eliminating Δ between Eqs. (3.8.6a,b) we obtain

$$\frac{D}{Dt}(\sigma \, \delta A) = 0, \tag{3.8.7a}$$

which expresses conservation of mass per unit θ for the volume depicted in Fig. 3.3. Likewise, Eqs. (3.8.6a,c) give

$$\frac{D}{Dt}(\tilde{\zeta} \, \delta A) = 0: \tag{3.8.7b}$$

this is the Kelvin–Bjerknes circulation theorem for the circuit C and can be thought of as a fluid-dynamical generalization of angular momentum conservation for a rigid body. Since $\sigma \, \delta A$ and $\tilde{\zeta} \, \delta A$ are both conserved following the motion when $X = Y = Q = 0$, their ratio $P = \tilde{\zeta}/\sigma$ is similarly conserved [this also follows from Eqs. (3.8.6b,c)]. Thus conservation of Ertel's potential vorticity is a consequence of the circulation theorem and mass conservation.

A useful alternative version of Eq. (3.8.5) is

$$(\sigma P)_t + (a \cos \phi)^{-1}(\sigma P u - Y + Q v_\theta)_\lambda$$
$$+ (a \cos \phi)^{-1}[(\sigma P v + X - Q u_\theta) \cos \phi]_\phi = 0, \tag{3.8.8}$$

which follows from Eqs. (3A.5) and (3.8.2). This can be regarded either as a vorticity equation [in view of Eq. (3.8.4a)] or as an equation for the mass-weighted potential vorticity σP. The second and third terms in Eq. (3.8.8) form the divergence of a vector flux $\mathbf{L}$, which has components

$$\mathbf{L} = (\sigma Pu - Y + Qv_\theta, \ \sigma Pv + X - Qu_\theta, \ 0) \tag{3.8.9}$$

in (λ, ϕ, θ) space. Integration of Eq. (3.8.8) over the whole atmosphere and use of the divergence theorem then shows that sources and sinks for the mass integrated potential vorticity $\iiint \sigma Pa^2 \cos \phi \, d\lambda \, d\phi \, d\theta$ can only occur at the ground, and not within the atmosphere itself; frictional and diabatic terms can only help redistribute potential vorticity. [This result can also be demonstrated using the z-coordinate equation, Eq. (3.1.5).] Even more remarkably, the fact that $\mathbf{L}$ has an identically zero θ component implies that there can be no net transport of potential vorticity across any isentrope, even in the presence of friction and diabatic heating.

3.8.3 Relationships between θ Coordinates and z Coordinates

The primitive equations in θ coordinates bear some formal resemblances to the quasi-geostrophic equations in z coordinates: note for instance the partial analogy between the quasi-geostrophic potential vorticity equation, Eq. (3.2.14), and the Ertel potential vorticity equation, Eq. (3.8.5). However, in several respects the analogy is not complete; for example, D_g involves advection by the geostrophic wind, while $\tilde{D}$ involves advection by the divergent flow (u, v). Another important difference is that the θ-coordinate primitive equations do not require that the potential temperature be always close to the reference profile $\theta_0(z)$ (cf. Section 3.2.3).

A more precise relationship, involving derivatives of q_g and P, was derived by Charney and Stern (1962), who showed that when quasi-geostrophic scaling holds,

$$\left(\frac{\partial P}{\partial s}\right)_{\theta = \text{const.}} \approx \frac{\theta_{0z}}{\rho_0} \left(\frac{\partial q_g}{\partial s}\right)_{z = \text{const.}} \tag{3.8.10}$$

where $s = t$, λ, or ϕ (or x or y on a beta-plane); a proof is given in Appendix 3A. In studies of planetary waves (Sections 4.5 and 5.3) and barotropic or baroclinic instability (Section 5.5), the isobaric gradient of q_g plays a central role, and one important use of Eq. (3.8.10) is to suggest interpretations of these phenomena in terms of the distribution of Ertel's potential vorticity on isentropic surfaces.

A major drawback to the use of θ coordinates for prognostic purposes in the troposphere is that the lower boundary condition is generally complicated: the "ground" is not usually an isentrope, and its position in (λ, ϕ, θ)

space is one of the unknowns. (The same difficulty arises in p or z coordinates, but is somewhat less acute unless large-amplitude topography is present.) For modeling the middle atmosphere it may often be sufficient to choose as a lower boundary an isentrope that never intersects the ground (say, the $\theta = 350\ \mathrm{K}$ surface, near the tropopause) and specify suitable conditions there from observations or in idealized form. This parallels the use of a constant-pressure surface as a lower boundary in z coordinates, as mentioned in Section 3.1.2.

3.9 The Zonal-Mean Equations in Isentropic Coordinates

The Eulerian zonal-mean equations in isentropic coordinates have a number of useful features: among other things they bear a close formal resemblance to the *transformed* Eulerian-mean equations in log-pressure coordinates and share the advantages of that set (see Sections 3.5 and 3.6). Moreover, some of the similarities between the primitive equations in θ coordinates and the quasi-geostrophic equations in z coordinates, mentioned in the previous section, carry over to the zonal-mean case, and these can be useful for extending certain quasi-geostrophic results to the primitive equations: an example is given in Section 7.5.

We consider first the zonal momentum equation. This can be derived in "flux form" from Eqs. (3.8.1a,c):

$$(\sigma u)_t + (a\cos\phi)^{-1}[(\sigma u^2)_\lambda + (\sigma uv\cos\phi)_\phi] - \sigma(f + ua^{-1}\tan\phi)v$$

$$+ (a\cos\phi)^{-1}\sigma M_\lambda = \sigma X - (\sigma Qu)_\theta. \tag{3.9.1}$$

Using Eqs. (3.8.1e) and (3.8.1d), the term σM_λ can be rewritten as follows:

$$\sigma M_\lambda = -g^{-1}p_\theta M_\lambda = -g^{-1}(pM_\lambda)_\theta + g^{-1}pM_{\theta\lambda}$$

$$= -g^{-1}(pM_\lambda)_\theta + g^{-1}pp_\lambda\frac{d\Pi}{dp}$$

$$= -g^{-1}(pM_\lambda)_\theta + \left[g^{-1}\int_{p_s}^{p} p_1\frac{d\Pi(p_1)}{dp_1}\,dp_1\right]_\lambda. \tag{3.9.2}$$

We define an average around a latitude circle on an isentropic surface:

$$\bar{A}(\phi, \theta, t) \equiv (2\pi)^{-1}\int_0^{2\pi} A(\lambda, \phi, \theta, t)\,d\lambda$$

for any field A [contrast the z-coordinate version of Eq. (3.3.1a)] and a deviation $A' \equiv A - \bar{A}$. Substituting from Eq. (3.9.2) into Eq. (3.9.1) and averaging, we obtain

$$(\overline{\sigma u})_t + (a \cos^2 \phi)^{-1}(\overline{\sigma v u} \cos^2 \phi)_\phi - f\overline{\sigma v}$$
$$-(ga \cos \phi)^{-1}(\overline{p'M'_\lambda})_\theta = \overline{\sigma X} - (\overline{\sigma Q u})_\theta, \tag{3.9.3}$$

where the two terms in Eq. (3.9.1) involving $\sigma u v$ have been combined and the result $\overline{pM_\lambda} = \overline{pM'_\lambda} = \overline{p'M'_\lambda}$ has been used. Likewise, the zonal average of Eq. (3.8.1c) is

$$\bar{\sigma}_t + (a \cos \phi)^{-1}(\overline{\sigma v} \cos \phi)_\phi = -(\overline{\sigma Q})_\theta. \tag{3.9.4}$$

We now introduce a mass-weighted zonal mean for any field A,

$$\bar{A}^* \equiv (\overline{\sigma A})/\bar{\sigma}, \tag{3.9.5}$$

and put

$$\sigma v = \overline{\sigma v} + (\sigma v)' = \bar{\sigma}\bar{v}^* + (\sigma v)',$$
$$\sigma Q = \overline{\sigma Q} + (\sigma Q)' = \bar{\sigma}\bar{Q}^* + (\sigma Q)',$$

so that

$$\overline{\sigma v u} = \bar{\sigma}\bar{v}^*\bar{u} + \overline{(\sigma v)'u'}, \tag{3.9.6a}$$

$$\overline{\sigma Q u} = \bar{\sigma}\bar{Q}^*\bar{u} + \overline{(\sigma Q)'u'}. \tag{3.9.6b}$$

On substituting $\overline{\sigma u} = \bar{\sigma}\bar{u} + \overline{\sigma'u'}$ in Eq. (3.9.3), subtracting $\bar{u} \times$ Eq. (3.9.4), using Eqs. (3.9.5) and (3.9.6), and dividing by $\bar{\sigma}$, we obtain

$$\bar{u}_t + \bar{v}^*[(a \cos \phi)^{-1}(\bar{u} \cos \phi)_\phi - f] + \bar{Q}^*\bar{u}_\theta - \bar{X}^*$$
$$= -\bar{\sigma}^{-1}(\overline{\sigma'u'})_t + (\bar{\sigma}a \cos \phi)^{-1}\tilde{\boldsymbol{\nabla}} \cdot \tilde{\mathbf{F}} \tag{3.9.7a}$$

where $\tilde{\mathbf{F}} = (0, \tilde{F}^{(\phi)}, \tilde{F}^{(\theta)})$ is the Eliassen–Palm flux in isentropic coordinates; its components are

$$\tilde{F}^{(\phi)} = -a \cos \phi \overline{(\sigma v)'u'}, \tag{3.9.8a}$$

$$\tilde{F}^{(\theta)} = g^{-1}\overline{p'M'_\lambda} - a \cos \phi \overline{(\sigma Q)'u'}, \tag{3.9.8b}$$

and its isentropic divergence is

$$\tilde{\boldsymbol{\nabla}} \cdot \tilde{\mathbf{F}} = (a \cos \phi)^{-1}\frac{\partial}{\partial \phi}(\tilde{F}^{(\phi)} \cos \phi) + \frac{\partial \tilde{F}(\theta)}{\partial \theta}.$$

The analogy with the z-coordinate TEM equation [Eq. (3.5.2a)] and the definitions in Eq. (3.5.3) should be noted. For small-amplitude disturbances a generalized Eliassen–Palm theorem can be derived, relating $\tilde{\boldsymbol{\nabla}} \cdot \tilde{\mathbf{F}}$ to wave transience, nonconservative effects, and nonlinearity (cf. Section 3.6). Under nonacceleration conditions it can be shown that $\tilde{\boldsymbol{\nabla}} \cdot \tilde{\mathbf{F}} \equiv 0$ at finite amplitude as well. Incidentally, the contribution $g^{-1}\overline{p'M'_\lambda}$ to $\tilde{F}^{(\theta)}$ equals the zonal component of the "pressure torque" (or $a \cos \phi$ times the "form drag" force) per unit horizontal area, exerted by the fluid above an isentrope on that below: cf. Section 3.7.2.

It is straightforward to derive the remaining mean-flow equations in the form

$$\bar{u}(f + \bar{u}a^{-1}\tan\phi) + a^{-1}\bar{M}_\phi = \tilde{G}, \tag{3.9.7b}$$

$$\bar{\sigma}_t + (a\cos\phi)^{-1}(\bar{\sigma}\bar{v}^*\cos\phi)_\phi + (\bar{\sigma}\bar{Q}^*)_\theta = 0, \tag{3.9.7c}$$

$$\bar{M}_\theta - \Pi(\bar{p}) = S \equiv \overline{\Pi(p)} - \Pi(\bar{p}), \tag{3.9.7d}$$

$$\bar{\sigma} = -g^{-1}\bar{p}_\theta. \tag{3.9.7e}$$

Here $\tilde{G}$ represents those terms that lead to departures from gradient-wind balance in Eq. (3.9.7b) [cf. the term G in Eq. (3.5.2b)] and S is a mean quantity of second order in wave amplitude. Analogies with the TEM set [Eq. (3.5.2)] should again be noted.

An alternative form of the mean zonal momentum equation [Eq. (3.9.7a)] that is useful for some purposes (see, e.g., Section 7.5) is

$$\bar{u}_t + \bar{v}^*\{(a\cos\phi)^{-1}(\bar{u}\cos\phi)_\phi - f\} = \bar{\sigma}\overline{\hat{v}\hat{P}}^* + \bar{X} - \overline{Qu_\theta}, \tag{3.9.9}$$

where

$$\hat{A} \equiv A - \bar{A}^* \tag{3.9.10}$$

and P is Ertel's potential vorticity [Eq. (3.8.4)]; the proof is given in Appendix 3A. This equation includes on its right-hand side a force per unit mass proportional to a mean northward advective eddy flux of Ertel's potential vorticity $\overline{\hat{v}\hat{P}}^*$; it is thus analogous to the quasi-geostrophic z-coordinate equation [Eq. (3.5.5a)] with $\rho_0^{-1}\nabla \cdot \mathbf{F}$ replaced by $\overline{v'q'}$, using Eq. (3.5.10). Comparison of Eqs. (3.9.7a) and (3.9.9) and use of Eqs. (3.9.5) and (3.9.10) gives the following relationship between $\overline{\hat{v}\hat{P}}^*$ and the isentropic EP flux divergence:

$$\bar{\sigma}^2\overline{\hat{v}\hat{P}}^* = (a\cos\phi)^{-1}\tilde{\nabla} \cdot \bar{\mathbf{F}} - \overline{(\sigma'u')}_t + \overline{\sigma'X'} + \bar{\sigma}^2\overline{\hat{Q}(u_\theta/\sigma)}^* \tag{3.9.11}$$

(Tung, 1986); this is more complicated than the quasi-geostrophic relation in Eq. (3.5.10).

Appendix 3A   Derivation of Some Equations in Isentropic Coordinates

3A.1   The Primitive Equations, Eq. (3.8.1)

First, note that, from Eq. (3.8.3b),

$$\frac{\partial M}{\partial\theta} = \Pi(p) + \theta\frac{d\Pi}{dp}\frac{\partial p}{\partial\theta} + \frac{\partial\Phi}{\partial\theta}. \tag{3A.1}$$

But $\theta \, d\Pi/dp = \theta \kappa c_p p^{\kappa-1} p_s^{-\kappa} = RTp^{-1}$ using the definition of $\Pi(p)$ in Eq. (3.8.1d), the fact that $\kappa c_p = R$, and Eq. (1.1.9a). Moreover, $\partial \Phi/\partial \theta = (\partial \Phi/\partial z)(\partial z/\partial \theta) = (RT/H)(-Hp_\theta/p)$ using Eqs. (3.1.3c') and (3.1.1); Eq. (3.8.1d) follows, since the last two terms in Eq. (3A.1) then cancel. The formula for transforming derivatives gives

$$\left(\frac{\partial M}{\partial s}\right)_{\theta=\text{const.}} = \left(\frac{\partial M}{\partial s}\right)_{z=\text{const.}} - \left(\frac{\partial M}{\partial \theta}\right)\left(\frac{\partial \theta}{\partial s}\right)_{z=\text{const.}}$$

where $s = \lambda$ or ϕ. Using Eqs. (3.8.3a), (3.8.1d), and (1.1.9b), we obtain

$$\left(\frac{\partial M}{\partial s}\right)_{\theta=\text{const.}} = \left(\frac{\partial \Phi}{\partial s}\right)_{z=\text{const.}} + c_p\left(\frac{\partial T}{\partial s}\right)_{z=\text{const.}} - \Pi e^{\kappa z/H}\left(\frac{\partial T}{\partial s}\right)_{z=\text{const.}};$$

by the definition of Π in Eq. (3.8.1d) the last two terms cancel, giving

$$\left(\frac{\partial M}{\partial s}\right)_{\theta=\text{const.}} = \left(\frac{\partial \Phi}{\partial s}\right)_{z=\text{const.}}.$$

Thus Eqs. (3.8.1a,b) follow from Eqs. (3.1.3a,b) on using the fact that $D/Dt \equiv \tilde{D} + Q \, \partial/\partial \theta$ is a coordinate-independent operator. The proof of Eq. (3.8.1c) follows from the conservation of mass, together with the fact that $\delta m = \sigma a^2 \cos \phi \, \delta\lambda \, \delta\phi \, \delta\theta$ is a mass element in (λ, ϕ, θ) space (see the caption of Fig. 3.3). [The analogy between Eq. (3.8.1c) and the compressible-fluid form $\rho_t + \nabla^* \cdot (\rho \mathbf{u}^*) = 0$ in geometric coordinates is immediately evident.]

3A.2   The Potential Vorticity Equation, Eq. (3.8.5)

First define

$$M_1 \equiv M + \tfrac{1}{2}(u^2 + v^2), \tag{3A.2}$$

$$X_1 \equiv X - Qu_\theta, \tag{3A.3a}$$

$$Y_1 \equiv Y - Qv_\theta. \tag{3A.3b}$$

It is then easy to verify that Eqs. (3.8.1a,b) can be written

$$u_t - v\tilde{\zeta} + (a \cos \phi)^{-1} M_{1\lambda} = X_1, \tag{3A.4a}$$

$$v_t + u\tilde{\zeta} + a^{-1} M_{1\phi} = Y_1, \tag{3A.4b}$$

where $\tilde{\zeta}$ is defined in Eq. (3.8.4b). Eliminating M_1 by cross-differentiating Eqs. (3A.4a,b), and rearranging, we can obtain the vorticity equation in the form

$$\tilde{D}\tilde{\zeta} + (a \cos \phi)^{-1}\tilde{\zeta}[u_\lambda + (v \cos \phi)_\phi] = (a \cos \phi)^{-1}[-(X_1 \cos \phi)_\phi + Y_{1\lambda}];$$

(3A.5)

moreover, Eq. (3.8.1c) can be written

$$\tilde{D}\sigma^{-1} - (a \cos \phi)^{-1}\sigma^{-1}[u_\lambda + (v \cos \phi)_\phi] = \sigma^{-2}(\sigma Q)_\theta. \quad (3A.6)$$

But from Eq. (3.8.4a),

$$P = \tilde{\zeta}\sigma^{-1}; \tag{3A.7}$$

hence, using Eqs. (3A.5)–(3A.7) we obtain

$$\tilde{D}P = (\sigma a \cos \phi)^{-1}[-(X_1 \cos \phi)_\phi + Y_{1\lambda}] + P\sigma^{-1}(\sigma Q)_\theta. \quad (3A.8)$$

Using Eqs. (3.8.4) and (3A.3), the right-hand side of Eq. (3A.8) can be rearranged to give Eq. (3.8.5).

To transform Eq. (3.8.5) to obtain the z-coordinate version [Eq. (3.1.5)] it is helpful to observe that

$$\frac{\sigma}{\rho_0} = \frac{\partial z}{\partial \theta} = \frac{\partial(\lambda, \phi, z)}{\partial(\lambda, \phi, \theta)} \tag{3A.9}$$

and that, for example, $\partial Y/\partial \lambda$ at constant θ can be written $\partial(Y, \phi, \theta)/\partial(\lambda, \phi, \theta)$. The rule for multiplication of Jacobians then gives

$$\sigma^{-1}\left(\frac{\partial Y}{\partial \lambda}\right)_{\theta = \text{const.}} = \rho_0^{-1}\frac{\partial(Y, \phi, \theta)}{\partial(\lambda, \phi, z)} = \rho_0^{-1}\frac{\partial(Y, \theta)}{\partial(\lambda, z)}.$$

The other Jacobian terms in Eq. (3.1.5) are obtained in a similar manner.

3A.3 Charney and Stern's Relation, Eq. (3.8.10)

From Eq. (3A.7) and the left-hand equality in Eq. (3A.9) we have

$$P = \rho_0^{-1}\theta_z\tilde{\zeta}; \tag{3A.10}$$

then from the usual formula for transforming derivatives,

$$\left(\frac{\partial P}{\partial s}\right)_{\theta = \text{const.}} = P_s - \theta_z^{-1}P_z\theta_s$$

(where the subscript s here denotes a derivative at constant z), it can easily be verified that

$$\left(\frac{\partial P}{\partial s}\right)_{\theta=\text{const.}} = \theta_z (P\theta_z^{-1})_s + P^2 \theta_z^{-1} (P^{-1}\theta_s)_z.$$

Using Eq. (3A.10) this yields

$$\left(\frac{\partial P}{\partial s}\right)_{\theta=\text{const.}} = \frac{\theta_z}{\rho_0}\left[\tilde{\zeta}_s + \frac{\tilde{\zeta}^2}{\rho_0}\left(\frac{\rho_0\theta_s}{\tilde{\zeta}\theta_z}\right)_z\right], \tag{3A.11}$$

since $\rho_0 = \rho_0(z)$. Equation (3A.11) is exact, but on using the leading approximations $\theta_z \approx \theta_{0z}$, $\tilde{\zeta}_s \approx \zeta_{gs}$, $\tilde{\zeta} \approx f_0$ we obtain the quasi-geostrophic result

$$\left(\frac{\partial P}{\partial s}\right)_{\theta=\text{const.}} \approx \frac{\theta_{0z}}{\rho_0}\left[\zeta_{gs} + \frac{f_0}{\rho_0}\left(\frac{\rho_0\theta_s}{\theta_{0z}}\right)_z\right],$$

whence Eq. (3.8.10) follows on using Eqs. (3.2.5), (3.2.15a), and the fact that ρ_0 and θ_0 are independent of s.

3A.4   The Mean Momentum Equation in the Form of Eq. (3.9.9)

The zonal mean of Eq. (3A.4a) is

$$\bar{u}_t - \overline{v\tilde{\zeta}} = \bar{X}_1 \tag{3A.12}$$

and

$$\overline{v\tilde{\zeta}} = \overline{\sigma vP} = \bar{\sigma}(\overline{vP})^* \tag{3A.13}$$

by Eqs. (3A.7) and (3.9.5). Now using Eqs. (3.9.5) and (3.9.10) it can be verified that $\overline{AB}^* = \bar{A}^* \bar{B}^* + \overline{\hat{A}\hat{B}}^*$ for any fields A, B. Thus

$$\bar{\sigma}(\overline{vP})^* = \bar{\sigma}\bar{v}^*\bar{P}^* + \bar{\sigma}\overline{\hat{v}\hat{P}}^* = \bar{v}^*\bar{\tilde{\zeta}} + \overline{\sigma\hat{v}\hat{P}}^* \tag{3A.14}$$

by Eqs. (3.9.5) and (3A.7). Then Eq. (3.9.9) follows from Eqs. (3A.12)–(3A.14), (3.8.4b), and (3A.3a). Note incidentally that Eq. (3A.12) can be written as $\bar{u}_t = \overline{L^{(\phi)}}$, by Eqs. (3A.3a) and (3A.13), where $L^{(\phi)}$ is the ϕ component of the potential vorticity flux vector defined in Eq. (3.8.9).

Appendix 3B   Boundary Conditions on the Residual Circulation

We present here examples of the kind of boundary conditions that may apply to $\bar{v}^*$ and $\bar{w}^*$.

3B.1 Side Boundaries

In a channel with vertical sidewalls, $v = 0$ on these walls, as in Section 3.1.2.c. Thus $\bar{v} = v' = 0$ on these walls, and hence $\bar{v}^* = 0$ there, by Eq. (3.5.1a).

3B.2 Lower Boundary

For simplicity, we concentrate on the case in which the log-pressure on a lower material boundary is specified, and use Cartesian coordinates. Thus,

$$w = \frac{D\zeta}{Dt} \quad \text{at} \quad z = \zeta(x, y, t) \tag{3B.1}$$

by Eq. (3.1.6c), when ζ is given. We suppose that $\zeta'(x, y, t)$ is $O(\alpha)$, where $\alpha \ll 1$, and that $\bar{\zeta}(y, t)$ is $O(\alpha^2)$. Consistent with these, we take $\bar{u} = O(1)$, $\bar{v}, \bar{w} = O(\alpha^2)$ and $(u', v', w') = O(\alpha)$. Then at $O(\alpha)$, Eq. (3B.1) gives

$$w' = \bar{D}\zeta' \quad \text{at} \quad z = \bar{\zeta}, \tag{3B.2}$$

where $\bar{D} = \partial/\partial t + \bar{u}\,\partial/\partial x$. Note that this applies at the "mean" value $\bar{\zeta}$ of the log-pressure altitude of the boundary.

At $O(\alpha^2)$, Eq. (3B.1) gives

$$\bar{w} + \overline{\zeta' w'_z} = \bar{\zeta}_t + \overline{u'\zeta'_x} + \overline{v'\zeta'_y} \quad \text{at} \quad z = \bar{\zeta} \tag{3B.3}$$

after zonal averaging; the second term on the left comes from Taylor-expanding $w(x, y, \zeta, t)$ about $z = \bar{\zeta}$. Now the Cartesian form of the continuity equation, Eq. (3.4.2d), can be written

$$u'_x + v'_y + w'_z = w'/H, \tag{3B.4}$$

since $\rho_{0z}/\rho_0 = -H^{-1}$. Rearranging Eq. (3B.3) and using Eqs. (3B.4), (3B.2), and the fact that $\overline{(\cdots)}_x \equiv 0$, we obtain

$$\bar{w} = \bar{\zeta}_t + (\overline{v'\zeta'})_y - (2H)^{-1}(\overline{\zeta'^2})_t \quad \text{at} \quad z = \bar{\zeta}; \tag{3B.5}$$

using the Cartesian form of Eq. (3.5.1b), this gives

$$\bar{w}^* = \bar{\zeta}_t + \overline{[v'(\zeta' + \theta'/\bar{\theta}_z)]}_y - (2H)^{-1}(\overline{\zeta'^2})_t \quad \text{at} \quad z = \bar{\zeta}. \tag{3B.6}$$

From the definition in Eq. (3.6.8) [or Eq. (3.7.1b)] of the northward parcel displacement η', we have,

$$\overline{v'(\zeta' + \theta'/\bar{\theta}_z)} = \overline{[\eta'(\zeta' + \theta'/\bar{\theta}_z)]}_t - \overline{\eta'\bar{D}(\zeta' + \theta'/\bar{\theta}_z)},$$

$$= \overline{[\eta'(\zeta' + \theta'/\bar{\theta}_z)]}_t - \overline{\eta'(Q' - v'\bar{\theta}_y)}/\bar{\theta}_z \tag{3B.7}$$

at $z = \bar{\zeta}$, by the Cartesian form of Eq. (3.4.2e), and Eq. (3B.2). Thus, from Eqs. (3B.6), (3B.7), and (3.6.8),

$$\bar{w}^* = \bar{\zeta}_t + [\overline{\eta'(\zeta' + \theta'/\bar{\theta}_z)}]_{yt} + \tfrac{1}{2}[\overline{\eta'^2 \bar{\theta}_y/\bar{\theta}_z}]_{yt} - (\overline{\eta'Q'/\bar{\theta}_z})_y$$

$$- (2H)^{-1}(\overline{\zeta'^2})_t + O(\alpha^3) \qquad \text{at} \quad z = \bar{\zeta}. \tag{3B.8}$$

In particular, if $\bar{\zeta}_t = 0$, then $\underline{\bar{w}^* = O(\alpha^3)}$ at $z = \bar{\zeta}$ for steady, adiabatic disturbances; however, $\bar{w} = (\overline{v'\zeta'})_y$ in this case, and is generally $O(\alpha^2)$. Thus $\bar{w}^*$ is negligible but $\bar{w}$ is generally not negligible at $z = \bar{\zeta}$, to second order in wave amplitude.

References

3.1. Comprehensive accounts of the principles of fluid dynamics are given in many texts, such as that of Batchelor (1967). Careful discussions of the dynamical and geometric arguments involved in the derivation of the primitive equations are given by Phillips (1973) and Gill (1982), Section 4.12. Hoskins *et al.* (1985) present a historical account of the use of Ertel's potential vorticity in meteorology, and mention several modern applications. The derivation of the kinematic boundary condition at a material surface is explained by Batchelor (1967), p. 73.

3.2. Formal derivations of the quasi-geostrophic equations on a beta-plane are given for example by Pedlosky (1979) and Gill (1982). The omega equation is discussed by Hoskins, Dragici and Davies (1978).

3.5. The transformation [Eq. (3.5.1)] leading to the TEM equations [Eqs. (3.5.2)] was introduced by Andrews and McIntyre (1976a, 1978a) and Boyd (1976). Alternative definitions of the residual circulation are given by Andrews and McIntyre (1978a) and Holton (1981). The identity of Eq. (3.5.10) is due to Bretherton (1966a).

3.6. The significance of conservation laws such as Eq. (3.6.2) is discussed by McIntyre (1980a, 1981). Successive generalizations of the original Charney–Drazin nonacceleration theorem were derived by Dickinson (1969), Holton (1974, 1975), Boyd (1976), and Andrews and McIntyre (1976a, 1978a).

3.7. Finite-amplitude versions of the generalized Eliassen–Palm theorem are discussed by Edmon *et al.* (1980), Andrews (1983) and Killworth and McIntyre (1985). The GLM theory is described in detail by Andrews and McIntyre (1978b,c); a useful introductory survey is that of McIntyre (1980a). Modified versions of this theory are outlined by McIntyre (1980a,b). For the relationship between the "form drag" and the EP flux see Bretherton (1969) and Andrews and McIntyre (1976a).

3.8. The primitive equations in θ coordinates are derived for example by Dutton (1976). Implications of the form of Eq. (3.8.8) of the potential vorticity conservation law are discussed by Haynes and McIntyre (1987).

3.9. The approach adopted here is a development of that due to Andrews (1983); see also Tung (1986).

Chapter 4 | Linear Wave Theory

4.1 Introduction and Classification of Wave Types

One of the most important dynamical properties of the atmosphere is its ability to support wave motions. These waves are of many different types, and in this chapter we shall concentrate on those that are of greatest importance for the large-scale behavior of the middle atmosphere. Thus we shall exclude discussion of acoustic waves, which have frequencies comparable to or larger than the buoyancy frequency and violate the hydrostatic relation of Eq. (3.1.3c); we shall also omit reference to waves modified by electrodynamic effects, which are important in the thermosphere.

The restoring effects necessary for the existence of the waves to be considered here are provided by the stable density or entropy *stratification* of the atmosphere, as represented by positive values of N^2 or θ_{0z}, and by the *rotation* of the earth, as represented by the Coriolis parameter $f = 2\Omega \sin \phi$ and its latitudinal derivative $\beta = 2\Omega a^{-1} \cos \phi$ (cf. Section 3.2).

In this chapter we shall concentrate exclusively on *linear* waves, assumed to be of small enough amplitude for the equations of Section 3.4 to apply. Even in this linear case, it is difficult to formulate a precise definition of a "wave"—indeed, no one definition is likely to satisfy all meteorologists. However, some sort of quasi periodicity is usually implied by the term, as well as the ability to transfer "information" over large distances without the corresponding transport of fluid parcels: all the waves to be discussed here possess these two attributes.[1]

Atmospheric waves can be classified in various ways, according to their physical or geometrical properties. In the first place, they can be categorized

---

[1] Incidentally, the departure from the zonal mean, defined by Eq. (3.3.1b), or any of its zonal Fourier components, is often called a "wave," despite the fact that it need not generally be "wave-like" in form.

according to their restoring mechanisms: thus buoyancy, or internal gravity, waves (often called "gravity waves" for short) owe their existence to stratification, while inertio–gravity waves result from a combination of stratification and Coriolis effects. Planetary, or Rossby, waves result from the beta-effect or, more generally, from the northward potential vorticity gradient [Eq. (3.4.8)].

A second type of classification is to distinguish *forced* waves, which must continually be maintained by an excitation mechanism of given phase speed and wave number, from *free* waves, which are not so maintained. Examples of forced waves include thermal tides, which are induced by the diurnal fluctuations in solar heating (see Section 4.3), while examples of free waves include global normal modes (Section 4.4).

A further classification results from the fact that some waves can propagate in all directions, while others may be trapped (or evanescent) in some directions. Thus under some circumstances horizontally propagating planetary waves can be trapped in the vertical (Section 4.5), while equatorial waves can propagate vertically and zonally but are evanescent with increasing distance from the equator (Section 4.7).

Waves can also be separated into *stationary* waves, whose surfaces of constant phase are fixed with respect to the earth, and *traveling* waves, whose phase surfaces move. Since information propagates with the group velocity (Section 4.5) and not with the phase speed, propagation can still occur in stationary waves. (We here use the adjective "steady" to denote waves whose amplitudes are independent of time, and "transient" for waves whose amplitudes are time-varying; see Section 3.6. Some authors use these terms as synonyms for our "stationary" and "traveling," while another definition of "transient" is mentioned in Section 5.1. What we have called stationary waves are sometimes also known as standing waves; however, the latter name is best reserved for waves with fixed nodal surfaces as typified, say, by a velocity disturbance $u' \propto \cos kx \cos \omega t$.)

The final general form of classification that we shall mention distinguishes waves that do not lead to any mean-flow acceleration from those that do. The former category includes waves that are linear, steady, frictionless, and adiabatic (see Section 3.6), while the latter usually includes any wave that is transient or nonconservative; however, nonlinear waves can sometimes satisfy nonacceleration conditions if they are steady and conservative.

4.2 Wave Disturbances to a Resting Spherical Atmosphere

When a stratified spherical atmosphere, at rest with respect to the rotating planet, undergoes small disturbances, an important class of wave motions results. The study of such a system originated with Laplace in the early nineteenth century, and has led to many insights into atmospheric behavior; in particular it underlies the theory of tides (Section 4.3) and global normal modes (Section 4.4).

We start with the linearized equations [Eqs. (3.4.2)], and set the basic flow $\bar{u}$ to zero; thus $\bar{\theta}_\phi$ also vanishes, by Eq. (3.4.1c). Moreover, we use Eq. (3.4.2c) to substitute for θ' in Eq. (3.4.2e) and use the definition $N^2(z) \equiv H^{-1}R\bar{\theta}_z e^{-\kappa z/H}$ [cf. Eq. (3.2.13)] to replace $\bar{\theta}_z$ in Eq. (3.4.2e). There result

$$u'_t - fv' + (a \cos \phi)^{-1}\Phi'_\lambda = X', \tag{4.2.1a}$$

$$v'_t + fu' + a^{-1}\Phi'_\phi = Y', \tag{4.2.1b}$$

$$(a \cos \phi)^{-1}[u'_\lambda + (v' \cos \phi)_\phi] + \rho_0^{-1}(\rho_0 w')_z = 0, \tag{4.2.1c}$$

$$\Phi'_{zt} + N^2 w' = \kappa J'/H, \tag{4.2.1d}$$

where the diabatic term $\kappa J'/H$ also equals $H^{-1}RQ'e^{-\kappa z/H}$; see Section 3.1.1. If the lower boundary is flat, and located at $z^* = 0$, the linearized version of Eq. (3.1.6a) becomes

$$\Phi'_t + w'\bar{\Phi}_z = 0 \qquad \text{at} \quad z = 0;$$

using Eq. (3.4.1b) and the relation between temperature and potential temperature, this can be written

$$\Phi'_t + \frac{R\bar{T}(0)w'}{H} = 0 \qquad \text{at} \quad z = 0. \tag{4.2.2}$$

The upper boundary condition depends on the problem in hand: see Section 4.3.3.

In this section we consider the conservative or unforced case $X' = Y' = J' = 0$, and use the method of separation of variables to investigate the solutions of Eq. (4.2.1). A first step is to separate the vertical dependence from the horizontal and time dependence; it turns out that the natural way to do this is by setting

$$(u', v', \Phi') = e^{z/2H}U(z)[\tilde{u}(\lambda, \phi, t), \tilde{v}(\lambda, \phi, t), \tilde{\Phi}(\lambda, \phi, t)] \tag{4.2.3a}$$

and

$$w' = e^{z/2H}W(z)\tilde{w}(\lambda, \phi, t). \tag{4.2.3b}$$

Substitution of Eq. (4.2.3a) into Eqs. (4.2.1a,b) (with $X' = Y' = 0$) and cancellation of the $e^{z/2H}U$ factor yields the following two equations, which involve only the (λ, ϕ, t) dependence:

$$\tilde{u}_t - f\tilde{v} + (a \cos \phi)^{-1}\tilde{\Phi}_\lambda = 0, \tag{4.2.4a}$$

$$\tilde{v}_t + f\tilde{u} + a^{-1}\tilde{\Phi}_\phi = 0, \tag{4.2.4b}$$

while substitution of Eqs. (4.2.3a,b) into Eqs. (4.2.1c,d) (with $J' = 0$) yields the following pair of equations, involving z dependence as well:

$$U(a \cos \phi)^{-1}[\tilde{u}_\lambda + (\tilde{v} \cos \phi)_\phi] + \left(W_z - \frac{W}{2H}\right) \tilde{w} = 0, \quad (4.2.5a)$$

$$\left(U_z + \frac{U}{2H}\right) \tilde{\Phi}_t + N^2 W \tilde{w} = 0. \quad (4.2.5b)$$

Examination of the expressions of Eqs. (4.2.3) and the z-dependent terms of Eq. (4.2.5a) shows that we can choose

$$U = \frac{dW}{dz} - \frac{W}{2H} \quad (4.2.6a)$$

without loss of generality. Substitution of Eq. (4.2.6a) into Eq. (4.2.5b) yields

$$-\left(W_{zz} - \frac{W}{4H^2}\right) \Big/ N^2 W = \tilde{w}/\tilde{\Phi}_t = (gh)^{-1} \quad (4.2.6b)$$

say, where $(gh)^{-1}$ is a separation constant and h has dimensions of length. Then Eqs. (4.2.6a,b) and Eq. (4.2.5a) give

$$(a \cos \phi)^{-1}[\tilde{u}_\lambda + (\tilde{v} \cos \phi)_\phi] + (gh)^{-1}\tilde{\Phi}_t = 0, \quad (4.2.4c)$$

while the outer terms in Eq. (4.2.6b) yield

$$\frac{d^2 W}{dz^2} + \left(\frac{N^2}{gh} - \frac{1}{4H^2}\right) W = 0. \quad (4.2.7a)$$

The three equations [Eqs. (4.2.4a,b,c)] for the horizontal and time structure are called *Laplace's tidal equations*, and are identical to those for small disturbances to a thin layer of fluid on a sphere, where the fluid has mean depth h much less than the radius a of the sphere, $\tilde{u}$ and $\tilde{v}$ are the velocity components, and $g^{-1}\tilde{\Phi}$ is the departure of the fluid depth from its mean value. The separation constant h is called the *equivalent depth*. Equation (4.2.7a) is called the *vertical structure equation*; from Eqs. (4.2.2), (4.2.3), and (4.2.6), the lower boundary condition can be written

$$\frac{dW}{dz} + \left(\frac{R\bar{T}(0)}{gh} - \frac{1}{2}\right) \frac{W}{H} = 0 \quad \text{at} \quad z = 0. \quad (4.2.7b)$$

It should be noted that h is not the vertical scale of $W(z)$. For example, when N is constant and $0 < h < 4N^2H^2/g$, Eq. (4.2.7a) has particular solutions that are sinusoidal in z with vertical wavelength $\lambda_v = 2\pi[(N^2/gh) - 1/4H^2]^{-1/2}$; the variation of λ_v with h is shown in Fig. 4.1.

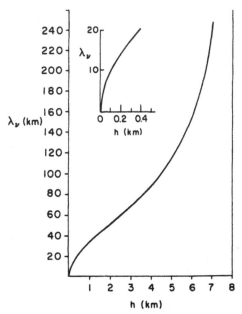

Fig. 4.1. The wavelength λ_v as a function of the equivalent depth h for $0 < h < 4N^2H^2/g$ in an isothermal atmosphere with $\bar{T} = 240$ K. In this case $H = 7$ km and $4N^2H^2/g = 4\kappa H = 8$ km by Eq. (3.2.13) and the definition $H = RT_s/g$, together with $\kappa = 2/7$. The inset gives an expanded view of the region near the origin.

The standard procedure for finding solutions to Laplace's tidal equations [Eqs. (4.2.4)] is to pose the sinusoidal forms

$$\{\tilde{u}, \tilde{v}, \tilde{\Phi}\} = \text{Re}\{[\hat{u}(\phi), \hat{v}(\phi), \hat{\Phi}(\phi)] \exp i(s\lambda - 2\Omega\sigma t)\} \qquad (4.2.8)$$

with zonal wavenumber s (an integer) and period $2\pi/2\Omega\sigma$ [or $(2\sigma)^{-1}$ in days]. Solving Eqs. (4.2.4a,b) for $\hat{u}$ and $\hat{v}$ and substituting in Eq. (4.2.4c), we obtain *Laplace's tidal equation,*

$$\mathcal{L}\hat{\Phi} + \gamma\hat{\Phi} = 0. \qquad (4.2.9)$$

Here $\gamma \equiv 4\Omega^2 a^2/gh$ is called Lamb's parameter and

$$\mathcal{L} \equiv \frac{d}{d\mu}\left[\frac{(1-\mu^2)}{(\sigma^2-\mu^2)}\frac{d}{d\mu}\right] - \frac{1}{\sigma^2-\mu^2}\left[\frac{-s(\sigma^2+\mu^2)}{\sigma(\sigma^2-\mu^2)} + \frac{s^2}{1-\mu^2}\right] \qquad (4.2.10)$$

is a second-order ordinary differential operator in the variable

$$\mu \equiv \sin\phi \qquad (-1 \leq \mu \leq 1) \qquad (4.2.11)$$

and depends on s and σ. The appropriate boundary conditions are that $\hat{\Phi}$ is bounded at the poles, $\mu = \pm 1$.

Given these boundary conditions, Eq. (4.2.9) is an eigenvalue problem and can be solved numerically. For example, if s and σ are specified, as in the theory of thermally forced tides to be treated in Section 4.3, a set of eigenvalues $\gamma_n^{(\sigma,s)}$ [or equivalent depths $h_n^{(\sigma,s)}$] and corresponding eigenfunctions $\Theta_n^{(\sigma,s)}$ can be found, which are bounded at the poles and satisfy

$$\mathscr{L}\Theta_n^{(\sigma,s)} + \gamma_n^{(\sigma,s)}\Theta_n^{(\sigma,s)} = 0 \qquad (4.2.12)$$

for integer values of n. The Θ_n are called *Hough functions* and, together with γ_n, have been extensively tabulated (e.g., by Longuet-Higgins, 1968). It should be noted that for some choices of σ, s, and n, $\gamma_n^{(\sigma,s)}$ and thus the equivalent "depth" $h_n^{(\sigma,s)}$ can be negative; moreover, the frequencies of the modes with $\gamma < 0$ all satisfy

$$|\sigma| < 1. \qquad (4.2.13)$$

Some examples of the variation of $|\gamma|^{-1/2}$ with σ and n at fixed zonal wave number s are shown in Fig. 4.2, for both positive and negative γ. The modes of eastward phase speed in Fig. 4.2a can all be regarded as (inertio-) gravity waves, except in the limit $\gamma^{-1/2} \to 0$ (to the left of the diagram), when the lowest two curves represent the Kelvin wave and mixed planetary-gravity (or Rossby-gravity) wave, respectively. The modes of westward phase speed in Fig. 4.2b fall into three classes: the central curve represents the mixed planetary-gravity wave, those above it (of higher frequency) represent gravity waves, while those below (of lower frequency) represent planetary waves.

Longuet-Higgins gives a more precise system of classification for these modes, and for those of negative equivalent depth in Figs. 4.2c,d. He also considers the limits of large and small $|\gamma|^{-1/2}$ in detail: for example, as $\gamma^{-1/2} \to \infty$ (i.e., large positive equivalent depths, on the right of the diagrams) the gravity waves, indexed by n_G in that limit, obey the dispersion relation

$$2\Omega\sigma \sim \pm[n_G(n_G + 1)gh]^{1/2}a^{-1}, \qquad (4.2.14a)$$

(see Section 4.6.1), and the corresponding Hough functions reduce to the associated Legendre polynomials $P_n^s(\mu)$. It can be shown that these limiting modes are divergent but irrotational—that is, $\tilde{\Phi} \not\equiv 0$ but the relative vorticity $(a\cos\phi)^{-1}[\tilde{v}_\lambda - (\tilde{u}\cos\phi)_\phi] \equiv 0$. On the other hand, the planetary waves and mixed mode, indexed by n_R, become nondivergent ($\tilde{\Phi} \equiv 0$) but rotational as $\gamma^{-1/2} \to \infty$, with relative vorticity proportional to P_n^s and dispersion relation

$$2\Omega\sigma \sim \frac{-2\Omega s}{n_R(n_R + 1)}; \qquad (4.2.14b)$$

see the end of Section 4.5.2.

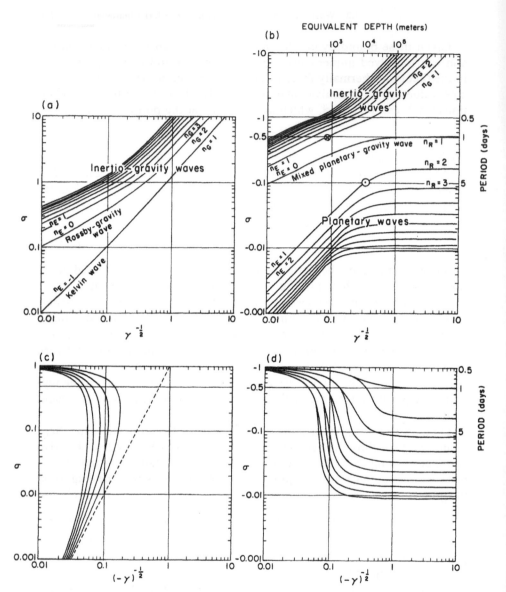

Fig. 4.2. Plots of $|\gamma|^{-1/2} \equiv (g|h|)^{1/2}/2\Omega a$ versus frequency σ for zonal wave number $s = 1$. Positive equivalent depths ($\gamma > 0$): (a) eastward phase speed ($\sigma > 0$), (b) westward phase speed ($\sigma < 0$). Negative equivalent depths ($\gamma < 0$): (c) eastward phase speed ($\sigma > 0$), (d) westward phase speed ($\sigma < 0$). In (b) and (d) some selected periods in days are given, and a horizontal line is drawn at the diurnal period ($\sigma = -\frac{1}{2}$). In (b) some equivalent depths are given (for $g = 9.8 \text{ m s}^{-1}$, $\Omega = 7.3 \times 10^{-5} \text{ s}^{-1}$, $a = 6.4 \times 10^6 \text{ m}$). The cross in (b) indicates the main propagating diurnal tidal mode (Section 4.3.3), and the circle indicates the "5-day wave" (Section 4.4). The indices n_G and n_R on the right-hand ends of some of the curves appear in Eqs. (4.2.14a,b). The indices n_E on the left-hand ends correspond to the index n used in Section 4.7 for equatorially trapped modes. [Adapted from Longuet-Higgins (1968).]

As $\gamma^{-1/2} \to 0$ (small positive equivalent depths, on the left of the diagrams), all modes become trapped near the equator: these equatorial waves are more readily studied using beta-plane geometry, and will be considered in Section 4.7. Asymptotic forms are also available for $\gamma < 0$; these include polar-trapped modes as $(-\gamma)^{-1/2} \to 0$, and nondivergent planetary waves with dispersion relation Eq. (4.2.14b) as $(-\gamma)^{-1/2} \to \infty$.

A different approach is used in the theory of free global normal modes, to be discussed in Section 4.4. Here h is sought as an eigenvalue of the vertical structure equation [Eq. (4.2.7a)], subject to the lower boundary condition of Eq. (4.2.7b) and an appropriate upper boundary condition. The corresponding $\gamma \equiv 4\Omega^2 a^2 / gh$, together with the zonal wave number s, is substituted into Laplace's tidal equation [Eq. (4.2.9)], which is now solved as an eigenvalue problem for σ, giving a set of eigenfrequencies (which can be read off from diagrams such as Figs. 4.2a,b) and corresponding Hough functions.

4.3 Atmospheric Thermal Tides

Atmospheric tides are global-scale daily oscillations, which are primarily forced by diurnal variations of the heating due to absorption of solar ultraviolet radiation by atmospheric water vapor and ozone. The solar and lunar gravitational forcing that produces ocean tides is much less important for the atmosphere, and we shall concentrate on thermally driven tides here. We shall further restrict attention to *migrating* tides, which move with the sun. Nonmigrating tides are associated, for example, with topography and geographically fixed tropospheric heat sources; they have received comparatively little theoretical attention. Migrating tides, on the other hand, have been studied extensively; they can sometimes propagate through great depths of the atmosphere, and can attain large amplitudes at some heights, especially in the thermosphere. Even in the mesosphere tidal temperature amplitudes can easily exceed 10 K, and can for example lead to problems in the interpretation of satellite data sampled once per day.

4.3.1 Summary of the Main Results of Atmospheric Tidal Theory

The basic ideas of tidal theory are quite simple, although some of the mathematical details are a little complicated. Thus, before going into the analysis in depth, we shall summarize the main results.

The first point to notice is that the solar heating is only active during the day, with a time variation which might look something like that shown in Fig. 4.3 at a point in midlatitudes. A Fourier analysis of this curve will

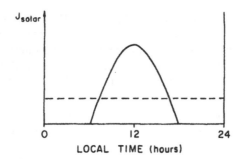

LOCAL TIME (hours)

Fig. 4.3. Schematic diagram of the typical daily variation of the solar heating J_{solar} (heavy curve) at a point in midlatitudes. The horizontal dashed line represents the diurnal or zonal average of J_{solar}, which is assumed to be balanced by a zonal-mean infrared cooling, so that the net $\bar{J} = 0$. The term J' is represented by the departure of the heavy curve from the dashed curve.

include a steady component, a diurnal component (with a 24-hr period), a somewhat smaller semidiurnal component (with a 12-hr period), and so on. The response of the atmosphere to this heating can likewise be decomposed into a steady part, and diurnal, semidiurnal and higher-frequency oscillations. An old puzzle, raised by Kelvin in 1882, was to explain why the *semidiurnal* surface pressure oscillation (about 1 mb in amplitude) is larger and more regular than the *diurnal*. The answer is best explained by describing the response of the whole middle and lower atmosphere to the diurnal and semidiurnal heating, as currently understood from theory and observation.

It transpires that the forced *semidiurnal* tidal oscillation has a large vertical wavelength (greater than 100 km). This allows it to be excited rather efficiently by the deep ozone heating region which is present around the stratopause. Moreover it can easily propagate to the ground and hence show up in surface pressure fluctuations. It is regular because the ozone heating is regular, and its latitudinal structure is fairly uniform.

The *diurnal* tide, on the other hand, has a more complex behavior. Between 30°N and 30°S it can propagate vertically, with a wavelength of about 28 km. Polewards of 30°, however, it is trapped in the vertical, close to its forcing region. As a result, the vertically propagating modes forced by the deep ozone heating tend to interfere destructively, and thus have small amplitude at the ground, while the trapped modes forced by the ozone never reach the ground at all. However, the comparatively shallow water-vapor heating region in the troposphere can excite this tide quite effectively, although this heating is somewhat intermittent in space and time. The resulting surface pressure oscillation is also intermittent. On the other hand, at the stratopause and above, the diurnal oscillation can be as strong as the semidiurnal.

4.3.2 Observations of Tides in the Middle Atmosphere

The surface pressure record is by far the largest existing atmospheric tidal data set, and it was the predominance of the semidiurnal component in this record that prompted Kelvin's question of 1882 and some of the earlier theories of atmospheric tides; data on other tropospheric tidal parameters, such as surface temperature and wind, are also very extensive now. Observations of tides in the middle atmosphere, on the other hand, are sparser, although a number of measurement techniques are currently available; these techniques will be briefly discussed here.

In the lower stratosphere, up to about 30 km altitude, wind observations from the tracking of meteorological balloons have provided data for large areas of the world and over many years. These data have been analyzed to provide estimates of the diurnal and semidiurnal tides, and show considerable vertical and horizontal structure and the presence of both migrating and nonmigrating components: see Fig. 4.4.

Above 30 km, a number of measurement techniques based on rockets and radars can be used. Rocket observations are mainly confined to altitudes below about 60 km, although there are a few important observations between 60 and 100 km. The radar techniques (summarized in Section 1.5) primarily provide wind measurements in the altitude range 70–110 km. The rocket and radar data have provided amplitudes and vertical wavelengths of both the diurnal and semidiurnal tides above selected locations on the globe and for selected periods of time: see Figs. 4.5 and 4.6.

These rocket and radar observations are necessarily restricted in geographical, and sometimes temporal, coverage, and this can be a drawback when comparison with theoretical calculations of global tidal structures is attempted. A recent advance has been the employment of simultaneous measurements from two Stratospheric Sounding Units on different NOAA satellites to obtain near-global tidal observations over long periods of time at several levels in the upper stratosphere. Although these observations have limited vertical resolution, satellite techniques of this kind promise a great expansion of the observational basis of our knowledge of tides in the middle atmosphere, thus providing a more critical test of theory. Such comparisons between theory and observations as are available at present are mentioned briefly in Section 4.3.4.

4.3.3 Classical Tidal Theory

We shall now outline the "classical" theory of tides, as given for example by Chapman and Lindzen (1970), who consider linear disturbances to an

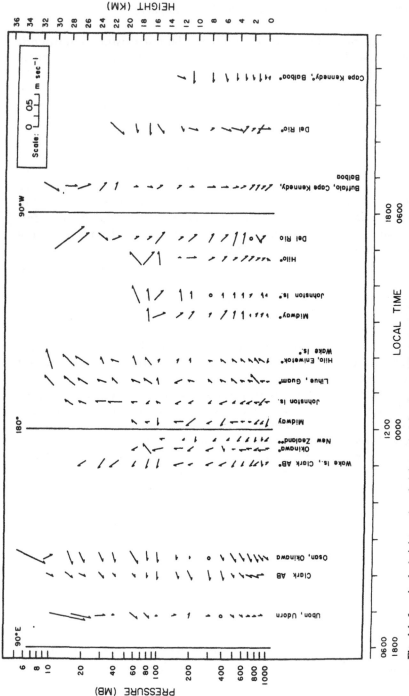

Fig. 4.4. Longitude–height or time–height section for the semidiurnal wind component for June–July, based on several years of radiosonde data. Arrows are centered on the longitude of the stations or the mean longitudes of groups of stations, and their directions indicate the flow direction, with upward corresponding to poleward. Note the roughly uniform rotation of the wind vectors with longitude and time, the increase in amplitude with height, and the lack of phase variation with height. [After Wallace and Tadd (1974).]

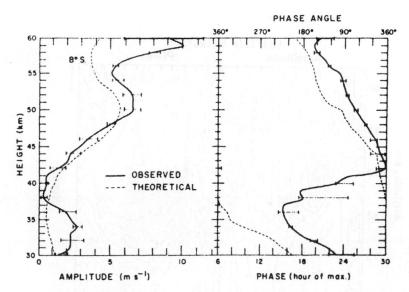

Fig. 4.5. Vertical variation of the amplitude and phase of the diurnal northerly wind component, as determined from rockets launched at Ascension Island (8°S). Also shown are theoretical calculations by Lindzen (1967), based on classical tidal theory (see Section 4.3.3). [After Reed *et al.* (1969). American Meteorological Society.]

atmosphere that is basically at rest. The set of Eqs. (4.2.1) is therefore used, together with suitable boundary conditions and a specified thermal forcing term $\kappa J'/H$ on the right of Eq. (4.2.1d); the mechanical forcing or dissipation terms X' and Y' are set to zero. The restriction to migrating tides means that all variables are taken to depend only on local time, latitude, and height; thus, for example, $J' = J'(\lambda + \Omega t, \phi, z)$, since $t + \lambda/\Omega$ is the local time at longitude λ if t is Greenwich Mean Time. (In accordance with the notation of Section 3.3, J' denotes the deviation of J from the zonal mean net heating $\bar{J}$. As seen from Fig. 4.3, the zonal-mean solar heating is nonzero; however, it will be assumed to be balanced by a zonal-mean infrared cooling, so that $\bar{J} = 0$.)

Since J' is periodic, it can be expanded in Fourier harmonics: thus

$$J' = \operatorname{Re} \sum_{s=1}^{\infty} J^{(s)}(\phi, z) e^{is(\lambda + \Omega t)} \tag{4.3.1}$$

where the coefficients $J^{(s)}$ are assumed known from a radiative–photochemical calculation. (This may not be a very accurate assumption: for example, the ozone distribution, which partly determines the heating, will depend to some extent on the tidal response to the heating.) The $s = 1$ component is the *diurnal* heating, and corresponds to $\sigma = -\frac{1}{2}$ in the notation

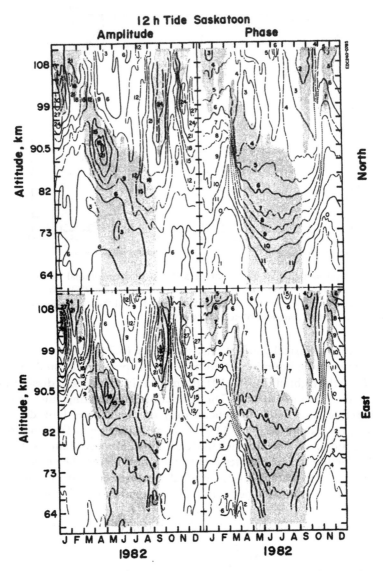

Fig. 4.6. Amplitude (m s^{-1}) and phase (local time of maximum) of the semidiurnal tidal wind components in the upper mesosphere and lower thermosphere, as measured by a partial reflection radar technique at Saskatoon (52°N) during 1982. The shaded region indicates that the time-mean zonal wind is easterly, the unshaded region that it is westerly. [After Manson and Meek (1986).]

of Eq. (4.2.8); the $s = 2$ component is the *semidiurnal* heating and corresponds to $\sigma = -1$; higher terms are seldom considered.

The next step is to expand each $J^{(s)}(\phi, z)$ in the Hough functions introduced in Section 4.2. (This procedure is analogous to that adopted in the theory, say, of the forced vibration of a stretched string, in which the forcing is expanded in terms of the eigenmodes of the unforced string.) In particular, $J^{(1)}$ is expanded in diurnal Hough functions $\Theta_n^{(-1/2,1)}$ which we shall call $\Theta_n^{(1)}$ for short, while $J^{(2)}$ is expanded in semidiurnal Hough functions $\Theta_n^{(-1,2)} \equiv \Theta_n^{(2)}$. Since modes of negative equivalent depth exist in the diurnal case ($\sigma = -\frac{1}{2}$), these must be included in the expansion of $J^{(1)}$. However, Eq. (4.2.13) shows that such modes are absent for semidiurnal and higher frequencies. Figures 4.2b,d for $s = 1$ include horizontal lines at the diurnal frequency, whose intersections with the plotted curves indicate the relevant modes and their equivalent depths. For positive equivalent depths (Fig. 4.2b) the relevant Hough functions all represent gravity modes, in the sense of Section 4.2, except for the gravest diurnal mode, which is of mixed type, has an infinite equivalent depth ($\gamma^{-1/2} \to \infty$), and is antisymmetric about the equator.

As an example, we shall consider the diurnal forcing $J^{(1)}(\phi, z)e^{i(\lambda + \Omega t)}$ and the atmospheric response to it. We write

$$J^{(1)}(\phi, z) = \sum_n J_n^{(1)}(z)\Theta_n^{(1)}(\mu) \tag{4.3.2}$$

where $\mu = \sin \phi$ as before; thus $J_n^{(1)}$ is the projection of the diurnal heating onto the nth diurnal Hough function. We look for a response with a similar migrating form, say, $w^{(1)}(\phi, z)e^{i(\lambda + \Omega t)}$ for the vertical velocity, where

$$w^{(1)}(\phi, z) = \sum_n e^{z/2H}W_n^{(1)}(z)\Theta_n^{(1)}(\mu), \tag{4.3.3}$$

by analogy with Eqs. (4.3.2) and (4.2.3b) (and with the vibrating string example, where the response to the forcing is also expanded in eigenmodes). We return to Eqs. (4.2.1), set $X' = Y' = 0$, and give all primed variables the diurnal $e^{i(\lambda + \Omega t)}$ dependence. The diurnal velocity amplitudes $u^{(1)}$ and $v^{(1)}$ can be expressed in terms of $\Phi^{(1)}$, using Eqs. (4.2.1a,b), and then expressed in terms of $w^{(1)}$ and $J^{(1)}$ by means of Eq. (4.2.1d). On substitution into Eq. (4.2.1c) we obtain a complicated partial differential equation in z and μ for $w^{(1)}$ in terms of $J^{(1)}$; however, on using Eqs. (4.3.2) and (4.3.3), and Laplace's tidal equation [Eq. (4.2.12)] for $\Theta_n^{(1)}$, the μ dependence separates out, leaving the inhomogeneous vertical structure equation [cf. Eq. (4.2.7a)]

$$\frac{d^2W_n^{(1)}}{dz^2} + \left[\frac{N^2}{gh_n^{(1)}} - \frac{1}{4H^2}\right]W_n^{(1)} = \frac{\kappa J_n^{(1)}(z)e^{-z/2H}}{gHh_n^{(1)}}, \tag{4.3.4}$$

with homogeneous lower boundary condition [cf. Eq. (4.2.7b)]

$$\frac{dW_n^{(1)}}{dz} + \left[\frac{R\bar{T}(0)}{gh_n^{(1)}} - \frac{1}{2}\right]\frac{W_n^{(1)}}{H} = 0 \qquad \text{at} \quad z = 0, \tag{4.3.5}$$

where $h_n^{(1)}$ is the nth diurnal equivalent depth.

The upper boundary condition needs careful consideration, and we follow the approach outlined in Section 3.1.2b. We suppose that above some height z_2 the forcing $J_n^{(1)}$ vanishes and the mean temperature $\bar{T}(z)$ becomes constant, T_∞ say, so that $N^2 = R\kappa T_\infty/H^2 = g\kappa T_\infty/HT_s \equiv N_\infty^2 =$ constant there. Thus for $z > z_2$, Eq. (4.3.4) becomes

$$\frac{d^2 W_n^{(1)}}{dz^2} + \left[\frac{N_\infty^2}{gh_n^{(1)}} - \frac{1}{4H^2}\right]W_n^{(1)} = 0,$$

with solutions that are exponential or sinusoidal in z according as the expression in square brackets is negative or positive. If it is negative, that is, if $h_n^{(1)} < 0$ or $h_n^{(1)} > 4H^2 N_\infty^2/g = 4\kappa HT_\infty/T_s$, then the decaying solution, with W_n proportional to

$$\exp - \left[\frac{1}{4H^2} - \frac{N_\infty^2}{gh_n^{(1)}}\right]^{1/2} z,$$

is clearly the one to be selected, and this provides the required upper boundary condition. Using Eqs. (4.2.3a), (4.2.6a), and (4.2.7a) and the fact that $\rho_0 \propto e^{-z/H}$, it can be shown that the mean wave energy per unit volume, $\frac{1}{2}\rho_0(\overline{u'^2} + \overline{v'^2} + \overline{\Phi_z'^2}/N^2)$ [see Eq. (3.6.3)], decays with height for this solution. However, if $0 < h_n^{(1)} < 4H^2 N_\infty^2/g$, then propagating solutions exist, with $W_n^{(1)} \propto e^{imz}$ for $z > z_2$ where

$$m = \pm\left[\frac{N_\infty^2}{gh_n^{(1)}} - \frac{1}{4H^2}\right]^{1/2} \tag{4.3.6}$$

is the vertical wave number. The wave energy per unit volume is independent of z for either sign of m, and the appropriate choice has to be made by computing the vertical component of the group velocity, $c_g^{(z)}$, and demanding that it be positive. This is the "radiation condition," which states that information must be traveling upward, and not downward, at great heights: see Section 3.1.2. We have $c_g^{(z)} \equiv \partial\omega/\partial m$, where $\omega = 2\Omega\sigma$ is the frequency. Thus

$$c_g^{(z)} \equiv \frac{\partial\omega}{\partial m} = 2\Omega\frac{\partial\sigma}{\partial m} = 4\Omega m\frac{\partial\sigma}{\partial m^2}$$

$$= 4\Omega m\left(\frac{\partial\sigma}{\partial h}\right)\bigg/\left(\frac{\partial m^2}{\partial h}\right) = -4\Omega m\left(\frac{\partial\sigma}{\partial h}\right)\bigg/\left(\frac{N_\infty^2}{gh^2}\right) \tag{4.3.7}$$

by Eq. (4.3.6), where indices on h have been dropped for clarity. Referring to Fig. 4.2b, we see that for all the relevant modes of finite, positive h (which are in fact gravity waves), $\partial(-\sigma)/\partial h > 0$, so that $c_g^{(z)}$ has the same sign as m, by Eq. (4.3.7). To satisfy the radiation condition, the positive sign must be chosen in Eq. (4.3.6), thus giving the upper boundary condition in the propagating-wave case.

The numerical solution of the vertical structure equation [Eq. (4.3.4)] for $W_n^{(1)}(z)$, subject to the appropriate boundary conditions, is straightforward. The series of Eq. (4.3.3) is then summed to obtain $w^{(1)}(\phi, z)$, and $u^{(1)}$, $v^{(1)}$, and $\Phi^{(1)}$ can be found from Eq. (4.2.1); a precisely similar procedure can be used for the semidiurnal response $w^{(2)}$, etc. Full details are given by Chapman and Lindzen (1970).

Although the global forms of the tidal structure are quite complicated, some general features can be deduced from basic properties of the vertical structure equation and Laplace's tidal equation. First note from Eq. (4.3.4) that if $0 < h < 4N^2H^2/g$ a tidal mode is roughly sinusoidal in the vertical, with wavelength given locally by

$$\lambda_v = 2\pi \left[\frac{N^2(z)}{gh} - \frac{1}{4H^2} \right]^{-1/2}$$

(The modal index n has again been omitted.) In the *diurnal* case some modes have $h < 0$, and the mixed mode of Fig. 4.2b has $h = \infty$; these modes are trapped in the vertical near to the forcing region. However, the diurnal gravity waves all propagate: the leading symmetric one, marked with a cross in Fig. 4.2b, has $h \approx 690$ m and $\lambda_v \approx 28$ km. Investigation of the horizontal Hough function structure (see Longuet-Higgins, 1968 and Chapman and Lindzen, 1970) shows that the propagating modes are confined equatorward of 30° latitude. (This is the latitude at which $|f| = 2\Omega|\sigma|$ or $|\mu| = |\sigma|$ for the diurnal frequency $\sigma = -\frac{1}{2}$, where Laplace's tidal equation [Eq. (4.2.9)] has apparent singularities: see also Section 4.6.3.) The mixed mode is global in extent with $\Theta \propto \sin\phi \cos\phi$, and the modes of negative equivalent depth are trapped poleward of 30°. Some calculations of the total diurnal response are given in Fig. 4.7; note the strong latitudinal variation, the fact that the amplitude of $T^{(1)}$ is several kelvins in the mesosphere, the vertical wavelength of about 28 km near the equator, and the weak phase variation at high latitudes.

The equivalent depths for the *semidiurnal* tide are all positive, all modes are of the gravity-wave class, and the dominant mode has $h \approx 7.85$ km. For this value, $N^2/gh \approx 1/4H^2$, implying a very large vertical wavelength λ_v, as mentioned above (see also Fig. 4.1), the vertical structure equation [Eq. (4.3.4)] is replaced by an equation of the form

$$\frac{d^2W}{dz^2} \approx F(z) \equiv \kappa J_1^{(2)}(z) e^{-z/2H} / gH h_1^{(2)} \tag{4.3.8}$$

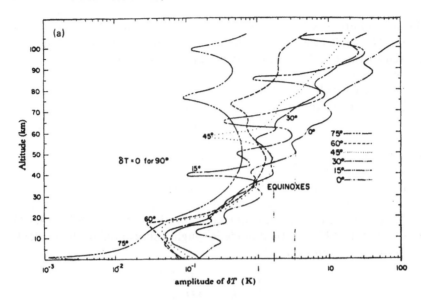

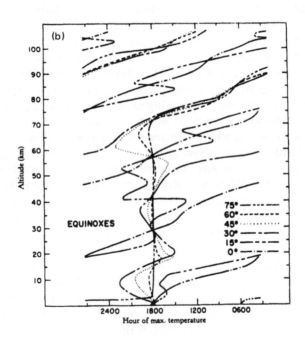

Fig. 4.7. (a) Amplitude and (b) phase of solar diurnal component of T' at various latitudes for equinox. [After Lindzen (1967).]

and the lower boundary condition [Eq. (4.3.5)] can be approximated by

$$\frac{dW}{dz} + \frac{W}{2H} \approx 0, \tag{4.3.9}$$

since $RT(0)/gh_1^{(2)} \approx 1$. If the semidiurnal excitation were all concentrated at one level, z_e say, we would have $F(z) = F_e \, \delta(z - z_e)$ and the solution to Eqs. (4.3.8) and (4.3.9) bounded at great heights would take the form $W = (2H - z)F_e$ for $z < z_e$, that is, *below* the excitation level. The corresponding solution for U would be

$$U = -(2 - z/2H)F_e \qquad \text{for} \quad z < z_e$$

by Eq. (4.2.6a), implying that U would change sign at $z = 4H \approx 28$ km if the excitation level z_e were above this height; thus the semidiurnal phases of u' and v' would shift by π at $z \approx 4H$, by Eq. (4.2.3a). Detailed calculations show that a similar behavior occurs in the more accurate theory when several semidiurnal Hough modes are included, when $F(z)$ is distributed in the vertical but confined mostly to altitudes above 28 km, and when Eqs. (4.3.4) and (4.3.5) are used rather than Eqs. (4.3.8) and (4.3.9). Some results are given in Fig. 4.8; they show a phase shift in the southward wind at about 28 km altitude, and quite large temperature amplitudes in the mesosphere. They also demonstrate the dominance of ozone in forcing this tide.

4.3.4 More Detailed Theory

In the last decade or so, various improvements on the "classical" theory have been made. Mean winds $\bar{u}(\phi, z)$ and latitudinally varying mean temperatures $\bar{T}(\phi, z)$, for equinox and solstice conditions, have been included, as have Newtonian cooling and (in the thermosphere) molecular viscosity, thermal conductivity, and ion drag. Better parameterizations of ozone and water-vapor heating have been used, and the effects of heating by molecular oxygen included. Such calculations have been performed for height ranges from the ground up to several hundred kilometers.

The theory as it stands now generally agrees fairly well with observations, although, as mentioned in Section 4.3.2, the latter are fairly scanty in the stratosphere and mesosphere. The main discrepancies are between the calculated and observed semidiurnal tides; theory predicts a phase shift in the horizontal winds of 180° at about 28 km altitude (see Fig. 4.8) which is not observed, and the calculated surface pressure phase is 30–60 min later than observed. Recent studies suggest that these differences may be due to the omission of a semidiurnal contribution from latent heating in the tropics. There is also a discrepancy between calculated and observed diurnal tides

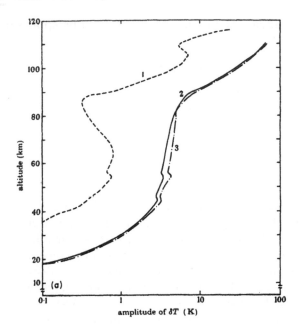

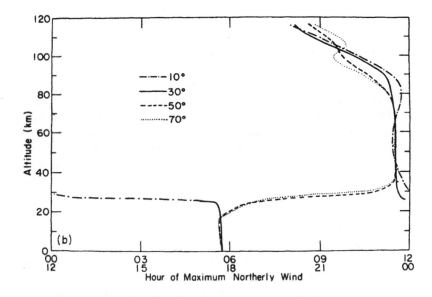

Fig. 4.8. (a) Amplitude of semidiurnal component of T' as forced by (1) H_2O only, (2) H_2O and O_3, (3) O_3 only (Lindzen 1968), and (b) phase of semidiurnal component of $(-v')$ at various latitudes. [After Chapman and Lindzen (1970).]

in the high latitude mesosphere; the observations show a strong seasonal variation in wavelength and phase that is not predicted by theory. As yet, the satellite measurements mentioned in Section 4.3.2 have only been compared with radiance calculations based on classical tidal theory; fair qualitative agreement is found, but some quantitative discrepancies appear, and the reasons for these are at present unresolved.

Another recent development has been the study of nonlinear processes associated with tides: possible implications of such processes for the mean circulation of the middle atmosphere are mentioned briefly in Section 7.3.

4.4 Free Traveling Planetary Waves

A well-documented group of atmospheric waves is a class of zonally traveling structures of global scale, which have periods of a few days. Such waves are thought to be examples of free traveling planetary (Rossby) waves or global normal modes and, unlike the thermally driven tides described in the previous section, they are apparently not maintained by traveling forcing effects. They appear for example in surface pressure data, in standard analyses of upper tropospheric radiosonde data, and in satellite data from the stratosphere. Space and time filtering can help to distinguish them from other atmospheric phenomena.

The most prominent mode of this class is the *5-day wave*, a westward-traveling disturbance whose period is close to 5 days and that is approximately sinusoidal in the east–west direction with zonal wave number $s = 1$. The geopotential disturbance Φ' or temperature disturbance T' associated with this wave is symmetric about the equator, and peaks in midlatitudes. Examples of middle atmosphere observations are presented in Figs. 4.9 and 4.10. The wave has little phase tilt in the vertical, and the observed temperature disturbance grows with height. At the ground the observed pressure fluctuation is about 0.5 mb; in the upper stratosphere at about 40 km altitude (≈ 2 mb) the temperature fluctuation is approximately 0.5 K. Numerous other traveling modes of this type have also been observed, including a wave-number 1, 16-day wave and a wave-number 3, 2-day wave (Fig. 4.11). More details are given in Section 5.4.

The simplest theory of free modes of this type involves unforced, global-scale, linearized disturbances to an atmosphere that is basically at rest. Equations (4.2.1) are used, with $X' = Y' = J' = 0$; then solutions are sought in the form

$$w' = e^{z/2H} \, \text{Re}[\, W(z)\hat{w}(\phi) \exp i(s\lambda - 2\Omega\sigma t)], \qquad (4.4.1)$$

for example [cf. Eqs. (4.2.3b) and (4.2.8)], where s is specified and σ is to

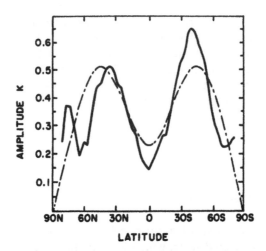

Fig. 4.9. Solid curve: brightness temperature amplitude of the 5-day wave as a function of latitude at about 42 km altitude for November 1973, as measured by the Selective Chopper Radiometer on the *Nimbus 5* satellite. [After Rodgers (1976a).] Broken curve: latitudinal structure of the Hough function corresponding to the 5-day wave ($s = 1$, $\sigma \approx -0.1$, $h \approx 10^4$ m). (Courtesy of P. J. Valdes.)

be determined. The method for calculating σ proceeds in two stages: first the vertical structure equation [Eq. (4.2.7a)] is solved as an eigenvalue problem for the equivalent depth h, subject to the lower boundary condition [Eq. (4.2.7b)] and the requirement that $W \to 0$ (and hence the wave energy per unit volume tends to zero) as $z \to \infty$. [Note that the radiation condition of Section 4.3 is not needed here, since the required modes must be evanescent at great heights. If they were to propagate upward at great heights, a continual forcing mechanism would be needed to maintain their amplitudes at low levels. Thus an eigenvalue h must be such that $N^2/gh < (1/4H^2)$ above some altitude z_1, say.]

The second stage of the calculation of σ takes the value or values of h found in this way, substitutes them into Laplace's tidal equation for the horizontal structure,

$$\mathscr{L}\hat{w} + (4\Omega^2 a^2/gh)\hat{w} = 0 \qquad (4.4.2)$$

[cf. Eq. (4.2.9)], and solves the latter as an eigenvalue equation for σ. [Recall that $\mathscr{L}$ depends on s and σ: see Eq. (4.2.10).] The eigenfunctions $\hat{w}$ are given by Hough functions. (Note how this procedure differs from that used in Section 4.3: see also the end of Section 4.2.)

As a simple example, consider an isothermal basic atmosphere, with $\bar{T} = T_s = $ constant; then $N^2 = R\kappa T_s H^{-2} = g\kappa H^{-1}$ by Eq. (3.2.13) and the definition $H = RT_s/g$. In this case a solution $W(z)$ of Eq. (4.2.7a) can only

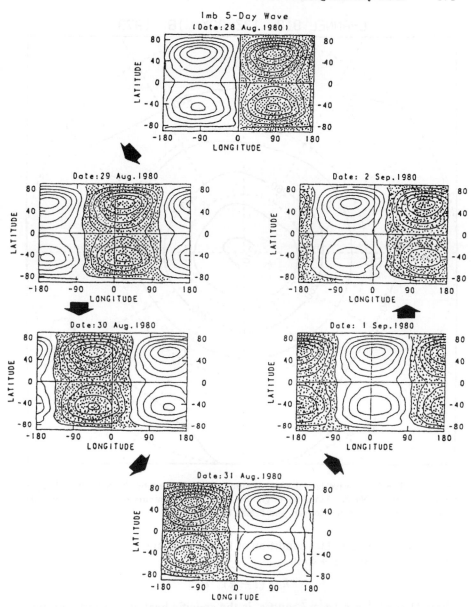

Fig. 4.10. The 5-day wave at 1 mb as observed by the Stratospheric Sounding Unit on the *TIROS-N* satellite, for 6 successive days in August–September 1980. The wave-number 1 Fourier component of the geopotential height anomaly has been bandpass-filtered to select periods of 5–6 days (eastward and westward). Note the clear westward-traveling pattern except south of 50°S. Shaded regions denote negative anomalies; the contour interval is 20 m. [After Hirota and Hirooka (1984). American Meteorological Society.]

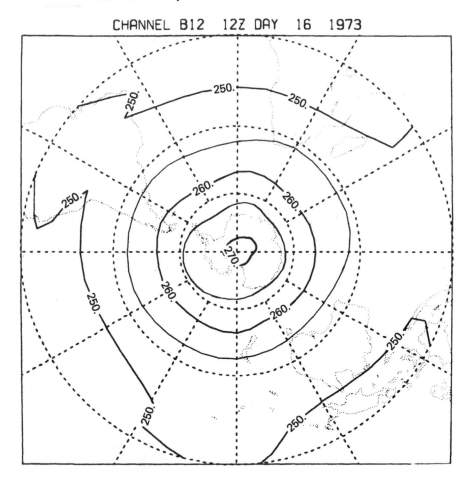

CHANNEL B12   12Z DAY   16   1973

Fig. 4.11. The 2-day wave (zonal wave number 3) as observed by the Selective Chopper Radiometer on the *Nimbus* 5 satellite: brightness temperature contours at about 42 km altitude on January 16, 1973. [After Rodgers and Prata (1981).]

satisfy the lower boundary condition of Eq. (4.2.7b) and the upper boundary condition $W \to 0$ as $z \to \infty$ if

$$h = (1 - \kappa)^{-1} H,$$

where $(1 - \kappa)^{-1} = c_p/c_v \approx \frac{7}{5}$, and c_v is the specific heat at constant volume. Then $W(z) \propto \exp[(\kappa - \frac{1}{2})z/H]$ and hence

$$w' \propto e^{\kappa z/H} \tag{4.4.3}$$

by Eq. (4.4.1). Using Eqs. (4.2.6) and (3.1.3c′) it can be shown that Φ', u', v', and T' have a similar z dependence. The wave thus has no phase tilt in the vertical, and although the velocity and temperature fields grow with height, the corresponding energy density $\frac{1}{2}\rho_0(u'^2 + v'^2 + \Phi_z'^2 N^{-2})$ [cf. Eq. (3.6.3)] decays with height since $\rho_0 \propto e^{-z/H}$. This is an example of an "external" or "edge" wave, being trapped against the lower boundary. It can be verified that the geometric vertical velocity Dz^*/Dt, as opposed to the "log-pressure" velocity $w \equiv Dz/Dt$, vanishes everywhere. Such a vertical structure is also exhibited by an acoustic wave known as the Lamb wave. [Incidentally, no such wave exists in a model with either of the artificial lower boundary conditions $\Phi' = 0$ or $w' = 0$ at $z = 0$: cf. Eqs. (3.1.6b,c).]

The planetary-wave dynamics of this wave enter through the horizontal structure equation [Eq. (4.4.2)]. Taking $T_s = 240$ K, so that $H = 7$ km, we obtain $h \approx 10$ km, and on consulting Fig. 4.2b for $s = 1$ we find that the gravest symmetric westward-traveling planetary mode with this equivalent depth (indicated by a circle) has a period of about 5 days. The corresponding latitudinal structure of $\hat{w}$ (and Φ') is given in Fig. 4.9. This theoretical solution is close in period and in horizontal and vertical structure to the observed 5-day wave.

This simple type of theory can be extended in several ways. Allowance for a nonisothermal basic temperature $\bar{T}(z)$ does not appear to permit vertical structures significantly different from the external mode form [Eq. (4.4.3)], except possibly for a "ducted" mode peaking near the stratopause. However, the latter is highly susceptible to dissipative processes and is therefore unlikely to have any analogue in the real atmosphere. Inclusion also of horizontally varying mean winds $\bar{u}(\phi)$ still gives a mathematically separable problem, with a modified horizontal structure equation, which can be solved numerically. If the mean wind $\bar{u}$, and thus the mean temperature $\bar{T}$, depends on latitude and height, the problem is no longer separable in ϕ and z, but numerical methods are still available for solving the linear equations of Eqs. (3.4.2), assuming all disturbance variables to have the $\exp i(s\lambda - 2\Omega\sigma t)$ form of Eq. (4.4.1) and given appropriate boundary conditions. Such a method has been used for example to show that the "5-day" wave period is relatively insensitive to the background wind structure and thus to partially explain why it is such a ubiquitous feature of observations. Details of this approach and further applications to models of the 16-day wave and the 2-day wave are given in Section 5.4.

Observations have suggested the existence of many other global free traveling waves of various periods. Not all of these modes may be even approximately sinusoidal in the zonal direction. For example, a sequence of synoptic maps (Fig. 4.12) shows that a feature that has been identified as a "4-day wave" by Fourier analysis of time series is in reality a localized warm pool that circles the Southern Hemisphere stratosphere in about 4 days.

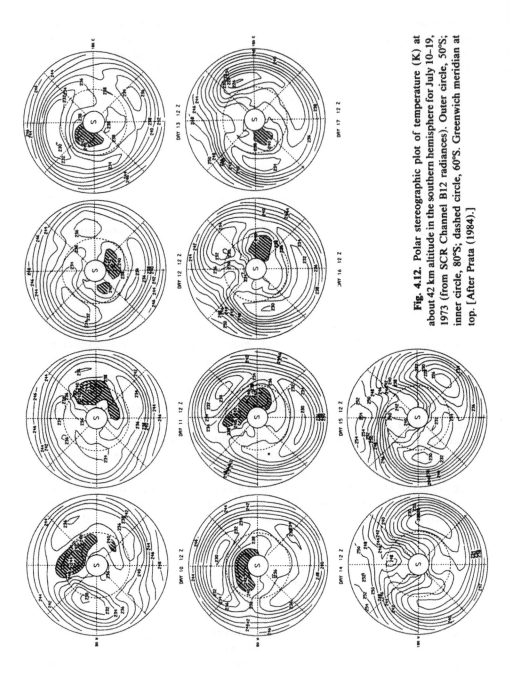

Fig. 4.12. Polar stereographic plot of temperature (K) at about 42 km altitude in the southern hemisphere for July 10–19, 1973 (from SCR Channel B12 radiances). Outer circle, 50°S; inner circle, 80°S; dashed circle, 60°S. Greenwich meridian at top. [After Prata (1984).]

Although the waves discussed in this section are not sustained by traveling forcing effects, some mechanism must still excite them initially, and perhaps repeatedly, so as to overcome dissipation and vertical leakage. Possibilities include stochastic forcing by latent heating, random disturbances in the atmosphere, or fluctuations of the mean winds. The waves may also be coupled to barotropic or baroclinic instabilities of matching phase speeds (see Section 5.5.2), although this coupling process has not yet been studied in detail.

4.5 Forced Planetary Waves

Some of the most important large-scale wave disturbances to be observed in the middle atmosphere are examples of forced planetary waves. The behavior of these waves is significantly different from that of the free planetary waves described in Section 4.4 and the propagating solar-forced tidal gravity waves of Section 4.3. In this section we present the basic theory of these forced planetary waves; observational details and interpretation are given in Chapter 5.

For simplicity we use beta-plane, rather than spherical, geometry and work with quasi-geostrophic theory, which is obeyed by these waves to a good approximation. We suppose the waves to be propagating on a basic zonal flow $[\bar{u}(y, z), 0, 0]$; the linearized potential vorticity equation is then

$$\left(\frac{\partial}{\partial t} + \bar{u}\frac{\partial}{\partial x}\right) q' + v'\bar{q}_y = Z', \tag{4.5.1}$$

where Z' here denotes the nonconservative terms on the right of Eq. (3.4.5),

$$q' = \psi'_{xx} + \psi'_{yy} + \rho_0^{-1}(\rho_0 \varepsilon \psi'_z)_z \tag{4.5.2}$$

is the disturbance quasi-geostrophic potential vorticity [cf. Eq. (3.4.7)],

$$v' = \psi'_x \tag{4.5.3}$$

is the northward geostrophic wind, and

$$\bar{q}_y = \beta - \bar{u}_{yy} - \rho_0^{-1}(\rho_0 \varepsilon \bar{u}_z)_z \tag{4.5.4}$$

is the basic northward quasi-geostrophic potential vorticity gradient [which by Eq. (3.8.10) can also be related to the basic northward isentropic gradient of Ertel's potential vorticity].

Most planetary (or Rossby) waves in the stratosphere and mesosphere appear to propagate upward from forcing regions in the troposphere, and a useful way to model their middle atmosphere behavior is to consider quasi-geostrophic disturbances forced from below by fluctuations in the height of some isobaric surface, $p = p_0$ say, which could be located in the upper troposphere or lower stratosphere (e.g., $p = 100$ mb). For simplicity, we take $p_s = p_0$ in this section (rather than $p_s = 1000$ mb: see Section 1.1.1), so that $z_0 \equiv -H \ln(p_0/p_s) = 0$. The relevant lower boundary condition is the disturbance part of Eq. (3.1.6b), which can be written in terms of $\psi' = f_0^{-1}\Phi'$ as

$$\psi' = \psi_0'(x, y, t) \qquad \text{at} \quad z = 0, \tag{4.5.5}$$

where the forcing function ψ_0' is prescribed. For simplicity we for the moment consider a "channel" between vertical walls at $y = 0, L$, with lateral boundary conditions:

$$v' = \psi_x' = 0 \qquad \text{at} \quad y = 0, L. \tag{4.5.6}$$

As in Section 4.3.3, the upper boundary condition is examined case by case.

4.5.1 $\bar{u}$ Depends on z Alone

A natural case to study first is that in which $\bar{u} = \bar{u}(z)$ and thus $\bar{q}_y = \beta - \rho_0^{-1}(\rho_0 \varepsilon \bar{u}_z)_z = \bar{q}_y(z)$. We take the nonconservative term $Z' = 0$, and suppose that the forcing is given by

$$\psi_0' = \text{Re } \hat{\psi}_0 e^{ik(x-ct)} \sin ly. \tag{4.5.7}$$

This has the form of a wavy pattern of zonal wavelength $2\pi k^{-1}$ and meridional wavelength $2\pi l^{-1}$, moving zonally with phase speed c (eastward if $c > 0$, westward if $c < 0$). Note that k is related to the spherical integer zonal wavenumber s by $s = ka \cos \phi$; moreover, $\omega \equiv ck = 2\Omega\sigma$ [cf. Eq. (4.2.8)]. For consistency with the forcing, we look for solutions to Eq. (4.5.1) of the form

$$\psi' = \text{Re } \hat{\psi}(z)e^{ik(x-ct)} \sin ly, \tag{4.5.8}$$

which satisfies the lower boundary condition of Eq. (4.5.5) if $\hat{\psi}(0) = \hat{\psi}_0$ and the lateral boundary condition of Eq. (4.5.6) if $lL\pi^{-1}$ is an integer. [Alternatively, we can replace $\sin ly$ by 1 in Eqs. (4.5.7) and (4.5.8).] Substitution of Eq. (4.5.8) into Eq. (4.5.1) and use of Eqs. (4.5.2,3) lead to the following second-order ordinary differential equation for $\hat{\psi}(z)$:

$$\rho_0^{-1}(\rho_0 \varepsilon \hat{\psi}_z)_z + \left[\frac{\bar{q}_y(z)}{\bar{u}(z) - c} - (k^2 + l^2)\right]\hat{\psi} = 0 \tag{4.5.9}$$

with lower boundary condition

$$\hat{\psi} = \hat{\psi}_0 \qquad \text{at} \quad z = 0 \tag{4.5.10}$$

and an upper boundary condition to be determined.

If $\bar{u}(z) = c$ at some level $z = z_c$—the *critical level*—the coefficient of $\hat{\psi}$ in Eq. (4.5.9) is infinite there if $\bar{q}_y(z_c) \neq 0$. In such a case it is known from the theory of differential equations that $\hat{\psi}$ is generally logarithmically infinite at z_c if $\bar{u}_z(z_c) \neq 0$; extra physics, such as transience, dissipation, or non-linearity, must be included in the theory if this infinity is to be avoided. Critical levels are the subject of much current research, and are mentioned again in Section 4.5.4 and also in Section 5.6.

4.5.2 Constant $\bar{u}$ and N

To gain some insight into the nature of solutions of Eq. (4.5.9), we specialize still further, by taking $\bar{u}$ constant, so that $\bar{q}_y = \beta = $ constant and N constant, so that ε is constant. Then Eq. (4.5.9) can be written

$$\hat{\psi}_{zz} - H^{-1}\hat{\psi}_z + B\hat{\psi} = 0$$

where

$$B = \varepsilon^{-1}\left[\frac{\beta}{\bar{u} - c} - (k^2 + l^2)\right], \qquad (4.5.11)$$

since $\rho_0 = \rho_s e^{-z/H}$. Looking for solutions $\hat{\psi} \propto e^{\Lambda z}$, we find

$$\Lambda = \frac{1}{2H} \pm \left(\frac{1}{4H^2} - B\right)^{1/2}.$$

Two possibilities arise, according as the term inside the square root is positive or negative:

1. $(1/4H^2) - B \equiv \nu^2 > 0$

In this case $\Lambda = (2H)^{-1} \pm \nu$ where $\nu = +[(1/4H^2) - B]^{1/2}$, and $\hat{\psi} = \hat{\psi}_0$ $\exp[(z/2H) \pm \nu z]$ or $\rho_0^{1/2}\hat{\psi} = \rho_s^{1/2}\hat{\psi}_0 e^{\pm \nu z}$. The wave activity density $A = \frac{1}{2}\rho_0 q'^2/\bar{q}_y$, noted in Section 3.6 as a natural measure of wave amplitude when $\bar{q}_y > 0$ will then vary with height as $e^{\pm 2\nu z}$: the wave energy per unit volume, as given by the term $\frac{1}{2}\rho_0(\overline{u'^2} + \overline{v'^2} + \overline{\Phi_z'^2}/N^2)$ in Eq. (3.6.3), will do likewise. Clearly the appropriate upper boundary condition is that quantities like these should be bounded as $z \to \infty$, and the lower sign must be selected. (In fact A and the wave-energy density then vanish as $z \to \infty$.) Thus

$$\psi' = \psi_0'(x, y, t) \exp\left(\frac{1}{2H} - \nu\right)z;$$

this is an example of a trapped, or edge, wave, and has no phase tilt with height. Note that ψ', u', v', etc., actually grow with height if $\nu < 1/2H$, that is, if $B > 0$; however, the decreasing basic density ρ_0 more than compensates for this in the wave-activity or wave-energy density.

2. $(1/4H^2) - B \equiv -m^2 < 0$

In this case we can put $\Lambda = (2H)^{-1} + im$, where

$$m = \pm\left(B - \frac{1}{4H^2}\right)^{1/2} \qquad (4.5.12)$$

and

$$\psi' = \text{Re } \hat{\psi}_0 \exp\left[\frac{z}{2H} + i(kx + mz - kct)\right] \sin ly. \qquad (4.5.13)$$

The presence of imz in the exponential here indicates vertical wave propagation, and phase lines tilt with height. The solution, Eq. (4.5.13), represents a Rossby wave propagating vertically and zonally. Eliminating B from Eqs. (4.5.11,12), we obtain the equation

$$\bar{u} - c = \beta[k^2 + l^2 + \varepsilon(m^2 + 1/4H^2)]^{-1}; \qquad (4.5.14)$$

in this case where $0 < m^2 < \infty$ and β, k^2, l^2, ε, and H^2 are all positive, we obtain the criterion

$$0 < \bar{u} - c < \bar{u}_c \equiv \beta(k^2 + l^2 + \varepsilon/4H^2)^{-1} \qquad (4.5.15)$$

(Charney and Drazin, 1961). For waves whose phase is stationary with respect to the ground, with $c = 0$, we have

$$0 < \bar{u} < \bar{u}_c \qquad (4.5.16)$$

for vertical propagation; thus for this case of constant $\bar{u}$ and N, "stationary" vertically propagating Rossby waves can only exist in winds that are westerly (eastward) and not too strong.

It should be noted that the limiting speed $\bar{u}_c$ depends on the zonal and meridional wave numbers of the mode in question. For a typical stratospheric static stability ($N^2 = 5 \times 10^{-4} \text{ s}^{-2}$), and choosing $l = \pi/(10{,}000 \text{ km})$, we find at 60°N

$$\bar{u}_c \approx 110/(s^2 + 3) \quad \text{m s}^{-1}$$

where the integer $s = ka \cos \phi$ is the spherical zonal wave number, as above. Thus on the basis of this very simple model, wave number 1 ($s = 1$) propagates in westerlies weaker than about 28 m s^{-1}, wave number 2 in westerlies weaker than about 16 m s^{-1}, and so on. Of course this model cannot hope to represent very faithfully the propagation of such global-scale modes in realistic mean zonal shear flows, but it does illustrate the fact that the "window" for propagation given by Eq. (4.5.16) becomes smaller as the zonal wave number s increases. This is in broad agreement with observations, which show that stationary disturbances tend to be composed only of the "ultralong" Fourier components $s = 1, 2, 3$ in the winter westerlies and tend to be absent in the stratospheric easterlies (see Chapter 5). More sophisticated theories, involving more general basic states and forms of disturbance, will be mentioned in Sections 4.5.4 and 5.3.

It remains to determine the sign of m in Eq. (4.5.12); here the radiation condition must be used, as in Section 4.3.3, since the wave-activity density and wave-energy density are both constant with height, irrespective of the sign of m. We again compute the vertical group velocity $c_g^{(z)}$ and choose it to be positive, in accordance with the general belief that planetary waves normally propagate upward from the troposphere into the stratosphere, rather than downwards.

To calculate $c_g^{(z)}$ we rearrange Eq. (4.5.14) so as to obtain the Rossby wave dispersion relation in the form

$$\omega \equiv ck = k\bar{u} - \beta k[k^2 + l^2 + \varepsilon(m^2 + 1/4H^2)]^{-1} \qquad (4.5.17)$$

so that

$$c_g^{(z)} \equiv \frac{\partial \omega}{\partial m} = 2\varepsilon\beta km[k^2 + l^2 + \varepsilon(m^2 + 1/4H^2)]^{-2}. \qquad (4.5.18)$$

Taking $k > 0$ by convention we see that $c_g^{(z)}$ is positive if $m > 0$, and so the upper sign must be chosen in Eq. (4.5.12). The phase surfaces, $kx + mz - kct = $ constant, thus tilt westward with height, as observed for planetary waves in the stratosphere.[2] For stationary waves, we put $c = \omega = 0$ after calculating $\partial \omega / \partial m$; although the phase surfaces for such waves are fixed in space, the waves still transfer information vertically.

An alternative method of motivating the choice of m is to include small dissipative terms. A simple model uses Newtonian cooling and Rayleigh friction with equal, constant, rate coefficients $\delta > 0$, so that $(X', Y', Q') = \delta(u', v', \theta')$; it can then be shown that $Z' = -\delta q'$ in Eq. (4.5.1). In the stationary-wave case, for example, it follows that $\bar{u} \, \partial/\partial x$ is replaced by $\bar{u} \, \partial/\partial x + \delta$, and we find $\hat{\psi} \propto e^{\Lambda z}$, where

$$\Lambda \approx \frac{1}{2H} \pm \left(\frac{1}{4H^2} - B - \frac{i\delta\beta}{\varepsilon k\bar{u}^2} \right)^{1/2}$$

[2] Note that if k is chosen negative then the group velocity condition implies that m must also be negative, and the phase tilt is still westward with height. All other physical properties of the wave solution are likewise unaltered.

if $\delta \ll |k\bar{u}|$ and B is as defined before. In case (2) above, this yields

$$\Lambda \approx \frac{1}{2H} - \frac{\delta\beta}{2\varepsilon k\bar{u}^2 m} + im$$

where m is given by Eq. (4.5.12) and

$$\psi' = \text{Re } \hat{\psi}_0 \exp\left[\frac{z}{2H} - \frac{\delta\beta z}{2\varepsilon k\bar{u}^2 m} + i(kx + mz)\right] \sin ly.$$

Thus small dissipation produces an extra factor $\exp(-\delta\beta z/2\varepsilon k\bar{u}^2 m)$, which implies that the wave-activity or wave-energy density decreases with z if $m > 0$, given the convention $k > 0$ as before. Letting $\delta \to 0$ we obtain the same choice of sign of m for the conservative case as given by the group velocity argument presented above.

We note finally that when N is constant, the vertical wave number m used here is given in terms of the equivalent depth h used in Sections 4.2 and 4.3 by

$$m^2 = (N^2/gh) - 1/4H^2 \tag{4.5.19}$$

[cf. Eq. (4.3.6)]; thus m^2 equals the coefficient of W in the vertical structure equation [Eq. (4.2.7a)]. When $\bar{u} = 0$, Eq. (4.5.17) can be written

$$\frac{\omega}{k} = -\beta(k^2 + l^2 + f_0^2/gh)^{-1}$$

using Eqs. (4.5.19) and (3.2.16), and as $h \to \infty$ this approaches the spherical version of Eq. (4.2.14b) for the same limit, if the definitions $\beta = 2\Omega a^{-1}\cos\phi$ and $s = ka\cos\phi$ are used and $k^2 + l^2$ replaced by its spherical analog $n_R(n_R + 1)a^{-2}$.

4.5.3 Fluid Parcel Orbits and the Stokes Drift for Rossby Waves

As well as being a useful idealized model of a propagating planetary wave in the winter stratosphere, the solution of Eq. (4.5.13) is also convenient for didactic purposes. We use it here to illustrate some of the Lagrangian concepts mentioned in Section 3.7.

For simplicity we consider stationary waves ($c = 0$), so that the uniform basic flow $\bar{u}$ is westerly and satisfies Eq. (4.5.16); we also choose the width of the channel L to equal πl^{-1}, so that $\sin ly$ has a single peak in midchannel. If $\hat{\psi}_0$ is taken to be real, Eq. (4.5.13) becomes

$$\psi' = \hat{\psi}_0 e^{z/2H} \cos(kx + mz) \sin ly. \tag{4.5.20}$$

The geostrophic velocities are

$$u' = -\psi'_y = -l\hat{\psi}_0 e^{z/2H} \cos(kx + mz) \cos ly, \qquad (4.5.21a)$$

$$v' = \psi'_x = -k\hat{\psi}_0 e^{z/2H} \sin(kx + mz) \sin ly \qquad (4.5.21b)$$

[cf. Eq. (3.5.9)]. Using Eqs. (3.2.11) and (3.2.12) with $Q = 0$ we obtain the adiabatic quasi-geostrophic relationship

$$w_a = -D_g\left(\frac{f_0\psi_z}{N^2}\right),$$

which on linearization in the present case gives

$$w' = -\frac{\bar{u}f_0\psi'_{zx}}{N^2}, \qquad (4.5.22)$$

since the waves are stationary ($\partial/\partial t = 0$) and since $\bar{\psi}_{zy} = -\bar{u}_z = 0$. We drop the subscript a on w' for convenience.

The definitions in Eqs. (3.7.1) of parcel displacements (ξ', η', ζ') reduce to

$$\bar{u}\frac{\partial}{\partial x}(\xi', \eta', \zeta') = (u', v', w') \qquad (4.5.23a,b,c)$$

here, and so

$$\xi' = -\frac{l}{k\bar{u}}\hat{\psi}_0 e^{z/2H} \sin(kx + mz) \cos ly, \qquad (4.5.24a)$$

$$\eta' = \frac{1}{\bar{u}}\hat{\psi}_0 e^{z/2H} \cos(kx + mz) \sin ly. \qquad (4.5.24b)$$

From Eqs. (4.5.22), (4.5.23c), and (4.5.20), we obtain

$$\zeta' = -\frac{f_0\psi'_z}{N^2}$$

$$= -\frac{f_0}{N^2}\hat{\psi}_0 e^{z/2H}\left[\frac{1}{2H}\cos(kx + mz) - m\sin(kx + mz)\right]\sin ly.$$

$$(4.5.24c)$$

The "constants" of x integration have been set to zero, in accordance with Eq. (3.7.2). Trajectories or orbits of parcels correct to the linear approximation, as viewed moving with the basic zonal flow $\bar{u}$, are shown in Fig. 4.13; their projections in the xy and yz planes are ellipses.

Knowledge of the parcel displacements of Eqs. (4.5.24) allows us to calculate quantities like the Stokes drift velocities ($\bar{u}^S$, $\bar{v}^S$, $\bar{w}^S$). From Eq.

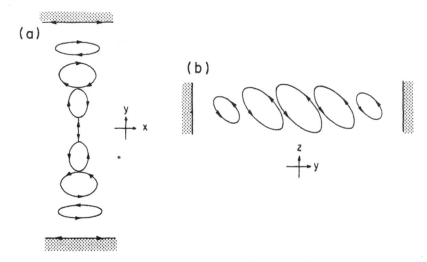

Fig. 4.13. Orbits of fluid parcels disturbed by a Rossby wave, correct to first order in wave amplitude, as viewed moving with the mean flow. (a) Projection of orbits in the xy plane. (b) Projection of orbits in the yz plane. [Adapted from Matsuno (1980), Birkhäusen Verlag AG, Basle, Switzerland.]

(3.7.7) we have $\bar{u}^S = \overline{\boldsymbol{\xi}' \cdot \boldsymbol{\nabla} u'} + O(\alpha^3)$, since $\bar{u}$ is constant to $O(\alpha^0)$; thus, using Eq. (3.7.4) and the fact that zonal averages are independent of x, we obtain

$$\bar{u}^S = \rho_0^{-1}\boldsymbol{\nabla} \cdot (\rho_0 \overline{\boldsymbol{\xi}' u'}) = (\overline{\eta' u'})_y + \rho_0^{-1}(\rho_0 \overline{\zeta' u'})_z. \qquad (4.5.25a)$$

In a rather similar manner we have

$$\bar{v}^S = (\overline{\eta' v'})_y + \rho_0^{-1}(\rho_0 \overline{\zeta' v'})_z, \qquad (4.5.25b)$$

$$\bar{w}^S = (\overline{\eta' w'})_y + \rho_0^{-1}(\rho_0 \overline{\zeta' w'})_z. \qquad (4.5.25c)$$

Using Eqs. (4.5.21), (4.5.22), and (4.5.24), we then obtain for the present case

$$\bar{u}^S = -\frac{l^2}{2\bar{u}} \hat{\psi}_0^2 e^{z/H} \cos 2ly, \qquad \bar{v}^S = 0,$$

$$\bar{w}^S = \frac{f_0 klm}{2N^2} \hat{\psi}_0^2 e^{z/H} \sin 2ly,$$

after a short calculation, which is expedited by using results like $\overline{\eta' v'} = \bar{u}\overline{\eta'\eta'_x} = \bar{u}(\frac{1}{2}\overline{\eta'^2})_x = 0$, from Eq. (4.5.23b). Thus the zonal Stokes drift $\bar{u}^S$ is directed eastward in midchannel ($L/4 < y < 3L/4$) and westward elsewhere, while the vertical Stokes drift $\bar{w}^S$ is upward in the southern half of the channel ($0 < y < L/2$) and downward in the northern half ($L/2 < y < L$).

It must be recalled, however, that the Stokes drift represents the difference between the Lagrangian-mean flow and the Eulerian-mean flow. Under the present "nonacceleration conditions," we can use the Charney–Drazin theorem of Section 3.6 to infer that $\bar{v}^* = \bar{w}^* = 0$ under appropriate boundary conditions, and hence to show that $\bar{v} = 0$ and $\bar{w} = -\bar{w}^S$ by Eq. (3.5.1) and manipulations similar to those used above. Thus $\bar{v}^L = \bar{w}^L = 0$ here, consistent with the GLM nonacceleration theorem mentioned in Section 3.7.1. However, the computation of $\bar{u}^L$ to second order in the wave amplitude α involves consideration of how the flow is initially set up, and requires a much more detailed analysis than the calculations based on the steady linear solutions that have been presented here.

4.5.4 Steady Planetary Waves in More General Basic States

For detailed comparison of linear theory with the observed middle atmosphere, the restriction to mean zonal flows $\bar{u}$ that are constant, or depend on z alone, is rather unsatisfactory. We therefore now consider basic flows $\bar{u}$ that depend on y and z, although we shall still take N, and thus ε, to be constant. The latter restriction can easily be relaxed, at the expense of a little extra algebra.

Returning to Eqs. (4.5.1)–(4.5.3), setting $Z' = 0$ and substituting the steady-wave form

$$\psi' = e^{z/2H} \operatorname{Re}\left[\Psi(y, z)e^{ik(x-ct)}\right] \tag{4.5.26}$$

we obtain

$$\Psi_{yy} + \varepsilon\Psi_{zz} + n_k^2\Psi = 0, \tag{4.5.27}$$

where

$$n_k^2(y, z) = (\bar{u} - c)^{-1}\bar{q}_y - k^2 - \varepsilon/4H^2, \tag{4.5.28}$$

(Dickinson, 1968). The quantity n_k^2 is the square of the *refractive index*, for zonal wave number k and phase speed c. In "stretched" coordinates $(\tilde{y}, \tilde{z}) \equiv (y, \varepsilon^{-1/2}z)$, Eq. (4.5.27) is identical to the equation for two-dimensional sound, or light, waves in a medium of varying refractive index, and we can use insights from the theory of acoustics or optics. In particular, we expect waves to propagate in regions where $n_k^2 > 0$ and to avoid regions where $n_k^2 < 0$. Note that n_k^2 depends on the two-dimensional structure of $(\bar{u} - c)^{-1}\bar{q}_y$, as well as on k^2, so that propagation generally depends on more complex criteria than the simple Charney–Drazin condition [Eq. (4.5.15)] that applies when $\bar{u}$ and $\bar{q}_y$ (and thus n_k^2) are constant. Note also that n_k^2 generally becomes infinite at a *critical line* (or *critical surface*) where $\bar{u}(y, z) = c$; as for the critical level, mentioned above for the case $\bar{u} = \bar{u}(z)$, such surfaces need special attention.

Given suitable boundary conditions, Eq. (4.5.27) or its spherical-geometry equivalent is readily amenable to numerical solution for realistic flows $\bar{u}(y, z)$, and a number of studies of this sort have been carried out, following Matsuno (1970). Some examples of calculations, and comparison with observed planetary waves, will be mentioned in Section 5.3; we shall here just discuss some semianalytical methods that can be used to solve special cases of Eq. (4.5.27).

The first of these is applicable when $\bar{u} = \bar{u}(y)$, so that n_k^2 depends on y alone; separable solutions $\Psi^{(l)}(y, z) = \Psi_l(y)e^{im_l z}$ to Eq. (4.5.27) can be sought, where

$$\frac{d^2\Psi_l}{dy^2} + [n_k^2(y) - \varepsilon m_l^2]\Psi_l = 0, \qquad l = 1, 2, \ldots$$

is solved as an eigenvalue problem, subject to suitable lateral boundary conditions, say $\Psi_l = 0$ at $y = 0, L$. The eigenvalues m_l^2 give the vertical structure—propagating if $m_l^2 > 0$, trapped if $m_l^2 < 0$. The analysis is similar to that of Section 4.5.2, and the response to a given lower boundary forcing can be found in terms of a sum of modes $\Psi^{(l)}$.

The assumption $\bar{u} = \bar{u}(y)$ and the imposition of vertical walls at $y = 0, L$ are, however, rather unrealistic, since they tend to confine wave propagation to the vertical, and thus inhibit latitudinal propagation. Observations of middle atmosphere planetary waves, on the other hand, suggest that latitudinal propagation is very significant in many cases: see Section 5.3. A popular approach for analyzing such cases, with n_k^2 varying with y and z, is to use an approximate theory, analogous to "geometric optics," to investigate Eq. (4.5.27). It is called the *WKBJ* or Liouville–Green method; a brief outline of the theory is given in Appendix 4A. In the present case one supposes that

$$\Psi = \hat{\Psi}(y, z) \exp i\chi(y, z)$$

where the phase χ is real, and varies much more rapidly with y and z than do the basic-flow quantities $\bar{u}$, $\bar{q}_y$, or n_k^2, the amplitude $\hat{\Psi}$, or the derivatives of χ. This is equivalent to looking for locally sinusoidal solutions whose y and z wavelengths are much less than typical y and z scales of the basic flow. The theory can be formalized, if desired, in terms of a small "*WKBJ* parameter," say μ_w, characterizing the ratio of these wavelengths to the meridional scales of the basic flow; it requires that $n_k^2 L_0^2 \geqslant O(\mu_w^{-2})$, where L_0 is a typical horizontal mean-flow scale or $\varepsilon^{-1/2}$ times a typical vertical mean-flow scale.

Under these assumptions, one can *define* local meridional wave numbers

$$l \equiv \partial\chi/\partial y, \qquad m \equiv \partial\chi/\partial z,$$

and to leading order in μ_w it is found that $\Psi_{yy} \approx -l^2\Psi$, $\Psi_{zz} \approx -m^2\Psi$. Thus Eqs. (4.5.27) and (4.5.28) yield an approximate dispersion relation for $\omega = ck$, analogous to Eq. (4.5.17):

$$\omega = \Delta(k, l, m; y, z) \equiv k\bar{u}(y, z) - k\bar{q}_y(y, z)[k^2 + l^2 + \varepsilon(m^2 + 1/4H^2)]^{-1}$$

(4.5.29)

and local group velocity components

$$c_g^{(y)} \equiv \partial\Delta/\partial l, \qquad c_g^{(z)} \equiv \partial\Delta/\partial m, \tag{4.5.30}$$

giving the local paths of propagation of information—the *rays*—in the yz plane. Standard *ray-tracing equations* can then be used to compute how the local wave numbers vary along a ray; for example, it can be shown that the variation of the vertical wavenumber along a ray is given by

$$\left[c_g^{(y)} \frac{\partial}{\partial y} + c_g^{(z)} \frac{\partial}{\partial z} \right] m = -\frac{\partial\Delta}{\partial z} \tag{4.5.31}$$

[where the z derivative on the right acts only on the terms $\bar{u}$ and $\bar{q}_y$ in Eq. (4.5.29) where z appears explicitly]. Using numerical methods, and given suitable "initial" conditions, the ray-tracing equations can be solved to find the paths of propagation through realistic stratospheric and mesospheric wind structures $\bar{u}(y, z)$. An extension of the approach also allows the calculation of amplitude variations along the rays: further details are mentioned in Appendix 4A. Because of the large latitudinal excursion of many of the rays, such calculations are best carried out with full allowance for the spherical geometry of the earth, rather than on a beta-plane. A typical example of a ray-tracing calculation is shown in Fig. 4.14a and the corresponding $\bar{u}$ and $n_k^2(y, z)$ are contoured in Figs. 4.14b,c. The most prominent feature of the rays is their tendency to be deflected up the gradient of n_k^2, and thus equatorward. This gradient depends partly on the variation of $\beta = 2\Omega a^{-1} \cos\phi$ with latitude ϕ, and is present even when $\bar{u}$ (or, rather, the angular velocity, on the sphere) is constant; however, it is greatly enhanced by variations in $(\bar{u} - c)^{-1}$ when the latter becomes large near a critical line, such as that present in low latitudes in Figs. 4.14b,c. Whether rays are absorbed or reflected at these critical lines in still a matter for investigation: see Section 5.6.

Despite the approximations involved, this *WKBJ* theory has provided useful insights into the propagation of planetary waves in the observed middle atmosphere and in more complex models such as the detailed linear numerical models described in Sections 5.3 and 6.3 and the general circulation models outlined in Chapter 11. It is also possible that an extension of the method might be of value for the investigation of disturbances to basic states that are not zonally symmetric.

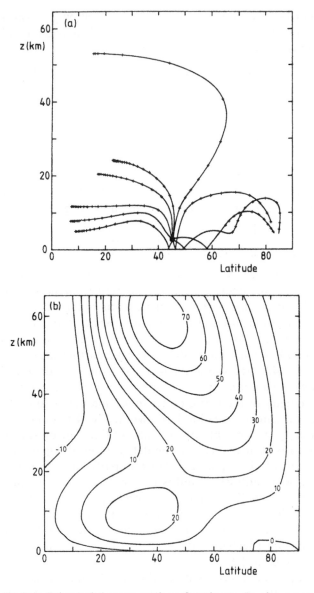

Fig. 4.14. Calculations of the propagation of stationary Rossby waves in an idealized northern hemisphere winter basic state. (a) Rays for zonal wave number 1, starting at 45°N and $z \approx 3$ km, with crosses marked at daily time intervals. (b) Basic zonal wind $\bar{u}(\phi, z)$ in m s^{-1}. (c) Spherical analog of $a^2 n_k^2 \cos^2 \phi$ for the wind field given in (b), where k corresponds to zonal wave number $s = 1$. Region of negative n_k^2 is shaded, and solid contours are spaced at unit intervals in the quantity $(a^2 n_k^2 \cos^2 \phi + 1)^{1/2}$. The closely packed contours at low latitudes indicate the presence of a critical line. [Adapted from Karoly and Hoskins (1982).]

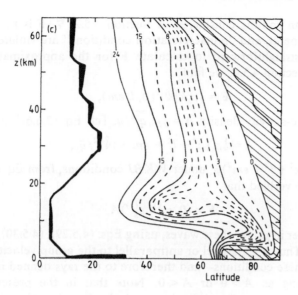

Fig. 4.14 (*continued*)

4.5.5 The Eliassen–Palm Flux for Forced Rossby Waves

The Eliassen–Palm (EP) flux was introduced in Sections 3.5 and 3.6, where its importance in the theory of wave, mean-flow interaction was described. It is of interest to calculate this quantity for the planetary waves discussed above; to this end it is convenient to rewrite the quasi-geostrophic beta-plane version of $\mathbf{F}$,

$$\mathbf{F} = (0, -\rho_0 \overline{v'u'}, \rho_0 f_0 \overline{v'\theta'} / \theta_{0z})$$

[cf. Eq. (3.5.6)], in the form

$$\mathbf{F} = (0, \rho_0 \overline{\psi'_x \psi'_y}, \rho_0 \varepsilon \overline{\psi'_x \psi'_z}), \tag{4.5.32}$$

using Eqs. (3.2.3), (3.2.12), and (3.2.16).

We first consider the simple solutions of Section 4.5.2, where $\bar{u}$ and N are constant, and the stream function has "modal" latitudinal structure, proportional to $\sin ly$. In the absence of dissipation it is easy to verify that in the "trapped" case (1) $\mathbf{F} \equiv \mathbf{0}$, while in the "vertically propagating" case (2) $\mathbf{F} = (0, 0, \frac{1}{2}\rho_s \varepsilon m k |\hat{\psi}_0|^2 \sin^2 ly)$ and points vertically upward. In each case $\nabla \cdot \mathbf{F} = \partial F^{(z)} / \partial z = 0$, as is to be expected from the Eliassen–Palm theorem [Eq. (3.6.1)], since these waves are steady, conservative, and linear; they thus induce no mean-flow acceleration (see Sections 3.6 and 4.1). On the other hand, if dissipation is included, such as the weak Rayleigh friction and Newtonian cooling considered in Section 4.5.2, $\nabla \cdot \mathbf{F}$ is found to be negative in general, and "nonacceleration conditions" are violated.

It is also straightforward to calculate $\mathbf{F}$ for the approximate *WKBJ* solutions of Section 4.5.4; we find

$$\mathbf{F} \approx \tfrac{1}{2}\rho_s k |\hat{\Psi}|^2 (0, l, \varepsilon m), \tag{4.5.33}$$

while the wave-activity density $A = \tfrac{1}{2}\rho_0 \overline{q'^2} / \bar{q}_y$ [cf. Eq. (3.6.6)] is given by

$$A \approx \tfrac{1}{4}\rho_s (k^2 + l^2 + \varepsilon m^2)^2 |\hat{\Psi}|^2 / \bar{q}_y, \tag{4.5.34}$$

since $q' \approx -(k^2 + l^2 + \varepsilon m^2)\psi'$ under *WKBJ* conditions, from Eq. (4.5.2). It can further be verified that

$$\mathbf{F} = (0, c_g^{(y)}, c_g^{(z)})A \tag{4.5.35}$$

to leading order in μ_w for these waves, using Eqs. (4.5.29), (4.5.30), (4.5.33), and (4.5.34). Thus $\mathbf{F}$ is parallel or antiparallel to the group velocity at each point under these conditions—and therefore to the rays defined in Section 4.5.4—according as $A > 0$ or $A < 0$. Note that in the present quasi-geostrophic case A has the same sign as $\bar{q}_y$, by Eqs. (3.6.6) or (4.5.34)—this is usually positive in the midlatitude stratosphere and mesosphere.

Following standard practice, the group velocity and the concept of a ray have been defined above only in the context of a *WKBJ* theory of almost-sinusoidal waves. However, Eq. (4.5.35) suggests that, in more general situations where the *WKBJ* approximations fail, one possible extended definition of the group velocity is

$$\mathbf{C} \equiv \mathbf{F}/A, \tag{4.5.36}$$

where $\mathbf{F}$ and A are given by Eqs. (3.5.6) and (3.6.6), respectively, or their primitive-equation equivalents. Generalized "rays" can then be defined as curves parallel to $\mathbf{C}$ at every point. These ideas will be used in Chapters 5 and 6 for interpreting the behavior of planetary waves in the atmosphere and in models.

4.6   Gravity Waves

It was mentioned in Section 4.1 that pure internal gravity waves owe their existence to buoyancy restoring forces, while inertio–gravity waves are due to the combined effects of buoyancy and Coriolis forces. The "gravity modes" identified in Figs. 4.2a,b, which include the main propagating tidally forced modes of Section 4.3, are, strictly speaking, examples of inertio–gravity waves, since they are generally affected to some extent by the rotation of the earth. In the present section we consider some rather simpler gravity waves, restricting attention to waves of comparatively small scale (tens to hundreds of kilometers in horizontal wavelength), so that the complications of spherical geometry can be avoided.

Gravity waves of this scale appear to be common in the upper mesosphere, where they have been detected by radars and other instruments; for example, Fig. 4.15 shows some radar observations of internal gravity waves in this region. The periods of waves of this type are typically a few minutes to an hour or so, and vertical wavelengths range from 5 to 15 km. Although direct measurements are difficult, the waves are thought to have horizontal wavelengths of up to about 100–200 km, and horizontal phase speeds of up to 80 m s^{-1}. Inertio-gravity waves, with periods approaching the local "inertial period" $2\pi f^{-1}$ and vertical wavelengths on the order of 10 km, have been detected in the upper mesosphere and also in the lower stratosphere, and these can have horizontal wavelengths of over a thousand kilometers: some observations are shown in Fig. 4.16. Internal gravity waves are also likely to be common in the lower mesosphere and the stratosphere, but observations at these levels are sparse at present.

4.6.1 Pure Internal Gravity Waves

We start by considering small wave disturbances about a basic state of rest, whose frequencies are large compared to the local inertial frequency $f = 2\Omega \sin \phi$. We therefore neglect the effects of the earth's rotation by putting $f = 0$ in Eq. (4.2.1), set $X' = Y' = J' = 0$, and employ Cartesian coordinates x and y. We obtain

$$u'_t + \Phi'_x = 0, \qquad v'_t + \Phi'_y = 0, \qquad (4.6.1a,b)$$

$$u'_x + v'_y + \rho_0^{-1}(\rho_0 w')_z = 0, \qquad \Phi'_{zt} + N^2 w' = 0. \qquad (4.6.1c,d)$$

For simplicity we take N to be constant; we then substitute

$$(u', v', w', \Phi') = e^{z/2H} \, \mathrm{Re}[(\hat{u}, \hat{v}, \hat{w}, \hat{\Phi}) \exp i(kx + ly + mz - \omega t)] \qquad (4.6.2)$$

into Eq. (4.6.1), where $\hat{u}$ etc. are constant, and obtain the following equations:

$$\hat{u} = \frac{k}{\omega} \hat{\Phi}, \qquad \hat{v} = \frac{l}{\omega} \hat{\Phi}, \qquad \hat{w} = -\frac{\omega}{N^2}\left(m - \frac{i}{2H}\right)\hat{\Phi}, \qquad (4.6.3a,b,c)$$

and the dispersion relation

$$\omega^2 = \frac{N^2(k^2 + l^2)}{m^2 + 1/4H^2}. \qquad (4.6.4)$$

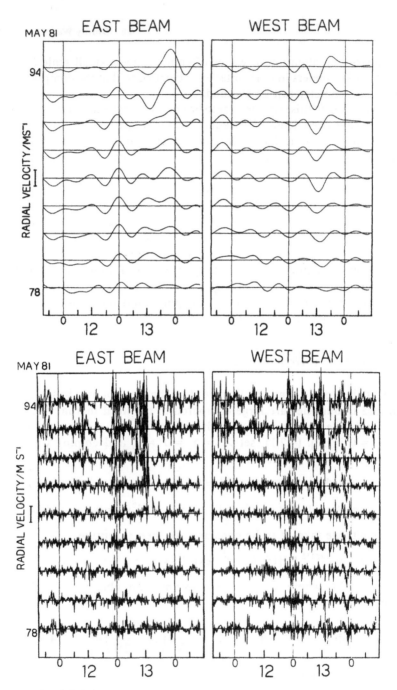

Fig. 4.15. High-frequency radar measurements of line-of-sight velocities at heights between 78 and 94 km in the upper mesosphere measured in two directions, equally inclined at small angles to the vertical. Top panels show data filtered to include only periods longer than 8 hr and bottom panels show data filtered to include only periods from 8 min to 8 hr. The data were collected during May 11–14, 1981. [After Vincent and Reid (1983).]

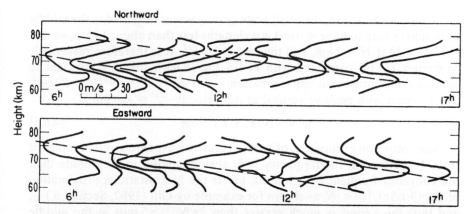

Fig. 4.16. Time sequence of northward and eastward velocity components as measured by very-high-frequency (VHF) radar during an unusually quasi-sinusoidal inertio–gravity wave event in the mesosphere, measured October 11, 1981, in Poker Flat, Alaska (1-hr average values). Velocity scale is shown in the lower left corner of the upper panel. Dashed lines indicate approximate height of velocity extrema and show downward phase propagation. Note that the left-to-right profile placements are not precisely uniform in time. [After Balsley *et al.* (1983). American Meteorological Society.]

Note that in terms of the equivalent depth h this can be written $\omega^2 = gh(k^2 + l^2)$, using Eq. (4.5.19); this is consistent with the "nonrotating" limit $\gamma^{-1/2} \to \infty$ for gravity waves given by Eq. (4.2.14a) if $k^2 + l^2$ is replaced by its spherical equivalent $n_G(n_G + 1)a^{-2}$.

Using methods analogous to those of Section 4.5.3, it can be shown that particles move in elliptical orbits (with tilted axes) in vertical planes perpendicular to the horizontal vector $(-l, k, 0)$.

It should be noted from Eqs. (4.6.2) and (4.6.3) that

$$\rho_0 \overline{u'w'} = \tfrac{1}{2}\rho_0 \, \mathrm{Re}(e^{z/H}\hat{u}\hat{w}^*) = \tfrac{1}{2}\rho_s \, \mathrm{Re}\left[\frac{k}{\omega}\hat{\Phi}\left(\frac{-\omega}{N^2}\right)\left(m + \frac{i}{2H}\right)\hat{\Phi}^*\right]$$

$$= -\tfrac{1}{2}\rho_s \frac{mk}{N^2}|\hat{\Phi}|^2$$

where an asterisk denotes the complex conjugate. Thus $\rho_0 \overline{u'w'}$ is independent of z, so that

$$\rho_0^{-1}(\rho_0 \overline{u'w'})_z = 0,$$

which is the Eliassen–Palm theorem [Eq. (3.6.1)] for the present steady, conservative, linear case, since $\mathbf{F} = (0, -\rho_0 \overline{v'u'}, -\rho_0 \overline{w'u'})$ here [cf. Eq. (3.5.3)], and depends only on z.

As mentioned above, the small-scale gravity waves observed in the middle atmosphere tend to have vertical wavelengths less than about 15 km, so that $4H^2m^2 \gtrsim 34$ if $H = 7$ km, and thus $m^2 \gg 1/4H^2$. It is therefore reasonable to neglect $1/4H^2$ compared with m^2 in Eq. (4.6.4); this is equivalent to making the "Boussinesq approximation," and gives $\omega^2 = N^2(k^2 + l^2)/m^2$. The solution with positive vertical group velocity $c_g^{(z)} \equiv \partial\omega/\partial m$ is

$$\omega = -N(k^2 + l^2)^{1/2}/m. \tag{4.6.5}$$

Numerous other properties of internal gravity waves are given in standard texts. In particular, the frequency ω is always smaller in magnitude than the buoyancy frequency N; indeed, under the hydrostatic approximation of Eq. (3.1.3c), $|\omega| \ll N$, as shown for example by Gill (1982, Section 6.14), and thus the period is much greater than $2\pi N^{-1}(\approx 5$ min in the middle atmosphere). The same inequality implies that the horizontal wavelength $2\pi(k^2 + l^2)^{-1/2}$ must be much larger than the vertical wavelength $2\pi|m|^{-1}$, by Eq. (4.6.5). Another interesting property is that if $i/2H$ is neglected in Eq. (4.6.3c)—a fair approximation for waves of vertical wavelength less than 15 km, for which $2H|m| \gtrsim 5.8$—then Eqs. (4.6.3) and (4.6.5) give $(\hat{u}, \hat{v}, \hat{w}) \cdot (k, l, m) \approx 0$, so that the velocity (u', v', w') is perpendicular to the wave vector $\mathbf{k} \equiv (k, l, m)$ and thus lies in planes of constant phase, $kx + ly + mz = $ constant. In this case parcel orbits are straight lines, also perpendicular to the wave vector.

It is often convenient to choose the horizontal axes so that $\mathbf{k} = (k, 0, m)$ and $l = 0$: this is possible since no preferred horizontal direction is imposed by the motionless basic state considered here. Then $\hat{v} = 0$ by Eq. (4.6.3b), and Eq. (4.6.5) becomes

$$\omega = -Nk/m, \tag{4.6.5'}$$

given the convention $k > 0$, as before. The vector group velocity is then

$$\mathbf{c_g} = \left(\frac{\partial\omega}{\partial k}, 0, \frac{\partial\omega}{\partial m}\right) = \frac{N}{m^2}(-m, 0, k), \tag{4.6.6}$$

and the tangent of the angle it makes with the horizontal has magnitude

$$|c_g^{(z)}/c_g^{(x)}| = |k/m| = |\omega/N|.$$

The foregoing theory can be extended to allow for a basic flow $[\bar{u}(z), \bar{v}(z), 0]$ and buoyancy frequency $N(z)$ that vary with height (although in this case one cannot generally chooose axes such that $l = 0$). When these quantities vary only on height scales much greater than a vertical wavelength, WKBJ methods analogous to those of Appendix 4A and Section 4.5.4 can be used, and ray-tracing can be carried out. Once again critical levels, at which $\omega = k\bar{u} + l\bar{v}$, need special attention.

4.6.2 A Simple Model of Breaking Gravity Waves

Owing to the presence of the $e^{z/2H}$ factor, proportional to $\rho_0^{-1/2}$, in Eq. (4.6.2), the linear, nondissipative theory of Section 4.6.1 predicts velocity and geopotential disturbances that grow with altitude; at some height the nonlinear terms that have been neglected will become important, and the linear theory will break down.

A physical picture of this breakdown can be obtained by considering a set of material surfaces at various levels, which are undulating as an internal gravity wave propagates vertically through them; Fig. 4.17 is a schematic diagram of this situation. In the lower mesosphere, say, the material surface (a) has a gentle sinusoidal variation, as predicted by linear theory. For gravity waves of period much less than a day the effects of radiative relaxation are small, and in the absence of other diabatic processes we can use an isentrope (a surface of constant θ) as the material surface (a). In the middle mesosphere the material surface (b) is also sinusoidal, but of larger amplitude than (a); linear theory still holds, and (b) can also be taken as an isentrope. In the upper mesosphere, however, nonlinear effects become important, leading to the rapid and irreversible deformation of material contours such as (c), followed by turbulence, small-scale mixing,

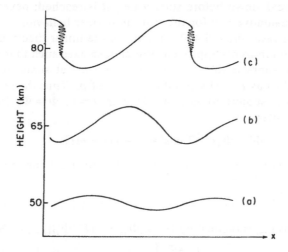

Fig. 4.17. Schematic diagram illustrating the breaking of vertically propagating internal gravity waves in the mesosphere. The curves labeled (a), (b), and (c) denote material surfaces. At the level of (a) and (b) the linear nondissipative theory of Section 4.6.1 is approximately valid. At the level of (c) nonlinear effects are important, with irreversible deformation of previously wavy material surfaces, and turbulence near the wave crests, presumably followed by small-scale mixing and dissipation.

and dissipation. Isentropes are no longer material surfaces, owing to the excitation of diabatic effects.

The process described here is known as *gravity-wave breaking*, by analogy with the overturning and breaking of oceanic surface waves on a shelving beach. It also has points in common with the phenomenon of planetary-wave breaking, described in Section 5.2.3. It will tend to limit the $e^{z/2H}$ growth of gravity wave amplitudes with height, and this has important consequences for the large-scale flow in the middle atmosphere, as will be seen below.

A simple model of this breaking process was suggested by Lindzen (1981), who considered linearized disturbances to a basic zonal flow $\bar{u}(z)$, and used a *WKBJ* method to generalize the solutions of Eq. (4.6.2). He then defined the *breaking level*, z_b, to be that altitude at which the isentropes first become vertical, with $\partial\theta/\partial z = 0$, thus implying a loss of static stability and the onset of turbulence and mixing. From Eqs. (3.2.13) and (3.4.2c)

$$\theta_z = \bar{\theta}_z + \theta_z' = HR^{-1}e^{\kappa z/H}[N^2 + \Phi_{zz}' + \kappa H^{-1}\Phi_z'], \qquad (4.6.7)$$

and since Φ', as calculated by linear theory, grows exponentially with z [cf. Eq. (4.6.2)], we expect that $-(\Phi_{zz}' + \kappa H^{-1}\Phi_z')$ will at a sufficient altitude become large enough to cancel N^2 at some values of x, y, and t, at the "crests" of the waves. At such points $\theta_z = 0$, and breaking occurs, in Lindzen's sense. This approach is not strictly self-consistent, since the linear solutions will break down before such a height is reached; nevertheless it should give a qualitative feel for the fully nonlinear behavior.

In the special case where $\bar{u} = 0$ we can use the linear theory of Section 4.6.1 to illustrate Lindzen's approach. We choose axes such that $l = 0$ and thus $v' = 0$, and suppose that $\Phi' = \Phi_0 \cos k(x - ct)$ at some lower level, which can be taken as $z = 0$ by suitable choice of p_s. Thus the solutions of Eq. (4.6.2) apply, subject to Eq. (4.6.3) with $l = 0$, $\hat{\Phi} = \Phi_0$ (real), and $\omega = ck$; in particular,

$$\Phi' = \Phi_0 e^{z/2H} \cos[k(x - ct) + mz]. \qquad (4.6.8)$$

Assuming once more that vertical wavelengths are small enough to ensure that $|m| \gg 1/2H$, we have

$$\Phi_{zz}' + \kappa H^{-1}\Phi_z' \approx -m^2\Phi' = -m^2\Phi_0 e^{z/2H} \cos[k(x - ct) + mz].$$

The breaking level z_b is defined by $\{\max|\Phi_{zz}' + \kappa H^{-1}\Phi_z'|\}_{z=z_b} = N^2$, and so

$$z_b \approx 2H \ln\left|\frac{N^2}{m^2\Phi_0}\right| = 2H \ln|c^2\Phi_0^{-1}| \qquad (4.6.9)$$

by Eqs. (4.6.8) and (4.6.5') with $c = \omega/k$. Note that z_b depends on the phase speed c and the initial geopotential amplitude Φ_0 of the wave; it increases as Φ_0 decreases because waves of small amplitude need to penetrate to greater heights before they have grown enough to break than do waves of larger amplitude.

The next step is to describe what happens above the breaking level. The turbulence that presumably sets in leads to diffusion of heat and momentum, and a crude way of parameterizing this diffusion is to modify Eqs. (4.6.1a,d) thus:

$$u'_t + \Phi'_x = Ku'_{zz}, \qquad \Phi'_{zt} + N^2 w' = K\Phi'_{zzz}, \qquad z > z_b, \qquad (4.6.10a,b)$$

where K is a constant diffusion coefficient; above z_b the continuity equation, Eq. (4.6.1c), still holds. On seeking solutions of the form

$$\Phi' = e^{z/2H} \, \mathrm{Re}[\Phi_1 \exp i(kx + m_1 z - kct)]$$

above z_b, chosen to match Eq. (4.6.8) at z_b, we find

$$\omega + im_1^2 K = -Nk/m_1 \qquad (4.6.11)$$

by analogy with (4.6.5'), where the convention $k > 0$ is retained. Thus if K is chosen so small that

$$m^2 K \ll |\omega|, \qquad (4.6.12)$$

it follows that

$$m_1 \approx -\frac{Nk}{\omega} + \frac{iKN^3 k^3}{\omega^4} = m + \frac{iKN^3}{c^4 k}$$

and so

$$\Phi' = \Phi_0 \exp\left[\frac{z}{2H} - \frac{KN^3}{c^4 k}(z - z_b)\right] \cos[k(x - ct) + mz], \qquad z > z_b, \qquad (4.6.13)$$

to satisfy continuity with Eq. (4.6.8) at z_b. The extra exponential factor involving K results from the postulated diffusive damping of the waves due to breaking. Lindzen hypothesizes that above z_b the waves are *saturated*, or just on the verge of breaking; thus $\max|\Phi'_{zz} + \kappa H^{-1}\Phi'_z| = N^2$ for all $z \geq z_b$. Hence the diffusive decay must be such as to exactly balance the $e^{z/2H}$ growth, and K must be chosen such that the coefficient of z in the exponential term in Eq. (4.6.13) vanishes, that is,

$$K = c^4 k/2HN^3. \qquad (4.6.14)$$

Note that this implies

$$m^2 K = \frac{N^2 k^2}{\omega^2}\left(\frac{c^4 k}{2HN^3}\right) = \frac{c^2 k}{2HN} = \left|\frac{\omega}{2Hm}\right|$$

by Eq. (4.6.5′), and thus Eq. (4.6.12) is consistent with our previous assumption that $|m| \gg 1/2H$. Substitution of Eq. (4.6.14) into Eq. (4.6.13) gives

$$\Phi' = \Phi_0 e^{z_b/2H} \cos[k(x - ct) + mz] \qquad \text{for} \quad z \geqslant z_b, \qquad (4.6.15a)$$

while Eqs. (4.6.10a,b) yield

$$(u', w') = (1, -k/m)c^{-1}\Phi_0 e^{z_b/2H} \cos[k(x - ct) + mz] \qquad \text{for} \quad z \geqslant z_b$$
$$(4.6.15b,c)$$

at leading order in the small parameter $m^2 K/\omega$. Thus Lindzen's saturation hypothesis implies no further growth of wave amplitude above z_b: this is roughly consistent with observed gravity-wave amplitudes in the mesosphere. Note, incidentally, that $\max|u'| = |c|$ for $z \geqslant z_b$, from Eqs. (4.6.15b) and (4.6.9).

We now calculate the quantity

$$\bar{X}_1 \equiv -\rho_0^{-1}(\rho_0 \overline{u'w'})_z; \qquad (4.6.16)$$

using Eqs. (4.6.15b,c), (4.6.9), (4.6.5′), and (4.6.14) it follows that

$$\bar{X}_1 = c^3 k/2NH = \frac{N^2 K}{c} \qquad \text{for} \quad z > z_b, \qquad (4.6.17)$$

which is a nonzero constant in general. Thus if $c > 0$ there is a zonal-mean vertical wave momentum flux convergence or Eliassen–Palm flux divergence above the breaking level (and vice versa if $c < 0$), and this implies a contribution to the wave-induced forcing of the zonal-mean flow. Lindzen also postulates that the momentum and heat diffusion represented by K acts on the zonal-mean flow, as well as on the gravity waves.

As mentioned above, Lindzen's method is more general than that given here, in that it includes a slowly varying mean flow $[\bar{u}(z), 0, 0]$ and nonzero meridional wave number l. The generalization of Eq. (4.6.17) is

$$\bar{X}_1 = \frac{(c - \bar{u})^3 k}{2N} \left[\frac{1}{H} + \frac{3 \, d\bar{u}/dz}{c - \bar{u}} \right] = \frac{N^2 K}{c - \bar{u}} \qquad (4.6.18a,b)$$

when $l = 0$ but $\bar{u} \neq 0$. On substituting typical wave parameters, such as those given at the beginning of Section 4.6, into these formulas, it is found that Lindzen's parameterization of the possible frictional effects due to breaking gravity waves implies a strong forcing of the zonal-mean circulation of the upper mesosphere, in accordance with observations that indicate that $\bar{X}_1$ may be on the order of several tens of meters per second per day (see Fig. 4.18). This topic will be discussed further in Sections 7.3 and 8.5.

It is also possible to allow for the radiative damping of gravity waves in the above theory. Since damping tends to decrease the amplitude of the

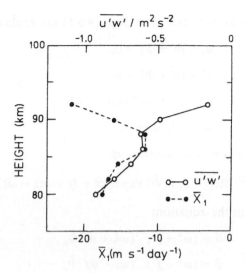

Fig. 4.18. Height profiles of $\overline{u'w'}$ and $\bar{X}_1$, in the upper mesosphere, derived from double-beam radar measurements in May 1981. [After Vincent and Reid (1983).]

upward-propagating waves, the breaking level tends to be raised; indeed, waves of sufficiently small intrinsic phase speed may not break at all. However, if they do break, the radiative damping has little effect on the resulting values of $\bar{X}_1$.

It has been emphasized that Lindzen's model is highly simplified, being essentially based on a linear theory of monochromatic gravity waves. Much more observational and theoretical work needs to be done to investigate the validity of the model and to understand the detailed nonlinear dynamics of a complex spectrum of breaking gravity waves (for which a "breaking amplitude" is more appropriate than a "breaking level") and their effects on the mean flow.

4.6.3 Inertio–Gravity Waves

On somewhat larger space and time scales (horizontal wavelengths ~1000 km and periods of several hours) than those considered in Section 4.6.1, gravity waves will be influenced by the rotation of the earth, and the theory given there will need modification. As a simple example that avoids the complexity of the spherical geometry used in Section 4.2, we consider small disturbances to a state of rest on an "f-plane," in which the Coriolis parameter f is taken as constant. Then Eqs. (4.6.1) are replaced by

$$u'_t - fv' + \Phi'_x = 0, \tag{4.6.19a}$$

$$v'_t + fu' + \Phi'_y = 0, \tag{4.6.19b}$$

$$u'_x + v'_y + \rho_0^{-1}(\rho_0 w')_z = 0, \tag{4.6.19c}$$

$$\Phi'_{zt} + N^2 w' = 0. \tag{4.6.19d}$$

Taking N constant again, and substituting

$$(u', v', w', \Phi') = e^{z/2H} \ \mathrm{Re}[(\hat{u}, \hat{v}, \hat{w}, \hat{\Phi}) \exp i(kx + ly + mz - \omega t)] \tag{4.6.20}$$

once more, we obtain the equations

$$\hat{u} = (\omega^2 - f^2)^{-1}(\omega k + ilf)\hat{\Phi}, \tag{4.6.21a}$$

$$\hat{v} = (\omega^2 - f^2)^{-1}(\omega l - ikf)\hat{\Phi}, \tag{4.6.21b}$$

$$\hat{w} = -\frac{\omega}{N^2}\left(m - \frac{i}{2H}\right)\hat{\Phi}, \tag{4.6.21c}$$

and the dispersion relation

$$\omega^2 = f^2 + \frac{N^2(k^2 + l^2)}{m^2 + 1/4H^2}. \tag{4.6.22}$$

These clearly reduce to Eqs. (4.6.3) and (4.6.4) when $f = 0$.

Some important properties of inertio–gravity waves follow from these results. Note first from Eq. (4.6.22) that the existence of propagating waves, with k, l, m all real, requires that

$$|f| \leqslant |\omega| \ll N, \tag{4.6.23}$$

where the right-hand inequality results from the hydrostatic approximation, as in Section 4.6.1. The left-hand inequality shows that the frequency of inertio–gravity waves is greater than the Coriolis parameter; this is a qualitative explanation of the confinement of propagating diurnal tides ($|\omega| = \Omega$) to the region where $|f| \equiv 2\Omega|\sin \phi| \leqslant \Omega$, that is, equatorward of 30° latitude (see Section 4.3.3). On the other hand, since the present analysis does not take account of the global variation of f, a detailed comparison of Eq. (4.6.22) with the dispersion curves plotted in Figs. 4.2a,b is difficult, except in the "nonrotating" limit $\gamma^{-1/2} \to \infty$ mentioned in Section 4.6.1. (A better comparison between the full spherical-geometry theory and a simpler analysis occurs as $\gamma^{-1/2} \to 0$ and the modes become equatorially trapped: see Section 4.7.)

From Eq. (4.6.22) we can obtain the group velocity; we choose axes such that $l = 0$ as before, and use the approximation $m^2 \gg 1/4H^2$. Then

$$\omega = \pm(f^2 + N^2 k^2 / m^2)^{1/2} \qquad (4.6.24)$$

and

$$\mathbf{c}_g = \frac{N^2 k}{m^3 \omega}(m, 0, -k); \qquad (4.6.25)$$

moreover,

$$|c_g^{(z)}/c_g^{(x)}| = |k/m| = (\omega^2 - f^2)^{1/2}/N, \qquad (4.6.26)$$

and this is generally smaller than for pure internal gravity waves [cf. Eq. (4.6.6)] since k is smaller, while m is about the same. Thus inertio-gravity waves will tend to propagate more horizontally than do pure internal waves, other things being equal.

One way of identifying the sign of $c_g^{(z)}$ from measurements of an inertio-gravity wave is to determine the sense of rotation of the horizontal velocity vector $(u', v', 0)$ with height. Taking $l = 0$ in Eqs. (4.6.21a,b) and setting $\hat{u}$ real (by a suitable choice of the time origin, say) we obtain

$$\tan \xi \equiv v'/u' = f\omega^{-1} \tan(kx + mz - \omega t),$$

where ξ is the angle between $(u', v', 0)$ and the positive x axis. Then, differentiating with respect to z,

$$\sec^2 \xi \frac{\partial \xi}{\partial z} = f\omega^{-1} m \sec^2(kx + mz - \omega t)$$

so that $\partial \xi/\partial z$ has the same sign as $f\omega^{-1}m$; by Eq. (4.6.25) it follows that

$$\mathrm{sgn}(\partial \xi/\partial z) = -\mathrm{sgn}(fc_g^{(z)}).$$

Thus the horizontal velocity vector rotates anticyclonically with height (clockwise in the northern hemisphere, anticlockwise in the southern hemisphere) for upward-propagating waves $(c_g^{(z)} > 0)$, and cyclonically for downward-propagating waves $(c_g^{(z)} < 0)$. [The same result can be proved when $l \neq 0$ by defining ξ as the angle between $(u', v', 0)$ and the horizontal wave vector $(k, l, 0)$.] In practice, this method is usually applied to measurements of inertio-gravity-wave spectra, rather than single waves.

The orbits of fluid parcels in inertio-gravity waves can be calculated by the methods of Section 4.5.3. In particular, when $|m| \gg 1/2H$ parcels move anticyclonically in ellipses in planes perpendicular to the wave vector (k, l, m): see Fig. 4.19.

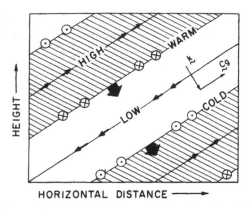

HORIZONTAL DISTANCE ——————▶

Fig. 4.19. Vertical section in a plane containing the wave vector **k** showing the phase relationships between velocity, geopotential, and temperature ($\propto \Phi'_z$) fluctuations in an upward-propagating inertio–gravity wave, with $|m| \gg 1/2H$, $m < 0$, $\omega > 0$ in the northern hemisphere ($f > 0$). The thin sloping lines denote the surfaces of constant phase (perpendicular to the wave vector) and thick arrows show the direction of phase propagation. The wave vector and group velocity are also shown. The same diagram applies to the pure internal gravity waves of Section 4.6.1 and the Kelvin waves of Section 4.7.1 if the arrows indicating velocity components into and out of the page are omitted, and if "horizontal" is taken to mean "eastward" in the Kelvin wave case.

4.7 Equatorial Waves

It has been known for many years that large-scale, equatorially confined wave motions propagate vertically and zonally through the middle atmosphere. These waves have periods of a few days and are of planetary scale in the zonal direction, but are trapped within about 15° north and south of the equator. The earliest observations of such waves were from radiosondes in the lower stratosphere, and more recently rocketsondes and satellites have detected them in the upper stratosphere and mesosphere as well; these observations will be discussed in Section 4.7.5. Waves of this kind are particularly significant for the dynamics of the middle atmosphere, since they are believed to play a central role in forcing the equatorial quasi-biennial oscillation (QBO) and semiannual oscillation (SAO): see Chapter 8.

A natural starting point for a theory of these equatorial waves is the fact, mentioned in Section 4.2, that as $\gamma^{-1/2} \equiv (gh)^{1/2}/2\Omega a \to 0$, all the Hough modes become confined near the equator; this is evident for example in Figs. 7–13 of Longuet-Higgins (1968). The latitudinal confinement suggests that a beta-plane approximation may be satisfactory for studying these modes, and we shall adopt such an approach in this section, so as to avoid the complexities of spherical geometry. The beta-plane will be centered at the equator so that $f_0 = 0$ and $f = \beta y$, where $\beta = 2\Omega a^{-1}$ [cf. Eq. (3.2.1f)] and y is distance north of the equator. The vanishing of f at the equator suggests that the quasi-geostrophic equations will not generally be valid: for example, the condition of Eq. (3.2.8a) on the Rossby number will usually be violated. We therefore work with the primitive equations, which, when linearized about a basic zonal flow $\bar{u}(y, z)$, become

$$\bar{D}u' + (\bar{u}_y - \beta y)v' + \bar{u}_z w' + \Phi'_x = X', \qquad (4.7.1a)$$

$$\bar{D}v' + \beta y u' + \Phi'_y = Y', \qquad (4.7.1b)$$

$$\Phi'_z = H^{-1}R\theta' e^{-\kappa z/H}, \qquad (4.7.1c)$$

$$u'_x + v'_y + \rho_0^{-1}(\rho_0 w')_z = 0, \qquad (4.7.1d)$$

$$\bar{D}\theta' + \bar{\theta}_y v' + \bar{\theta}_z w' = Q', \qquad (4.7.1e)$$

where $\bar{D} = \partial/\partial t + \bar{u}\,\partial/\partial x$. These equations follow from Eqs. (3.4.2) on setting $f = \beta y$, $\cos \phi = 1$, $\tan \phi = 0$, and using Cartesian coordinates (x, y), where the x axis points eastward along the equator. The thermal wind equation for the basic flow,

$$\beta y \bar{u}_z = -H^{-1}R\bar{\theta}_y e^{-\kappa z/H}, \qquad (4.7.2)$$

follows similarly from Eq. (3.4.1c).

Aspects of the roles of the frictional and diabatic terms X', Y', and Q' in forcing and dissipating equatorial waves will be discussed in Sections 4.7.3 and 4.7.4. However, the simplest way of deriving the basic theoretical equatorial wave structures is to ignore such nonconservative processes for the moment by setting $X' = Y' = Q' = 0$, and also to neglect the complicating effects of the basic wind shear by setting $\bar{u} = 0$ and thus $\bar{\theta}_y = 0$, by Eq. (4.7.2). (The effects of a constant nonzero $\bar{u}$ will be mentioned later.) The disturbances then satisfy

$$u'_t - \beta y v' + \Phi'_x = 0, \qquad (4.7.3a)$$

$$v'_t + \beta y u' + \Phi'_y = 0, \qquad (4.7.3b)$$

$$u'_x + v'_y + \rho_0^{-1}(\rho_0 w')_z = 0, \qquad (4.7.3c)$$

$$\Phi'_{zt} + N^2 w' = 0, \qquad (4.7.3d)$$

where Eqs. (4.7.1c,e) have been combined to give Eq. (4.7.3d), as in Eq. (4.2.1), and $N^2(z) = H^{-1}R\bar{\theta}_z e^{-\kappa z/H}$ as before. For simplicity we set $N =$ constant, and then seek solutions in the form

$$(u', v', w', \Phi') = e^{z/2H}\, \mathrm{Re}\{[\hat{u}(y), \hat{v}(y), \hat{w}(y), \hat{\Phi}(y)]\exp i(kx + mz - \omega t)\}.$$

$$(4.7.4)$$

We obtain

$$\hat{w} = -\frac{\omega}{N^2}\left(m - \frac{i}{2H}\right)\hat{\Phi} \qquad (4.7.5)$$

from Eq. (4.7.3d) and

$$-i\omega\hat{u} - \beta y\hat{v} + ik\hat{\Phi} = 0, \qquad (4.7.6a)$$

$$-i\omega\hat{v} + \beta y\hat{u} + \hat{\Phi}_y = 0, \qquad (4.7.6b)$$

$$ik\hat{u} + \hat{v}_y - i\omega m^2 N^{-2}\hat{\Phi} = 0, \qquad (4.7.6c)$$

from Eqs. (4.6.3a,b,c) and (4.7.5). In Eq. (4.7.6c) a factor $m^2 + (1/4H^2)$ [which by Eq. (4.5.19) equals the N^2/gh of classical tidal theory] has been replaced by m^2. This "Boussinesq" approximation is a reasonable one for many of the observed equatorial waves that, like the gravity waves of Section 4.6, have vertical wavelengths less than about 15 km, so that $4H^2m^2 \geqslant 34$. The approximation is more doubtful for the "fast" Kelvin waves observed in the upper stratosphere, which may have vertical wavelengths of about 40 km; however, the theory is easily reworked with the inclusion of the $1/4H^2$ term.

It will be observed that two y derivatives appear in Eqs. (4.7.6): calculation of the latitudinal wave structures therefore generally involves the solution of a second-order ordinary differential equation in y. This equation, Eq. (4.7.11), is discussed in Section 4.7.2; we first consider a special case in which only a first-order ordinary differential equation need be solved.

4.7.1 The Kelvin Wave

Among the equatorially trapped waves that are observed in the stratosphere and mesosphere (see Section 4.7.5) is a class of modes with small meridional wind component v'. As a first attempt to model these waves we look for solutions to Eqs. (4.7.6) in which $\hat{v}$ is identically zero; thus $\hat{u}$ and $\hat{\Phi}$ must satisfy

$$-\omega\hat{u} + k\hat{\Phi} = 0, \qquad \beta y\hat{u} + \hat{\Phi}_y = 0, \qquad k\hat{u} - \omega m^2 N^{-2}\hat{\Phi} = 0. \qquad (4.7.7a,b,c)$$

If $\hat{u}$ and $\hat{\Phi}$ do not vanish, Eqs. (4.7.7a,c) immediately give $\omega = \pm Nkm^{-1}$, with vertical group velocity $c_g^{(z)} \equiv \partial\omega/\partial m = \mp Nkm^{-2}$. Anticipating that the root with positive $c_g^{(z)}$ will be the relevant one for the middle atmosphere, corresponding to a wave that propagates upward from the troposphere, we therefore have

$$\omega = -Nk/m \qquad (4.7.8a)$$

and

$$c_g^{(z)} = Nk/m^2 = \omega^2/Nk \qquad (4.7.8b)$$

if k is chosen positive, by our usual convention. The dispersion relation, Eq. (4.7.8a), is identical to that for pure internal gravity waves with $l = 0$; see Eq. (4.6.5′).

The meridional structure of this wave solution can be found by eliminating $\hat{u}$ from Eqs. (4.7.7a,b) to obtain the first-order ordinary differential equation

$$\hat{\Phi}_y + k\beta\omega^{-1}y\hat{\Phi} = 0,$$

which has the solution

$$\hat{\Phi}(y) = \hat{\Phi}_0 \exp(-\beta k y^2/2\omega), \tag{4.7.9}$$

where $\hat{\Phi}_0$ is a constant. If $\hat{\Phi}$ and $\hat{u}$ are to be bounded far from the equator, where $|y|$ becomes large, it is necessary that the coefficient of y^2 in the exponent in Eq. (4.7.9) be negative, and thus that $c = \omega/k$ be positive: this mode therefore has an eastward zonal phase speed and, by Eq. (4.7.8a), a negative value of m. The surfaces of constant phase $(kx + mz - \omega t)$ thus tilt eastward with height and move downward with time, as indicated in Fig. 4.19; a plan view of the meridional structure is shown in Fig. 4.20. The wave solution is called an *equatorial Kelvin wave*; its structure in the xz plane is analogous to that of an internal gravity wave, and the y variation given by Eq. (4.7.9) allows geostrophic balance to hold (exceptionally) right up to the equator.

A similar analysis can be performed in the case when $\bar{u}$ is constant and nonzero: the same formulas hold except that the absolute frequency ω is replaced by the intrinsic, or Doppler-shifted, frequency

$$\omega^+ \equiv \omega - k\bar{u}. \tag{4.7.10}$$

Use of this crude allowance for a basic zonal wind gives quite good agreement between theory and observation, as will be mentioned in Section 4.7.5. In particular $\omega^+/k > 0$, so that the absolute phase speed ω/k must be eastward with respect to the basic flow, and this is in accord with observations of Kelvin waves.

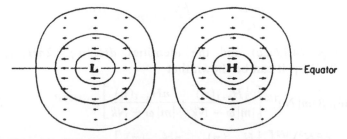

Fig. 4.20. Schematic illustration of geopotential and horizontal wind fluctuations for the Kelvin wave. [Adapted from Matsuno (1966).]

4.7.2 Modes with Nonzero Meridional Velocity

We now consider the case $\hat{v} \neq 0$, and return to Eq. (4.7.6). On eliminating $\hat{u}$ and $\hat{\Phi}$ from Eqs. (4.7.6a,c) and substituting in Eq. (4.7.6b), we obtain

$$\left[\frac{d^2}{dy^2} + \left(\frac{m^2\omega^2}{N^2} - k^2 - \frac{k\beta}{\omega}\right) - \frac{\beta^2 m^2}{N^2} y^2\right]\hat{v} = 0, \qquad (4.7.11)$$

provided that

$$m^2\omega^2 \neq N^2 k^2. \qquad (4.7.12)$$

The substitutions

$$\eta \equiv \left(\frac{\beta|m|}{N}\right)^{1/2} y, \qquad M \equiv \frac{N}{\beta|m|}\left(\frac{m^2\omega^2}{N^2} - k^2 - \frac{k\beta}{\omega}\right), \qquad (4.7.13a,b)$$

where $|m|$ is assumed nonzero, allow Eq. (4.7.11) to be reduced to the dimensionless form

$$\left(\frac{d^2}{d\eta^2} + M - \eta^2\right)\hat{v} = 0, \qquad (4.7.13c)$$

which also arises in the theory of the quantum harmonic oscillator and has the solutions

$$\hat{v} = \hat{v}_0 e^{-(1/2)\eta^2} H_n(\eta)$$

if $M = 2n + 1$, where n is a nonnegative integer, the H_n are the Hermite polynomials ($H_0 = 1$, $H_1 = 2\eta$, $H_2 = 4\eta^2 - 2$, etc.) and $\hat{v}_0$ is constant. Thus Eq. (4.7.11) has solutions

$$\hat{v} = \hat{v}_0 \exp(-\beta|m|y^2/2N) H_n[(\beta|m|N^{-1})^{1/2}y], \qquad (4.7.14a)$$

provided that

$$\frac{m^2\omega^2}{N^2} - k^2 - \frac{\beta k}{\omega} = (2n + 1)\frac{\beta|m|}{N}. \qquad (4.7.15)$$

Using Eqs. (4.7.6a,c) and the identities $dH_n/d\eta = 2nH_{n-1}$, $H_{n+1} = 2\eta H_n - 2nH_{n-1}$, it can be shown that

$$\hat{u} = i\hat{v}_0(\beta|m|N)^{1/2}\left[\frac{\frac{1}{2}H_{n+1}(\eta)}{|m|\omega - Nk} + \frac{nH_{n-1}(\eta)}{|m|\omega + Nk}\right]e^{-(1/2)\eta^2}, \qquad (4.7.14b)$$

$$\hat{\Phi} = i\hat{v}_0\left(\frac{\beta N^3}{|m|}\right)^{1/2}\left[\frac{\frac{1}{2}H_{n+1}(\eta)}{|m|\omega - Nk} - \frac{nH_{n-1}(\eta)}{|m|\omega + Nk}\right]e^{-(1/2)\eta^2}, \qquad (4.7.14c)$$

and $\hat{w}$ follows from Eqs. (4.7.5) and (4.7.14c). These solutions are trapped near the equator, with a latitudinal decay scale of order $(2N/\beta|m|)^{1/2}$; for a vertical wavelength of 10 km, this is approximately 1660 km or 15 degrees of latitude. [The same scale $(2N/\beta|m|)^{1/2}$ applies for Kelvin waves, as can be seen from Eqs. (4.7.9) and (4.7.8a) and the fact that $m < 0$ in this case.] The "turning point," at which $\hat{v}$ changes from oscillatory behavior to exponential decay, is seen from Eq. (4.7.13c) to occur at $\eta = \pm M^{1/2}$, that is, $y = \pm[(2n + 1)N/\beta|m|]^{1/2}$. Note that the solutions, Eqs. (4.7.9) and (4.7.14c), are the equatorial beta-plane analogs of the Hough functions of Section 4.2 in the limit $\gamma^{-1/2} \to 0$.

The gravest of the modes with nonzero $\hat{v}$ is of particularly simple form: setting $n = 0$ in Eq. (4.7.15) we obtain

$$(|m|\omega - Nk)(|m|\omega + Nk) = \beta\omega^{-1}N(|m|\omega + Nk);$$

but by Eq. (4.7.12), $|m|\omega + Nk \neq 0$, so this factor can be canceled to give

$$|m| = \frac{N}{\omega^2}(\beta + \omega k) \tag{4.7.16}$$

as the dispersion relation for the $n = 0$ mode; since $|m|$ is positive it follows that

$$c \equiv \omega/k > -\beta/k^2. \tag{4.7.17}$$

Equation (4.7.16) can be written $m = \pm N\omega^{-2}(\beta + \omega k)$, so that

$$c_g^{(z)} = \frac{\partial\omega}{\partial m} = \left(\frac{\partial m}{\partial\omega}\right)^{-1} = \frac{\mp\omega^3}{N(2\beta + \omega k)}. \tag{4.7.18}$$

The denominator of the last term in Eq. (4.7.18) is positive, by Eq. (4.7.17), and thus the choice of sign for upward group velocity $c_g^{(z)}$ depends on the sign of ω. If $-\beta/k < \omega < 0$ the upper sign applies, while if $\omega > 0$ the lower sign applies; the dispersion relation [Eq. (4.7.16)] therefore becomes

$$m = -\text{sgn}(\omega)\frac{N}{\omega^2}(\beta + \omega k). \tag{4.7.16'}$$

Since $H_0 = 1$, the solution is found to be

$$(\hat{u}, \hat{v}, \hat{\Phi}) = \hat{v}_0\left(\frac{i|m|\omega y}{N}, 1, i\omega y\right)\exp\left(\frac{-\beta|m|y^2}{2N}\right); \tag{4.7.19}$$

this solution is called the *Rossby-gravity wave*; its structures in the xz and xy planes are illustrated in Figs. 4.21 and 4.22. It approximately corresponds to an observed stratospheric wave disturbance if allowance is again made for a basic flow $\bar{u}$ by replacing ω by $\omega^+ = \omega - k\bar{u}$. The observed wave has

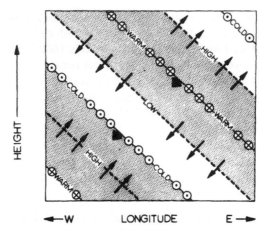

← W   LONGITUDE   E →

Fig. 4.21. Longitude–height section at a latitude north of the equator, showing geopotential, temperature, and wind fluctuations in the Rossby-gravity wave of westward phase speed $(-\beta/k < \omega < 0)$ when $|m| \gg 1/2H$. Northward winds are indicated by arrows into the page and southward winds by arrows out of the page. Thick arrows indicate direction of phase propagation. [After Holton (1975). American Meteorological Society.]

a westward phase speed with respect to the mean flow, so that $\omega^+/k < 0$: see Section 4.7.5.

The equatorial modes for $n \geq 1$ have more complex meridional structure than the Kelvin and Rossby-gravity modes. Their dispersion relation, Eq. (4.7.15), is a quadratic in $|m|$ if $k(>0)$, ω and n are given, and the solutions fall into two categories. First, there is a set of high-frequency *equatorial inertio-gravity waves*, with dispersion relations

$$m = -\mathrm{sgn}(\omega)N\beta\omega^{-2}\{(n + \tfrac{1}{2}) + [(n + \tfrac{1}{2})^2 + \omega k\beta^{-1}(1 + \omega k\beta^{-1})]^{1/2}\}, \quad (4.7.20)$$

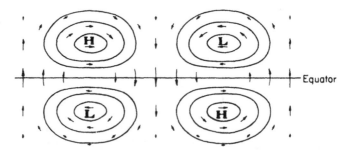

Fig. 4.22. Schematic illustration of geopotential and horizonal wind fluctuations for the Rossby-gravity wave of westward phase speed. [Adapted from Matsuno (1966).]

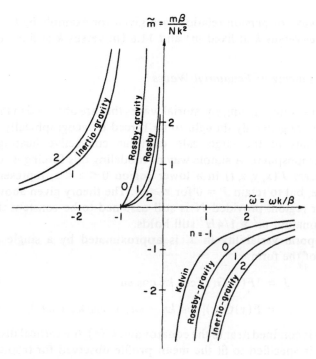

Fig. 4.23. Dispersion curves for upward-propagating equatorial waves. See text for details.

which occur for all values of ω. Second, there is a set of low-frequency *equatorial Rossby waves*, with dispersion relations

$$m = N\beta\omega^{-2}\{(n + \tfrac{1}{2}) - [(n + \tfrac{1}{2})^2 + \omega k\beta^{-1}(1 + \omega k\beta^{-1})]^{1/2}\}, \qquad (4.7.21)$$

which only occur for $-\beta k^{-2} < c \equiv \omega/k < 0$. (This condition on the phase speed also happens to be a corollary of the Charney–Drazin condition [Eq. (4.5.15)] for midlatitude Rossby waves when $\bar{u} = 0$, since $\bar{u}_c < \beta k^{-2}$.) In Eqs. (4.7.20) and (4.7.21) we have given only the solutions corresponding to an upward group velocity. Dispersion curves are presented in Fig. 4.23 in terms of the dimensionless parameters $\tilde{m} \equiv m\beta/Nk^2$ and $\tilde{\omega} \equiv \omega k/\beta$, for several values of $n \geq 1$. Also plotted are the dispersion curves for the upward-propagating Rossby-gravity wave ($n = 0$) with the dispersion relation of Eq. (4.7.16') and Kelvin wave (often designated by the index $n = -1$) with the dispersion relation of Eq. (4.7.8a).[3] Other ways of plotting the

---

[3] The index n, which has been used here to distinguish the various equatorial wave modes, corresponds to n_E appearing at the left-hand end of the curves in Figs. 4.2a,b.

equatorial-wave dispersion relations are given for example by Gill (1982), Figs. 11.1 (ω versus k at fixed m) and 11.8 (m versus k at fixed ω).

4.7.3 The Forcing of Equatorial Waves

The upward-propagating equatorial waves that are observed in the middle atmosphere are generally thought to be forced by geographically confined time variations in the large-scale cumulus convective heating in the equatorial troposphere. A simple way of modeling this forcing is to include a heating term $J'(x, y, z, t)$ in a lower region $0 \leq z < z_1$ representing the troposphere, but to retain $J' = 0$ for $z \geq z_1$. The theory given above relates to the latter region, provided N is still assumed to be constant there and the approximation $m^2 \gg 1/4H^2$ still holds.

The temporal variability in J' is approximated by a single standing oscillation of the form

$$J' = 2F(y)G(z) \cos kx \cos \omega t \qquad (4.7.22a)$$

$$= F(y)G(z)[\cos(kx - \omega t) + \cos(kx + \omega t)], \qquad (4.7.22b)$$

where $F(y)$ is confined near to the equator and $G(z)$, the vertical distribution of heating, is specified to fit the mean profile observed for tropical cloud clusters (Fig. 4.24); it vanishes for $z \geq z_1$. Equation (4.7.22b) shows that the standing oscillation can be represented as a superposition of eastward and westward moving forcing terms. If $\kappa J'/H$ is now included on the right of Eq. (4.7.3d), Eqs. (4.7.3) are analogous to Eqs. (4.2.1) of classical tidal theory, except that the equatorial beta-plane is used, rather than spherical geometry. By linearity, the responses to the eastward and westward traveling forcing in Eq. (4.7.22b) can be found separately, and then summed. The method is similar to that of tidal theory: for the specified k and ω, the

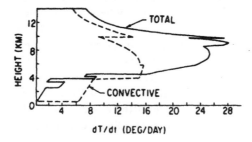

Fig. 4.24. Observed vertical profile of total large-scale diabatic heating by mature cloud clusters in the tropical troposphere (solid curve). Heating by convective towers alone (dashed curve) is shown for comparison. [After Houze (1982).]

relevant modes and their vertical wave numbers m can be read off from a diagram like Fig. 4.23. The function $F(y)$ is then expanded in the geopotential eigenfunctions associated with these modes; since the latter are limiting cases of the Hough functions, they form a complete set. The response, say in $\hat{w}$, is similarly summed, and the z-dependent amplitudes of the eigenfunctions found by solving a vertical structure equation like Eq. (4.3.4), but with $N^2/gh_n - (1/4H^2)$ replaced by m^2 for the relevant mode. In detailed calculations, the non-Boussinesq version of the theory might be needed if values of m^2 arise that are not much larger than $1/4H^2$.

A simple model of this form has been used by Chang (1976) to explain the observed distribution of Kelvin waves. In general, however, a more detailed calculation is probably needed, to examine the mechanism of the excitation of equatorial waves. For example, Holton (1972) used a model involving a longitudinally localized heat source and a basic zonal shear flow $\bar{u}(z)$ to account for the observed Rossby-gravity waves. Further details of these studies are given in Section 4.7.5. Strong support for the hypothesis that both Kelvin and Rossby-gravity waves are indeed forced by tropical convective heating has been provided by Hayashi and Golder (1978) by means of controlled experiments with a sophisticated nonlinear general circulation model.

4.7.4 WKBJ Theory of Dissipating Equatorial Waves in a Shear Flow

As mentioned above, the theory given in Sections 4.7.1 and 4.7.2 immediately extends to the case in which a constant basic zonal flow is present but dissipation is neglected. However, the influence of basic shear, particularly vertical shear $\partial\bar{u}/\partial z$, is believed to account for several important features of the observed Kelvin and Rossby-gravity waves. Together with thermal and perhaps mechanical dissipation, it is also an essential component of the models of the quasi-biennial oscillation mentioned in Chapter 8.

The most detailed treatments of the effects of a basic shear flow $\bar{u}(y, z)$ on the propagation of equatorial waves in the presence of dissipation necessarily involve numerical solution of Eqs. (4.7.1). For many purposes, though, a *WKBJ* approach (cf. Appendix 4A) is sufficient for gaining physical insight into the effects of shear. This theory is still quite complicated, and we shall only quote some basic results here.

For simplicity we consider the case where $\bar{u}$ depends on z alone. The *WKBJ* assumption then requires that the basic shear is weak, in the sense that the height scale on which $\bar{u}$ varies is much greater than the vertical wavelength $2\pi m^{-1}$ of the waves under consideration. It also requires that dissipative effects are small. These conditions can be formalized by introducing a small *WKBJ* parameter μ_w, as in Section 4.5.4 and Appendix 4A.

At leading order in μ_w, some of the results given above still hold locally, at each z. For example, upward-propagating Kelvin waves of absolute frequency ω and zonal wave number k have a local vertical wavenumber $m(z)$ given by

$$\omega - k\bar{u}(z) = -Nk/m(z) > 0, \qquad (4.7.23)$$

[cf. Eqs. (4.7.8a) and (4.7.10)] and a local vertical group velocity $c_g^{(z)}(z)$ given by

$$c_g^{(z)}(z) = Nk[m(z)]^{-2} = [\omega - k\bar{u}(z)]^2/Nk. \qquad (4.7.24)$$

(The same formulas also hold when N varies slowly with z.) In terms of the absolute zonal phase speed $c \equiv \omega/k$, these imply that Kelvin waves only exist in regions where $\bar{u}(z) < c$, since $k > 0$ by convention, and that the vertical wavelength $2\pi m^{-1}$ and $c_g^{(z)}$ both become small as $\bar{u}(z)$ approaches c, that is, as a wave approaches a critical level. Intuitively one expects the waves to become more and more susceptible to dissipation as the vertical group velocity decreases, and this can be confirmed theoretically, as mentioned below.

Similar results hold for Rossby-gravity waves: in particular, for the modes with westward phase speed with respect to the mean flow $(-\beta/k < \omega - k\bar{u} < 0)$, we have

$$m(z) = N[\omega - k\bar{u}(z)]^{-2}\{\beta + [\omega - k\bar{u}(z)]k\} \qquad (4.7.25)$$

and

$$c_g^{(z)}(z) = -[\omega - k\bar{u}(z)]^3 N^{-1}\{2\beta + [\omega - k\bar{u}(z)]k\}^{-1}. \qquad (4.7.26)$$

Once again the vertical wavelength and group velocity decrease as a critical level is approached, and the waves are expected to be strongly dissipated there.

At the next order in μ_w, the effects of the weak shear and dissipation on the variation of wave amplitude appear. The amplitude factors $\hat{\Phi}_0$ and $\hat{v}_0$ of Eqs. (4.7.9) and (4.7.19) now vary slowly with height and time. The simplest way of determining this variation is to substitute the lowest-order solutions [e.g., Eqs. (4.7.9) or (4.7.19)] into the full form of the generalized Eliassen–Palm theorem (cf. Section 3.6) and integrate in y: see Appendix 4A for a general outline of this approach. This calculation confirms that the waves become strongly dissipated as $c_g^{(z)} \to 0$. It also gives the slow height dependence of the latitudinally integrated vertical component of the Eliassen–Palm flux $\int_{-\infty}^{\infty} F^{(z)}\, dy$, associated with the waves in question. This quantity depends, among other things, on the form and magnitude of the dissipation that is present in the middle atmosphere, and is an essential ingredient for some of the models of the QBO to be presented in Chapter 8.

As with the *WKBJ* theory for midlatitude planetary waves, described in Section 4.5.4, the conditions for the strict validity of the results given in this section are frequently not satisfied. Nevertheless, the theory still provides a valuable qualitative, and often quantitative, model of the behavior of the observed equatorial waves in the middle atmosphere.

4.7.5 Observed Equatorial Waves

The existence of Kelvin and Rossby-gravity waves in the equatorial lower stratosphere has been verified by a number of time-series analyses based on radiosonde data. The Kelvin waves are primarily of zonal wave number 1 and 2 with periods in the range 10–20 days. The Rossby-gravity waves are primarily of wave number 4, with westward phase propagation and with periods of 4–5 days. The observed characteristics of these modes are summarized in Table 4.1. In both cases the structures are in approximate

Table 4.1

Characteristics of the Dominant Observed Planetary-Scale Waves in the Equatorial Lower Stratosphere

Theoretical description	Kelvin wave	Rossby-gravity wave
Discovered by	Wallace and Kousky (1968)	Yanai and Maruyama (1966)
Period (ground-based) $2\pi\omega^{-1}$	15 days	4–5 days
Zonal wave number $s = ka\cos\phi$	1–2	4
Vertical wavelength $2\pi m^{-1}$	6–10 km	4–8 km
Average phase speed relative to ground	$+25\ \mathrm{m\ s^{-1}}$	$-23\ \mathrm{m\ s^{-1}}$
Observed when mean zonal flow is	Easterly (max. $\approx -25\ \mathrm{m\ s^{-1}}$)	Westerly (max. $\approx +7\ \mathrm{m\ s^{-1}}$)
Average phase speed relative to maximum zonal flow	$+50\ \mathrm{m\ s^{-1}}$	$-30\ \mathrm{m\ s^{-1}}$
Approximate observed amplitudes		
u'	$8\ \mathrm{m\ s^{-1}}$	2–$3\ \mathrm{m\ s^{-1}}$
v'	0	2–$3\ \mathrm{m\ s^{-1}}$
T'	2–3 K	1 K
Approximate inferred amplitudes		
Φ'/g	30 m	4 m
w'	$1.5 \times 10^{-3}\ \mathrm{m\ s^{-1}}$	$1.5 \times 10^{-3}\ \mathrm{m\ s^{-1}}$
Approximate meridional scales $\left(\dfrac{2N}{\beta\|m\|}\right)^{1/2}$	1300–1700 km	1000–1500 km

agreement with theoretical expectations when doppler shifting by mean winds is considered.

As will be shown in Chapter 8, these waves are primarily responsible for driving the so-called quasi-biennial oscillation of the zonal mean winds in the lower stratosphere. The Kelvin waves appear in the data most prominently during periods when mean easterly winds exist at the base of the equatorial stratosphere. The zonal wind and temperature oscillations

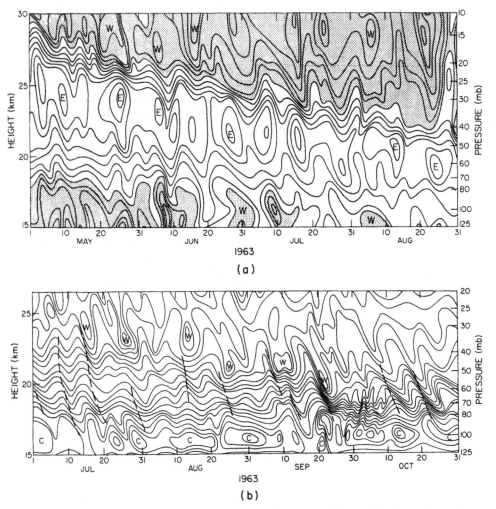

Fig. 4.25. Time–height sections for the equatorial lower stratosphere, showing evidence of Kelvin-wave activity. (a) Zonal wind and (b) temperature at Canton Island (3°S). Note the westerly phase of the QBO encroaching from upper levels in (a): see Chapter 8. [From Giu (1982).]

associated with the Kelvin wave can be quite dramatic, as indicated by the time–height sections of Fig. 4.25. The figure clearly indicates that the temperature fluctuations lag velocity fluctuations by one-fourth cycle and that phase propagates downward, consistent with Kelvin wave dynamics. Cross correlation with stations at other longitudes has confirmed the eastward propagation and long zonal wavelength.

It was noted in Section 4.7.3 that both the Kelvin and Rossby-gravity modes are thought to be generated by heating occurring in large-scale convective complexes (cloud clusters) in the equatorial zone. The model of Chang (1976) showed that randomly distributed sources most efficiently excite the longest zonal-scale Kelvin waves, and that the preferred vertical wavelength of the excited waves is about twice the vertical scale of the heat source, which from Fig. 4.24 is about 6 km for mature cloud clusters. This theory appears to account satisfactorily for the observed spectral distribution of Kelvin waves in the lower stratosphere.

A similar mechanism for frequency selection does not seem to operate for the Rossby-gravity mode. However, there is a rather distinct period of

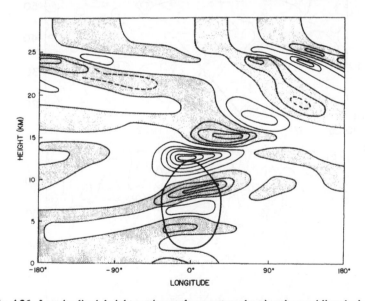

Fig. 4.26. Longitudinal–height section at the equator, showing the meridional wind disturbance excited by an antisymmetric heat source, at the time of maximum heating north of the equator. The mean zonal flow $\bar{u}(z)$ is westerly between 12 and 25 km altitude and easterly elsewhere. The heavy line encloses the region where the amplitude of the diabatic heating exceeds 4 K day^{-1} at the latitude where it is a maximum. Isopleths are at 2-m s^{-1} intervals, with shading indicating southerly winds. [After Holton (1972). American Meteorological Society.]

4–5 days in equatorial tropospheric convection. The model of Holton (1972, 1973) showed that a localized tropical heat source antisymmetric about the equator with a 5-day period can generate a Rossby-gravity response that is dominated by wave number 4 in the lower stratosphere. Figure 4.26, which shows the response to a standing heat source, also illustrates the eastward group velocity of the Rossby-gravity wave.

The relatively slow-moving Kelvin and Rossby-gravity modes observed in the lower stratosphere are effectively damped out by thermal dissipation by about the 10-mb level. Above that level the wave spectrum is dominated by more rapidly propagating Kelvin waves that were first reported in an analysis of rocketsonde data by Hirota (1978). Kelvin waves of several distinct frequency bands have been detected in the LIMS satellite data

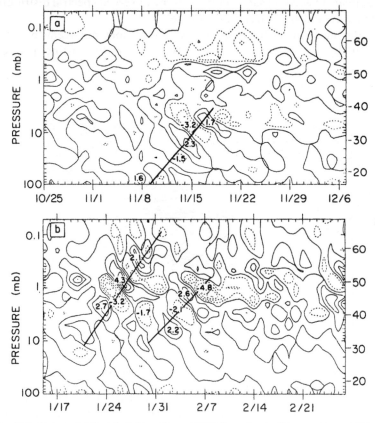

Fig. 4.27. Equatorial time-height sections of LIMS zonal wave 2 temperature at 0° longitude for the periods (a) October 25 to December 7, 1978, and (b) January 15 to February 27, 1979. Contour interval of 1 K. [After Coy and Hitchman (1984).]

(Salby *et al.*, 1984). These waves appear to be excited by isolated "events" in the troposphere and propagate into the middle atmosphere as distinct wave packets, as shown by the example of Fig. 4.27. Such waves are believed to provide the westerly accelerations necessary to maintain the semiannual mean zonal wind oscillation (see Section 8.5).

Appendix 4A   Ray-Tracing Theory and Wave Action in a Slowly Varying Medium

We present here a brief account of the application of "*WKBJ*" or "Liouville–Green" methods to problems involving the propagation of linearized waves in slowly-varying background states. We start by supposing that each disturbance quantity, such as the disturbance zonal velocity u', can be written in the form

$$u' = \text{Re } \hat{u}(\mathbf{x}, t) \exp i\chi(\mathbf{x}, t) \qquad (4A.1)$$

where $\mathbf{x} \equiv (x, y, z)$ and the phase χ is real. We *define* a local wave number vector $\mathbf{k} = (k, l, m)$ and frequency ω in terms of derivatives of χ:

$$k \equiv \frac{\partial \chi}{\partial x}, \qquad l \equiv \frac{\partial \chi}{\partial y}, \qquad m \equiv \frac{\partial \chi}{\partial z}, \qquad \omega \equiv -\frac{\partial \chi}{\partial t}. \qquad (4A.2)$$

We now make the crucial assumption that $\hat{u}$, k, l, m, and ω, as well as the background medium, all vary much more slowly in time and space than does the phase. This can be formalized by requiring that their time and space scales are $O(2\pi\omega^{-1}\mu_w^{-1})$ and $O[2\pi(k^2 + l^2 + m^2)^{-1/2}\mu_w^{-1}]$, respectively, where μ_w is a small "*WKBJ* parameter," and thus are large compared with the wave period and wavelength, respectively. We also assume that a dispersion relation holds at each point in time and space:

$$\omega = \Delta(\mathbf{k}; \mathbf{x}, t). \qquad (4A.3)$$

This relation follows, at leading order in μ_w, from substitution into the linearized equations of motion [e.g., Eqs. (3.4.2)] of Eq. (4A.1) and similar expressions for the other disturbance variables; it thus contains dynamical information. Note that ω can depend on $\mathbf{x}$ both through the dependence of $\mathbf{k}$ on $\mathbf{x}$ and through the "explicit" $\mathbf{x}$ dependence of Δ due to $\mathbf{x}$ variations of the background state. (For example, if the background flow velocity depends on z, then Δ will contain explicit z dependence.)

We define the group velocity $\mathbf{c_g}$ by

$$\mathbf{c_g}(\mathbf{k}; \mathbf{x}, t) = (c_g^{(x)}, c_g^{(y)}, c_g^{(z)}) = \left(\frac{\partial \Delta}{\partial k}, \frac{\partial \Delta}{\partial l}, \frac{\partial \Delta}{\partial m}\right), \qquad (4A.4)$$

and introduce the time rate-of-change as measured by an observer moving with the local group velocity:

$$\frac{d_g}{dt} \equiv \frac{\partial}{\partial t} + \mathbf{c}_g \cdot \nabla. \tag{4A.5}$$

An important set of identities follows from the definitions of Eq. (4A.2) using relations like $\partial^2 \chi / \partial x \, \partial t \equiv \partial^2 \chi / \partial t \, \partial x$: for example,

$$\frac{\partial k}{\partial t} = -\frac{\partial \omega}{\partial x}, \qquad \frac{\partial k}{\partial y} = \frac{\partial l}{\partial x}, \qquad \frac{\partial k}{\partial z} = \frac{\partial m}{\partial x}. \tag{4A.6}$$

Thus

$$\frac{d_g k}{dt} \equiv \frac{\partial k}{\partial t} + \mathbf{c}_g \cdot \nabla k = -\frac{\partial \omega}{\partial x} + \frac{\partial \Delta}{\partial k} \frac{\partial k}{\partial x} + \frac{\partial \Delta}{\partial l} \frac{\partial l}{\partial x} + \frac{\partial \Delta}{\partial m} \frac{\partial m}{\partial x}, \tag{4A.7}$$

using Eqs. (4A.5), (4A.6), and (4A.4); but from Eq. (4A.3),

$$\frac{\partial \omega}{\partial x} = \frac{\partial \Delta}{\partial k} \frac{\partial k}{\partial x} + \frac{\partial \Delta}{\partial l} \frac{\partial l}{\partial x} + \frac{\partial \Delta}{\partial m} \frac{\partial m}{\partial x} + \frac{\partial \Delta}{\partial x} \tag{4A.8}$$

using the chain rule, and so Eqs. (4A.7) and (4A.8) imply

$$\frac{d_g k}{dt} = -\frac{\partial \Delta}{\partial x}. \tag{4A.9a}$$

In a similar manner it can be shown that

$$\frac{d_g l}{dt} = -\frac{\partial \Delta}{\partial y}, \qquad \frac{d_g m}{dt} = -\frac{\partial \Delta}{\partial z}, \qquad \frac{d_g \omega}{dt} = \frac{\partial \Delta}{\partial t}. \tag{4A.9b,c,d}$$

Thus the rate-of-change, following the group velocity, of the local wave number or frequency depends on the "explicit" space or time variation of Δ due to inhomogeneities of the background state.

We now define a *ray* as the trajectory $\mathbf{x}(t)$ of a point moving with the local group velocity, that is,

$$\frac{d_g \mathbf{x}}{dt} = \mathbf{c}_g[\mathbf{k}(\mathbf{x}(t), t); \mathbf{x}(t), t]. \tag{4A.10}$$

Equations (4A.9) and (4A.10) are called the *ray-tracing equations*: given the dispersion relation [Eq. (4A.3)] and appropriate initial conditions on $\mathbf{x}$ and $\mathbf{k}$, they can in principle be solved for the ray paths and the variation of local wave number and frequency along the rays. Physically, Eq. (4A.10) can be regarded as giving the trajectory of a "wave packet" whose wave number $\mathbf{k}$ and frequency ω vary according to Eq. (4A.9).

Examples of the use of ray-tracing theory are discussed in Section 4.5.4 for planetary waves in a zonal shear flow $\bar{u}(y, z)$ and mentioned in Section 4.6.2 for internal gravity waves in a zonal shear flow $\bar{u}(z)$; the latter case is particularly amenable to analytical treatment (e.g., Bretherton, 1966b). Further details of the technique are given, for example, by Lighthill (1978) and Gill (1982).

The ray-tracing equations give no information about the amplitude variation of wave packets; for this purpose a more complex theory is required, and we shall only quote the main results here. For simplicity, we specialize to the case of waves of slowly varying amplitude in a basic unforced flow $[\bar{u}(y, z), 0, 0]$ that is zonal, independent of x (so that $\partial\Delta/\partial x = 0$) and t, and slowly varying in y and z. Since this flow is zonal, unforced, and x-independent, the generalized Eliassen–Palm theorem [Eq. (3.6.2)] holds; using the slowly varying properties of the waves and mean flow it can be shown [using methods analogous to those of Andrews and McIntyre (1976a,b, 1978c)] that

$$\mathbf{F} = \mathbf{c}_g A \tag{4A.11}$$

to leading order in μ_w. (A special case of this result, for planetary waves, is mentioned in Section 4.5.5.) It can also be shown that

$$A = -k\left(\frac{E}{\omega - k\bar{u}}\right) \tag{4A.12}$$

to leading order in μ_w, where E is the wave-energy density $\frac{1}{2}\rho_0(\overline{u'^2} + \overline{v'^2} + \overline{\Phi_z'^2}/N^2)$ [cf. Eq. (3.6.3)]. The expression $E/(\omega - k\bar{u})$ is called the *wave-action density*; using Eqs. (3.6.2), (4A.11), (4A.12), and (4A.9a), together with the fact that $\partial\Delta/\partial x = 0$ here, one obtains the wave-action equation

$$\frac{\partial}{\partial t}\left(\frac{E}{\omega - k\bar{u}}\right) + \mathbf{\nabla}\cdot\left(\frac{\mathbf{c}_g E}{\omega - k\bar{u}}\right) = \text{nonconservative effects} + O(\alpha^3)$$

$$\tag{4A.13}$$

for slowly varying waves (Bretherton and Garrett, 1968; Andrews and McIntyre, 1978c). (Here $\alpha \ll 1$ is a dimensionless wave amplitude, as in Chapter 3.) Equation (4A.13) can be rewritten

$$\frac{d_g}{dt}\left(\frac{E}{\omega - k\bar{u}}\right) + \left(\frac{E}{\omega - k\bar{u}}\right)\mathbf{\nabla}\cdot\mathbf{c}_g = \text{nonconservative effects} + O(\alpha^3)$$

and, if the nonconservative terms are known, this can be used to find the variation of wave action (and thence of wave energy and other measures of wave amplitude) along a ray, to $O(\alpha^2)$. Note that, in general, information on a bundle of adjacent rays is required for calculation of $\mathbf{\nabla}\cdot\mathbf{c}_g$ here.

For equatorial waves (Sections 4.7 and 8.3) the assumption of slow variation of wave amplitude in y is no longer valid; however, in the case where $\bar{u}$ depends on z alone it can be shown that

$$\int F^{(z)} \, dy = c_g^{(z)} \int A \, dy, \tag{4A.14}$$

where the integrals are taken from $-\infty$ to ∞. From Eq. (3.6.2) it then follows that

$$\frac{\partial}{\partial t} \int A \, dy + \frac{\partial}{\partial z} \left(c_g^{(z)} \int A \, dy \right) = -2\gamma(z) \int A \, dy \tag{4A.15}$$

to $O(\alpha^2)$, where $2\gamma(z) \equiv -\int D \, dy / \int A \, dy$ and D is the nonconservative term in Eq. (3.6.2). If X', Y', Q' [see Eq. (3.4.2)] represent dissipative effects, it can be shown that $\gamma(z)$ is a positive inverse relaxation time. [For example, if $X' = -\gamma_1 u'$, $Y' = -\gamma_2 v'$, $Q' = -\gamma_3 \theta'$, then γ is a weighted mean of γ_1, γ_2, γ_3 at each z whose precise form generally depends on the detailed equatorial wave solutions: cf. Andrews and McIntyre (1976b); however, if $\gamma_1 = \gamma_2 = \gamma_3$, then $\gamma = \gamma_1$.] For steady waves, Eqs. (4A.14) and (4A.15) give

$$\int F^{(z)} \, dy = \int F^{(z)} \, dy \bigg|_{z=z_0} \exp \left\{ -\int_{z_0}^{z} \frac{2\gamma(z') \, dz'}{c_g^{(z)}(z')} \right\} \tag{4A.16}$$

[cf. Eqs. (8.3.1) and (8.3.6)]. Note that the exponential decay factor in Eq. (4A.16) tends to be enhanced by small values of the vertical group velocity $c_g^{(z)}$, implying strong attenuation of the waves: see Sections 4.7.4 and 8.3.2 for some possible consequences of this result.

References

4.1. General treatments of linear waves in rotating, stratified fluids are given in many texts: for example Gill (1982) and Pedlosky (1979) both discuss a variety of waves of atmospheric and oceanic interest. Lighthill (1978) gives a thorough account of the general principles of wave propagation in fluids, and includes several geophysical applications.

4.2. The material summarized in this section is treated in more detail by Chapman and Lindzen (1970), for example. Some mathematical properties of Hough functions are given by Lindzen (1971); questions of their completeness are discussed by Holl (1970).

4.3. A basic source on atmospheric tidal theory and observation is Chapman and Lindzen (1970). Recent reviews include Lindzen (1979), Forbes and Garrett (1979) and Forbes (1984). Rocket observations are discussed by Reed (1972) and Groves (1976). Radar observations of mesospheric tides are given by Manson et al. (1985). Recent satellite measurements of middle atmosphere tides, and comparisons with simple theory, are presented by Brownscombe *et al.* (1985). The simple, approximate theory for the vertical structure of the semidiurnal tide given near the end of Section 4.3.3 follows Green (1965). The contributions of latent heating to the excitation of the semidiurnal tide are investigated by Hamilton (1981a).

4.4. Recent reviews of the theory and observation of free traveling planetary waves are given by Madden (1979), Walterscheid (1980), and Salby (1984). The "ducted" mode peaking near the stratopause was investigated theoretically by Salby (1979, 1980). The sensitivity of the 5-day wave to the background wind structure was studied by Geisler and Dickinson (1976). Observational evidence for the excitation of the 5-day wave by tropical latent heating is presented by Hamilton (1985).

4.5. The theory of tropospherically forced planetary waves propagating into the middle atmosphere was initiated by Charney and Drazin (1961) and is to be found in several standard texts. The calculation of parcel orbits in Section 4.5.3 follows Matsuno (1980). Fundamental work on the propagation of planetary waves in basic flows depending on latitude and height, including consideration of critical lines and ray-tracing, was performed by Dickinson (1968). The propagation of planetary waves in mean flows $\bar{u}(y)$, containing only meridional shear, is considered, for example, by Boyd (1982a,b).

4.6. Fritts (1984) gives an excellent review of middle atmosphere gravity wave observations and theory. Radar observations are discussed in some detail by Vincent (1984). For the Boussinesq approximation see, for example, Gill (1982). The effects of radiative damping on gravity-wave breaking are examined by Fels (1984, 1985) and Holton and Zhu (1984). The possible importance of inertio-gravity waves in the middle atmosphere is considered by Dunkerton (1984). Evidence for inertio-gravity waves in rocketsonde winds is presented by Hirota and Niki (1985).

4.7. The foundations of the modern theory of atmospheric equatorial waves were laid by Matsuno (1966). Further details on the *WKBJ* theory of Section 4.7.4 are given by Lindzen (1971, 1972), Andrews and McIntyre (1976a,b), and Boyd (1978a,b).

Chapter 5 | Extratropical Planetary-Scale Circulations

5.1 Introduction

This chapter will discuss observational and theoretical aspects of some of the planetary-scale "climatological" features of the extratropical stratosphere and mesosphere, which vary slowly from month to month during the annual cycle and recur regularly from year to year. Such features include the westerly zonal-mean winds that occur in the winter hemisphere, the easterly zonal-mean winds that occur in summer, and the geographically fixed planetary-scale wave patterns observed primarily in the northern winter.

A natural way to isolate features of this type is to perform time averages for individual calendar months in a record extending over many years. Thus, from 10 years of data, say, one can average the 10 Januaries to obtain a "mean January" field, and so on. The resulting climatological monthly mean flow patterns tend to take a wavy form in the Northern-Hemisphere winter: these patterns can, if desired, be separated into zonal-mean and zonally varying parts, as in Section 3.3. Their zonally varying parts are known as "stationary waves" and can be further separated into zonal Fourier components. The time-dependent departures from the climatological average are often known as "transient eddies." (Note that this definition of "transient" differs from that of Sections 3.6 and 4.1.) Space–time spectral analysis of these transient components sometimes reveals the presence of large-scale, coherent, zonally propagating "traveling waves."

5.2 The Observed Annual Cycle

It was mentioned in Section 1.5 that global observation of stratospheric temperature began in earnest in the late 1960s with the advent of satellite-borne infrared radiometers. Using instruments of this kind, together with conventional radiosonde measurements up to about the 10-mb level in the lower stratosphere, our knowledge of the planetary-scale structure of the middle atmosphere has gradually expanded over the last 20 years. There now exists a data base potentially large enough to construct fairly stable climatologies of temperature, geopotential height, geostrophic wind, and other quantities up to mesopause levels in the extratropics. So far, little detailed analysis of this climatological data set has been performed, but some general remarks about the gross planetary-scale structure of the middle atmosphere can be made.

We first describe the annual cycle of certain monthly mean fields. These can be displayed in several different ways: for example, the mean geopotential fields, $\langle \Phi \rangle$, say (where $\langle \ldots \rangle$ here denotes the climatological monthly average), for each month can be contoured on polar stereographic charts at various log-pressure levels. Alternatively, the monthly mean geopotential field can be expanded in zonal Fourier harmonics up to some zonal wave number S representing the limit of resolution of the data,

$$\langle \Phi \rangle = A_0(\phi, z) + \sum_{s=1}^{S} A_s(\phi, z) \cos[s\lambda + \alpha_s(\phi, z)], \qquad (5.2.1)$$

and the amplitude A_s and phase α_s of each harmonic contoured in the (ϕ, z) plane for each month. We present examples of both of these approaches for various atmospheric fields.

5.2.1 The Zonal-Mean Flow

The most basic monthly mean fields are the zonal means, corresponding to the term A_0 in Eq. (5.2.1). Schematic solstice cross sections of zonal-mean temperature and zonal-mean zonal wind in the middle atmosphere were presented in Figs. 1.3 and 1.4, which were based primarily on Northern-Hemisphere data.

To give some idea of the seasonal cycle and interhemispheric differences, Figs. 5.1 and 5.2 show monthly mean data for January, April, July, and October, from the 1986 COSPAR International Reference Atmosphere (CIRA) compilation, based on approximately 5 years of satellite data. Figure 5.1 shows the zonal-mean temperature for these months and, like Fig. 1.3,

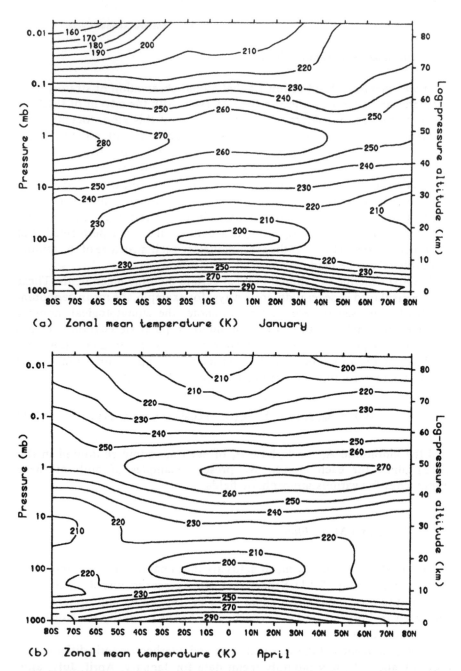

Fig. 5.1. Monthly and zonally averaged temperature (K) for altitudes up to approximately 90 km, based on about 5 years of data from the *Nimbus* 5 and 6 satellites (January 1973–December 1974 and July 1975–June 1978, respectively) above 30 mb; data supplied by Berlin Free University at 30 mb and Oort's (1983) climatology for 50 mb and below. (a) January, (b)

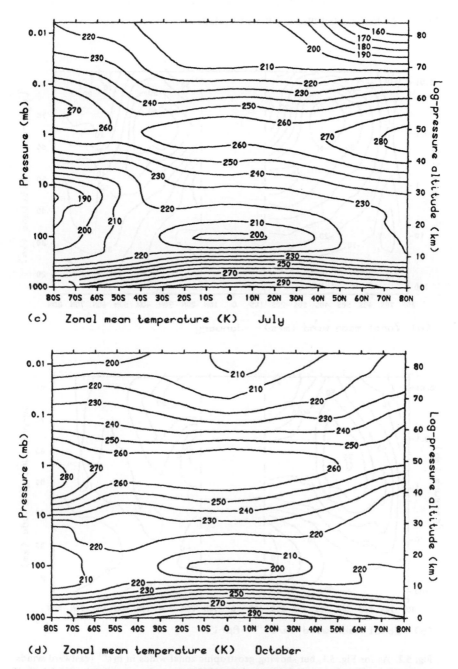

(c)   Zonal mean temperature (K)   July

(d)   Zonal mean temperature (K)   October

April, (c) July, (d) October. From the 1986 CIRA compilation, courtesy of J. J. Barnett and
M. Corney, Department of Atmospheric Physics, Oxford University. (See also Barnett and
Corney, 1985a.)

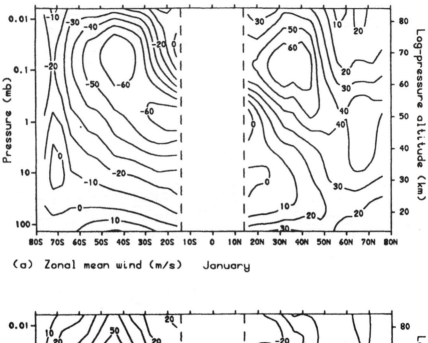

(a) Zonal mean wind (m/s)   January

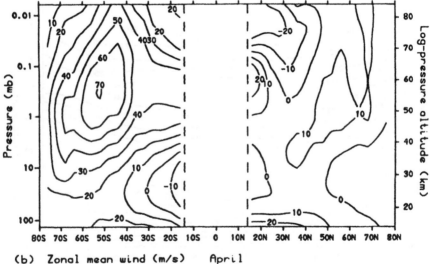

(b) Zonal mean wind (m/s)   April

Fig. 5.2. As for Fig. 5.1, but showing geostrophic zonal winds in m s^{-1} (eastward winds positive, westward winds negative), above about 15 km.

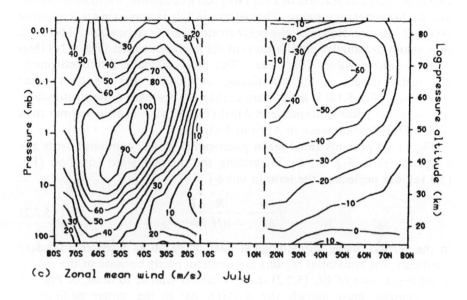

(c) Zonal mean wind (m/s)   July

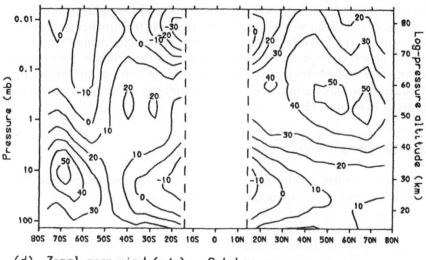

(d) Zonal mean wind (m/s)   October

Fig. 5.2 (continued)

illustrates the basic features of a cold tropical tropopause, warm stratopause, and cold summer mesopause. The pattern in July is approximately a mirror image of January. However, the polar stratosphere is somewhat colder (and thus nearer to radiative equilibrium) in the southern winter (Fig. 5.1c) than in the northern winter (Fig. 5.1a). The most dramatic interhemispheric asymmetry occurs at the equinoxes; in October the southern polar stratopause (Fig. 5.1d) is much warmer than the northern polar stratopause, or than either polar stratopause in April (Fig. 5.1b); it is even warmer than the equatorial stratopause in April or October.

Figure 5.2 presents zonal-mean geostrophic winds, computed from the temperatures in Fig. 5.1 by integrating the thermal wind equation [Eq. (3.4.1c), but neglecting the term in tan ϕ]

$$f\frac{\partial \bar{u}}{\partial z} = -\frac{R}{aH}\frac{\partial \bar{T}}{\partial \phi} \tag{5.2.2}$$

in the vertical, using climatological data at 30 mb as a lower boundary condition. The equatorial regions are omitted, partly because of the anticipated breakdown of Eq. (5.2.2) there. Features similar to those of Fig. 1.4 are apparent; these include the westerly jets in the winter midlatitude mesosphere (stronger in the Southern Hemisphere than the Northern), which extend down to the "polar night jets" in the winter polar stratosphere. Easterlies of somewhat weaker magnitude appear in the summer mesosphere and stratosphere. Away from the subtropics, equinoctial winds tend to be westerly or weakly easterly in both hemispheres. In Chapter 7 we discuss the dynamical reasons for some of these observed features of the climatological zonal-mean circulation. Interannual variability in the stratosphere is discussed in Section 12.5.

5.2.2 Stationary Waves

We next consider zonally asymmetric aspects of the monthly mean data. The most graphic way of depicting these is by means of maps of the various fields at different levels. For example, Fig. 5.3 shows monthly mean temperatures at 10 mb for January, April, July, and October in the form of polar stereographic charts for each hemisphere. (These maps give the *total* field, and include the zonal mean as well as the zonally asymmetric components.) Some clear differences between the seasons and between the hemispheres are immediately evident: thus, the Northern-Hemisphere winter 10-mb temperature field (January) is characterized by strong departures from zonal symmetry, indicating the presence of stationary waves, while the Southern-Hemisphere winter field (July) is more symmetric about the

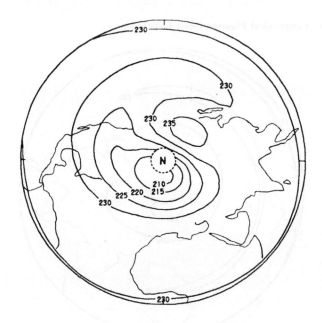

TEMPERATURE (K) AT 10 MB : JANUARY

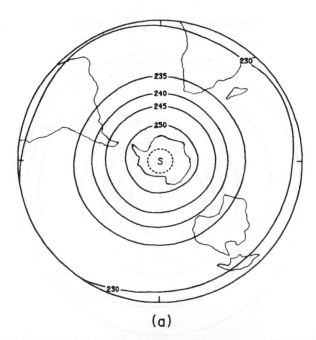

(a)

Fig. 5.3. Polar stereographic maps of monthly averaged temperature (K) at 10 mb (approximately 30 km altitude) for (a) January, (b) April, (c) July, (d) October, for the Northern Hemisphere (above) and the Southern Hemisphere (below). Outer circle, equator; inner circle; 80° latitude. (Courtesy of J. J. Barnett and M. Corney, Department of Atmospheric Physics, Oxford University. See also Barnett and Corney, 1985b.) *Figure continues.*

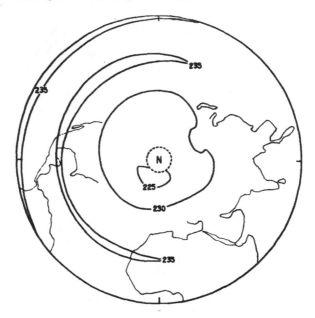

TEMPERATURE (K) AT 10 MB : APRIL

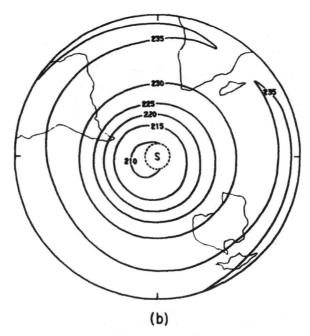

(b)

Fig. 5.3 (continued)

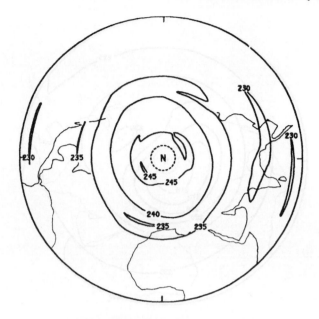

TEMPERATURE (K) AT 10 MB : JULY

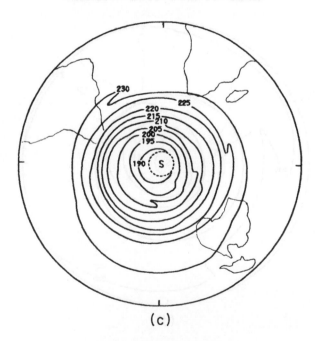

(c)

Fig. 5.3 (*figure continues*)

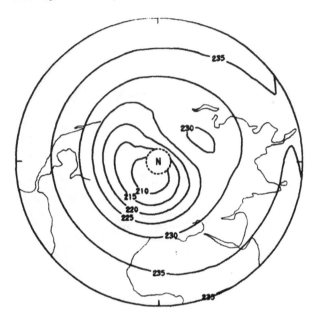

TEMPERATURE (K) AT 10 MB : OCTOBER

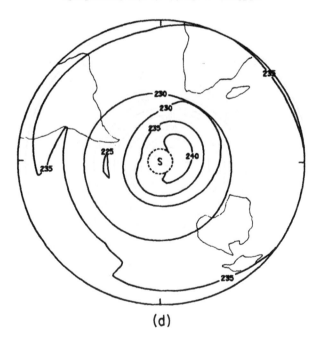

(d)

Fig. 5.3 (continued)

pole and exhibits strong latitudinal temperature gradients, as in Fig. 5.1c. Both summer hemispheres are fairly zonally symmetric, with quite weak temperature gradients. The spring equinoctial fields (Northern Hemisphere in April, Southern Hemisphere in October) are close to isothermal, but the autumn equinoxes have noticeable temperature gradients and (in the Northern Hemisphere) significant departures from zonal symmetry.

An alternative method of examining zonal asymmetries is by means of meridional cross sections of amplitude A_s and phase α_s [see Eq. (5.2.1)] for various zonal wave numbers $s \geq 1$. Examples of the geopotential amplitude and (negative) phase for the mean January and July $s = 1$ and 2 components are shown in Fig. 5.4. The summer hemispheres have been omitted from this figure since amplitudes are much smaller there than in the winter hemispheres. It will be seen that winter amplitudes are greater in the Northern than in the Southern Hemisphere. These differences, of course, reflect the differences in asymmetry observed in Fig. 5.3. The lines of constant phase are fairly horizontal in the extratropical lower stratosphere, but tilt equatorward–downward in the upper stratosphere and mesosphere. The longitudes of the ridges of each Fourier component ($-\alpha_1$ for wave number 1 and $-\frac{1}{2}\alpha_2$ and $-\frac{1}{2}\alpha_2 + 180°$ for wave number 2) generally progress westward with increasing height and decreasing latitude. (Note that the phase lines in Fig. 5.4 are labeled with $-\alpha_s$.)

As discussed in Section 5.3 below, there have been many attempts to model the stationary waves appearing, for example, in Fig. 5.3 in terms of the linear theory of forced planetary waves on a zonally symmetric basic state. For this reason another diagnostic that is often applied is the Eliassen–Palm flux vector $\mathbf{F}$ (see Sections 3.5, 3.6, and 4.5.5). A quasi-geostrophic version of $\mathbf{F}$ in spherical geometry is

$$\mathbf{F} = [0, -\rho_0 a(\cos \phi)\overline{v'u'}, \rho_0 a(\cos \phi)\overline{fv'\theta'}/\theta_{0z}] \qquad (5.2.3)$$

[cf. Eqs. (3.5.3) and (3.5.6)], where $f = 2\Omega \sin \phi$. Using the following approximate geostrophic formulas in spherical coordinates,

$$u' = -(fa)^{-1}\Phi'_\phi, \qquad v' = (fa \cos \phi)^{-1}\Phi'_\lambda, \qquad (5.2.4)$$

[cf. Eqs. (3.2.2)–(3.2.4)], together with Eqs. (3.1.3c) and (3.2.13), this reduces to

$$\mathbf{F} = \rho_0(0, \overline{\Phi'_\phi\Phi'_\lambda}/f^2 a, \overline{\Phi'_z\Phi'_\lambda}/N^2), \qquad (5.2.5)$$

where Φ' here represents the departure of the climatological mean geopotential from its zonal mean value [cf. Eq. (4.5.32)]. (This expression, like quasi-geostrophic theory itself, will not be valid near the equator, where f

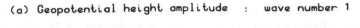

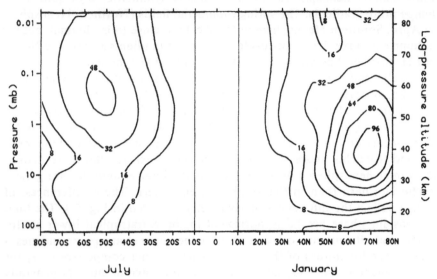

July   January

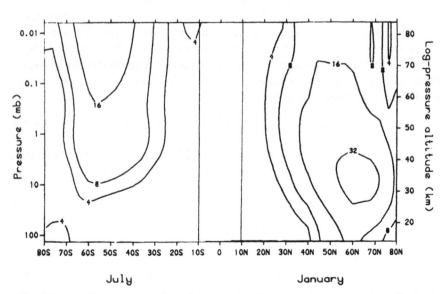

July   January

Fig. 5.4. Meridional cross section of the negative phase $-\alpha_s$ (degrees) and amplitude A_s (decameters) of the monthly mean geopotential height $\langle \Phi/g \rangle$ for wave numbers $s = 1$ and 2, from the same data source as Fig. 5.1. Only the winter hemispheres are shown: (a) $s = 1$ amplitude, (b) $s = 2$ amplitude, (c) $s = 1$ phase, (d) $s = 2$ phase.

(c) Geopotential height phase (deg. E) : wave number 1

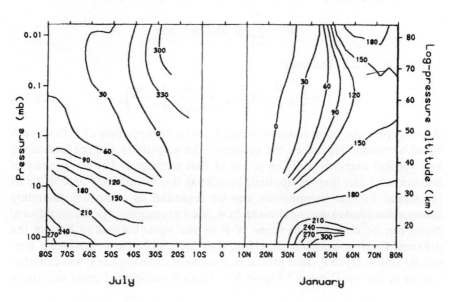

July   January

(d) Geopotential height phase (deg. E) : wave number 2

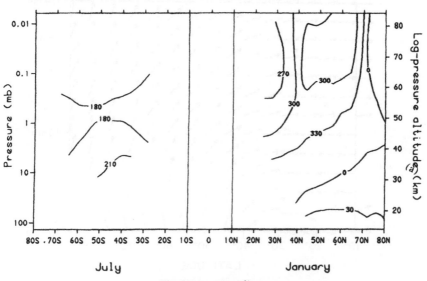

July   January

Fig. 5.4 (*continued*)

is small.) Employing a Fourier decomposition as in Eq. (5.2.1), we have

$$\Phi' = \sum_{s=1}^{S} A_s \cos(s\lambda + \alpha_s)$$

so that Eq. (5.2.5) becomes

$$\mathbf{F} = \frac{1}{2}\rho_0 \sum_{s=1}^{S} sA_s^2 \left(0, \frac{1}{f^2 a}\frac{\partial \alpha_s}{\partial \phi}, \frac{1}{N^2}\frac{\partial \alpha_s}{\partial z}\right), \tag{5.2.6}$$

showing how the contributions to the ϕ and z components of $\mathbf{F}$ from each zonal harmonic depend on the square of the amplitude and the latitudinal and vertical derivatives of the phase of that harmonic. The philosophy of Section 4.5.5 can then be applied; insofar as the disturbance field Φ' or its individual Fourier components can be regarded as stationary planetary waves embedded in the zonal-mean flow, their propagation in the meridional plane can be described in terms of $\mathbf{F}$ or the contributions to $\mathbf{F}$ from the different Fourier components. (Note that these different components contribute additively to $\mathbf{F}$, since products of terms of differing zonal wavenumber vanish in the zonal mean.) Figure 5.5 shows a meridional cross section in

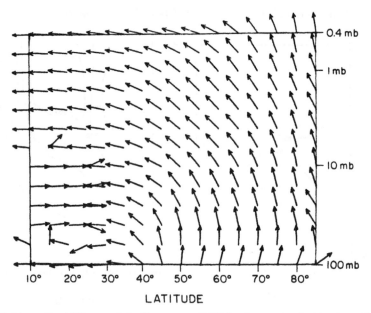

Fig. 5.5. Normalized Eliassen–Palm flux vectors $\mathbf{F}/|\mathbf{F}|$ for January in the Northern Hemisphere stratosphere, based on 4 years of data. The vertical coordinate is z. The vectors are normalized to avoid plotting problems associated with the rapid decrease of $|\mathbf{F}|$ with z. [After Hamilton (1982b). American Meteorological Society.]

which **F** is represented by suitably scaled arrows, which suggest propagation of wave activity from the winter troposphere up into the stratosphere and toward the equator; this is consistent with the ray-tracing calculations of Section 4.5.4 and also with the more detailed numerical models of Sections 5.3 and 11.2. The flux **F** tends to be dominated by $s = 1$ in these cases, since $A_1^2 \gtrsim 2A_2^2$, and the directions of the **F** arrows in Fig. 5.5 tend to be up the $s = 1$ phase gradient in Fig. 5.4c, consistent with Eq. (5.2.6).

5.2.3 Transient Eddies

As mentioned in Section 5.1, the term "transient eddies" (or "transient waves") is often used to describe departures from a time-mean flow such as the zonally asymmetric climatological monthly mean considered above. These departures can take many different forms, and careful space–time filtering, or examination of individual cases of transient development, may be needed to distinguish between the different phenomena. Transient disturbance fields can always be decomposed into zonal Fourier components, including in general a zonal-mean contribution. Whether such a decomposition helps to illuminate the dynamics will depend on circumstances.

The most spectacular transient phenomenon to be observed in the middle atmosphere is the major stratospheric sudden warming; however, this does not occur every year, and its discussion will be postponed to Chapter 6. More ubiquitous large-scale transient features include traveling planetary waves and breaking planetary waves, and these will be mentioned briefly here.

Some observations of traveling planetary waves, such as the 5-day wave and the 2-day wave, were mentioned in Section 4.4, and the simplest theory of the 5-day wave was described there. Detailed space–time spectral analysis of the type used by Mechoso and Hartmann (1982) has revealed a number of coherent, traveling, planetary-scale wave structures in the middle atmosphere. These traveling waves may vary from season to season and between hemispheres; both westward-moving and eastward-moving disturbances are found. Further theoretical attempts to model some of these structures are discussed in Sections 5.4 and 5.5.2.

A phenomenon that has recently been identified from stratospheric satellite data, and that may well turn out to be a common transient process, is the breaking planetary wave. A case study is presented by McIntyre and Palmer (1983, 1984), who use isentropic maps of Ertel's potential vorticity, as well as isobaric charts of geopotential height, from January and February 1979 to discuss the dynamics of the event. The time development is illustrated in Fig. 5.6: on January 26 there is an off-centered cyclonic vortex in the middle stratosphere, as revealed by the low geopotential at 10 mb over Scandinavia in Fig. 5.6a and the corresponding high potential vorticity on the 850-K isentropic surface (also near 10 mb, or 30 km altitude) in Fig. 5.6b. The latter diagram shows a "tongue" of high potential vorticity emanating from the main vortex and extending westward over North America. A sequence of isentropic potential vorticity maps for the days immediately preceding January 26 suggests that this tongue represents material that has been dragged out of the main cyclonic vortex by the flow associated with the secondary "Aleutian anticyclone" in Fig. 5.6a. This argument is based on the facts that potential temperature θ and Ertel's potential vorticity P are both conserved by air parcels in adiabatic, frictionless flow (see Sections 3.1 and 3.8); in the middle atmosphere they should represent quasi-conservative tracers over periods of a few days. The sugges-

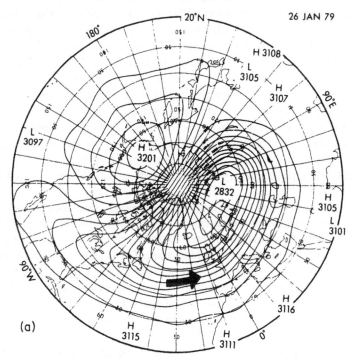

Fig. 5.6. Polar stereographic maps (outer circle: 20°N) of (a) NMC-based analysis of the 10 mb geopotential height (decameters) on January 26, 1979. (b) Coarse-grain estimate of Ertel's potential vorticity divided by gH/p_0 (where $H = 6.5$ km and $p_0 = 1000$ mb) on the 850-K isentropic surface on January 26, 1979, in 10^{-4} K m^{-1} s^{-1}. Values greater than 4 units are lightly shaded, and those greater than 6 units are heavily shaded. The dashed circle shows the position of a local maximum of just under 4 units. (c) As for (b), but for January 27, 1979. [From McIntyre and Palmer (1984), with permission.]

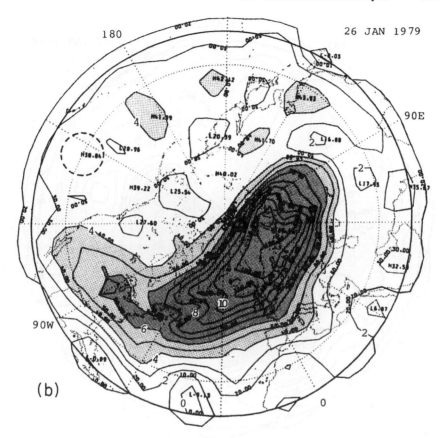

Fig. 5.6 (*figure continues*)

tion is confirmed by calculations of air-parcel trajectories and by independent observations of ozone, which is also quasi-conservative at these altitudes and latitudes in winter (Leovy *et al.*, 1985).

The sequence of events discussed by McIntyre and Palmer is characterized by rapid and irreversible deformation of material contours, as represented for example by the isopleths of P on the $\theta = 850$ K surface (which in small-amplitude wave motions would merely undulate back and forth). For this reason the authors describe the process as "planetary-wave breaking," by analogy with the breaking of ocean waves on a beach (see also the discussion of breaking internal gravity waves in Section 4.6.2). They suggest that isentropic mixing associated with events of this kind may be responsible for eroding the main winter polar vortex, to produce the region of uniform potential vorticity (or "surf zone") that is often observed to surround it in the northern hemisphere. However, Clough *et al.* (1985) find that other

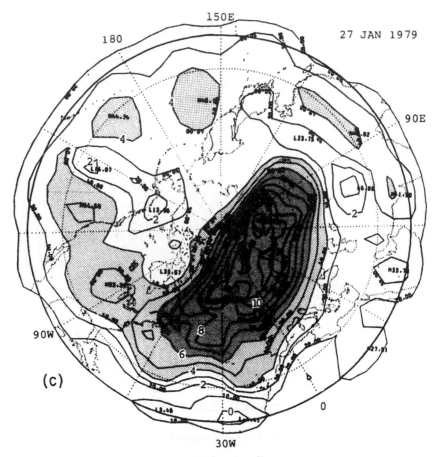

Fig. 5.6 (*continued*)

wave-breaking events, such as that of December 1981, appear to exhibit much less isentropic mixing than that described by McIntyre and Palmer. The reasons for these differences in behavior are not yet understood.

It should be noted that on January 27, 1979 the potential vorticity tongue appears to start breaking up into "blobs": if real, these features could indicate the presence of barotropic or baroclinic instability (see Section 5.5.2).

We have shown in this section that a variety of transient planetary-scale eddy phenomena can occur in the middle atmosphere. Through the time-averaged quadratic eddy flux terms, such transient eddies may exert an important influence on aspects of the time-mean flow, including the stationary eddies. The stationary eddies, in turn, may strongly control the behavior of the transient eddies. These effects are difficult to interpret and evaluate, since a comprehensive theory for the interaction of transient eddies and a zonally asymmetric time-mean flow (comparable to that of Sections 3.3–3.6, for the interaction of eddies with the zonal-mean flow) is still lacking.

5.3 Detailed Linear Models of Stationary Planetary Waves in the Middle Atmosphere

The modeling of stationary planetary waves in the *troposphere* began with the paper by Charney and Eliassen (1949), who used a simple linear, barotropic beta-plane model. The waves were forced by a uniform eastward basic wind blowing over the surface orography. Later studies (e.g., Smagorinsky, 1953) also included the influence of thermal sources in forcing the waves.

The first major study of stratospheric planetary waves, including stationary waves, was performed by Charney and Drazin (1961), using quasi-geostrophic theory on a beta-plane. Their largely analytical methods were outlined in Section 4.5, and their main results were discussed there.

Detailed quantitative investigation of vertically propagating stationary planetary waves in the stratosphere began with the work of Matsuno (1970). Rather than studying the precise generation mechanism for the waves, he concentrated on the hypothesis that the stationary waves in the Northern Hemisphere winter stratosphere are forced from the troposphere; he therefore imposed 500-mb heights based on observation as a lower boundary condition.

Matsuno's study used a linearized quasi-geostrophic potential vorticity equation in spherical coordinates, modified to include an ageostrophic term (namely, the "isallobaric" contribution to the northward disturbance wind). The equation takes the form

$$\left(\frac{\partial}{\partial t} + \frac{\bar{u}}{a \cos \phi}\frac{\partial}{\partial \lambda}\right)q'_{(M)} + a^{-1}\bar{q}_\phi v' = 0, \qquad (5.3.1)$$

where

$$v' = (fa \cos \phi)^{-1}\Phi'_\lambda, \qquad (5.3.2)$$

$$q'_{(M)} = \frac{1}{fa^2}\left[\frac{\Phi'_{\lambda\lambda}}{\cos^2 \phi} + \frac{f^2}{\cos \phi}\left(\frac{\cos \phi}{f^2}\Phi'_\phi\right)_\phi + \frac{f^2 a^2}{\rho_0}\left(\frac{\rho_0 \Phi'_z}{N^2}\right)_z\right], \qquad (5.3.3)$$

and

$$\bar{q}_\phi = 2\Omega \cos \phi - \left[\frac{(\bar{u} \cos \phi)_\phi}{a \cos \phi}\right]_\phi - \frac{a}{\rho_0}\left(\frac{\rho_0 f^2}{N^2}\bar{u}_z\right)_z. \qquad (5.3.4)$$

Here $\bar{u}(\phi, z)$ is the basic zonal flow, Φ' is the geopotential disturbance, and $f = 2\Omega \sin \phi$. The ageostrophic modification ensures energetic consistency and also a generalized Eliassen–Palm theorem of the form of Eq. (3.6.2). On posing a stationary wave solution of zonal wave number s,

$$\Phi' = e^{z/2H} \operatorname{Re} \Psi(\phi, z) e^{is\lambda}, \qquad (5.3.5)$$

and taking N = constant, for simplicity, the equation

$$\frac{f^2}{a^2 \cos \phi} \left(\frac{\cos \phi}{f^2} \Psi_\phi \right)_\phi + \frac{f^2}{N^2} \Psi_{zz} + n_s^2 \Psi = 0 \qquad (5.3.6)$$

is obtained, where

$$n_s^2 = \frac{\bar{q}_\phi}{a\bar{u}} - \frac{s^2}{a^2 \cos^2 \phi} - \frac{f^2}{4N^2 H^2} . \qquad (5.3.7)$$

Equations (5.3.6) and (5.3.7) are the spherical analogs of Eqs. (4.5.27) and (4.5.28) respectively; in particular, n_s^2 is the squared refractive index.

   In Section 4.5.4 some cases were mentioned in which the basic flow $\bar{u}$ is simple enough that Eq. (5.3.6), or rather its beta-plane analog [Eq. (4.5.27)], can be solved semianalytically, under certain approximations. Matsuno studied more general flows $\bar{u}(\phi, z)$, representative of the observed zonal-mean wind in the northern winter stratosphere, for which a numerical method of solution was necessary. He considered only the "ultralong" stationary-wave components $s = 1$–3, which by the Charney–Drazin criterion [Eq. (4.5.16)] include those expected to propagate into the winter stratosphere, and for each s forced the model from below by the appropriate zonal Fourier component of the observed monthly mean 500-mb height for January 1967, using a boundary condition of the form of Eq. (3.1.6b). By linearity, the complete solution is the sum of the responses in each Fourier component. A radiation condition was imposed at the top of the model, $z = z_1$, say (near 60 km altitude), by assuming that $\bar{u}$ is independent of z above z_1. This allows separable solutions to be found above z_1, and the upward-propagating solutions can be identified as in Section 4.5.2; these then supply the required upper boundary condition. Only the Northern Hemisphere was considered; the lateral boundary conditions were that $\Psi = 0$ at the pole, to ensure bounded solutions there, and that $\Psi = 0$ at the equator. The latter is somewhat artificial, but can be rationalized by the fact that a critical line (a "zero wind line" $\bar{u} = 0$, for these stationary waves; see Section 5.6) is present near the equator in the chosen basic flow (see Fig. 5.7a). On the equatorward side of this line, n_s^2 is large and negative, and the waves are evanescent there, so that the solution is insensitive to the actual equatorial boundary condition. The singularity at the zero wind line was removed by replacing $\bar{u}$ in the denominator of the first term on the right of (5.3.7) by $\bar{u} - i\gamma a^{-1} \cos \phi$, where γ is a small damping coefficient.

The basic zonal flow used by Matsuno is shown in Fig. 5.7a; it is reasonably similar to the January monthly mean Northern Hemisphere geostrophic wind of Fig. 5.2a. In Fig. 5.7b is shown the quantity $a^2 n_0^2$, where

$$n_0^2 = \frac{\bar{q}_\phi}{a\bar{u}} - \frac{f^2}{4N^2 H^2} = n_s^2 + \frac{s^2}{a^2 \cos^2 \phi} \; . \tag{5.3.8}$$

Plots of the calculated amplitude and phase of Ψ for $s = 1$ and 2 are shown

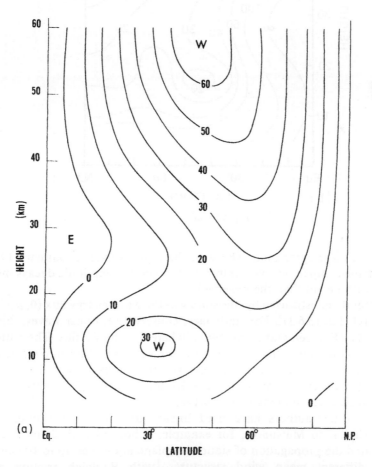

Fig. 5.7. (a) Basic wind distribution $\bar{u}(\phi, z)$ and (b) $a^2 n_0^2$, used by Matsuno in the study of the propagation of stationary planetary waves into the stratosphere. The refractive index squared for zonal wave number s can be obtained from Eq. (5.3.8). [After Matsuno (1970). American Meteorological Society.] *Figure continues.*

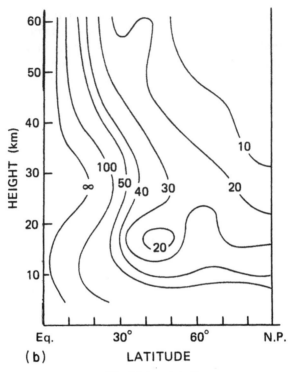

Fig. 5.7 (*continued*)

in Fig. 5.8; these agree quite well with the observations for January 1967, although the computed wave number 2 geopotential amplitude decays more rapidly with height than the observed.

Another diagnostic used by Matsuno was the wave-energy flux $(0, \overline{\rho_0 v' \Phi'}, \overline{\rho_0 w' \Phi'})$ [cf. Eq. (3.6.3)]. For stationary, conservative, linear waves, this is equal to $\bar{u}\mathbf{F}$ (Eliassen and Palm, 1961), where $\mathbf{F}$ is the EP flux. The pattern of arrows representing the wave-energy flux vectors in Matsuno's calculations shows the characteristic upward and then equatorward orientation noted in the observed stationary-wave EP fluxes in Fig. 5.5, as well as in the ray-tracing calculations of Fig. 4.14a.

Several other authors have used linear quasi-geostrophic numerical models similar to Matsuno's; for example, Schoeberl and Geller (1977) investigated the propagation of stationary planetary waves up to 100 km in several different mean wind structures, with Rayleigh friction and Newtonian cooling present. They found considerable sensitivity of the geopotential amplitude of the waves to the strength of the polar night jet and the magnitude of the Newtonian cooling. They interpreted their results

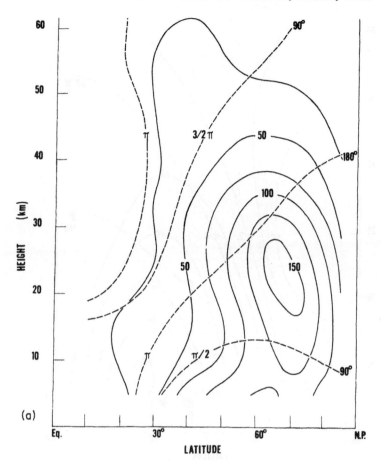

Fig. 5.8. Computed amplitude $|\Psi|$ (in meters, times a factor 1.43: solid lines) and phase of Ψ (longitude of ridges: dashed lines) for (a) $s = 1$, (b) $s = 2$, forced by the observed mean 500-mb height for January 1967. [After Matsuno (1970). American Meteorological Society.] *Figure continues.*

in terms of the vertical propagation of the gravest horizontal modes $\Theta_n(\phi)$ in an expansion of the solution Ψ of Eq. (5.3.6) in the form

$$\Psi = \sum_n \Psi_n(z)\Theta_n(\phi);$$

the Θ_n approximate the planetary-wave Hough functions discussed in Section 4.2.

The most comprehensive linear calculation to date of the winter-mean stationary planetary waves throughout the Northern-Hemisphere stratosphere is that of Lin (1982). He used a hemispheric primitive-equation model

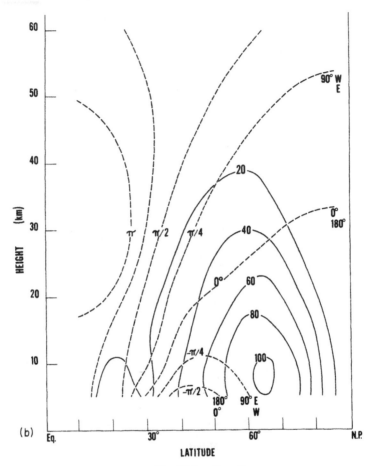

Fig. 5.8 (*continued*)

of the atmosphere from the ground up to 60 km, solving a set of equations equivalent to Eqs. (3.4.2), with various representative forms of $\bar{u}(\phi, z)$. Dissipation in the forms of Rayleigh friction and Newtonian cooling were incorporated, with specified thermal forcing, and topographic forcing included in the lower boundary condition.

Lin's results confirm those of Matsuno and of Schoeberl and Geller in most respects; in particular, they show that the vertical propagation of the stationary waves into the stratosphere is sensitive to the zonal-mean wind structure, and especially its latitudinal curvature and the latitude of the polar night jet. They also indicate that only the $s = 1$ and 2 components can propagate into the stratosphere, in qualitative agreement with the Charney and Drazin criterion. With topographic, but not thermal, forcing, the solution simulates quite well the observed winter-mean stationary waves in the troposphere and stratosphere. However, inclusion of diabatic heating degrades the results somewhat. The reason for this is not entirely clear, but the good simulation of tropospheric waves when topographic forcing alone is used in this linear model may be partly fortuitous. This is because experiments using *nonlinear* tropospheric general circulation models (see, e.g., Held, 1983) indicate that topographic and thermal forcing produce roughly comparable contributions to the tropospheric stationary waves.

Linear models of the type discussed here do not include the time-averaged nonlinear effects of the transient eddies on the stationary waves. Recent studies suggest that these effects may be important in a number of tropospheric phenomena; the same may also be true in the stratosphere and mesosphere, although no studies have yet investigated this possibility in detail. However, general circulation models of the middle atmosphere do perform nonlinear simulations of the stationary planetary waves, and these are discussed in Section 11.2.

5.4   Detailed Linear Models of Free Traveling Planetary Waves in the Atmosphere

In Section 4.4 we presented a simple linear theory of the "5-day wave," the most prominent observed free traveling Rossby wave or global normal mode. This theory took the basic zonal flow $\bar{u}$ to be zero. In the present section we outline the theoretical methods that can be used to search for free traveling waves in more realistic zonal-mean wind structures $\bar{u}(\phi, z)$.

The mathematical problem amounts *in principle* to seeking eigensolutions of the form

$$\Phi' = \text{Re}[\hat{\Phi}(\phi, z)e^{i(s\lambda - \omega t)}] \qquad (5.4.1)$$

to the linearized primitive equations of Eqs. (3.4.2) on the sphere, subject to an upper boundary condition of decaying wave-energy density (as in Section 4.4) and to the lower boundary condition [Eq. (3.1.6a)]. The linearized version of the latter can be applied at $z = 0$, as well as $z^* = 0$, when topography is absent $(h = 0)$; it takes the form

$$\Phi'_t + \frac{\bar{u}}{a \cos \phi}\Phi'_\lambda + \frac{v'}{a}\bar{\Phi}_\phi + w'\bar{\Phi}_z = 0 \qquad \text{at} \quad z = 0. \qquad (5.4.2)$$

In practice, the easiest way of finding these solutions is to adopt the method of Geisler and Dickinson (1976) and add a forcing term

$$\text{Re}[\hat{W}(\phi)e^{i(s\lambda - \omega t)}]$$

to the right of Eq. (5.4.2), corresponding to g times an assumed nonzero vertical velocity imposed at $z^* = 0$. The integer zonal wave number s and (real) frequency ω are varied until a large, quasi-resonant response occurs in the model atmosphere. (The restriction to real ω means that only stable modes are considered: unstable modes, with complex ω, are discussed in Section 5.5.) An infinite, truly resonant response would correspond to a mode that is a true free mode of the unforced problem. In practice, weak damping in the form of Newtonian cooling and Rayleigh friction is usually included in Eqs. (3.4.2), and this ensures that the response remains finite. Moreover, some of the quasi-resonant modes may in fact be weakly propagating (or "leaky"), rather than evanescent, as $z \to \infty$, and a radiation condition may be required.[1] It is reasonable to suppose that those theoretical modes with the largest response to the imposed forcing may be good candidates for representing free traveling modes in the atmosphere, which are perhaps excited by random or other forcing effects.

Geisler and Dickinson used their method to search for theoretical waves of zonal wave number 1 with periods close to 5 days in zonal-mean winds that are representative of the middle atmosphere at solstice. They found that the period and low-level structure of the quasi-resonant mode are not very sensitive to the zonal wind configuration, and resemble those of the simple solution of Section 4.4 quite closely. In the upper stratosphere and mesosphere, however, the amplitude and phase of the model 5-day wave become strongly asymmetric about the equator, with relatively large geopotential amplitude in the summer mesosphere. Geisler and Dickinson also included Newtonian cooling with a 10-day relaxation time: the main effect of this was to halve the summer mesosphere maximum.

Another free traveling mode to receive much attention is the 2-day wave (Fig. 4.11), which may perhaps be identified with the gravest antisymmetric (in w') mode for $s = 3$ in the absence of mean winds. Here the inclusion of a mean flow $\bar{u}(\phi, z)$ in numerical calculations (e.g., by Salby, 1981a,b) leads to larger amplitudes in the summer hemisphere than in the winter hemisphere, in agreement with observation. The observed 16-day wave can likewise be identified with the second symmetric $s = 1$ mode; this mode is also fairly sensitive to the details of the basic wind field. Some observations of the 16-day wave in the upper stratosphere are presented in Fig. 5.9, and some theoretical calculations are shown in Fig. 5.10. Note that both show roughly equatorially symmetric amplitude structure at equinox, but much

---

[1] One can also regard the (real) quasi-resonant frequencies as approximations to the true (complex) eigenfrequencies of the dissipative problem. From this viewpoint, the distinction between the free modes considered here and the unstable modes of Section 5.5 may become blurred.

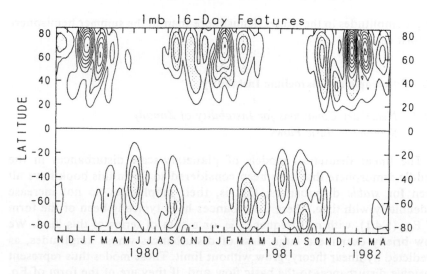

Fig. 5.9. The observed "16-day wave," as revealed by a latitude–time section of the geopotential height amplitude of $s = 1$ westward-traveling waves at 1 mb, band-passed to include periods between 12 and 24 days. (The results are insensitive to the precise bandwidth.) Values are averaged over 10 days. The contour interval is 50 m, and stippling indicates values greater than 100 m. [After Hirooka and Hirota (1985). American Meteorological Society.]

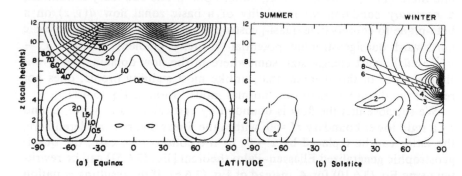

Fig. 5.10. Geopotential height amplitude (arbitrary units) of the second symmetric mode for $s = 1$ in idealized background wind $\bar{u}(\phi, z)$ and temperature $\bar{T}(\phi, z)$ fields for (a) equinox (period = 16.4 days) and (b) solstice (period = 15.7 days). [After Salby (1981b). American Meteorological Society.]

larger amplitudes in the winter hemisphere than in the summer hemisphere at solstice.

5.5   Barotropic and Baroclinic Instability

5.5.1   Necessary Conditions for Instability of Zonally Symmetric Basic Flows

The linear theoretical models of planetary-scale disturbances in the middle atmosphere that have been considered so far in this book have all been for *stable* disturbances; that is, their amplitudes do not increase indefinitely with time. These disturbances have typically been of the form of Eq. (5.4.1), with real frequency ω, or a sum of terms of this kind. We now briefly examine a large class of disturbances whose amplitudes, as predicted by linear theory, grow without limit. These modes thus represent *unstable* disturbances to the basic flow and, if they are of the form of Eq. (5.4.1), their frequencies ω are complex, with positive imaginary parts.

Two types of large-scale instability that may be important in the middle atmosphere are barotropic and baroclinic instability; these are both described by quasi-geostrophic theory. Barotropic instability depends on large horizontal curvature of the basic flow profile, while baroclinic instability depends, roughly speaking, on vertical curvature. Combined barotropic-baroclinic instability can occur in a basic flow that varies both horizontally and vertically.

A more precise statement of some of the necessary conditions for barotropic or baroclinic instability is provided by the theorem of Charney and Stern (1962). This states that, under appropriate boundary conditions, a *necessary* condition for instability of a basic zonal flow $\bar{u}(y, z)$ on a beta-plane to conservative quasi-geostrophic disturbances is that the basic northward quasigeostrophic potential vorticity gradient $\bar{q}_y \equiv \beta - \bar{u}_{yy} - \rho_0^{-1}(\rho_0 \varepsilon \bar{u}_z)_z$ must change sign somewhere in the flow domain. We give a simple proof of this theorem that, unlike most stability proofs, does not restrict the disturbances to the "normal mode" form of Eq. (5.4.1).

We suppose that the flow is bounded by vertical walls at $y = 0, L$, and by a rigid lower boundary $z^* = 0$; within the linear theory considered here, the latter can be replaced by $z = 0$. In the conservative case, the quasi-geostrophic generalized Eliassen–Palm theorem [Eq. (3.6.5)] can be rewritten using Eq. (3.6.10) for A, instead of Eq. (3.6.6). If the resulting equation is integrated over y and z we obtain

$$\frac{\partial}{\partial t} \int_0^\infty \int_0^L \frac{1}{2} \rho_0 \bar{q}_y \overline{\eta'^2} \, dy \, dz = \int_0^L \rho_0 f_0 \overline{v'\theta'} / \theta_{0z}|_{z=0} \, dy \qquad (5.5.1)$$

to second order in amplitude, using the sidewall boundary conditions $v' = 0$ at $y = 0, L$ and assuming that $\rho_0 \overline{v' \theta'} / \theta_{0z} \to 0$ as $z \to \infty$; here η' represents the northward parcel displacement, defined for example in Eq. (3.6.8).

The linearized lower boundary condition [Eq. (5.4.2)] can be used to simplify the right hand side of Eq. (5.5.1). In the quasigeostrophic beta-plane case, this boundary condition becomes

$$\bar{D}\psi' + v'\bar{\psi}_y + w'f_0^{-1}\bar{\Phi}_z = 0 \qquad \text{at} \quad z = 0, \tag{5.5.2}$$

where $\bar{D} = \partial/\partial t + \bar{u}\, \partial/\partial x$ and $\psi = f_0^{-1}[\Phi - \Phi_0(z)]$ by Eq. (3.2.4). The linearized quasi-geostrophic disturbance potential temperature equation is

$$\bar{D}\theta' + v'\bar{\theta}_y + w'\theta_{0z} = 0 \tag{5.5.3}$$

[see Eqs. (3.2.9d) and (3.4.2e)]. Now w' is not the geometric disturbance vertical velocity and does not vanish at $z = 0$; however, it can be eliminated between Eqs. (5.5.2) and (5.5.3). Equations (3.2.3), (3.2.5'), (3.2.13), (3.5.5d), and (3.6.8) then give

$$\bar{D}\left\{ \frac{R}{f_0 H} \theta' - \eta'(\bar{u}_z - B\bar{u}) - B\psi' \right\} = 0 \qquad \text{at} \quad z = 0, \tag{5.5.4}$$

where

$$B \equiv N^2 H / R\bar{T} = (T_s/\bar{T})N^2/g.$$

Integrating Eq. (5.5.4), given suitable initial conditions, we obtain

$$\theta' = f_0 H R^{-1}[\eta'(\bar{u}_z - B\bar{u}) + B\psi'] \qquad \text{at} \quad z = 0,$$

and hence

$$f_0 \overline{v'\theta'}/\theta_{0z} = \varepsilon(\bar{u}_z - B\bar{u})\overline{v'\eta'} = \varepsilon(\bar{u}_z - B\bar{u})(\tfrac{1}{2}\overline{\eta'^2})_t, \qquad \text{at} \quad z = 0, \tag{5.5.5}$$

by Eqs. (3.2.13), (3.2.16), and (3.6.8). Substitution of Eq. (5.5.5) into Eq. (5.5.1) yields

$$\frac{\partial}{\partial t}\int_0^\infty \int_0^L \frac{1}{2}\rho_0 \bar{q}_y \overline{\eta'^2}\, dy\, dz - \frac{\partial}{\partial t}\int_0^L \frac{1}{2}\rho_0 \varepsilon(\bar{u}_z - B\bar{u})\overline{\eta'^2}\bigg|_{z=0} dy = 0, \tag{5.5.6}$$

or, more compactly,

$$\frac{\partial}{\partial t}\int_{0_-}^\infty \int_0^L \frac{1}{2}\rho_0 \tilde{q}_y \overline{\eta'^2}\, dy\, dz = 0, \tag{5.5.7}$$

where

$$\tilde{q}_y \equiv \bar{q}_y - \varepsilon\, \delta(z)(\bar{u}_z - B\bar{u})|_{z=0}, \tag{5.5.8}$$

$\delta(z)$ is the Dirac delta function, and the z integration extends from just below $z = 0$. The term $B\bar{u}|_{z=0}$ in Eq. (5.5.8) is known as a "non-Doppler" term. It is not invariant under a Galilean transformation $\bar{u} \to \bar{u} - u_0$, although it is often negligible compared to $\bar{u}_z$ at $z = 0$.

If $\tilde{q}_y$ is positive everywhere, the integral in Eq. (5.5.7) is positive definite and can be taken as a global measure of disturbance amplitude (see the end of Section 3.6). Equation (5.5.7) states that this quantity is constant in time; in particular it does not grow, and if it is initially small it will remain small. This accords with the usual properties of a globally *stable* disturbance. A similar result holds if $\tilde{q}_y$ is negative everywhere.

On the other hand, these considerations do not apply if $\tilde{q}_y$ takes both positive and negative values. The parcel displacement η' may then perhaps grow indefinitely with time, indicating an *unstable* disturbance, at least while linear theory remains valid. The change of sign of $\tilde{q}_y$ can be due to the interior potential vorticity gradient $\bar{q}_y$ changing sign, or to the boundary term somewhere having the opposite sign to the interior $\bar{q}_y$. An important example of the latter case often occurs when vertical shear—and thus a horizontal potential temperature gradient, by the thermal wind equation, Eq. (3.5.5d)—is present at $z = 0$. It should be recalled that $\bar{q}_y$ can be related to the northward gradient of Ertel's potential vorticity on an isentropic surface, by Eq. (3.8.10).

A stronger condition, due essentially to Fjørtoft (1950), states that if there exists a constant u_0 such that $(\bar{u} - u_0)\bar{q}_y < 0$ for all y and z, then the flow is stable to conservative quasi-geostrophic disturbances. This can be proved by a method similar to that given above for the Charney–Stern theorem, but using the quasi-geostrophic version of the wave-energy equation, Eq. (3.6.3), in addition to the generalized Eliassen–Palm theorem. The Charney–Stern theorem follows as a special case by choosing $u_0 > \max(\bar{u})$ if $\tilde{q}_y > 0$ everywhere and $u_0 < \min(\bar{u})$ if $\tilde{q}_y < 0$ everywhere.

It should be noted that these theorems only give *sufficient* conditions for *stability* or, conversely, *necessary* conditions for *instability*. They cannot by themselves tell us that a particular flow is definitely unstable; an explicit search for unstable modes will normally be necessary. The simplest cases occur when $\bar{u}$ and $\bar{q}_y$ depend only on y (leading to the possibility of barotropic instability) or when they depend only on z (leading to the possibility of baroclinic instability). Numerous calculations of unstable disturbances to such flows have been made.

5.5.2. Barotropic and Baroclinic Instability Calculations for Representative Middle Atmosphere States

In this section we describe some calculations of unstable modes that have been suggested as possible explanations for some of the observed traveling-wave structures mentioned in Sections 4.4 and 5.2.3, or at least for the initiation of such structures.

The first example is that of the localized eastward-moving "warm pool" documented by Prata (1984) in the Southern Hemisphere winter upper stratosphere and shown in Fig. 4.12. Hartmann (1983) has shown that, in the region where this phenomenon occurs, the spherical analog of $\bar{q}_y$ tends to change sign, owing to strong meridional curvature of the zonal-mean zonal wind on the poleward flank of the stratospheric jet (Fig. 5.11). He performed linear instability calculations for this basic flow, and found barotropically unstable disturbances of zonal wave numbers 1 and 2, which move eastward with periods of 3–4 days and 1.5–2 days, respectively, and which have e-folding times of a few days and geopotential amplitude maxima near 70°S. He suggested that the localized disturbances observed by Prata may represent a "phase-locking" of these wave-number 1 and 2 modes. However, linear theory is only able to describe the early, small-amplitude evolution of disturbances of this kind, and a full understanding will probably need to await nonlinear calculations, in which the unstable disturbances are allowed to grow to (and perhaps equilibrate at) finite amplitude.

The Southern Hemisphere zonal-mean wind structure shown in Fig. 5.11 also has a reversed potential vorticity gradient on the equatorward flank of the jet, and Hartmann (1983) showed that instabilities are associated with this also. These bear some resemblance to observed modes that move slowly

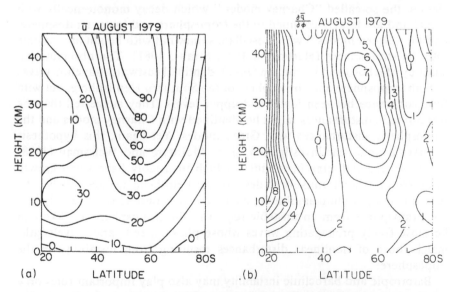

Fig. 5.11. (a) Basic wind distribution $\bar{u}(\phi, z)$ and (b) quasi-geostrophic potential vorticity gradient $\partial\bar{q}/\partial\phi$ (divided by Ω; regions of negative $\partial\bar{q}/\partial\phi$ stippled) for the month of August 1979 in the southern hemisphere. [After Hartmann (1983). American Meteorological Society.]

eastward in the Southern Hemisphere. However, Hartmann (1985) points out that the theoretical modes have a poleward EP flux ($-\rho_0 \overline{u'v'} < 0$), while the observed modes have an equatorward EP flux.

Plumb (1983) examined the baroclinic instability of profiles $\bar{u}(z)$, which broadly capture the vertical variation of the zonal wind in the Southern Hemisphere winter mesosphere. He found that a baroclinically unstable mode of zonal wave number 3 and 2-day period can develop when the westerly upper mesospheric shear exceeds $6 \text{ m s}^{-1} \text{ km}^{-1}$. He suggested that this mechanism may elucidate certain aspects of the observed 2-day wave that are not explained by the "free normal mode" theory of Salby (1981a) (see Section 5.4). Further observations and, again, a nonlinear treatment may be needed to resolve the question; it could be that the instability process describes the generation of the 2-day wave and Salby's calculations, despite being linear, describe some aspects of a finite-amplitude equilibrated state.

Plumb's example, like those of Hartmann (1983), depends on a change of sign of $\bar{q}_y$ within the atmosphere to violate the Charney–Stern stability criterion. Other studies have investigated possible baroclinic instabilities in the troposphere, stratosphere, and mesosphere that depend on temperature gradients (or, equivalently, vertical shear) at the ground, so that the delta-function term in Eq. (5.5.8) is opposite in sign to $\bar{q}_y$ in the interior. The resulting unstable modes for typical winter flows tend to fall into two broad classes: the so-called "Charney modes," which decay monotonically with height and are largely confined to the troposphere and lower stratosphere, and the "Green modes," which oscillate somewhat with height and radiate into the stratosphere (Hartmann, 1979; Straus, 1981). The observational study by Mechoso and Hartmann (1982) identified eastward-traveling waves in both troposphere and stratosphere of the Southern Hemisphere, but with little coherence between lower and upper levels. They suggested that the tropospheric disturbances might be identified with Charney modes and the stratospheric disturbances with Green modes. Yet again, this hypothesis deserves a nonlinear investigation, since McIntyre and Weissman (1978) point out that radiating instabilities, which invariably have small growth rates, are in general unlikely to describe far-field behavior accurately. The reason is that, at finite amplitude, the modes that radiate most readily into the stratosphere from an unstable region in the troposphere are likely to be those freely propagating waves whose phase speeds and wavelengths match those of nonlinear disturbances resulting from instability in the troposphere.

Barotropic and baroclinic instability may also play important roles on a smaller scale in the middle atmosphere—for example, in leading to the apparent break-up of tongues of potential vorticity into "blobs," as mentioned at the end of Section 5.2.3 and depicted in Figs. 5.6b,c.

5.6   Planetary-Wave Critical Layers

It was mentioned in Section 4.5 that the theory of linear, steady, conservative Rossby waves breaks down at "critical surfaces," on which the basic zonal wind $\bar{u}(\phi, z)$ matches the zonal phase speed c. This is because the factor $(\bar{u} - c)^{-1}$ in the wave equation [Eq. (4.5.9)] or, more generally, in the refractive index squared [Eq. (4.5.28)], becomes infinite there. The infinity is removed by the inclusion of further physical effects, namely, wave transience, dissipation, and nonlinearity (the effects that violate nonacceleration conditions), one or more of which must become important in a region called the "critical layer," which surrounds the critical surface. The detailed dynamical behavior within the critical layer depends on the relative importance of these different effects, and this behavior in turn can have a crucial influence on the wave structure far from the critical surface. An understanding of critical-layer dynamics is thus likely to be of great importance for the study of planetary-wave propagation in the presence of critical surfaces, and much theoretical research is currently being devoted to the phenomenon.

A convenient idealized model of critical layer behavior uses beta-plane geometry and takes the flow to be barotropic (i.e., independent of z); in this case the critical surface reduces to a "critical line." We suppose that a steady zonal-mean flow $\bar{u}(y)$ of the form sketched in Fig. 5.12 supports

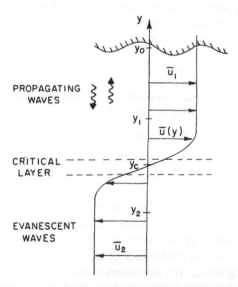

Fig. 5.12. Schèmatic diagram of a basic zonal shear flow $\bar{u}(y)$ containing Rossby waves of zero zonal phase speed forced by a corrugated northern boundary near $y = y_0$. See text for details.

small-amplitude Rossby waves of zonal wave number $k(>0)$ and zero zonal phase speed $(c = 0)$. For the purposes of the model, these waves can be regarded as generated, say, by the flow past a northern boundary near $y = y_0$ containing stationary sinusoidal corrugations. This configuration mimics aspects of the quasi-horizontal equatorward propagation of stationary planetary waves in the mid- and low-latitude stratosphere (Fig. 5.5). It is assumed that $\bar{u}$ is eastward and constant $(\bar{u} = \bar{u}_1 > 0)$ in the northern region $y_1 < y < y_0$; if the waves are essentially steady, linear, and conservative there, it can easily be shown by the methods of Section 4.5 that the disturbance stream function may be written in the form

$$\psi' = C \operatorname{Re}[e^{ikx}(e^{-ily} + Re^{ily})] \qquad \text{for} \quad y_1 < y < y_0 \qquad (5.6.1)$$

in that region. Here C is a real constant, R is a complex constant, and $l = +(\beta\bar{u}_1^{-1} - k^2)^{1/2}$, which is real provided that $\bar{u}_1$ is chosen to be less than βk^{-2}. It is easy to verify that the northward Eliassen–Palm flux component [see Eq. (3.5.6)] is given by

$$F^{(y)} \equiv -\rho_0 \overline{v'u'} \equiv \rho_0 \overline{\psi'_x \psi'_y}$$

$$= -\tfrac{1}{2}\rho_0 C^2 kl(1 - |R|^2) \qquad \text{for} \quad y_1 < y < y_0. \qquad (5.6.2)$$

Equation (5.6.1) thus represents the superposition of a southward-propagating wave of streamfunction amplitude C and EP flux $-\tfrac{1}{2}\rho_0 C^2 kl$ and a northward-propagating wave of amplitude CR and EP flux $\tfrac{1}{2}\rho_0 C^2 |R|^2 kl$, which is "reflected" by the shear layer south of y_1.

In the region south of y_2 it is assumed that $\bar{u}$ is westward and constant $(\bar{u} = \bar{u}_2 < 0)$ and so

$$\psi' \propto e^{ikx+\Lambda y} \qquad \text{for} \quad y < y_2, \qquad (5.6.3)$$

where $\Lambda = (\beta|\bar{u}_2|^{-1} + k^2)^{1/2}$. This represents a disturbance that is evanescent with decreasing latitude; it can be verified that

$$F^{(y)} \equiv 0 \qquad \text{for} \quad y < y_2. \qquad (5.6.4)$$

Thus far, the theory is simple; the difficulties arise when the wave solutions are sought in the intervening "shear zone" between y_1 and y_2, which includes the critical line, $y = y_c$ say, where $\bar{u} = c \equiv 0$. The mathematical theory describing the dynamics of the waves in the shear zone, and in the critical layer in particular, is quite complicated, and we only quote the main results here. For simplicity, we suppose that the curvature of the $\bar{u}(y)$ profile is so small that partial reflections of the waves from outside the critical layer can be neglected. We concentrate on the ways in which the details of the flow in the critical layer affect the "reflection coefficient" R, and thus the amplitude of the reflected waves in $y > y_1$, far from the critical line.

It has long been known that if the waves are steady, linear, and dissipative in the critical layer (as for example in Matsuno's model of Section 5.3), the layer is perfectly absorbing under the hypotheses mentioned above, with $R = 0$. A similar result was shown by Dickinson (1970) and Warn and Warn (1976) to hold at moderately large times under the long-wave approximation $k(y_1 - y_2) \ll 1$ (specifically, for $1 \ll t^* \ll \alpha^{-1/2}$, where $t^* = k\bar{u}_1 t$ and α is a small dimensionless wave-amplitude parameter) for linear, conservative, transient waves that are "switched on" at some initial instant $t = 0$ and maintained thereafter. However, at larger times ($t^* \sim \alpha^{-1/2}$), nonlinear effects must become important in the critical layer [which then has a thickness $\sim \alpha^{1/2}(y_1 - y_2)$]. Stewartson (1978) and Warn and Warn (1978) showed that the conservative, nonlinear critical layer then oscillates between partial absorption ($|R|^2 < 1$), reflection ($|R|^2 = 1$), and over-reflection ($|R|^2 > 1$), tending to a state of perfect reflection ($|R|^2 = 1$) at still larger times ($t^* \gg \alpha^{-1/2}$). It has recently been shown that this flow is, in fact, barotropically unstable (Killworth and McIntyre, 1985; Haynes, 1985); while, at first sight, the resulting mixing in the critical layer might be expected to introduce enhanced dissipative effects, and thus to bring about absorption, Killworth and McIntyre demonstrate that absorption does not necessarily occur: under fairly general conditions the critical layer remains a perfect reflector in a time-integrated sense.

A partial physical description of the reflection process is as follows. In the Stewartson-Warn-Warn model the flow within the critical layer takes the form of "Kelvin's cats'-eyes." As shown schematically in Fig. 5.13, this flow tends to wrap up the contours of absolute vorticity $\zeta \equiv f - u_y + v_x$ [to which the quasi-geostrophic potential vorticity q reduces in this barotropic model: see Eq. (3.2.15)]. The barotropic analog of Eq. (3.5.10) is

$$\overline{v'\zeta'} = \rho_0^{-1} \frac{dF^{(y)}}{dy};$$

integrating this over the shear layer and using Eqs. (5.6.2) and (5.6.4) we obtain

$$\int_{y_2}^{y_1} \overline{v'\zeta'} \, dy = -\tfrac{1}{2}C^2 kl(1 - |R|^2). \tag{5.6.5}$$

Since nonacceleration conditions hold outside the critical layer, $\overline{v'\zeta'} \equiv 0$ there, and the left of Eq. (5.6.5) can be replaced by

$$\int_{\text{critical layer}} \overline{v'\zeta'} \, dy. \tag{5.6.6}$$

Examination of the sequence of diagrams in Fig. 5.13 shows that Eq. (5.6.6) oscillates between negative and positive values of continually decreasing

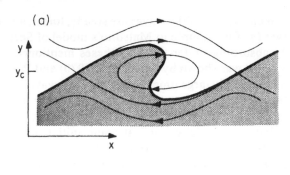

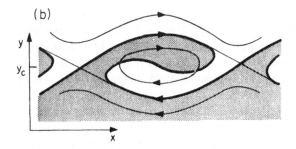

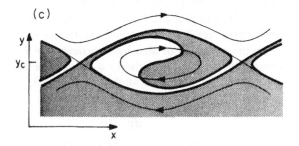

Fig. 5.13. Time-dependent analytical solution for a barotropic Rossby-wave nonlinear critical layer at non-dimensional times (a) $\alpha^{1/2}t^* = 2$, (b) $\alpha^{1/2}t^* = 4$, (c) $\alpha^{1/2}t^* = 6$ (Stewartson, 1978; Warn and Warn, 1978). The flow is periodic in x, and the y scale is greatly exaggerated: the initial critical line was at $y = y_c$. The thin lines indicate streamlines, and the "Kelvin's cats'-eyes" are the lens-shaped regions of closed streamlines; their width is of order $\alpha^{1/2}(y_1 - y_2)$. The thick line shows the successive positions of the material absolute vorticity contour, $\zeta = \zeta_c$ say, that initially lay along $y = y_c$. Thus $\zeta < \zeta_c$ in the stippled regions and $\zeta > \zeta_c$ in the unstippled regions. For this model, $\bar{v} \equiv 0$, so $\overline{v'\zeta'} = \overline{v'\zeta} = \overline{v'(\zeta - \zeta_c)} = \overline{v(\zeta - \zeta_c)}$. In (a) it can be seen that most of the stippled region within the cats'-eyes has $v > 0$, and most of the unstippled region has $v < 0$. Thus $\overline{v'\zeta'} = \overline{v(\zeta - \zeta_c)} < 0$ throughout most of the critical layer, Eq. (5.6.6) is negative, and $|R| < 1$ by Eq. (5.6.5), indicating partial absorption. Similar arguments show that in (b), $\overline{v'\zeta'} \approx 0$ and $|R| \approx 1$, indicating near-perfect reflection, and in (c) $\overline{v'\zeta'} > 0$ and $|R| > 1$, indicating overreflection. (Courtesy of P. H. Haynes.)

magnitude as the velocity field wraps up the vorticity contours more and more. Thus, by Eq. (5.6.5), $|R|^2$ oscillates between values less than and greater than 1, approaching 1 (perfect reflection) at large times. The picture is complicated by the fact that the Stewartson–Warn–Warn flow goes unstable, leading to complex small-scale features in the velocity and vorticity fields. Nevertheless, Killworth and McIntyre's general result shows that $|R|^2$ cannot depart systematically from 1 at large times in the model, even when this happens.

While great strides have recently been made in understanding these models of critical layers, the assumptions made in the theory (for example, that the waves are small-amplitude disturbances to an initially zonal flow) are of course still rather poor idealizations of actual atmospheric situations. Thus, the implications of the theory for the modeling of planetary waves and for the interpretation of atmospheric observations are not yet clear. It is, however, of interest that the time-dependent nonlinear theories predict behavior within the critical layer that bears some resemblance to the breaking-wave structures observed by McIntyre and Palmer (1983, 1984) in isentropic potential vorticity maps of the stratosphere (see Section 5.2.3). (The z-independent vertical absolute vorticity component in the barotropic theory plays the role of the potential vorticity in the more general case.) In particular, the theoretical flows exhibit the same kind of rapid and irreversible deformation of material contours (Fig. 5.13) as is suggested by the stratospheric maps of McIntyre and Palmer (Figs. 5.6b,c), and for this reason these authors propose that nonlinear critical-layer theory may model certain aspects of the observed breaking planetary wave events. [Rapid, irreversible deformation of material contours can also occur, for example, when vortices interact nonlinearly (see, e.g., Dritschel, 1986). Such flows, too, may turn out to provide useful idealizations of features of breaking planetary waves.]

References

5.2. Recent compilations of stratospheric monthly mean climatological data include those of Hamilton (1982a,b), Geller *et al.* (1983), and Wu *et al.* (1984). The most comprehensive collection is to be found in the 1986 COSPAR International Reference Atmosphere [Labitzke *et al.* (1985)]. Recent reviews of observed stationary and transient eddies in the *troposphere* are those of Wallace (1983) and Holopainen (1983), respectively.

5.3. The fundamentals of the theory and modeling of *tropospheric* stationary planetary waves are discussed by Held (1983). Developments of the Charney and Drazin semianalytical theory of stationary planetary waves in the stratosphere are presented by Dickinson (1968) and Simmons (1974), for example. Discussions of the self-consistency of Eqs. (5.3.1)–(5.3.4) are given by Matsuno (1970) and Palmer (1982). The paper by Alpert *et al.* (1983) gives a detailed study of the sensitivity of a linear quasi-geostrophic stationary planetary-wave model to differing representations of the topographic and thermal forcing. Jacqmin and Lindzen (1985) present calculations of stationary planetary waves in the troposphere and lower stratosphere derived from a high-resolution, linear, primitive-equation model. The time-averaged nonlinear effects of transient eddies on stationary waves in the troposphere are considered by Hoskins (1983).

5.5. The proof given in Section 5.5.1 is due essentially to Bretherton (1966a), who used ideas originally developed by Taylor (1915). "Non-Doppler" effects are considered for example by White (1982). The proof of Fjørtoft's (1950) result using methods similar to those of this section is given by Blumen (1978). Examples of barotropically and baroclinically unstable modes, and thorough discussions of their dynamics, are given by Pedlosky (1979) and Gill (1982), among others.

5.6. Killworth and McIntyre (1985) review the current understanding of the theory of planetary-wave critical layers and present some important new results.

Chapter 6 | Stratospheric Sudden Warmings

6.1 Introduction

It has been shown in Section 5.2.1 that the climatological zonal-mean zonal winds in the winter stratosphere are generally westerly and increase with height, peaking in the "polar night jet" vortex (see, for example, Figs. 5.2a,c). The zonal-mean climatological temperature fields (Figs. 5.1a,c) decrease towards the winter pole on each pressure surface in the stratosphere.

During some winters, however, this zonal-mean configuration is dramatically disrupted, with polar stratospheric temperatures increasing rapidly with time, leading to a poleward increase of zonal-mean temperature and, on occasion, a reversal of zonal-mean winds to an easterly direction. Such an event is called a *stratospheric sudden warming*. It is defined, somewhat arbitrarily, to be a *major warming* if at 10 mb or below the zonal-mean temperature increases poleward from 60° latitude and the zonal-mean zonal wind reverses. If the temperature gradient reverses there but the circulation does not, it is defined to be a *minor warming*. Major warmings occur on average about once every other winter in the Northern Hemisphere, but have not been observed in the Southern Hemisphere. Minor warmings in the upper stratosphere occur more frequently, and in both hemispheres. Table 6.1 gives a list of recent occurrences of major warmings in the Northern Hemisphere. The sudden and dramatic nature of stratospheric warmings is illustrated by the fact that temperatures near the pole may increase by 40–60 K in 1 week at 10 mb.

Although the definitions of major and minor stratospheric warmings are given in zonal-mean terms, these phenomena are far from zonally symmetric

259

Table 6.1

Occurrences of Major Stratospheric Warmings in Recent
Northern-Hemisphere Winters[a]

Year	Number	Month
1964–1965	0	
1965–1966	1	F
1966–1967	0	
1967–1968	1	J
1968–1969	0	
1969–1970	1	J
1970–1971	1	J
1971–1972	0	
1972–1973	1	J–F
1973–1974	0	
1974–1975	0	
1975–1976	0	
1976–1977	1	J
1977–1978	0	
1978–1979	1	F
1979–1980	0	
1980–1981	1	F
1981–1982	0	
1982–1983	0	
1983–1984	0	
1984–1985	1	J
1985–1986	0	

[a] Note: This table differs slightly from those given by
Labitzke (1982) and McInturff (1978), owing to differing
definitions of final warmings.

in form. It has been known for many years that the stratospheric circulation
is strongly zonally asymmetric before and during sudden warmings, and
satellite data are now providing complex three-dimensional pictures of the
time-development of these events and the variety of forms that they can
assume. Current theories, primarily stemming from the paper of Matsuno
(1971), suggest that tropospherically forced planetary waves play a crucial
role in the dynamics of sudden warmings. However, although the combination of satellite data with a hierarchy of models has done much in recent
years to elucidate the mechanisms of warmings, our understanding of all
the observed details of these events and the necessary conditions for their
occurrence is still by no means complete.

6.2. Observed Features of Sudden Warmings

The first observation of a stratospheric sudden warming was made by Scherhag in 1952, using radiosonde measurements over Berlin, and the radiosonde network has continued to provide much information on sudden warmings in the lower and middle stratosphere. This information was originally supplemented by rocket observations up to the high mesosphere at selected times and places. More recently, satellite measurements with good horizontal and temporal coverage and satisfactory vertical resolution at various levels in the stratosphere and mesosphere have given new insights into the three-dimensional morphology of sudden warmings.

As an example of the type of behavior that can be observed during a major stratospheric warming, we describe some basic features of the warming of February 1979, using a variety of diagnostics. This particular event has received much attention, since it occurred during a period of intensive observation of the atmosphere (the First GARP Global Experiment) and shortly after the launch of the *TIROS-N* and *Nimbus* 7 satellites. Although perhaps not "typical" in some respects, it nevertheless displays most of the features that are generally associated with major sudden warmings.

6.2.1   The Zonal-Mean Picture

During January and February 1979, three warming episodes took place in the northern polar stratosphere, with zonal-mean temperature maxima at 10 mb and 80°N, for example, occurring around January 25, February 6, and February 26 (see Fig. 6.1). The first two of these were minor warmings, but the third and largest was associated with a reversal of the zonal-mean wind $\bar{u}$ at and below 10 mb, corresponding to a major stratospheric warming. Latitude–height plots of $\bar{u}$ for December 8, 1978 (well before the warming

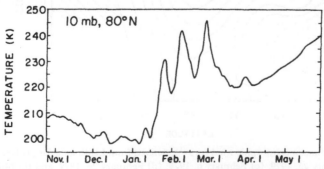

Fig. 6.1. Variation of zonal-mean temperature at 10 mb, 80°N, from October 1978 through May 1979, derived from LIMS data. [After Gille and Lyjak (1984).]

events), and January 25, and February 6 and 26, 1979 are shown in Figs. 6.2a–d, and give some idea of the time-variation of the mean zonal wind before and during the warmings. The broad region of easterlies that appears in the polar stratosphere and mesosphere at the time of the major warming is clearly visible in Fig. 6.2d. In early March the temperature falls again in the polar stratosphere (Fig. 6.1) before resuming its increase toward summer values, and winds revert to westerlies for a while (Fig. 6.2e) before the spring transition to summer easterlies.

6.2.2 Synoptic Description

As mentioned in Section 6.1, zonal-mean diagnostics give only a partial view of the rich three-dimensional structure of sudden warmings. Further observational details have traditionally been supplied by synoptic maps of temperature and geopotential at various levels in the stratosphere and by rocket soundings from isolated points on the globe. As an example, Fig. 6.3 presents polar stereographic charts of the 10-mb height and temperature fields for the days leading up to, and following, the major warming of late February 1979. On February 17 (Fig. 6.3a), the basic cyclonic polar vortex is elongated (as shown by the roughly elliptical height contours) and is

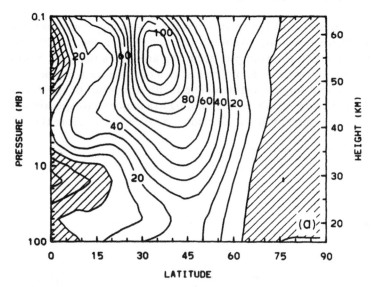

Fig. 6.2. Zonal-mean geostrophic wind (m s^{-1}), derived from LIMS data for (a) December 8, 1978, (b) January 25, 1979, (c) February 6, 1979, (d) February 26, 1979, and (e) March 3, 1979. Easterlies are shaded. [Adapted from Gille and Lyjak (1984); courtesy of Dr. John C. Gille.]

centered just off the pole, with an Aleutian High being present at about 180°W, 60°N. The temperature contours show an elongated cold region centered at about 20°E, 75°N and warm regions at about 100°E, 60°N and 80°W, 60°N. On February 19 (Fig. 6.3b), these structures have rotated slightly eastward and elongated further. By February 21 (Fig. 6.3c), a dramatic

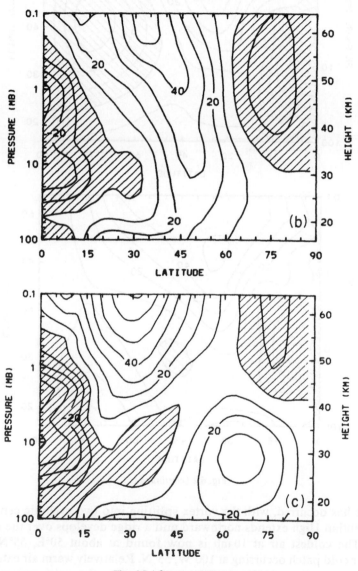

Fig. 6.2 (*figure continues*)

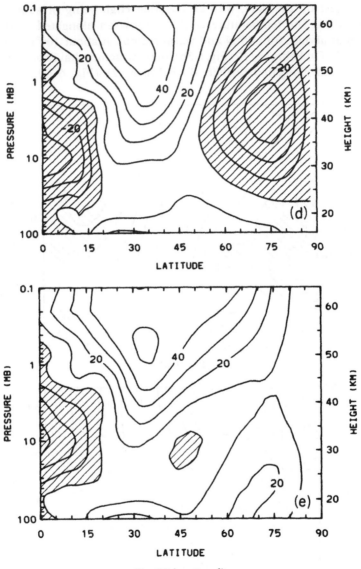

Fig. 6.2 (*continued*)

change has occurred, with the vortex splitting into two cyclonic centers as the Aleutian High extends northward and a ridge develops over the British Isles. The coldest air at 10 mb is now found at about 50°E, 55°N, with another cold patch occurring at 100°W, 55°N. Relatively warm air extending over the pole from 110°E to 60°W indicates that the polar stratospheric

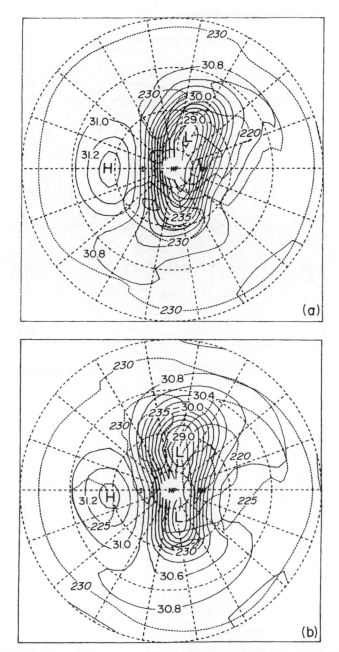

Fig. 6.3. Polar stereographic charts of 10 mb height (solid curves; contour interval 0.2 km) and temperature (dashed curves; contour interval 5 K) from LIMS data for the following days in 1979; (a) February 17, (b) February 19, (c) February 21, (d) February 26, (e) March 1, and (f) March 5. "GM" designates the Greenwich Meridian, which extends horizontally toward the right from the North Pole (NP). Latitude circles are shown at 20° intervals, with the outermost circle at 20°N. (*Figure continues.*)

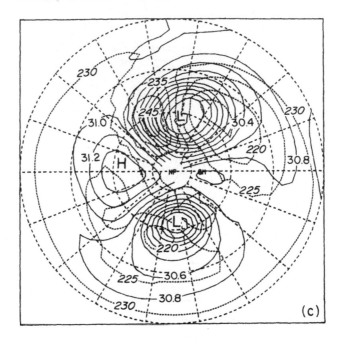

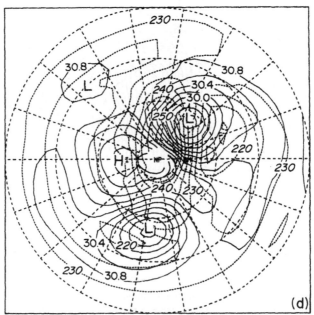

Fig. 6.3 (*continued*)

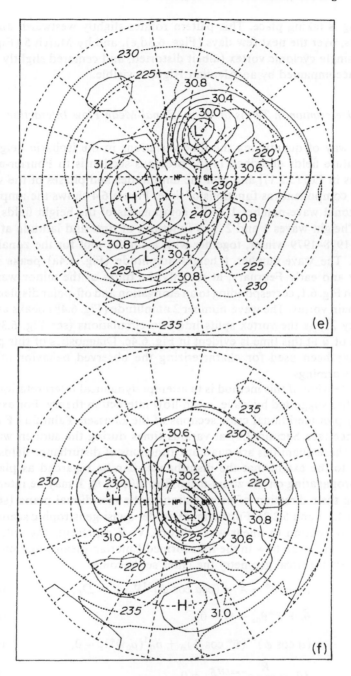

Fig. 6.3. (*continued*)

warming is taking place. This pattern rotates slightly westward, and then weakens, over the next few days (Figs. 6.3d,e), and by March 5 (Fig. 6.3f) only a single cyclonic vortex (albeit distorted, and centered slightly off the pole), accompanied by an Aleutian High, is visible.

6.2.3 Description in Terms of Wave, Zonal-Mean Flow Interaction

One way of quantifying the time-variations in the isobaric height and temperature fields described in the previous section is to Fourier-analyze them [as in Eq. (5.2.1)] and plot the amplitudes and phases of the various Fourier components as functions of time. Figure 6.4 shows the amplitudes of the zonal wave-number 1 and 2 components of the height fields (often called "height waves 1 and 2") as functions of time and latitude at 10 mb for the 1978–1979 winter, together with a similar plot for the zonal-mean wind $\bar{u}$. The wave-number 1 height amplitude (Fig. 6.4a) peaks in late January and early February, at roughly the times of the minor warmings shown in Fig. 6.1, corresponding to an elongation and off-polar displacement of the main vortex. The wave-number 2 amplitude (Fig. 6.4b) peaks at about February 21, as the vortex splits into two circulations (see Fig. 6.3c): the reversal of $\bar{u}$ at this time is evident in Fig. 6.4c. Diagnostics of this general type have been used for characterizing the observed behavior of many sudden warmings.

An extension of this method is to attempt dynamical interpretation using diagnostics suggested by wave, mean-flow interaction theory. For example, one may plot meridional cross sections of the Eliassen–Palm flux $\mathbf{F}$ and its divergence (see Section 3.5) at various times during the sudden warming period. These quantities are quadratic functions of disturbance fields u', v', θ', etc.; to the extent that the disturbances can be regarded as planetary waves propagating on the zonal-mean flow $\bar{u}(\phi, z, t)$, $\mathbf{F}$ may be interpreted as giving their direction of propagation in the meridional plane (see also Sections 4.5.5 and 5.2.2). Furthermore, under quasi-geostrophic scaling, but with no restriction to linear wave theory, $\nabla \cdot \mathbf{F}$ represents the sole eddy-forcing of the mean flow in the transformed Eulerian-mean equations [Eqs. (3.5.5)]. In spherical geometry the latter set become

$$\bar{u}_t - f\bar{v}^* - \bar{X} = (\rho_0 a \cos \phi)^{-1}\nabla \cdot \mathbf{F} \equiv D_F, \tag{6.2.1a}$$

$$\bar{\theta}_t + \bar{w}^*\theta_{0z} - \bar{Q} = 0, \tag{6.2.1b}$$

$$(a \cos \phi)^{-1}(\bar{v}^* \cos \phi)_\phi + \rho_0^{-1}(\rho_0 \bar{w}^*)_z = 0, \tag{6.2.1c}$$

$$f\bar{u}_z + \frac{R}{aH}e^{-\kappa z/H}\bar{\theta}_\phi = 0, \tag{6.2.1d}$$

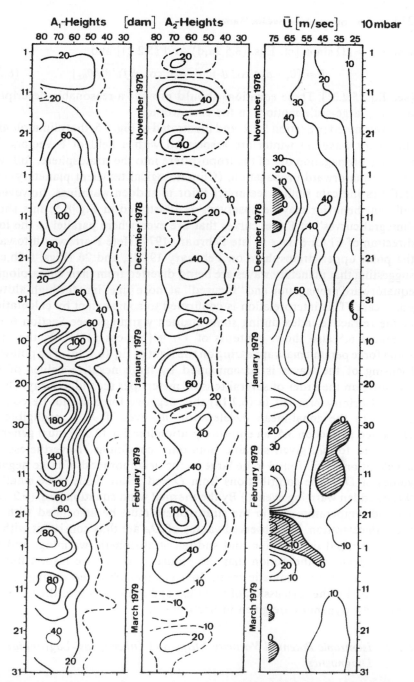

Fig. 6.4. Latitude–time sections at 10 mb from November 1978 through March 1979 for amplitudes (decameters) of the (a) wave-number 1 and (b) wave-number 2 components of the geopotential height field, together with (c) the mean zonal wind (m s^{-1}). [From Labitzke (1981a), with permission.]

where $f = 2\Omega \sin \phi$ [cf. Eqs. (3.5.2a,d) and (5.2.2)] and

$$\mathbf{F} = [0, -\rho_0 a \cos \phi \, \overline{v'u'}, \rho_0 a \cos \phi \overline{fv'\theta'}/\theta_{0z}] \tag{6.2.2}$$

[see Eq. (5.2.3)]. These equations should provide a reasonable description of slow, large-scale motions in the extratropics.

It was noted in Section 5.2.2, and illustrated in Fig. 5.5, that climatological stationary waves in winter are associated with a field of $\mathbf{F}$ vectors that indicate propagation out of the troposphere into the stratosphere, followed by an equatorward propagation. (Climatological transient planetary-wave statistics indicate similar features.) Prior to sudden warmings, however, a different behavior may be observed, as illustrated in Fig. 6.5; this shows "integral curves" of $\mathbf{F}$ (i.e. curves that are everywhere parallel to the local direction of $\mathbf{F}$) for 6 days in late February 1979. Here $\mathbf{F}$ arrows tilt towards the polar upper stratosphere on February 19, 21, and 26 (Figs. 6.5b,c,e), suggesting that planetary waves are being diverted from their climatological equatorward propagation and "focused" at those times into the high-altitude polar cap. (This interpretation is confirmed to some extent by calculations of the refractive index during some sudden warmings: see Section 6.3.2.)

Figure 6.5 also shows contours of $D_F \equiv (\rho_0 a \cos \phi)^{-1} \nabla \cdot \mathbf{F}$, the mean zonal force per unit mass appearing on the right of Eq. (6.2.1a). The poleward focusing of the waves is accompanied by large negative values of this quantity, on the order of several tens of meters per second per day. Owing to the Coriolis term $f\bar{v}^*$ in (6.2.1a) this negative force cannot be equated exactly to the zonal-mean deceleration $\bar{u}_t$, even if $\bar{X}$ is negligible: as explained in Section 3.5, an elliptic equation like Eq. (3.5.7) must generally be solved for $\bar{u}_t$, given suitable boundary conditions. Nevertheless, the deceleration is still found to be large in the neighborhood of large negative values of D_F. (The contributions from $\bar{Q}$ and $\bar{X}$ are likely to be small for these sudden warming events.) By the thermal wind equation [Eq. (6.2.1d)], a rapid temperature rise in the stratosphere must be associated with this rapid deceleration. The observed values of D_F are thus consistent with the observed rapid deceleration and sudden stratospheric warming, and this lends support to the notion that the planetary waves are responsible for bringing about the dramatic mean-flow changes observed during sudden warmings. Further discussion of this idea and more details of the wave and mean-flow dynamics are given in Section 6.3.

6.2.4   Isentropic Potential Vorticity Maps and Other Quasi-Lagrangian Diagnostics

Despite the fairly self-consistent dynamical picture offered by the wave mean-flow diagnostics discussed in the previous section, the separation of

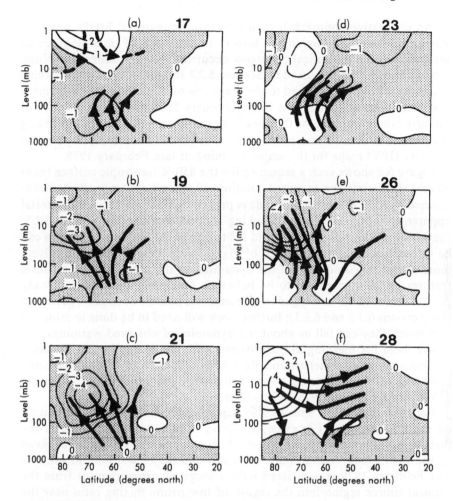

Fig. 6.5. Some integral curves of **F** and contours of D_F in units of $10^{-4}\,\mathrm{m\,s^{-2}}$ (negative values stippled) for (a) February 17, (b) February 19, (c) February 21, (d) February 23, (e) February 26, and (f) February 28, 1979. Dashed integral curves are dominated by wave-number 1 contributions to **F**, full curves by wave-number 2 contributions. [From Palmer (1981a), with permission.]

flow quantities into zonal-mean and wave contributions is a rather arbitrary process and may be an unnecessarily complicated (and even misleading) way of viewing the dynamics, especially when wave amplitudes are large. The synoptic maps described in Section 6.2.2 do not make this separation but are, however, purely descriptive, and difficult to interpret in terms of a simple, coherent, dynamical framework.

An alternative approach is to use isentropic maps of Ertel's potential vorticity P (a quasi-Lagrangian tracer) in the hope of obtaining deeper insights into the dynamical processes occurring during sudden warmings. This approach was discussed in Section 5.2.3 in connection with breaking planetary waves, and applied to the diagnosis of the strong cyclonic vortex that was centered off the pole in late January 1979 (in fact at the time of the first minor warming of that year). For comparison with the preceding sections, it is thus of interest to study a sequence of isentropic potential vorticity (IPV) maps for the major warming of late February 1979.

Figure 6.6 shows such a sequence for the 850-K isentropic surface (near 10 mb). It depicts the eastward rotation of the main vortex (here distinguished by high P values) on the days preceding February 17, with material apparently being shed by the breaking process discussed in Section 5.2.3; thereafter, the vortex elongates and splits into two. Similar information can be inferred from the height maps of Fig. 6.3, but the IPV maps may give a more sharply focused view of the dynamics, to the extent that small-scale features in these maps are to be believed. Maps like these have already proved useful in the investigation of numerical models of sudden warmings (see Sections 6.3.2 and 6.3.3); further work will need to be done to establish how much they can tell us about the dynamics of observed warmings and to use them for meaningful quantitative analysis. A step in this direction has been taken by Butchart and Remsberg (1986), who studied the time-evolution of the horizontal areas enclosed by the isopleths of P on the 850-K surface for the 1978–1979 winter.

Long-lived chemical species also supply quasi-Lagrangian information on sudden warmings. For example, Leovy et al. (1985), in a study of the three-dimensional transport of ozone during the 1978–1979 winter as given by the LIMS data, found that each minor and major warming in January and February 1979 is associated with a surge of ozone-rich air from the tropical source region into the region of low ozone mixing ratio near the North Pole (see Section 9.5).

6.2.5   Other Observational Aspects

We conclude this brief observational account by mentioning a few further features of observed sudden warmings.

Labitzke (1981a) has noted that a typical "precondition" of many warmings is that a pulse of wave number 1 geopotential amplitude precedes the pulse of wave number 2 (corresponding to a splitting of the main vortex) that occurs at the time of the mean zonal wind reversal. This wave-number 1 pulse may be accompanied by a poleward shifting of the zonal-mean polar night jet. Such a wave-one precursor did indeed occur in 1979, at the

FEB 1979   **850 K**

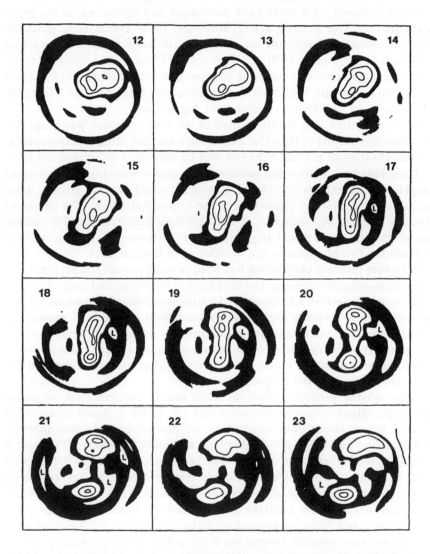

Fig. 6.6. Evolution of Ertel's potential vorticity P on the 850-K isentropic surface for part of February 1979. The north pole is at the center of each square; the sides of each square are tangent to 20°N, and no data are plotted south of that latitude. The contour values are given by $1.3 \times 10^{-4} \times n$ K kg^{-1} m^2 s^{-1} (corresponding approximately to $2 \times 10^{-4} \times n$ K m^{-1} s^{-1} in the scaled units of Figs. 5.6b and 6.12, where P is divided by gH/p_0), with $n = 0$, 2 (outer boundary of black strip), 3 (inner boundary of black strip), 5 and 7. Values increase monotonically towards the center of each plot unless an "L" is present, indicating a local minimum. [From Dunkerton and Delisi (1986), with permission.]

time of the minor warming of January 25 (see Fig. 6.4a and also Fig. 5.6), and was followed by a northward movement and tightening of the polar night jet (see also Section 6.3.3). However, these events took place an unusually long time before the major warming of late February. Other major warmings (e.g., January 1970 and January 1977) take yet a different form, in that no significant development of wave number 2 occurs. On the other hand, that of January 1985 is unusual in involving wave number 2 from the outset, with no significant wave-number 1 precursor.

Another feature that is often noted at the time of major warmings is a simultaneous cooling of the polar mesosphere and low-latitude stratosphere. Dynamical reasons for this behavior will be discussed in Section 6.3.2.

It has long been known that sudden stratospheric warmings tend to occur roughly concurrently with "blocking events," which involve strong, long-lasting, quasi-stationary distortions of the tropospheric flow. The causes of blocking are not well understood, still less their connections with sudden warmings: some theoretical suggestions for cause-and-effect relationships between the two phenomena are mentioned in Section 6.3.4.

A further possible correlation has been noted between the incidence of major stratospheric warmings and the phase of the equatorial quasi-biennial oscillation (see Section 8.2), in which the tropical stratospheric winds reverse from easterly to westerly and back, with a period of about 27 months. It was mentioned above that major warmings only occur about once in every two northern winters; more often than not, these are the winters when equatorial stratospheric winds are easterly. A speculative mechanism for this relationship is noted in Section 6.3.4.

Although the preceding sections have mostly concentrated on major warmings, minor stratospheric warmings (in which a zonal-mean wind reversal below 10 mb does not occur) are also of considerable interest. Examples include those of late January and early February 1979, and a Southern Hemisphere case that has received much attention is that of July 1974, in which considerable *midlatitude* temperature increases and wind decelerations (from an initially very strong westerly flow) were observed in the upper stratosphere. Labitzke (1981b) defines *Canadian warmings* as events that result from an anomalous strengthening of the Aleutian anti-cyclone and may possibly reverse the mean poleward temperature gradient north of 60°N. *Final warmings* are those events that are followed not by a reversion of stratospheric conditions to the usual winter pattern but by a transition to the summer structure of warm temperatures and easterly winds. Late winter in the Southern Hemisphere stratosphere is notable for a downward and poleward shifting of the polar night jet; the spring transition tends to occur later than in the Northern Hemisphere, and is accompanied by fluctuating wave-number 1 activity.

6.3 Theoretical Modeling of Sudden Warmings

6.3.1 Matsuno's Model

Early attempts to explain sudden warmings investigated the possibility that they may be primarily due to the baroclinic instability of the polar night jet in the zonally averaged winter stratospheric flow. A further development of this idea was the examination of the possible barotropic instability of the large-scale zonally asymmetric polar vortex. Although some unstable modes were found in these studies, their growth rates and spatial scales were too small to explain the rapid development and large horizontal and vertical scales of observed warmings.

It is now generally accepted that the essential dynamical mechanism reponsible for sudden warmings involves the upward propagation from the troposphere of planetary (Rossby) waves and their interaction with the mean stratospheric flow. This hypothesis was first proposed by Matsuno (1971), who tested it using a numerical model of the stratosphere. Most subsequent "mechanistic" models of sudden warmings have been generalizations of Matsuno's.

The dynamical reasoning on which Matsuno's model is based can be summarized as follows. He supposed that the mean-flow changes observed during sudden warmings, including the mean zonal flow deceleration and the mean temperature rise near the pole, are attributable to the nonlinear rectified effects of vertically propagating planetary waves forced in the troposphere by large-scale disturbances there. In one of the first applications of the Charney–Drazin nonacceleration theorem (see Section 3.6), he noted that steady, conservative, linear planetary waves are incapable of inducing such mean-flow changes, and therefore sought conditions under which the nonacceleration theorem is violated. One of these occurs when a transient packet of planetary waves first starts to propagate upward: at the leading edge of such a packet, wave amplitudes are growing in time. The effects of this wave growth can conveniently be described using the ideas presented in Section 3.6 (although these were not explicitly used by Matsuno); a simple schematic picture of the process is given in Fig. 6.7. If the wave-activity density

$$A \equiv \tfrac{1}{2}\rho_0 \overline{\eta'^2}\, \bar{q}_y$$

[a quadratic measure of wave amplitude: see Eq. (3.6.10)] is introduced, where η' is the northward parcel displacement and $\bar{q}_y$ is the northward mean potential vorticity gradient (assumed positive), then $\partial A/\partial t > 0$ at the leading edge of the packet. If the waves are considered to be linear and conservative, then by the generalized Eliassen–Palm theorem [Eq. (3.6.2)],

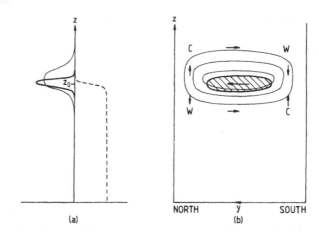

Fig. 6.7. Schematic description of the interaction with the zonal-mean flow of a transient, vertically propagating planetary-wave packet that is maintained by continual forcing at $z = 0$ in a zonal channel, to illustrate aspects of Matsuno's model of sudden warmings. (a) Midchannel height profiles of wave activity A (dashed), EP flux divergence $\nabla \cdot \mathbf{F} = -\partial A/\partial t$ (heavy line), and acceleration $\bar{u}_t$ (thin line); z_0 is the height reached by the leading edge of the packet, and is of order (vertical group velocity) × (time since packet started at $z = 0$). (b) Latitude-height cross section showing region where $\nabla \cdot \mathbf{F} < 0$ (hatched), contours of induced acceleration $\bar{u}_t$ (thin lines), and induced residual circulation ($\bar{v}^*$, $\bar{w}^*$) (arrows). Regions of warming (W) and cooling (C) are also indicated.

$\nabla \cdot \mathbf{F} = -\partial A/\partial t$, where $\mathbf{F}$ is the Eliassen–Palm flux, and a patch of negative $\nabla \cdot \mathbf{F}$ is therefore to be expected near the front of the packet. As noted in Section 6.2.3, this implies a negative force per unit mass on the zonal-mean flow, the response to which will include a westward mean-flow acceleration in the vicinity of the patch of negative $\nabla \cdot \mathbf{F}$. A mean temperature increase will occur on the lower poleward flank of this patch. Thus, the mean-flow changes associated with the leading edge of a transient planetary-wave packet are qualitatively similar to those observed during a sudden warming.[1]

Matsuno tested this qualitative mechanism with a quasi-geostrophic numerical model based on a "quasi-linear" separation of the waves and the mean flow. Thus, to represent the waves he used the linearized spherical-geometry potential vorticity equation [Eq. (5.3.1)] [see also Eqs. (5.3.2)–(5.3.4)], including the time-derivative terms, together with a suitable time-

---

[1] Matsuno also considered the mean-flow changes associated with vertically propagating planetary waves encountering a descending horizontal critical level (see Section 5.6), with all the transience being concentrated in a thin critical layer. However, it is now known that such horizontal critical levels do not normally appear in the stratosphere during sudden warmings. On the other hand, approximately vertical critical surfaces may encroach from low latitudes and play a somewhat different role in the dynamics of some sudden warmings: see Section 6.3.2.

dependent lower boundary condition at $z = 10$ km to represent the switch-on of the tropospheric forcing associated with some unspecified large-scale transient disturbance. The geopotential disturbance Φ' given by solving this system under a specified mean-flow configuration is then substituted into the quadratic wave-forcing of the mean flow. In terms of the formulation of Section 6.2.3, this amounts to evaluating $\mathbf{F}$ by means of Eq. (6.2.2) and substituting into Eq. (6.2.1a). The mean-flow equations [Eqs. (6.2.1)] can then be solved for the tendencies $\bar{u}_t$ and $\bar{\theta}_t$, and the mean flow can be updated after one time step. The process is then repeated by solving Eq. (5.3.1), given the new mean-flow configuration and the forcing at the new time step, and so on. This quasi-linear approach excludes the possibility of interactions between waves of different zonal wave numbers.

Matsuno performed a series of experiments in which transient disturb-ances of a single zonal wave number, $s = 1, 2$, or 3, were forced at the lower boundary and propagated into a model stratosphere whose initial zonal wind structure $\bar{u}$ resembled the observed climatology. Interaction between the waves and mean flow led to an evolution of the mean wind and temperature fields, which in some cases was reminiscent of that occurring during sudden warmings. For example, with an $s = 2$ forcing, the climato-logical westerlies gave way to easterlies after about 20 days of integration, followed by a rapid polar temperature increase (of over 80 K) below about 50 km and a decrease above that level: see Fig. 6.8. The corresponding geopotential height charts (Fig. 6.9) showed an elongation and splitting of the polar cyclonic vortex in the midstratosphere, somewhat similar to the February 1979 observations given in Figs. 6.3a–d. This was followed by an amalgamation of the model's subsidiary anticyclones to form a weak anti-cyclonic polar vortex that was not apparent in the 1979 observations of Figs. 6.3e,f.

6.3.2   The Use of Lagrangian and Eliassen–Palm Diagnostics in Models of Sudden Warmings

In recent years there have been numerous model studies of the sudden warming phenomenon, mostly based on quasi-linear models of the type discussed in the previous section. A variety of diagnostic techniques have been used for describing and interpreting the model dynamics. Some of these techniques have also been used in observational studies of warmings and have been mentioned briefly in Section 6.2; others are less easily applied to observational data and have mainly been restricted to model experiments.

An example of the latter type of diagnostic is the air-parcel trajectory. Accurate Lagrangian calculations of this kind are difficult to achieve for

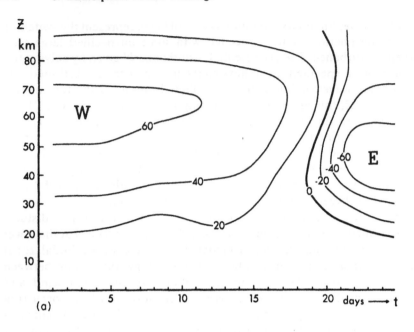

(a)

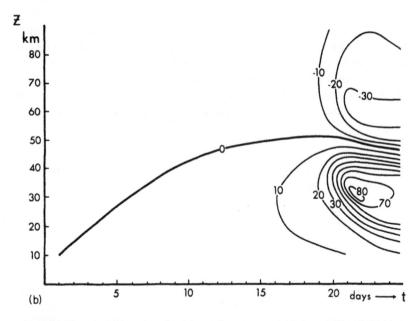

(b)

Fig. 6.8. Time–height sections for (a) zonal-mean zonal wind at 60°N and (b) temperature at the north pole, from the model wave-number 2 warming in Matsuno's (1971) model. [From Matsuno (1971). American Meteorological Society.]

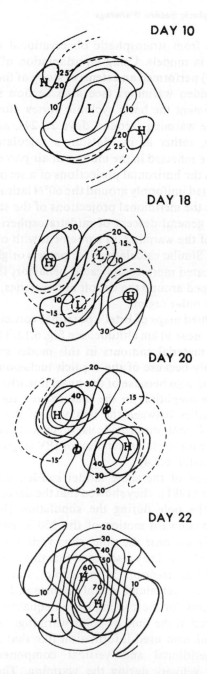

Fig. 6.9. Polar stereographic plots of the time evolution of the temperature (thin curves, marked in degrees Celsius) and height (thick curves, 500-m contour intervals) for the isobaric surface $z = 30$ km ($p \approx 13$ mb) for the same model warming as in Fig. 6.8. The contours cover the area north of 30°N. [From Matsuno (1971). American Meteorological Society.]

extended periods from stratospheric observational data, but can be performed routinely in models. Using an adaptation of the model of Holton (1976), Hsu (1980) performed an extensive study of the motion of air parcels during model sudden warmings (see also Section 9.7.4). In her "wavenumber 2" experiment the basic cyclonic vortex elongated, split into two at the peak of the warming (around days 22-24), and was replaced by a polar anticyclone, rather as in Matsuno's calculation (Fig. 6.9). These developments were reflected in the motion of air parcels, as depicted in Fig. 6.10, which shows the horizontal projections of a set of air parcels that were originally distributed uniformly around the 60°N latitude circle at $z = 32$ km. Figure 6.11 shows the meridional projections of the same set of parcels and demonstrates the general descent of midstratospheric air parcels near the pole at the peak of the warming, in agreement with observational evidence (Mahlman, 1969). Similar calculations for parcels originally located at 30°N, $z = 30.8$ km, indicated more complicated behavior; these low-latitude parcels became wrapped around the cut-off anticyclones, with some eventually penetrating to the polar region.

Hsu also presented maps of Ertel's potential vorticity P on the $\theta = 850$ K isentropic surface, near 30 km altitude: see Fig. 6.12. The isentropic isopleths of P differ from material contours in this model since θ and P are not conserved, not only because of the explicit inclusion of diabatic and frictional processes but also because of the quasi linearity, which neglects terms quadratic in wave amplitude in the disturbance equations. Nevertheless, qualitative similarities between the behavior of the strings of air parcels and the isentropic P contours provide useful cross-checks on the calculations and some guidance for the interpretation of IPV maps derived from observational data (cf. Section 6.2.4).

Further diagnosis of the same model sudden warming was provided by Dunkerton et al. (1981). They showed that the descent of midstratospheric air parcels near the pole during the simulation (Fig. 6.11) was closely followed by the downward motion of the 850-K isentrope relative to the isobars (or z surfaces) near the pole, thus verifying that the process was approximately adiabatic. They also noted that this polar descent was accompanied by ascent of the isentrope in the tropics, since the atmospheric mass beneath the isentrope remained roughly constant. Such a "see-saw" effect presumably accounts for the cooling of the equatorial stratosphere that is commonly observed at the time of polar warmings (see Section 6.2.5).

Dunkerton et al. also used information like that shown in Fig. 6.11 to calculate the meridional and vertical components $(\bar{v}^L, \bar{w}^L)$ of the Lagrangian-mean velocity during the warming. These are given by the motion of the center of mass, as viewed in the meridional plane, of the initially zonal ring of parcels (see Section 3.7.1). As expected from the

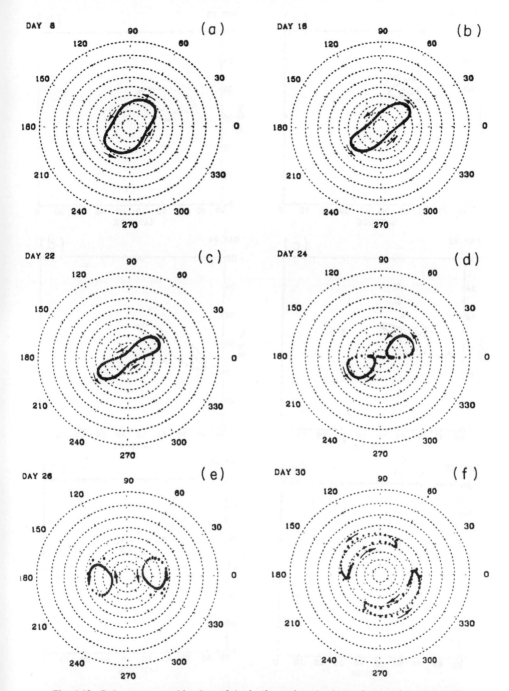

Fig. 6.10. Polar stereographic plots of the horizontal projections of a set of air parcels for 6 days during the model warming described by Hsu (1980). The parcels were distributed uniformly around the 60°N latitude circle at $z = 32$ km on day 0. Latitude circles, at 10° intervals, and longitudes, at 30° intervals, are shown dashed. [From Hsu (1980). American Meteorological Society.]

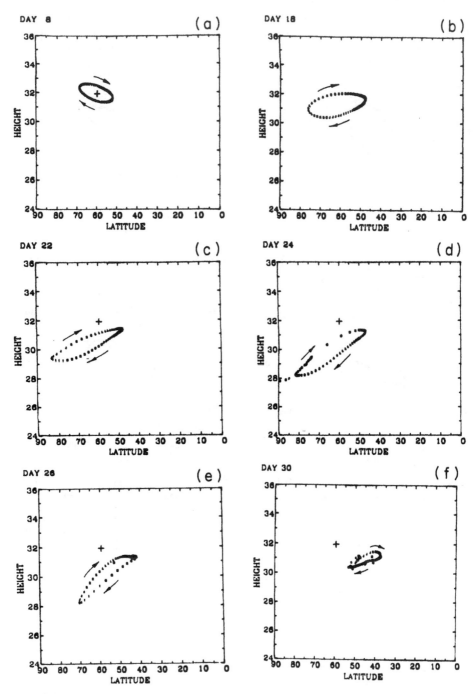

Fig. 6.11. As in Fig. 6.10, but for projections in the meridional (ϕ, z) plane. The initial position of the line of parcels is marked by a plus sign. [From Hsu (1980). American Meteorological Society.]

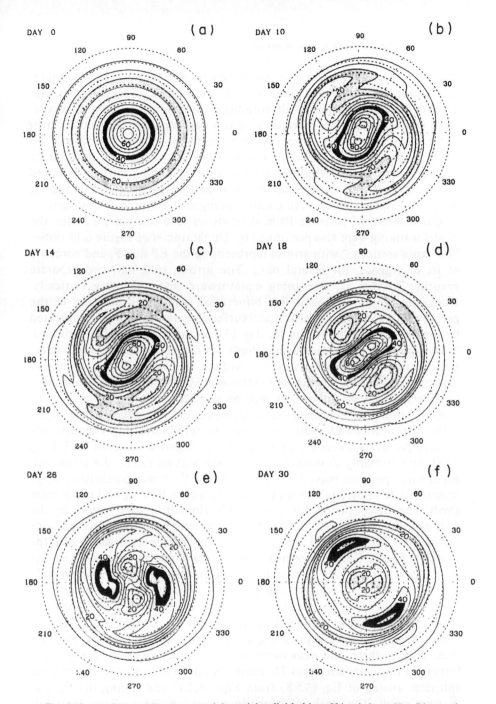

Fig. 6.12. Isopleths of Ertel's potential vorticity divided by gH/p_0 (where $H = 7$ km and $p_0 = 1000$ mb) on the 850-K isentropic surface for 6 days during the model warming described by Hsu (1980). The units are 10^{-5} K m^{-1} s^{-1} and the contour interval is 5 units. The area between the contours for 15 and 20 units is shaded lightly, and that between the contours for 40 and 45 units is shaded heavily. [From Hsu (1980). American Meteorological Society.]

previous discussion, Lagrangian-mean descent was found in high latitudes and ascent in low latitudes in the middle stratosphere. (This contrasts with an Eulerian-mean *ascent* in high latitudes.) Another feature of these calculations was a strong mass divergence of the Lagrangian-mean velocity over a deep layer, associated primarily with the north–south dispersion of air parcels. The authors also indicated ways in which the generalized Lagrangian-mean theory may fail at large wave amplitude, and they explored the possible use of an alternative "modified Lagrangian-mean" theory, based on averages around the quasi-Lagrangian isentropic P contours.

Calculations of Eliassen–Palm diagnostics (see Section 6.2.3) for the model warming were also performed by Dunkerton *et al.* Figure 6.13 shows "EP cross sections," with arrows representing the EP flux F, and contours of its divergence, for several days. The arrows split into two separate branches by day 22, one going equatorward and one more vertically. Dunkerton *et al.* attributed this bifurcation and polar focusing to the presence of a zero-wind line (a critical surface for stationary waves, indicated by the heavy curve in the plot for day 15) encroaching from low latitudes. They argued that the model may have produced a qualitatively correct reflection of the waves from such a surface, as expected from nonlinear Rossby-wave critical layer theory (Section 5.6). This reflection would tend to deflect EP arrows upward, as seen on day 22.

Dunkerton *et al.* examined the wave, mean-flow interaction in the model using the transformed Eulerian-mean equations [Eqs. (6.2.1)]. The vertically focused planetary waves exhibited significant negative values of $\nabla \cdot \mathbf{F}$ (Fig. 6.13), due primarily to wave transience (see Section 6.3.1); the associated zonal force per unit mass $D_F \equiv (\rho_0 a \cos \phi)^{-1} \nabla \cdot \mathbf{F}$ was particularly large in magnitude in the high-altitude polar cap where both ρ_0 and $\cos \phi$ were small. The effects of this force were redistributed to some extent by the residual circulation $(\bar{v}^*, \bar{w}^*)$, but nevertheless produced decelerations (leading to the poleward migration of low-latitude easterlies) and the associated polar temperature increases. Diabatic effects were small on the short timescales involved here, and the net acceleration $\bar{u}_t$ tended to resemble D_F as a function of latitude: see Fig. 6.14. (A quite different balance of terms occurs when long-time averages are taken: see Section 7.2.) The mass stream function for the residual circulation on day 27 (Fig. 6.15) showed a descent in stratospheric polar regions roughly similar to the Lagrangian-mean descent there, and weak ascent in the polar mesosphere. The primary forcing of this circulation was D_F again, as can be seen by constructing the spherical analog of Eq. (3.5.8) from Eqs. (6.2.1) and noting that $\bar{Q}$ and $\bar{X}$ were small. The weak mesospheric ascent implies cooling there: this may be analogous to the observed mesospheric cooling noted in Section 6.2.5.

DAY 11

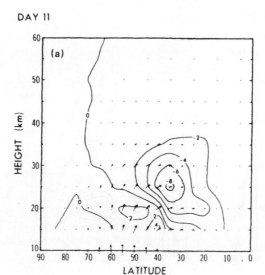

DAY 15

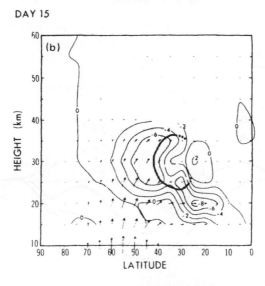

Fig. 6.13. Eliassen–Palm cross sections (in the ϕ, z plane) for (a) day 11, (b) day 15, and (c) day 22 of the evolution of Hsu's (1980) model. The heavy curve in (b) shows part of the zero-wind line for day 15. The contours represent values of $(\cos \phi)(\nabla \cdot \mathbf{F})$, in units of $a\rho_0(0) \times 10^{-7}$ m s^{-2}. The arrows represent the vector $(\cos \phi)\mathbf{F}$, scaled such that the distance occupied by $10°$ in ϕ represents a value $5.56\rho_0(0)a$ m^2 s^{-2} of $(\cos \phi)F^{(\phi)}$ and that occupied by 10 km in z represents a value $0.05\rho_0(0)a$ m^2 s^{-2} of $(\cos \phi)F^{(z)}$. [From Dunkerton *et al.* (1981). American Meteorological Society.] (*Figure continues.*)

DAY 22

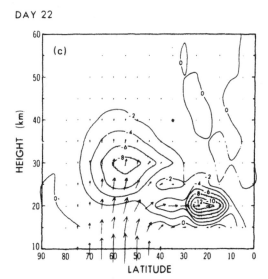

Fig. 6.13 (*continued*)

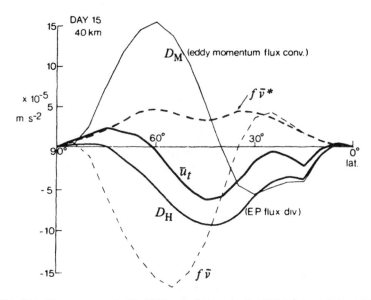

Fig. 6.14. Heavy curves: typical balance of terms in Eq. (6.2.1a) from the model warming described by Hsu (1980). Light curves: principal terms in the conventional momentum equation, Eq. (3.3.2a), where $D_M \equiv -(a \cos^2 \phi)^{-1} (\overline{v'u'} \cos^2 \phi)_\phi$. [From Dunkerton *et al.* (1981). American Meteorological Society.]

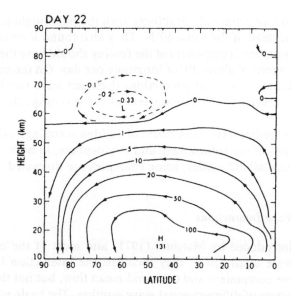

Fig. 6.15. Mass stream function $\bar{\chi}^*$ associated with the residual circulation on day 22 of the model warming described by Hsu (1980); $\bar{\chi}^*$ is defined such that $\partial\bar{\chi}^*/\partial\phi = \rho_0 a(\cos\phi)\bar{w}^*$ and $\partial\bar{\chi}^*/\partial z = -\rho_0(\cos\phi)\bar{v}^*$. Contour values are to be multiplied by $\rho_0(0)$ times $1\ m^2\ s^{-1}$. The contour interval is nonuniform. [From Dunkerton *et al.* (1981). American Meteorological Society.]

A more complex series of experiments was performed by Butchart *et al.* (1982). They first used a fully nonlinear model of the stratosphere, starting with the observed zonal-mean wind field for February 16, 1979, and forced at 100 mb by the observed height field, to simulate the observed warming of late February. Most of the main features of the observed warming were quite well reproduced. The model was then simplified in various ways, to try to isolate the essential dynamical behavior of the simulation, and hence of the observed warming. For example, runs were carried out using only the wave-number 2 Fourier component of the lower boundary forcing, or using idealized stationary or steadily progressing forcing there, and replacing the observed initial winds by their climatological counterparts. EP fluxes were used as diagnostics, as was the planetary-wave refractive index squared, n_k^2 (see Section 4.5.4). Although the strict derivation of the latter quantity assumes the waves to be steady, it nevertheless suggests interpretations of the bifurcation and polar focusing of transient-wave EP fluxes during sudden warmings, by showing the locations of possible barriers to propagation, where n_k^2 is small or negative.

Butchart *et al.* concluded that an important condition for the warming to occur was that the initial mean zonal wind should take its observed, "preconditioned," nonclimatological form, with the polar night jet at 10 mb at about 75°N, instead of the usual 60°N. They also found it necessary that the zonal wave-number 2 component of the forcing should have the observed eastward phase speed of about 10° of longitude per day. On the other hand, they concluded that wave–wave interactions (between wave numbers 1 and 2, say) were not crucial to the development of the warming after February 16.

Several other studies have been made of sudden warmings using global circulation models of the troposphere and stratosphere, in both "forecast" and "general circulation" modes; some examples are discussed in Section 11.3.

6.3.3   Wave–Wave Interactions

The model introduced by Matsuno (1971) and most of the others discussed above were quasi-linear: they included the interaction between a single zonal wave component and the zonal-mean flow, but not the interactions between waves of different zonal wave number. The basic model used by Butchart *et al.*, on the other hand, was fully nonlinear but, as mentioned above, they found that wave–wave interactions were not essential to the simulation of the major warming of late February 1979, given the observed nonclimatological flow as an initial condition on February 16.

Several model studies have sought to elucidate the role of wave–wave interactions in the overall development of sudden warmings. For example, Hsu (1981) found that inclusion of such interactions tended to result in increased wave activity, stronger polar warmings, and more rapid mean-flow decelerations. As expected, the Lagrangian behavior of air parcels was more complex than in the quasilinear cases depicted in Figs. 6.10 and 6.11.

Palmer and Hsu (1983) used the same nonlinear model to study aspects of the stratospheric behavior in late January and early February 1979. After the minor warming associated with the wave-number 1 "precursor" in late January (see Section 6.2.5), the polar temperatures dropped once more (see Fig. 6.1) in a "sudden cooling," and high-latitude easterlies in the upper stratosphere accelerated rapidly to form a tight westerly polar night jet near 75°N by February 3. (This jet later relaxed, then formed again by February 16, just before the major warming.) Palmer and Hsu concluded that nonlinear wave–wave interactions were important during the cooling period, a period that included the essentially nonlinear planetary wave breaking event studied by McIntyre and Palmer (1983, 1984; see also Section 5.2.3). Synoptically, the sudden cooling corresponds to the movement of the cyclonic vortex in Fig. 5.6a back to a more pole-centered position. The fact that the polar night jet was further poleward after this event than before may be a result of the erosion of potential vorticity while the vortex was in a displaced position.

Matsuno (1984) added wave–wave interactions to his earlier quasi-geostrophic model and carried out some experiments to study the differences between minor and major warmings. With lower boundary forcing in wave-number 1 of geopotential height amplitude 300 m at $z = 10$ km, on an already distorted flow, a wave-number 1 type minor warming ensued, while if the forcing amplitude was increased to 400 m, a major warming followed. The differences were diagnosed in terms of potential vorticity, among other things. In each case, low, tropical potential vorticity was advected into the polar stratosphere and high potential vorticity contours were displaced off the pole, as shown in Fig. 6.16 (cf. also Fig. 6.12). However, in the major warming irreversible mixing of potential vorticity occurred on a hemispheric scale, leading to polar easterlies, whereas in the minor warming irreversible mixing was restricted to low and middle latitudes, leading to a tight westerly polar jet. (In the latter case any mixing near the pole was reversible, being due to the temporary juxtaposition of air masses of differing potential vorticities that later returned toward their original latitudes.) This behavior was anticipated on theoretical grounds by McIntyre (1982), and is illustrated schematically in Fig. 6.17. In the model the irreversible mixing took place by the stretching-out of potential vorticity contours to such an extent that dissipation, whether explicitly represented or implicit in the model's zonal truncation, could take over: the inclusion of wave number 2 in addition to wave number 1 apparently expedited this process.

6.3.4 The Role of Instabilities

The theories discussed above have prescribed some form of planetary-wave forcing near the tropopause as a lower boundary condition. The temporal growth of this forcing leads to amplification of the planetary waves propagating into the stratosphere; these waves in turn interact with the mean stratospheric flow, and the warming follows. It has been assumed that the increased lower-boundary forcing is related in some way to large-scale tropospheric disturbances, such as blocking events, which have been taken to develop more or less independently of happenings in the stratosphere.

An alternative approach, which seeks to include an explanation of the planetary-wave amplification, rather than regarding it as imposed externally, was considered by Plumb (1981). Using both an asymptotic analytical theory and a simple quasi-linear model, which included a highly idealized troposphere and vertical sidewalls parallel to the x axis and no meridional wave propagation, he found that wave amplification and stratospheric warmings

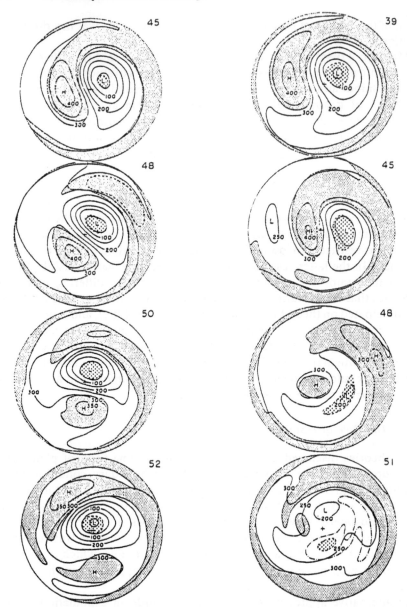

Fig. 6.16. Polar stereographic plots of the evolution of the isobaric height (solid curves, at 500-m intervals) and quasi-geostrophic potential vorticity $q'_{(M)}$ [see, e.g., Eq. (5.3.3)] at $z = 40$ km, for several days from the model experiments of Matsuno (1984). Left-hand diagrams, minor warming; right-hand diagrams, major warming. Small values of $q'_{(M)}$ ($<2 \times 10^{-4}$ s^{-1}) are shaded. Stippled areas indicate large values of $q'_{(M)}$ ($>2.5 \times 10^{-4}$ s^{-1} in minor warming and $>2 \times 10^{-4}$ s^{-1} in major warming). Day numbers marked next to each frame.

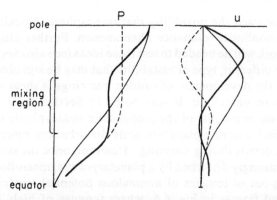

Fig. 6.17. Schematic latitudinal distributions of Ertel's potential vorticity P on an isentropic surface and corresponding polar night jet profiles u in the model of Matsuno (1984). Thin solid curves represent the initial states, and thick solid curves and dashed curves show distributions after minor and major warmings, respectively. [From McIntyre (1982) and Matsuno (1984).]

could occur through a nonlinear near-resonance between a topographically forced stationary wave and a free traveling wave in a zonal-mean flow $\bar{u}$ that depends on height.

Plumb's mechanism works essentially as follows. Suppose that under given mean-flow conditions a free normal mode has a small horizontal phase speed, so that it is nearly in resonance with the stationary, topographically forced wave. If this forced wave grows a little in amplitude, it can under certain conditions bring about mean-flow changes that make the free mode's phase speed even smaller and thus even nearer to resonance. This "self-tuning" leads to further stationary-wave growth, and a positive feedback is set up; as a result, large wave-amplification can occur, and large mean-flow changes resembling those associated with warmings. It is interesting that some observational studies, including those of the minor warming of January 1979, note a slowing-down of a traveling wave prior to the warming.

Plumb's "self-tuned resonant cavity instability" is analogous to the "topographic instability" studied by Charney and DeVore (1979) in another context. It has been discussed in detail by McIntyre (1982), who considered the likelihood of suitable resonant cavities occurring in the real atmosphere, perhaps aided by the presence of reflecting nonlinear critical layers (see Section 5.6) at low latitudes. For quasi-stationary waves, these critical layers surround zero-wind lines where $\bar{u} = 0$: the modulation of the positions of such lines by the equatorial quasi-biennial oscillation might affect the properties of any such cavity, and hence account for the roughly biennial occurrence of major warmings noted in Section 6.2.5 (but see Section 12.5.1).

McIntyre also discussed the possibility that tropospheric blocking may be part of the same nonlinear resonance phenomenon. Further observational and theoretical work will be needed to test these ideas (see also Section 12.4).

Another, quite different, type of instability that may be significant during certain phases of the development of sudden warmings is shear instability, either barotropic or baroclinic. It was noted in Section 6.3.1 that early theories of the shear instability of the global-scale stratospheric circulation were unable to find unstable planetary-scale disturbances resembling the wave-growth that occurs during warmings. However, once the stratospheric flow has become strongly disturbed by a planetary-wave, mean-flow interaction, the drawing-out of tongues of anomalous potential vorticity (as for example in several frames in Fig. 6.6, where tongues of high P become sandwiched between regions of low P) implies sign changes in $\partial P/\partial y$ on isentropic surfaces. If these occur over a sufficient spread of longitudes, for a roughly zonally aligned tongue, it is plausible that the flow may support *local* shear instabilities, owing to the local breakdown of the Charney–Stern criterion (Section 5.5.1).

6.4 Conclusions

This chapter has presented a variety of observational findings and theoretical ideas relating to stratospheric warmings. An attempt will now be made to synthesize these into an overall picture of the phenomenon.

A major sudden warming is initiated by an anomalous growth of a planetary-wave disturbance (mainly comprising wave-number 1 and 2 components) that propagates from the troposphere into the stratosphere and interacts strongly with the preexisting circulation there. The growth of the wave may be due to an independently generated forcing mechanism in the troposphere or some process (perhaps a self-tuned resonance) that disturbs the stratosphere and troposphere together. The relevant tropospheric disturbance usually manifests itself as a blocking event.

In either case, the stratosphere probably has to be in some suitable state if a warming is to follow. Thus, the zonal-mean polar night jet may need to be further poleward than its climatological position or, in synoptic terms, the basic winter polar cyclonic vortex may need to be tighter than its usual form and perhaps displaced off the pole. This "preconditioned" state can itself be time-varying, and may be the result of an earlier wave event that disrupts the climatological flow: such a disruption may be explicable in terms of wave mean-flow interaction theory or, equivalently, in terms of planetary-wave breaking mechanisms.

The major warming itself may likewise be diagnosed in terms of wave mean-flow concepts or in terms of synoptic maps. The former suggest and quantify a dynamical scenario in which the planetary waves are diverted from their climatological equatorward propagation by the anomalous refractive properties of the preconditioned flow and are focused into the high polar cap. Once there, they lead to mean-flow deceleration and temperature rises in the stratosphere, and the associated mean meridional circulations. The chain of cause and effect provided by this picture is borne out to some extent by experiments with quasi-linear models. It is, however, complicated by the presence of wave–wave interactions, which seem to be significant in some observational and modeling studies. Moreover, it is difficult to justify for large-amplitude disturbances, when the separation into zonal-mean and wave components may be especially artificial.

Synoptic maps, on the other hand, do not make this separation, and isentropic potential vorticity maps and maps of chemical tracers such as ozone give particularly vivid descriptions of the development of observed and modeled warmings. For example, they show how the main cyclonic vortex is distorted and displaced by the growth of the Aleutian High, and how it may split and move off the pole at the height of the warming. They also give strong qualitative hints of some of the dynamical processes involved. As yet, however, few methods have been devised for the quantitative dynamical analysis of such maps. In the long run, a judicious combination of these approaches and others (such as the consideration of the propagation of localized waves on strong cross-polar flows, for example) will probably be needed if the maximum insight is to be gained from observations and models of sudden warmings. Account will also need to be taken of aspects of warmings (such as the sharp "front-like" vertical temperature gradients that have long been observed in rocket data) that can be obscured by wave mean diagnostics and IPV maps at only a few levels.

There is much interannual variability between one stratospheric winter and another: major warmings only occur in about one in two Northern-Hemisphere winters, and when they do occur they take a variety of forms. Theoretical suggestions of possible explanations for some of this variability are now beginning to emerge and will require careful testing. The total absence of major warmings in the Southern-Hemisphere winter may be partly due to the fact that the winter vortex is stronger there than in the Northern Hemisphere, and thus more deceleration (and hence more wave forcing) would be required to reverse it. Another, related, factor may be the comparative lack of large-amplitude quasi-stationary waves there: see Sections 5.2.2 and 7.4.

The stratospheric sudden warming is perhaps the most dramatic large-scale dynamical event to occur in the middle atmosphere, and the elucidation of the processes involved remains a major challenge to meteorologists. Much further work will need to be done before a full understanding of the phenomenon is attained. This will call for further interaction between observational studies using data from satellites and other sources on the one hand, and experiments with a hierarchy of theoretical models on the other.

References

6.1. Review articles on sudden warmings include those by Quiroz *et al.* (1975), Schoeberl (1978), McInturff (1978), Holton (1980), Labitzke (1981b), and McIntyre (1982).

6.2 Synoptic descriptions of the evolution of several individual warmings are given in some of the articles cited for Section 6.1 and other papers referenced therein. Labitzke (1981b) summarizes the behavior of the wave-number 1 and 2 height field components and zonal-mean temperature field at 30 mb for all winters from 1964 to 1980. Eliassen–Palm diagnostics and refractive indices have been applied to sudden warmings by Palmer (1981a,b), O'Neill and Youngblut (1982), and Kanzawa (1982, 1984), among others. The use of potential vorticity for interpreting sudden warmings was suggested by Davies (1981).

Low-latitude stratospheric cooling at the time of a polar warming has been noted, for example, by Fritz and Soules (1972), and upper-mesospheric cooling by Hirota and Barnett (1977).

A recent statistical study of the observed association between sudden warmings and tropospheric blocking was performed by Quiroz (1986).

A possible correlation of sudden warmings with the equatorial quasi-biennial oscillation was noted by Labitzke (1982).

A major study of the Southern Hemispheric warming of July 1974 was performed by Al-Ajmi *et al.* (1985). Hartmann *et al.* (1984) and Shiotani and Hirota (1985) give detailed descriptions of wave mean-flow interaction in the southern winter, including some minor warming events.

6.3. References to numerous other developments of Matsuno's model of the sudden warming can be found in the reviews cited for Section 6.1.

The slowing-down of a traveling wave prior to a stratospheric warming was noted, for example, by Quiroz (1975) and Madden and Labitzke (1981). A recent theoretical study of the role of baroclinic instability of distorted stratospheric flows was carried out by Frederiksen (1982).

Chapter 7 | The Extratropical Zonal-Mean Circulation

7.1 Introduction

In Section 5.2.1 we presented some aspects of the observed climatology of the zonal-mean circulation of the middle atmosphere, its annual cycle, and its interhemispheric variations. For example, Fig. 5.1 showed the existence, in the zonal mean, of a cold tropical tropopause, a warm stratopause, and a cold summer mesopause, while the corresponding mean zonal geostrophic winds, illustrated in Fig. 5.2, are generally westerly in winter and easterly in summer, decreasing to small values near the mesopause. In the present chapter we examine the processes that maintain the climatological zonal-mean state in the extratropics. We concentrate especially on the ways in which dynamical phenomena can lead to large departures, in certain parts of the middle atmosphere, from a hypothetical climatology determined solely by radiative and photochemical effects.

To investigate the role played by dynamical processes in producing the observed middle atmosphere circulation, it is first useful to consider what form the circulation would take in the absence of dynamical processes, other than some representation of convection, and perhaps baroclinic wave activity, in the troposphere. The temperature field associated with such a circulation can be calculated from a radiative–photochemical model of the stratosphere and mesosphere, together with a radiative–convective model of the troposphere. An example for near-solstice conditions was given in Fig. 1.2, from the time-marched calculations decribed by Fels (1985). This shows strong latitude and height variations of the resulting zonally symmetric temperature, with a maximum of about 290 K at the summer

295

stratopause and temperatures below 180 K throughout the middle atmosphere at the winter pole, decreasing to 130 K at the winter mesopause. The temperature field, $T_r(\phi, z, t)$ say, calculated in this way will be referred to as the *radiatively determined temperature.*

A comparison of the calculated radiatively determined temperature field in Fig. 1.2 with the observed January-mean field of Fig. 5.1a reveals some overall similarities between the two, but also some striking differences. For example, the observed midlatitude summer stratopause temperature, at about 280 K, is close to T_r, but the observed north polar night is much warmer than the corresponding T_r (by about 30 K in the lower stratosphere, increasing to 100 K in the mesosphere), while the observed southern summer mesopause is much colder (by about 60 K). The July-mean zonally averaged observed temperature of Fig. 5.1c is roughly a mirror image of Fig. 5.1a, except that the southern winter polar midstratosphere, at about 180 K, is only just above the radiatively determined value.

Assuming thermal-wind balance,

$$\left(f + \frac{2u_{gr} \tan \phi}{a}\right) \frac{\partial u_{gr}}{\partial z} = -\frac{R}{aH} \frac{\partial T_r}{\partial \phi} \tag{7.1.1}$$

[cf. Eqs. (3.4.1c) and (5.2.2)], and a suitable lower boundary condition, one may calculate the zonal gradient wind, u_{gr} say, that would be associated with the radiatively-determined temperature T_r. Figure 7.1 shows the $u_{gr}(\phi, z)$ field corresponding to the temperature field of Fig. 1.2, given that u_{gr} equals the observed climatological values at 100 mb. This "radiatively determined" gradient wind exhibits extremely strong westerlies in the winter polar night, associated with the strong latitudinal gradients of T_r there, and quite strong easterlies in the summer hemisphere. The magnitudes of the winds increase with height throughout the stratosphere and mesosphere. In contrast, the observed zonal-mean geostrophic zonal winds for January (Fig. 5.2a) show more moderate growth with height, peaking near 60 km altitude in both hemispheres and decreasing to small values near the mesopause: the observed jets "close off" in the upper mesosphere. In July (Fig. 5.2c) the Southern Hemisphere winter westerlies are stronger than their Northern Hemisphere counterparts of January, but still decrease above 60 km altitude. (Note that observed zonal-mean geostrophic zonal winds are usually small enough to be a good approximation to the observed gradient winds.)

It is possible that some of the differences between the climatological observations of Figs. 5.1 and 5.2 and Fels's time-marched radiative–photochemical–convective calculations of Figs. 1.2 and 7.1 may be due to deficiencies in the radiative formulation of the latter. However, by far the most important reason for these differences is the presence of dynamical processes

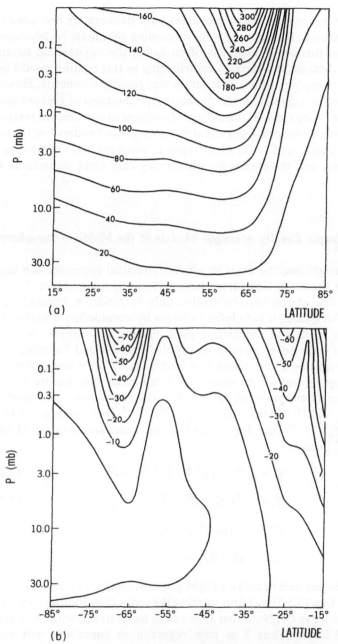

Fig. 7.1. Zonal gradient wind u_{gr} that is in thermal-wind balance with the temperature field T_r of Fig. 1.2 and equals the observed climatological zonal wind at 100 mb. (a) Northern Hemisphere (winter), (b) Southern Hemisphere (summer). (Courtesy of Dr. S. B. Fels.)

in the middle atmosphere; such processes were deliberately excluded from
Fels's calculations. The extra heating or cooling that must be provided by
the dynamical thermal transport is often called the "dynamical heating."
Some of the dynamical processes contributing to this heating would occur
in a middle atmosphere whose circulation was zonally symmetric. However,
simple arguments suggest that the dynamical phenomena of greatest importance for accounting for departures from T_r—in the extratropics at least—are
associated with deviations from zonal symmetry: the "eddies" or "waves."
The following sections discuss which types of wave are likely to be involved
in this process, and the means by which they may force departures of $\bar{T}$
from T_r.

7.2 Some Simple Zonally Averaged Models of the Middle Atmosphere

To gain insight into the ways in which dynamical processes can lead to
departures of the zonal-mean temperature from the temperature $T_r(\phi, z, t)$
of a hypothetical atmosphere controlled only by radiative, photochemical,
and convective effects, it is helpful to begin by considering a hierarchy of
rather simple models of the extratropical middle atmosphere. A suitable
starting point is the set of quasi-geostrophic "transformed Eulerian-mean"
(TEM) equations on a beta-plane (see Section 3.5). Since comparison with
observed temperatures will be made, the zonal-mean temperature $\bar{T}$ will be
used as a dependent variable, rather than the zonal-mean potential tem-
perature $\bar{\theta} = \bar{T}e^{\kappa z/H}$. Then, with $\bar{J}/c_p \equiv \bar{Q}e^{-\kappa z/H}$ (see Section 3.1.1) and
$N^2 \equiv H^{-1}R\theta_{0z}e^{-\kappa z/H}$ [see Eq. (3.2.13)], the quasi-geostrophic TEM set
[Eqs. (3.5.5)] becomes

$$\bar{u}_t - f_0\bar{v}^* = \rho_0^{-1} \nabla \cdot \mathbf{F} + \bar{X} \equiv \bar{G}, \tag{7.2.1a}$$

$$\bar{T}_t + N^2HR^{-1}\bar{w}^* = \bar{J}/c_p, \tag{7.2.1b}$$

$$\bar{v}_y^* + \rho_0^{-1}(\rho_0\bar{w}^*)_z = 0, \tag{7.2.1c}$$

$$f_0\bar{u}_z + H^{-1}R\bar{T}_y = 0. \tag{7.2.1d}$$

It is convenient here to make a slight physical distinction from the system
described by Eqs. (3.5.5) by reinterpreting the terms contributing to the zonal
force per unit mass $\bar{G}$. [Note that $\bar{G}$ is not to be confused with the quantity
G in Eq. (3.5.2b).] Thus $\mathbf{F}$ is now regarded as containing not only a
contribution

$$\mathbf{F}_{(p)} \equiv (0, -\rho_0\overline{u'v'}, \rho_0f_0\overline{v'\theta'}/\theta_{0z}) \tag{7.2.2a}$$

from planetary waves [cf. Eq. (3.5.6)], but also a contribution

$$\mathbf{F}_{(g)} \equiv (0, -\rho_0 \overline{u'v'}, -\rho_0 \overline{u'w'}) \tag{7.2.2b}$$

from small-scale gravity waves [cf. Eq. (3.5.3b)]. The term $\bar{X}$ now represents all further contributions to the mean zonal force per unit mass associated with gravity waves and other small-scale disturbances. The term $\bar{J}$ is the zonal-mean diabatic heating rate per unit mass, and will be assumed here to equal the zonal-mean net *radiative* heating rate: the "wave heating" term on the right of Eq. (3.5.2e) is negligible for quasi-geostrophic motions and also for gravity waves, and will be ignored here together with all other wave-induced and molecular contributions to $\bar{J}$. As in Section 3.5, $(\bar{v}^*, \bar{w}^*)$ is the residual circulation, defined by

$$\bar{v}^* \equiv \bar{v}_a - \rho_0^{-1}(\rho_0 \overline{v'\theta'}/\theta_{0z})_z, \qquad \bar{w}^* \equiv \bar{w}_a + (\overline{v'\theta'}/\theta_{0z})_y, \tag{7.2.3}$$

where $(\bar{v}_a, \bar{w}_a)$ is the Eulerian zonal-mean (ageostrophic) meridional circulation. It should be noted that $\overline{v'\theta'}/\theta_{0z}$ can alternatively be written as $\overline{v'T'}/N^2 H R^{-1}$ in Eqs. (7.2.2a) and (7.2.3). Several advantages of the set of Eqs. (7.2.1) over the conventional Eulerian-mean set were mentioned in Section 3.5.

7.2.1 Steady-State Model

We first consider a hypothetical steady-state atmosphere in which the seasonal cycle is absent; with time derivatives set to zero, Eqs. (7.2.1a,b) give

$$-f_0 \bar{v}^* = \bar{G}, \qquad N^2 H R^{-1} \bar{w}^* = \bar{J}/c_p, \tag{7.2.4a,b}$$

and substitution into Eq. (7.2.1c) yields

$$-\bar{G}_y + f_0 \rho_0^{-1}(\rho_0 \bar{J}\kappa/N^2 H)_z = 0 \tag{7.2.4c}$$

(since $\kappa = R/c_p$), showing how the diabatic heating rate $\bar{J}$ must be related to $\bar{G} \equiv \rho_0^{-1}\nabla \cdot \mathbf{F} + \bar{X}$ in this hypothetical state. [Note that Eq. (7.2.4c) is a steady-state version of the quasi-geostrophic potential vorticity equation; this can be seen by substituting Eq. (3.5.10) into Eq. (3.3.4).] If all eddy and small-scale effects are absent, so that $\bar{G}$ vanishes, then $\bar{v}^* = 0$ by Eq. (7.2.4a): the continuity equation [Eq. (7.2.1c)] then implies that $|\bar{w}^*|$ grows exponentially with z, in general. To prevent this unphysical behavior we can impose the boundary condition $\bar{w}^* = 0$ at a lower boundary $z = z_0$, say. Then $\bar{w}^* = 0$ everywhere and so $\bar{J} = 0$ everywhere, by Eq. (7.2.4b). The atmosphere is then in radiative equilibrium, under our assumption that $\bar{J}$ is the net *radiative* heating rate: thus, the temperature $\bar{T}$ must equal the the radiatively determined value $T_r(y, z)$, so that the long-wave cooling everywhere balances the solar heating. Moreover, $\bar{u} = u_r$, where u_r is the geostrophic radiatively determined wind, satisfying

$$f_0 \frac{\partial u_r}{\partial z} + H^{-1} R \frac{\partial T_r}{\partial y} = 0 \qquad (7.2.5)$$

[see Eq. (7.2.1d), but contrast Eq. (7.1.1)], and since $\bar{v}^* = \bar{w}^* = 0$ and eddies are absent, $\bar{v}_a = \bar{w}_a = 0$ as well, by Eq. (7.2.3).

7.2.2 Annually Varying Model with No Waves

We next include time-dependence by letting the solar heating rate take on an annual variation, $\bar{J}_s(y, z, t)$, say, but still assume that eddy and small-scale effects are absent, so that $\bar{G} = 0$. Further qualitative insight can be obtained by parameterizing $\bar{J}$ in terms of $\bar{T}$. As a simple example we use the Newtonian cooling form

$$\bar{J}/c_p = -[\bar{T} - T_r(y, z, t)]/\tau_r(z), \qquad (7.2.6)$$

where $T_r(y, z, t)$ is the temperature calculated from a time-dependent radiative–photochemical model (such as that from which Fig. 7.1 was obtained) with specified solar heating $\bar{J}_s(y, z, t)$, and $\tau_r(z)$ is a radiative relaxation time. This parameterization is not expected to be quantitatively accurate for large departures of $\bar{T}$ from T_r; however, it does contain the important physical feature of relating the net heating rate to departures of $\bar{T}$ from the radiatively determined temperature T_r.

Using Eqs. (7.2.1) with $\bar{G} = 0$ and $\bar{J}/c_p$ given by Eq. (7.2.6), it can be shown that

$$\left[\frac{\partial^2}{\partial y^2} + \frac{\partial}{\partial z} \left(\rho_0^{-1} \frac{\partial}{\partial z} \rho_0 \varepsilon \right) \right] \bar{T}_t + \left\{ \rho_0^{-1} \left[\frac{\rho_0 \varepsilon}{\tau_r} (\bar{T} - T_r) \right]_z \right\}_z = 0, \qquad (7.2.7)$$

where $\varepsilon(z) = f_0^2 / N^2(z)$, as in Eq. (3.2.16). In this equation the term in T_r (which depends on the solar heating $\bar{J}_s$) provides the *forcing*, to which $\bar{T}$ is the *response*. In general $\bar{T}$ will follow $T_r(y, z, t)$ but will be somewhat lagged in time (since the relaxation time τ_r is nonzero) and somewhat more smoothly distributed in space [because of the properties of the elliptic operator on the far left of Eq. (7.2.7)]: the zonally symmetric dynamics thus provide a kind of "inertia." Since $\bar{T}$ is not equal to T_r in general, Eq. (7.2.6) implies that the net radiative heating rate does not generally vanish; by Eqs. (7.2.1) the residual circulation $(\bar{v}^*, \bar{w}^*)$ does not vanish, either. This idealized model makes it clear that the nonvanishing of the *net* heating rate $\bar{J}$ is essentially due to the presence of the "dynamical inertia," and cannot be regarded as *imposed* by external radiative agencies. To put it another way, although the solar heating $\bar{J}_s$ has been specified in advance in this model, the "long-wave cooling" $\bar{J} - \bar{J}_s$ must be determined as part of the solution.

Assuming vertical scales of order H, where H is the scale height, and horizontal scales of order L, where $f_0^2 L^2 \sim N^2 H^2$, we can apply a simple order-of-magnitude argument to Eq. (7.2.7) to obtain

$$\frac{\partial \bar{T}}{\partial t} \sim (\bar{T} - T_r)/\tau_r \qquad (7.2.8)$$

approximately, in the present model. (This argument does not apply near the equator, where $f_0^2 L^2 \ll N^2 H^2$: see Section 8.1 for a discussion of the relevant time-dependent balance in the tropics.) If $\Delta \bar{T}(y, z)$ is the maximum annual variation of $\bar{T}$ in the model and τ is a seasonal timescale (say 3 months), an estimate of the temperature tendency is

$$\partial \bar{T}/\partial t \sim \Delta \bar{T}/\tau. \qquad (7.2.9)$$

However, typical radiative relaxation times are usually a few days ($\tau_r \ll \tau$), so that

$$\bar{T} - T_r \sim \frac{\tau_r}{\tau} \Delta \bar{T} \ll \Delta \bar{T} \qquad (7.2.10)$$

from Eqs. (7.2.8) and (7.2.9), and thus departures of $\bar{T}$ from the annually varying radiatively determined value $T_r(y, z, t)$ are much smaller in this model than the amplitude of the annual swing $\Delta \bar{T}$. It follows that these departures are also much smaller in magnitude than the annual swing ΔT_r in the radiatively determined temperature, which can be estimated by comparison of opposite hemispheres in Fig. 7.1. The model, therefore, predicts extratropical temperatures $\bar{T}(y, z, t)$ that are always "close" to the annually varying temperatures $T_r(y, z, t)$ determined from radiative-photo-chemical-convective considerations alone.[1]

On the other hand, an examination of the fields of T_r in Fig. 1.2 and the observed climatological $\bar{T}$ in Fig. 5.1a shows that in January at 0.01 mb (in the upper mesosphere, near 80 km altitude), $\bar{T} - T_r \approx 60$ K at 60°N and -45 K at 60°S, values comparable with that of $\Delta T_r \approx 50$ K (as obtained by subtracting T_r at 60°N from T_r at 60°S). Similarly, at 10 mb and 60°N, in the winter middle stratosphere, $\bar{T} - T_r \approx 45$ K, comparable with $\Delta T_r \approx 55$ K for that latitude and height. Clearly, the model described in the present section fails to predict these *large* departures of $\bar{T}$ from T_r in the upper

---

[1] The same conclusion holds if Eq. (7.2.6) is regarded as an order-of-magnitude estimate, rather than an equality: the argument does not depend on the precise form of the parameterization of $\bar{J}$, but only on the fact that the radiative timescale is much less than the seasonal timescale.

mesosphere and winter stratosphere. Additional effects must be included if a basic understanding of the observed annual variations of the temperature structure of these regions is to be obtained.

7.2.3   Inclusion of Wave-Forcing Effects

To obtain the next model in our hierarchy, we now suppose that $\bar{G}$ is nonzero. The set of Eqs. (7.2.1) with parameterization as in Eq. (7.2.6) yields the equation

$$\left[\frac{\partial^2}{\partial y^2} + \frac{\partial}{\partial z}\left(\rho_0^{-1}\frac{\partial}{\partial z}\rho_0\varepsilon\right)\right]\bar{T}_t + \left\{\rho_0^{-1}\left[\frac{\rho_0\varepsilon}{\tau_r}(\bar{T} - T_r)\right]_z\right\}_z + f_0HR^{-1}\bar{G}_{yz} = 0$$

$$\qquad\qquad [1] \qquad\qquad\qquad\qquad\qquad [2] \qquad\qquad\qquad\qquad\qquad [3] \qquad (7.2.11)$$

for $\bar{T}$ and, using Eq. (7.2.5), the equation

$$\left[\frac{\partial^2}{\partial y^2} + \rho_0^{-1}\frac{\partial}{\partial z}\left(\rho_0\varepsilon\frac{\partial}{\partial z}\right)\right]\bar{u}_t + \rho_0^{-1}\left[\frac{\rho_0\varepsilon}{\tau_r}(\bar{u} - u_r)_z\right]_z - \bar{G}_{yy} = 0 \quad (7.2.12)$$

for $\bar{u}$: the latter is a form of Eq. (3.5.7). [Note that the elliptic operator acting on $\bar{u}_t$ in Eq. (7.2.12) differs slightly from that acting on $\bar{T}_t$ in Eqs. (7.2.11) and (7.2.7).]

We now suppose that the wave forcing represented by $\bar{G}$ varies on a timescale τ_w: except for rapid events like sudden warmings we can expect τ_w to be comparable to the seasonal timescale τ and much greater than the radiative timescale τ_r. If rapid wave events like sudden warmings *are* present, the application of a running time average over a period $\tau_w = O(\tau)$ will remove shorter-period fluctuations in Eq. (7.2.11), leaving only smooth variations of timescale $O(\tau_w)$. In either case a scaling argument similar to the one that was applied to Eq. (7.2.7) gives the ratio of terms in Eq. (7.2.11) as

$$[1]:[2]:[3] = \frac{\Delta\bar{T}}{\tau}:\frac{\bar{T} - T_r}{\tau_r}:f_0LR^{-1}\Delta\bar{G} \qquad (7.2.13)$$

if $f_0^2L^2 \sim N^2H^2$ again (thus excluding equatorial regions once more) and $\Delta\bar{G}$ is a typical variation in $\bar{G}$ over a time τ_w. As mentioned in the previous section, it is generally found that $(T - \bar{T}_r)/\tau_r \gg \Delta T/\tau$ in the polar night and the summer upper mesosphere, so in these regions the term [1] in Eq. (7.2.11) is small, and the effects represented by $\bar{G}$ must be large enough to give a balance between terms [2] and [3]. Equivalently, the time derivatives in Eqs. (7.2.1a,b) are small, so that—according to this model—the balances expressed by Eq. (7.2.4) hold approximately, at each t, in those regions where $\bar{T}$ exhibits large departures from T_r. This is especially true near the solstices, when time derivatives are expected to be small, in any case; for example, $\bar{T}_t$ is then smaller than the estimate of Eq. (7.2.9).

This simple model suggests that dynamical effects contributing to the mean zonal force per unit mass $\bar{G} \equiv \rho_0^{-1} \nabla \cdot \mathbf{F} + \bar{X}$ are the primary agents responsible for maintaining the large departures of $\bar{T}$ from T_r that are observed in parts of the middle atmosphere, and thus for producing the net radiative heating rates $\bar{J}$ that occur there. At the same time, these dynamical processes also drive the residual circulation, as can be seen from the continuity equation [Eq. (7.2.1c)] together with the approximate equation Eq. (7.2.4a) or more generally from Eq. (3.5.8), which can be written as

$$\left[\frac{\partial^2}{\partial y^2} + \rho_0^{-1} \frac{\partial}{\partial z} \left(\rho_0 \varepsilon \frac{\partial}{\partial z} \right) \right] f_0 \bar{v}^* + \rho_0^{-1} (\rho_0 \varepsilon \bar{G}_z)_z + \rho_0^{-1} \left(\frac{\rho_0 f_0 \kappa \bar{J}}{N^2 H} \right)_{yz} = 0$$

(7.2.14)

in the present notation. [The transformation of Eq. (7.2.3) shows that the Eulerian-mean meridional circulation is also dynamically driven.] Conversely, the model also suggests that those regions that are observed to be close to radiative equilibrium, such as parts of the midlatitude lower stratosphere and the summer stratopause, are in such a state because of the absence of any dynamical effects that can lead to significant values of $\bar{G}$ there.

The residual circulation in the middle atmosphere can be calculated *diagnostically* using *observed* temperatures, an accurate radiative heating algorithm for $\bar{J}$, the spherical-geometry equivalents of Eqs. (7.2.4b) [or, more accurately, Eq. (7.2.1b)] and (7.2.1c), and appropriate boundary conditions. A pioneering calculation of this kind was performed by Murgatroyd and Singleton (1961), who attempted to estimate the Eulerian-mean meridional circulation $(\bar{v}, \bar{w})$ by making what is now known to be an unjustifiable approximation, namely that the eddy heating terms on the right of the Eulerian-mean thermodynamic equation [Eq. (3.3.2e)] are negligible. However, under this assumption Eq. (3.3.2e) becomes essentially isomorphic to Eq. (7.2.1b); the continuity equation [Eq. (3.3.2d)] is of the same form as Eq. (7.2.1c), and so the Murgatroyd and Singleton circulation, sometimes also called a "diabatic circulation" (see Section 9.3.1), is a close approximation to $(\bar{v}^*, \bar{w}^*)$. It is shown schematically in Fig. 7.2: generally rising motion takes place above 30 km in the summer hemisphere, with flow from the summer to the winter hemisphere in the upper stratosphere and mesosphere, and descent in the winter hemisphere. The lower stratospheric circulation is more symmetric about the equator, with rising motion at low latitudes and descent in high latitudes.

It must be emphasized that the Murgatroyd and Singleton diagnostic calculation assumes $\bar{T}(y, z, t)$ to be given, and then derives $(\bar{v}^*, \bar{w}^*)$. It does

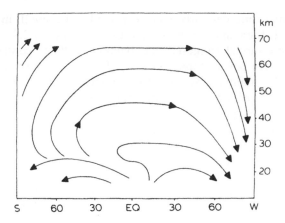

Fig. 7.2. Schematic streamlines of the solstice diabatic circulation in the middle atmosphere, as obtained from the Murgatroyd and Singleton (1961) calculation. S, summer pole; W, winter pole. [After Dunkerton (1978). American Meteorological Society.]

not require details of the eddy-forcing $\bar{G}$. A more predictive approach is to solve the complete set of mean-flow equations for $\bar{u}$, $\bar{T}$, $\bar{v}^*$, and $\bar{w}^*$, given some knowledge of $\bar{G}$. A classic study of this type was that of Leovy (1964b), who parameterized $\bar{G}$ in terms of $\bar{u}$ by assuming a linear "Rayleigh friction" drag,

$$\bar{G} = -\bar{u}/\tau_m, \tag{7.2.15}$$

where τ_m is a constant mechanical relaxation time, and used an expression of the form of Eq. (7.2.6) for $\bar{J}$. A variety of choices of τ_r and τ_m were examined: with $\tau_r = \tau_m \approx 15$ days, for example, several basic features of the seasonal cycle of $\bar{T}$ and $\bar{u}$ were simulated, although the details of the polar night stratosphere were not captured.

Current wave, mean-flow interaction theory leads us to expect that the parameterization of Eq. (7.2.15) is not likely to be very accurate, even with a relaxation time τ_m that depends on z. For some types of wave (e.g., gravity waves: see Sections 4.6.2 and 7.3), improved parameterizations are now available; for others, no satisfactory alternative parameterization has yet come to light.

A different, more qualitative, approach makes use of the result of Section 3.6 that the quantity $\nabla \cdot \mathbf{F}$ depends on wave transience, nonconservative wave effects, and wave nonlinearity. Thus, if waves are in some sense strongly transient, nonconservative, or nonlinear (or any combination of these) in a particular region, we can anticipate large local values of $\rho_0^{-1} \nabla \cdot \mathbf{F}$ there and perhaps large local departures of $\bar{T}$ from T_r. (Under the same conditions large values of $\bar{X}$ often tend to occur as well.) The identification of the mechanisms responsible for local deviations of the climatological temperature fields from the radiatively determined value thus amounts to a search for wave motions that significantly violate the "nonacceleration conditions" in the relevant regions of the middle atmosphere. Likely candidates in the mesosphere and in the winter stratosphere are discussed in the next two sections.

7.3 The Upper Mesosphere

As noted in Section 7.1, a comparison of Figs. 5.1a,c with Fig. 1.2 reveals that the observed climatological temperature field $\bar{T}$ at the solstices in the upper mesosphere is much warmer than the radiatively determined value T_r in midwinter and much colder in midsummer. In accordance with the arguments of Section 7.2.3 we consider in this section what wave motions could significantly break the nonacceleration conditions in this part of the atmosphere, in a climatological sense, and thus lead to the large climatological values of $\bar{G} = \rho_0^{-1} \nabla \cdot \mathbf{F} + \bar{X}$, on the order of $100 \text{ m s}^{-1} \text{ day}^{-1}$, required to account for such discrepancies.

It is now generally believed that gravity waves provide the main part of the necessary forcing $\bar{G}$ in the upper mesosphere: as discussed in Section 4.6.2, such waves grow in amplitude as they propagate from tropospheric source regions up into the rarefied mesosphere, and "break," leading to turbulence, small-scale mixing, and dissipation. Associated with these waves there is an Eliassen–Palm flux divergence $\nabla \cdot \mathbf{F}_{(g)} \approx -\rho_0^{-1} (\overline{\rho_0 u'w'})_z$ (called $\bar{X}_1$ in Section 4.6.2), representing a zonal force per unit mass on the zonal-mean flow: the mixing induced by the waves also leads to diffusion of mean-flow properties. Parameterizations of these processes, due to Lindzen (1981), are given in Eq. (4.6.18). The net contribution to the forcing $\bar{G}$ associated in this way with gravity waves tends to drag the mean flow toward the horizontal phase speed of the waves.

Since gravity waves are absorbed at or near critical levels, where the local horizontal wind speed equals their horizontal phase speed, the only gravity waves reaching the mesosphere from below are expected to have phase speeds outside the range of horizontal wind speeds occurring in the underlying stratosphere. Thus, when winter westerly winds are present in the stratosphere we can anticipate that gravity waves within a range of easterly phase speeds will occur in the mesosphere and break there. Conversely, when summer easterly winds are present in the stratosphere, gravity waves with westerly phase speeds would be expected to appear, and break, in the mesosphere. In winter, the breaking gravity waves of easterly phase speed will exert an easterly force on the westerly jet, and hence tend to close it off; similarly, the breaking gravity waves in summer will tend to close off the easterly jet. (An analogous "shielding effect" has also been postulated as an explanation of the mesospheric equatorial semiannual oscillation: see Section 8.5.2.)

A more detailed argument is as follows. Consider the solstices [see Figs. 5.2a,c]: in the winter mesosphere, above the peak of the westerly jet, the gravity-wave contribution to $\bar{G}$ is negative, while in the summer mesosphere, above the peak of the easterly jet, $\bar{G}$ is positive. By the approximate Eq. (7.2.4a) we expect $\bar{v}^*$ to be directed from summer to winter hemisphere at these altitudes: mass continuity then requires that $(\rho_0 \bar{w}^*)_z > 0$ in the winter mesosphere and $(\rho_0 \bar{w}^*)_z < 0$ in the summer. Assuming that $\bar{w}^*$ does not vary rapidly over a pressure scale-height H (or, alternatively, that $\bar{w}^* = 0$ at some level at or above the mesopause), this implies descent in the winter mesosphere and ascent in the summer (see Fig. 7.2). By Eqs. (7.2.4b) and (7.2.6), this means that $\bar{T}$ should be warmer than T_r in the winter upper mesosphere and cooler than T_r in the summer, in accordance with the observations. By the thermal wind relation of Eq. (7.2.1d), the reversed latitudinal gradients in $\bar{T}$ imply reversed vertical shears in $\bar{u}$ and hence a closing-off of the jets.

Experiments with Lindzen's parameterizations in fairly simple models of the middle atmosphere suggest that the gravity-wave drag and diffusion contributions to $\bar{G}$ can account satisfactorily for much of the departure of $\bar{T}$ from T_r in the mesosphere, given reasonable values of the wave parameters: see Fig. 7.3. The magnitude of the computed drag in such models

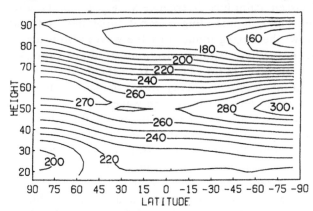

Fig. 7.3. Zonal-mean temperature (K) at the Northern-Hemisphere winter solstice, derived from a zonally symmetric model including a parameterization of the zonal drag associated with breaking gravity waves, but no representation of planetary waves. The reversed meridional temperature gradient in the upper mesosphere should be compared with the observations in Fig. 5.1a and contrasted with the radiatively determined calculations in Fig. 1.2. [After Holton (1983a). American Meteorological Society.]

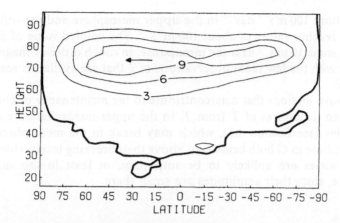

Fig. 7.4. As in Fig. 7.3, but for residual mean meridional wind $\bar{v}^*$ (m s^{-1}). [After Holton (1983a). American Meteorological Society.]

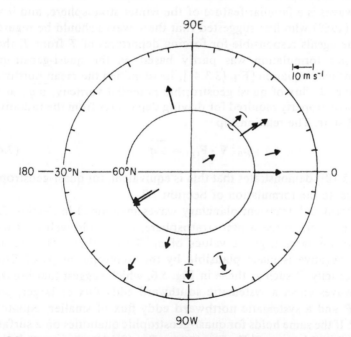

Fig. 7.5. Time-averaged north–south wind components near 90 km in June in the Northern Hemisphere, measured by various methods at different locations. Note the strong equatorward flow in all cases. [After Nastrom *et al.* (1982).]

peaks at about $100 \text{ m s}^{-1} \text{ day}^{-1}$ in the upper mesosphere and the diffusion coefficient reaches values of about $100 \text{ m}^2 \text{ s}^{-1}$. The implied value of $\bar{v}^*$ can amount to some 10 m s^{-1} near the mesopause, in rough order-of-magnitude agreement with the few radar observations of $\bar{v}$ that are available: see Figs. 7.4 and 7.5.

Other wave motions that may contribute to the maintenance of climatological-mean departures of $\bar{T}$ from T_r in the upper mesosphere are atmospheric tides (see Section 4.3), which may break in the mesosphere and could contribute to $\bar{G}$ both below and above their breaking levels. However, planetary waves are unlikely to be important, at least in the summer mesosphere, since their amplitudes are small there.

7.4   The Winter Polar Stratosphere

The other region of the middle atmosphere in which large climatological departures from the radiatively determined temperature are observed is the winter polar stratosphere. As discussed in Chapter 5, the occurrence of planetary waves is a familiar feature of the winter stratosphere, and it was Dickinson (1969) who first suggested that these waves should be regarded as the prime agents responsible for forcing departures of $\bar{T}$ from T_r there.

Dickinson's formulation was partly based on the quasi-geostrophic potential vorticity equation [Eq. (3.3.4)]; he identified the mean northward geostrophic eddy flux of quasi-geostrophic potential vorticity, $\overline{v'q'}$, as the essential wave property required for driving departures from the radiatively determined state. The relationship

$$\rho_0^{-1} \boldsymbol{\nabla} \cdot \mathbf{F}_{(p)} = \overline{v'q'} \tag{7.4.1}$$

[see Eq. (3.5.10)] demonstrates that this is equivalent, for quasi-geostrophic disturbances, to the formulation of Section 7.2.

If the "breaking" transient planetary waves mentioned in Section 5.2.3 are common events in the winter stratosphere, they could well contribute to systematically large negative values of $\rho_0^{-1} \boldsymbol{\nabla} \cdot \mathbf{F}_{(p)}$ there. The fact that the sign is negative is made plausible by the isentropic maps of Ertel's potential vorticity P, such as those in Fig. 5.6, which suggest that breaking planetary waves effect a systematic southward eddy flux of larger, polar values of P and a systematic northward eddy flux of smaller, equatorial values of P. If the same holds for quasi-geostrophic quantities on z surfaces, then $\overline{v'q'} < 0$ and hence $\rho_0^{-1} \boldsymbol{\nabla} \cdot \mathbf{F}_{(p)} < 0$ by Eq. (7.4.1): as expected, these waves are violating nonacceleration conditions. [The analogous result for isentropic eddy fluxes of P is expressed by Eq. (3.9.11): see Section 7.5.]

Dissipating stationary planetary waves (cf. Section 5.2.2) are also likely to be associated with negative values of $\nabla \cdot \mathbf{F}_{(p)}$, as in the simple example mentioned in Section 4.5.5. Radiative damping is one source of dissipation for such waves; it has also been suggested that breaking, small-scale waves may introduce nonconservative mechanical forcing or damping of the planetary waves, just as they do for the zonal-mean flow. However, one type of planetary wave that appears unlikely to induce significant forcing of the zonal-mean climatological flow is the global normal mode or free traveling planetary wave discussed in Sections 4.4 and 5.4. Except perhaps during their growth and decay phases, such waves do not strongly violate nonacceleration conditions.

Further research will be required to determine whether the values of $\rho_0^{-1} \nabla \cdot \mathbf{F}$, or equivalently of $\overline{v'q'}$, associated with planetary waves are large enough in magnitude and of the appropriate distribution in latitude and height to account for the observed departures of $\bar{T}$ from T_r in the winter polar stratosphere. [Note incidentally that the calculation of the *response* to such a zonal force is not entirely trivial: equations like Eqs. (7.2.11), (7.2.12), and (7.2.14) must be solved subject to appropriate boundary conditions.] However, support for some of the ideas presented in this section comes from the general circulation model experiments described in Section 11.2.1: with some versions of the "SKYHI" model, the polar night stratosphere temperature appears to be too close to T_r because the planetary waves are too weak. A similar picture emerges from experiments with the NCAR Community Climate Model, mentioned at the end of Section 11.1. Also relevant is the fact that the observed Southern-Hemisphere winter stratosphere is closer to the radiatively determined state than is the Northern-Hemisphere winter stratosphere: this may be due to the weaker southern winter planetary-wave activity mentioned above.

It should finally be mentioned that observational and model studies suggest that gravity-wave drag may also have some role to play in the winter stratosphere in contributing *directly* to the zonal force $\bar{G}$ as well as perhaps forcing or dissipating the planetary waves there.

7.5 Interpretation and Generalization

The models described in Section 7.2 are highly simplified and cannot be expected to give full quantitative agreement with observations. Their purpose is rather to provide qualitative insights into the physical mechanisms that determine the zonal-mean climatological state of the middle atmosphere. They suggest how eddy motions on various scales can keep certain parts of the middle atmosphere in a state far from that predicted by radiative–photochemical models; further, they shed light on some of the physical properties of the eddies that may be responsible for this maintenance process.

These models show how dynamical effects are associated with a "dynamical heating" [as represented, say, by the terms on the left of Eq. (7.2.1b)] which must be balanced by a net radiative heating. In particular, they make it clear that the residual circulation cannot be regarded as a flow that is purely *driven by* an externally imposed net radiative heating rate $\bar{J}$, and independent of the eddy forcing. Indeed, in the steady-state limit described by Eqs. (7.2.4)—which is shown in Section 7.2.3 to be a reasonable first approximation for some purposes—the residual circulation and the net radiative heating are both *entirely* eddy-driven under appropriate boundary conditions, and would both vanish if the eddy-forcing $\bar{G}$ were zero. Of course, the eddy-forcing $\bar{G}$ itself is not generally independent of the mean flow structure (this fact is made explicit, for example, by the parameterizations mentioned in Section 7.3). As a result, interesting and complex feedbacks may occur in the wave, mean-flow interaction and radiative-dynamical interaction processes.

Some of the results derived above for quasi-geostrophic flows on a beta plane can be extended to flows described by the primitive equations on the sphere; it is convenient for this purpose to use the isentropic coordinates introduced in Sections 3.8 and 3.9. For example, the analogs of the steady-state model equations [Eqs. (7.2.4a,b)] and the continuity equation [Eq. (7.2.1c)] in these coordinates are

$$a^{-1}\bar{v}^*\bar{m}_\phi + \bar{Q}^*\bar{m}_\theta = \bar{G}\cos\phi, \tag{7.5.1a}$$

$$(a\cos\phi)^{-1}(\bar{\sigma}\bar{v}^*\cos\phi)_\phi + (\bar{\sigma}\bar{Q}^*)_\theta = 0, \tag{7.5.1b}$$

from Eqs. (3.9.7a,c), where $a\bar{m} \equiv a^2\Omega\cos^2\phi + a\bar{u}\cos\phi$ is the zonal-mean absolute angular momentum per unit mass, $Q = (J/c_p)e^{\kappa z/H}$, and $\bar{G}$ now represents the terms $\bar{X}^* + (\bar{\sigma}a\cos\phi)^{-1}\tilde{\nabla}\cdot\tilde{F}$ in Eq. (3.9.7a). [Note that $\overline{(\sigma'u')}_t = 0$ because the waves are assumed steady.] Equation (7.5.1b) implies the existence of a stream function $\bar{\psi}^*$ such that

$$\bar{\sigma}a(\cos\phi)\bar{Q}^* = \bar{\psi}_\phi^*, \qquad \bar{\sigma}(\cos\phi)\bar{v}^* = -\bar{\psi}_\theta^* \tag{7.5.2a,b}$$

and substitution into Eq. (7.5.1a) then yields

$$\frac{\partial(\bar{\psi}^*, \bar{m})}{\partial(\phi, \theta)} = \bar{\sigma}a(\cos^2\phi)\bar{G}. \tag{7.5.3}$$

If eddy and frictional effects are absent, then $\bar{G} = 0$ and Eq. (7.5.3) implies that

$$\bar{\psi}^* = \Psi(\bar{m}), \tag{7.5.4}$$

say, for some function Ψ; that is, $\bar{\psi}^*$ is constant on surfaces of constant $\bar{m}$. The residual circulation ($\bar{v}^*, \bar{Q}^*$) flows along surfaces of constant zonal-mean absolute angular momentum, since it must conserve angular momentum when the right-hand side of Eq. (7.5.1a) vanishes. Now if $\bar{Q}^* = 0$ everywhere along some isentrope, $\theta = \theta_0$ say (perhaps a nominal lower boundary for the stratosphere), it follows from Eq. (7.5.2a) that $\bar{\psi}^*$ is constant on θ_0 and hence, by Eq. (7.5.4), is constant on all $\bar{m}$ surfaces that intersect θ_0. Thus, by Eq. (7.5.2), $\bar{v}^*$ and $\bar{Q}^*$ both vanish everywhere in the region threaded by such surfaces. Since $\bar{Q}^* = \bar{Q}$ in the absence of eddy motions, $\bar{Q} \equiv 0$ here and the region is in radiative equilibrium, with $\bar{v}^* \equiv 0$ also, by analogy with Section 7.2.1.[2]

The results of Sections 7.2.2 and 7.2.3 for the annually varying case can similarly be extended, but once again the scaling arguments that demonstrate the primacy of the eddy forcing in the extratropical middle atmosphere fail near the equator. Note incidentally that the mean zonal momentum equation in the form of Eq. (3.9.9) shows how Dickinson's argument concerning the role of the northward eddy potential vorticity flux generalizes to the primitive equations in isentropic coordinates, with Ertel's potential vorticity P replacing the quasi-geostrophic potential vorticity q.

References

7.1. The approach of considering how dynamical processes can lead to large departures from a hypothetical radiatively determined climatology was pioneered by Dickinson (1969). More recent use of the same ideas has been made by Kurzeja (1981), Plumb (1982), and Mahlman *et al.* (1984), among others.

7.2. The use of a hierarchy of models of the kind described here was suggested by Dr. S. B. Fels (1982 personal communication).

7.3. The original suggestion that gravity waves might by responsible for most of the forcing $\bar{G}$ in the upper mesosphere was made by Houghton (1978). Some indirect evidence for the "filtering" of gravity-wave phase speeds by the mean winds through which they propagate is provided by the radar measurements of Balsley *et al.* (1983). Simple models of the middle atmosphere involving parameterizations of gravity-wave drag in the mesosphere include those of Holton (1982, 1983a), Matsuno (1982), and Schoeberl *et al.* (1983). The effects of tides are considered, for example, by Lindzen (1981) and Miyahara (1984).

---

[2] Note that this argument does not exclude the possibility of regions containing steady, closed meridional circulations, in which angular momentum is conserved, so that Eq. (7.5.4) holds, but that are not in radiative equilibrium since the streamlines do not cut any isentropes on which $\bar{Q}^* = 0$. Such flows are useful idealizations of the low-latitude tropospheric Hadley cell, but their relevance for the middle atmosphere is not at present clear.

7.4. The forcing and dissipation of planetary waves by breaking gravity waves has been investigated by Schoeberl and Strobel (1984), Holton (1984a), and Miyahara (1985). Observations suggesting that direct gravity-wave drag on the zonal-mean flow may be important in the stratosphere are presented by Hamilton (1983) and Smith and Lyjak (1985).

7.5. Questions of wave, mean-flow feedbacks and radiative-dynamical feedbacks are considered for example by Fels *et al.* (1980) and Fels (1985). Models of angular-momentum-conserving Hadley circulations are discussed by Held and Hou (1980), Held and Hoskins (1985), and Tung (1986), among others.

Chapter 8 | Equatorial Circulations

8.1 Introduction

In Section 5.2 and Chapter 7 we discussed the global annual cycle in the mean zonal wind and temperature fields. This cycle is roughly antisymmetric with respect to the equator and, although dominant in the extratropics, has relatively little amplitude in equatorial latitudes. In the equatorial middle atmosphere above 35 km the seasonal variation is characterized primarily by a *semiannual* oscillation of the mean zonal wind. Well below 35 km, on the other hand, the equatorial stratospheric seasonal cycle is completely overwhelmed by a long-term oscillation that is not directly linked to the march of the seasons. This oscillation, which is of somewhat irregular period (averaging 27 months), is called the "quasi-biennial" oscillation (QBO).

The annual cycle in the mean zonal wind is fundamentally driven by the annual cycle in solar insolation. As shown in Chapter 7, in the extratropical stratosphere, eddy forcing acts primarily to warm the winter polar regions and hence damp the amplitude of the response to radiative driving. Since the sun passes overhead twice per year in the equatorial zone, radiative heating in that region has a significant semiannual component. Thus, it is not surprising that by analogy with the annual cycle efforts to explain the semiannual wind oscillation at first focused on thermal forcing. Despite the irregularity of the period of the QBO, early theoretical attempts to account for that oscillation also invoked thermal forcing mechanisms.

However, oscillating thermal forcing, in the absence of eddies, is far less effective in generating an oscillating mean zonal temperature and wind response in equatorial regions than at higher latitudes. For low latitudes, taking $f_0 = O(\beta L)$, where L is a typical meridional scale for the mean flow (~ 1200 km) and β takes its equatorial value $2\Omega/a$, we have

$$f_0^2 L^2 / N^2 H^2 = O(\beta^2 L^4 / N^2 H^2) = O(10^{-2}).$$

In this case the scaling argument described in Section 7.2.2 gives

$$\frac{\partial \bar{T}}{\partial t} \sim (\bar{T} - T_r)/\tau_e,$$

in place of Eq. (7.2.8), where $\tau_e \equiv (N^2 H^2 / \beta^2 L^4)\tau_r = O(10^2 \tau_r)$. If τ_r is a few days, then τ_e is much greater than the seasonal timescale $\tau \sim 3$ months. The amplitude $\Delta \bar{T}$ of the associated oscillation in $\bar{T}$ will thus be much smaller [by a factor $O(\tau/\tau_e)$] than the amplitude ΔT_r of the oscillation of T_r in a purely radiative model, rather than the same magnitude as at high latitudes. The amplitude of the corresponding oscillation in $\bar{u}$ can be estimated from Eq. (7.2.1d) as

$$\Delta \bar{u} \sim \left(\frac{R \Delta \bar{T}}{f_0 L} \right) \sim \left(\frac{R \Delta \bar{T}}{\beta L^2} \right),$$

which is $O(\tau/\tau_e)$ smaller than $\Delta u_r \sim R \Delta T_r / \beta L^2$, rather than the same order, as in the extratropics. This argument, which is supported by detailed calculations, indicates that it is very difficult to drive mean zonal temperature and wind oscillations in the equatorial region by thermal forcing alone. For the equatorial quasi-biennial and semiannual oscillations, wave driving must be considered the primary forcing.

8.2   The Observed Structure of the Equatorial Quasi-Biennial Oscillation

The equatorial QBO was originally discovered in 1960 by Reed and independently by Veryard and Ebdon (Reed et al., 1961; Veryard and Ebdon, 1961). Many of the essential dynamical aspects of the QBO are best displayed by time–height sections showing the departure of the monthly averaged mean zonal wind from its long-term monthly mean at equatorial stations. The example shown in Fig. 8.1 clearly illustrates the following features:

1. An alternating pattern of eastward and westward wind regimes that repeat at intervals varying from about 22 to 34 months, with an average period of about 27 months.

2. Downward propagation of successive regimes at an average rate of about 1 km/month, but with westerly shear zones descending more regularly and more rapidly than easterly shear zones.

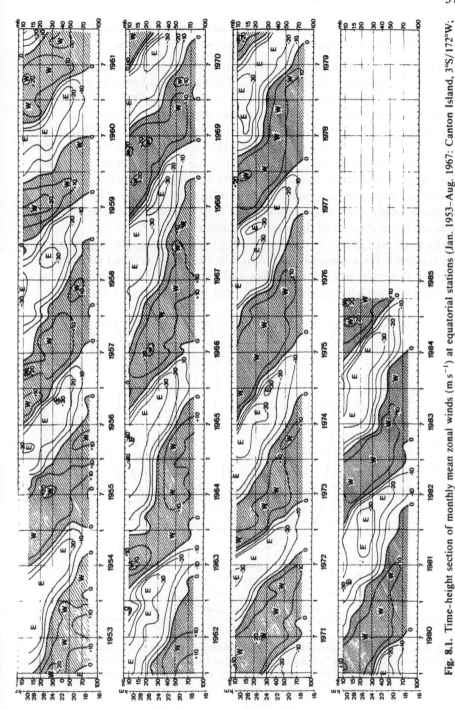

Fig. 8.1. Time–height section of monthly mean zonal winds (m s⁻¹) at equatorial stations (Jan. 1953–Aug. 1967: Canton Island, 3°S/172°W; Sept. 1967–Dec. 1975: Gan/Maldive Islands, 1°S/73°E; Jan. 1976–Apr. 1985: Singapore, 1°N/104°E). Isopleths are at 10-m s⁻¹ intervals. Note the alternating downward propagating westerly (W) and easterly (E) regimes. [From Naujokat (1986), with permission.]

3. Amplitude nearly constant in height between about 40 and 10 mb, but decreasing rapidly as regimes descend below the 50 mb level.

The last of these features is further illustrated by the latitude–height section of amplitude and phase shown in Fig. 8.2. The oscillation has an approximate Gaussian distribution about the equator with a latitudinal half-width of about 12 degrees. There is very little phase dependence on latitude.

A closer examination of the latitude–time dependence of the QBO does, however, reveal some remarkable asymmetries between the westerly and easterly acceleration phases. As shown in Fig. 8.3 at 30 mb, westerly accelerations appear first at the equator and spread to higher latitudes, while the easterly accelerations are rather more uniform in latitude. The westerly accelerations are more intense on average than the easterly, consistent with the more rapid descent of the westerly shear zone.

Although the QBO is definitely not a 2-year oscillation, there appears to be a tendency for a seasonal preference in the phase reversal so that, for example, the onset of both westerlies and easterlies occurs mainly during Northern Hemisphere summer at the 50 mb level (see Figs. 8.1 and 8.4).

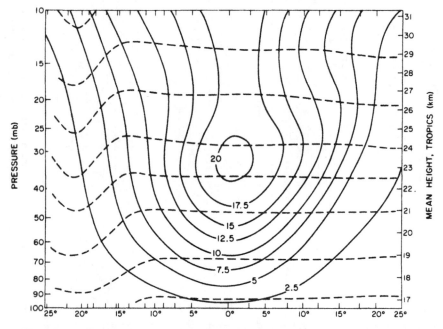

Fig. 8.2. Latitude–height distribution of the amplitude and phase of the zonal wind QBO. Amplitude (solid lines) in m s^{-1}, phase (dashed lines) at 1-month intervals with time increasing downward. [From Wallace (1973), with permission, adapted from an original drawing by R. J. Reed.]

30 mb

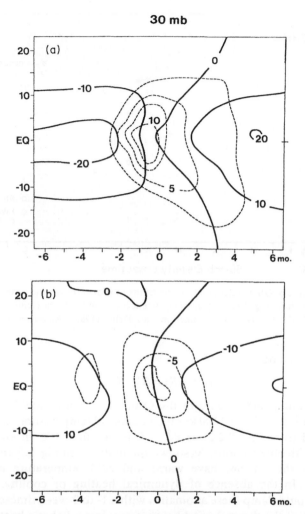

Fig. 8.3. Composite latitude–time section of zonal wind at 30-mb for westerly (upper) and easterly (lower) phases of the QBO. Zonal winds (solid lines) in m s^{-1}; acceleration (dashed lines) in m s^{-1}month^{-1}. [From Dunkerton and Delisi (1985). American Meteorological Society.]

Because of its long period and equatorial symmetry, the QBO in the mean zonal wind is in thermal wind balance throughout the equatorial region. For the equatorial β-plane this balance is given by Eq. (4.7.2), which in terms of $\bar{T}$ may be expressed as

$$\bar{u}_z = -R(H\beta y)^{-1}\bar{T}_y. \qquad (8.2.1)$$

Assuming equatorial symmetry so that $\bar{T}_y = 0$ at $y = 0$, L'Hôpital's rule

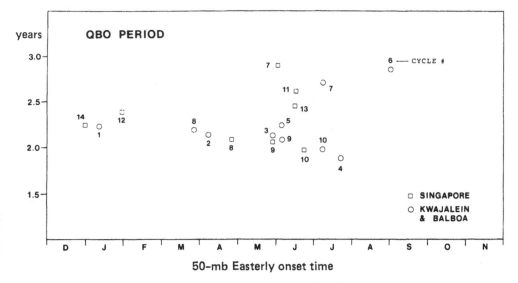

Fig. 8.4. Period of QBO cycles as a function of the time of easterly wind onset at the 50-mb level. Notice the clustering of easterly onset occurrences during Northern-Hemisphere summer. [Adapted by T. Dunkerton, from Dunkerton and Delisi (1985). American Meteorological Society.]

yields at the equator

$$\bar{u}_z = -R(H\beta)^{-1}\bar{T}_{yy}. \tag{8.2.2}$$

In the descending shear zones of the QBO, $\bar{u}_z \sim 5\ \mathrm{m\ s^{-1}\ km^{-1}}$, so that for a meridional scale of $L \sim 1200$ km the equatorial temperature anomaly can be estimated from Eq. (8.2.2) as $\delta\bar{T} \sim L^2 H\beta R^{-1}\bar{u}_z \sim 3$ K, in approximate agreement with observations. As shown qualitatively in Fig. 8.5, the westerly and easterly shear zones have warm and cold temperature anomalies, respectively. In the absence of dynamical heating or cooling, radiative relaxation would damp these anomalies with a 1- to 2-week timescale. Thus, maintenance of the thermal wind balance requires adiabatic heating (cooling) by a secondary residual meridional circulation with sinking (rising) at the equator in the westerly (easterly) shear zones as shown schematically in Fig. 8.5.

This mean meridional circulation is too weak to be directly observed. However, the vertical advection and the Coriolis torque associated with it (see Fig. 8.5) may account for some of the observed asymmetries between the easterly and westerly shear zones. The vertical advection is also consistent with the phase of the equatorial portion of the QBO observed in total ozone (Hasebe, 1983), although the QBO in ozone appears to have marked asymmetry with respect to the equator and substantial amplitude at high latitudes.

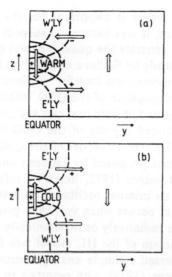

Fig. 8.5. Schematic latitude–height sections showing the mean meridional circulation associated with the equatorial temperature anomaly of the QBO. Solid contours show temperature anomaly isotherms, dashed contours are zonal wind isopleths. Plus and minus signs designate signs of the zonal wind accelerations driven by the mean meridional circulation. (a) Westerly shear zone, (b) easterly shear zone. [From Plumb and Bell (1982), with permission.]

This and other high latitude manifestations of the QBO will be discussed in Section 12.5.1.

8.3 Theory of the Quasi-Biennial Oscillation

Perhaps the three most remarkable features of the QBO that theory must explain are (1) the quasi-biennial periodicity, (2) the occurrence of zonally symmetric westerlies at the equator, and (3) the downward propagation without loss of amplitude. Early efforts to explain the periodicity were generally based on postulated biennial cycles either in the diabatic forcing in the stratosphere or in various elements of the tropospheric circulation. Aside from the fact that the observed oscillation is *not* biennial, but quite irregular in period, none of these theories could plausibly account for the structure of the observed oscillation. In particular, mechanisms based on advection by zonally symmetric inviscid circulations (which are angular-momentum-conserving) could not account for the production of a maximum in the absolute angular momentum at the equator during the westerly phase of the oscillation. Apparently zonally asymmetric wave forcing is required to explain this "superrotation." Since it is well known that lateral momentum transport by large-scale waves is important for the angular momentum budget of the troposphere, it was natural to suspect that planetary wave momentum fluxes might generate the quasi-biennial oscillation. However, the numerical modeling study by Wallace and Holton (1968) showed rather conclusively that lateral momentum transfer by planetary waves could not explain the downward propagation of the QBO without loss of amplitude.

The only process suggested to date that can successfully account for all three of the above-mentioned aspects of the QBO is vertical transfer of momentum by equatorial waves. A model of the QBO based on vertically propagating waves was first proposed by Lindzen and Holton (1968) and refined by Holton and Lindzen (1972, hereafter referred to as HL). HL argued that the QBO is an internal oscillation that results from the wave mean-flow interaction that occurs when vertically propagating Kelvin and Rossby-gravity waves are radiatively or mechanically damped in the lower stratosphere. The mechanism of the HL model was further elucidated by Plumb (1977), who discussed a simple analog nonrotating stratified fluid, and by Plumb and McEwan (1978), who reported an ingenious laboratory simulation of Plumb's analog to the QBO. Because it provides a simply understood prototype, it is useful to discuss Plumb's model before considering the HL model in detail.

8.3.1   A Two-Dimensional Analog of the QBO[1]

In his theoretical model, Plumb (1977) considered a vertically unbounded, nonrotating, stratified fluid subject to a standing wave forcing at its lower boundary of the form

$$\cos(kx)\cos(kct) = \tfrac{1}{2}[\cos k(x - ct) + \cos k(x + ct)].$$

Thus, the oscillation at the lower boundary is equivalent to two traveling waves of equal amplitude and oppositely directed phase speeds. This forcing generates vertically propagating internal gravity waves that propagate upward into the fluid, where Plumb supposed that they were dissipated by a weak Newtonian cooling with constant rate α.

Assuming that the mean flow is slowly varying so that the WKB approximation is valid for the waves, the EP flux $F^{(z)}$ [cf. Eq. (3.5.3b)] can be expressed as

$$F^{(z)} = \sum_{i=1}^{2} F_i(z) = \sum_{i=1}^{2} F_i(0) \exp\left[-\int_0^z g_i(z')\,dz'\right] \qquad (8.3.1a)$$

where

$$g_i(z) = \alpha N[k(\bar{u} - c_i)^2]^{-1} \qquad (8.3.1b)$$

[1] The discussion in this section follows Plumb (1984).

is just the Newtonian relaxation rate divided by the vertical component of the group velocity [see Eq. (4.6.6)]. Here $c_1 = -c_2 = c$ and $F_1(0) = -F_2(0)$.

The mean flow acceleration is given by the simple one-dimensional equation

$$\frac{\partial \bar{u}}{\partial t} = \frac{1}{\rho_0} \sum \frac{\partial F_i}{\partial z} + \nu \frac{\partial^2 \bar{u}}{\partial z^2} \qquad (8.3.2)$$

where ν designates a viscosity that is assumed to act on the mean flow. When $\bar{u} = 0$ the total EP flux vanishes for all z: $F_1 + F_2 = 0$, and the mean flow acceleration is zero. However, as shown by Plumb, this equilibrium is unstable for sufficiently small ν, since if $\bar{u}$ has a small westerly perturbation (for example), the westerly wave has lower vertical group velocity than the easterly wave and hence will be damped more rapidly, as shown in Fig. 8.6a (see Appendix 4A). Thus, a positive acceleration will occur at higher levels due to the "shielding" effect of the westerly $\bar{u}$. As shown in Fig. 8.6b, the initial perturbation in $\bar{u}$ will then amplify. Plumb (1977) showed that the maximum acceleration will occur below the maximum $\bar{u}$, and hence the jet maximum must descend in time.

That this instability can lead to a finite amplitude mean wind oscillation resembling the QBO is shown in Fig. 8.7. As the shear zone separating the easterly and westerly regimes descends, it becomes sufficiently narrow that

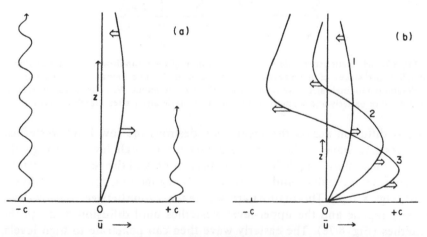

Fig. 8.6. Schematic representation of the instability of zonal flow in a stratified fluid with standing-wave forcing at a lower boundary. (a) Onset of instability from a small zonal flow perturbation. (b) Early stages of the subsequent mean-flow evolution. Broad arrows show locations and direction of maxima in mean wind acceleration. Wavy lines indicate relative penetration of wave components of positive and negative phase speeds c. [From Plumb (1982), with permission.]

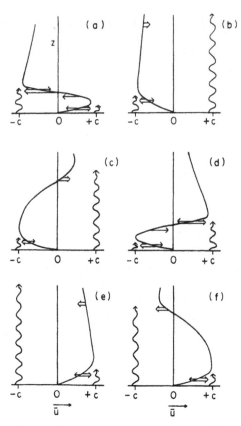

Fig. 8.7. Schematic representation of the evolution of the mean flow in Plumb's analog of the QBO. Six stages of a complete cycle are shown. Double arrows show wave-driven accelerations and single arrows show viscously driven accelerations. Wavy lines indicate relative penetration of easterly and westerly waves. See text for details. [After Plumb (1984).]

viscous diffusion across the shear zone destroys the low level westerlies. The westerly wave is then free to propagate to high levels through the easterly mean flow (Fig. 8.7b), where dissipation and the resulting westerly acceleration gradually build a new westerly regime that propagates downward (Figs. 8.7c,d). This in turn narrows the shear zone between the low-level easterly regime and the upper-level westerlies until diffusion destroys the easterlies (Fig. 8.7e). The easterly wave then can penetrate to high levels, where it initiates a new easterly phase of the oscillation (Fig. 8.7f). Thus, the nonlinear interaction between the mean flow and the upward-propagating dissipating waves generates a finite-amplitude oscillation in the mean flow whose period depends primarily on the amplitude of the EP flux and the vertical decay scale of the waves.

This mechanism was demonstrated in a laboratory experiment by Plumb and McEwan (1978). They used an annular vessel filled with salt-stratified water with a flexible membrane divided into a number of equal segments at the base (Fig. 8.8). This membrane was oscillated to produce a standing wave forcing that excited two vertically propagating internal gravity waves of equal but opposite horizontal phase speeds. Unlike the situation in the model described above, the waves in the experiment were dissipated by viscosity rather than Newtonian cooling. However, similar considerations apply; for a forcing exceeding some critical amplitude, the experiment generated a strong mean flow consisting of alternating downward-propagating easterly and westerly regimes. The period of the oscillation was very long compared to the period of the forced waves. This experiment provides convincing evidence that the HL mechanism contains the basic physics necessary to understand the atmospheric equatorial QBO.

8.3.2 The Holton and Lindzen (HL) Theory of the QBO

The eastward- and westward-propagating gravity waves in Plumb's model are symmetric and produce a mean wind oscillation with symmetric westerly and easterly phases. In the atmosphere, however, rotation introduces an important asymmetry between waves with westerly and easterly phase speeds. The Kelvin and Rossby-gravity modes have structures that are strongly dependent on the β effect. As described in Section 4.7, the Kelvin wave has a structure in the (x, z) plane that is identical to an eastward-propagating internal gravity wave, but has an exponential decay away from the equator. The Rossby-gravity wave, on the other hand, has a rather

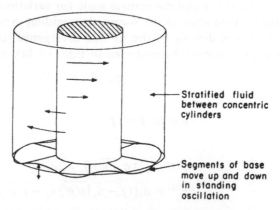

Stratified fluid between concentric cylinders

Segments of base move up and down in standing oscillation

Fig. 8.8. Schematic representation of the apparatus used in the Plumb and McEwan laboratory analog of the QBO.

complex three-dimensional structure. The total effect of this wave on the mean flow cannot be assessed by examining the wave momentum fluxes alone, but rather the complete EP flux must be considered. For the Kelvin wave the EP flux is downward at each latitude, while the Rossby-gravity wave (like the Rossby wave) has an upward-directed EP flux at each latitude. Thus for waves generated in the troposphere, wave damping causes a divergence (convergence) of the latitudinally integrated EP flux for the Kelvin (Rossby-gravity) wave as illustrated in Fig. 8.9. Hence the Kelvin (Rossby-gravity) wave can in principle provide the mean zonal force necessary to generate the acceleration leading to the westerly (easterly) phase of the QBO.

In the atmosphere, unlike Plumb's model, it is necessary to account for the meridional distribution of the oscillation as well as its vertical structure. The observations reviewed in Section 4.7 indicate that observed Kelvin and Rossby-gravity waves in the equatorial lower stratosphere have meridional scales comparable to that of the zonal wind QBO. However, the latitudinal distribution of the mean wind acceleration depends on the meridional profile of the EP flux divergence. For waves damped solely by radiative relaxation (Newtonian cooling), the EP flux convergence for the Rossby-gravity wave vanishes at the equator—just where the observed acceleration is a maximum. The meridional distribution of the EP flux convergence in the Rossby-gravity wave turns out to be very sensitive to the relative magnitudes of mechanical and thermal damping and to wave transience effects. Because of the complexity of dealing theoretically with the meridional dependent mean flow, HL used a meridionally averaged model.

The HL theory is best introduced by considering the TEM version of the governing equations, Eqs. (3.5.2). For the equatorial stratosphere where $\text{Ri} \equiv N^2/(\bar{u}_z)^2 \gg 1$ and the vertical scale for variation in $\bar{u}$ is much greater than the vertical scale of the waves, the residual mean meridional motion $(\bar{v}^*, \bar{w}^*)$ driven directly by the EP flux divergence [cf. Eq. (3.5.8)] can, to a first approximation, be neglected in Eq. (3.5.2a), which reduces then to

$$\bar{u}_t = \rho_0^{-1} \boldsymbol{\nabla} \cdot \mathbf{F} + \bar{X}, \qquad (8.3.3)$$

where

$$\mathbf{F} \equiv (0, F^{(y)}, F^{(z)})$$

and

$$F^{(y)} = \rho_0[\bar{u}_z \overline{v'\theta'}/\bar{\theta}_z - \overline{u'v'}],$$
$$F^{(z)} = \rho_0[(f - \bar{u}_y)\overline{v'\theta'}/\bar{\theta}_z - \overline{u'w'}].$$

Letting L designate the meridional scale of the QBO and noting that the QBO, the Kelvin wave, and the Rossby-gravity wave are all equatorially

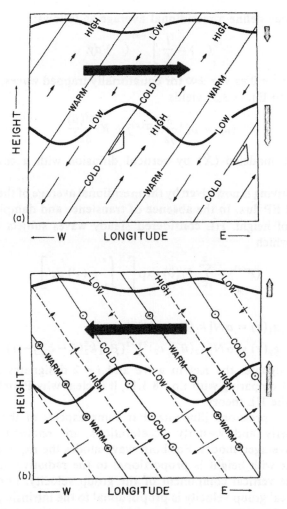

Fig. 8.9. (a) Longitude–height section along the equator for a thermally damped Kelvin wave; (b) longitude–height section along a latitude circle north of the equator for a thermally damped Rossby-gravity wave of westward phase speed; cf. Figs. 4.19 and 4.21. Both panels show geopotential, temperature and wind perturbations, with areas of high geopotential shaded. Heavy wavy lines indicate material surfaces, and open blunt arrows in (a) show phase propagation. Small arrows indicate zonal and vertical wind perturbations with length proportional to the wave amplitude, which decreases in height due to damping. Meridional wind perturbations are shown in (b) by arrows pointed into the page (northward) and out of the page (southward). Heavy vertical arrows to the right of the diagrams show the direction and an indication of the magnitude of the vertical component of the EP flux: this component equals the "form drag" on the indicated material surfaces (see Section 3.7.2). Heavy horizontal arrows indicate the latitudinal integral of the mean zonal force per unit mass associated with the EP flux divergence.

trapped, we may define a meridional average as follows:

$$\langle \ \rangle \equiv \frac{1}{L} \int_{-\infty}^{+\infty} (\ \ ) \, dy. \qquad (8.3.4)$$

Thus, since $F^{(y)} \to 0$ as $y \to \pm\infty$ for equatorially trapped waves, application of Eq. (8.3.4) to Eq. (8.3.3) yields

$$\langle \bar{u}_t \rangle = \rho_0^{-1} \frac{\partial}{\partial z} \langle F^{(z)} \rangle + K \frac{\partial^2 \langle \bar{u} \rangle}{\partial z^2}. \qquad (8.3.5)$$

Here we have modeled $\langle \bar{X} \rangle$ by vertical diffusion with a constant eddy viscosity K.

The wave driving is now given by the meridional average of the divergence of the vertical EP flux. In the absence of transience and damping, $\langle F^{(z)} \rangle$ is independent of height. HL considered steady waves subject to radiative damping for which

$$\langle F^{(z)} \rangle = \sum_{i=1}^{2} F_i(0) \exp\left[-\int_0^z g_i(z') \, dz' \right] \qquad (8.3.6)$$

with

$$g_1(z) = \alpha N [k_1(\bar{u} - c_1)^2]^{-1}$$

$$g_2(z) = \alpha N [k_2(\bar{u} - c_2)^2]^{-1} (\beta / [k_2^2(\bar{u} - c_2)] - 1).$$

Here $i = 1$ designates the Kelvin wave and $i = 2$ designates the Rossby-gravity wave. Comparing with Eq. (8.3.1), it is clear why Plumb's model is an analog to the HL theory.

The above expressions illustrate that some aspects of the asymmetry between easterly and westerly accelerations are retained even in a latitudinally averaged model. For both wave modes the exponential decay of the EP flux with height is proportional to the radiative damping rate divided by the vertical component of the group velocity. For the Kelvin wave the vertical group velocity is proportional to the intrinsic phase speed squared, while for the Rossby-gravity wave it is approximately proportional to the intrinsic phase speed cubed. The Kelvin wave is preferentially damped in westerly shear zones and the Rossby-gravity wave is damped in easterly shear zones. However, the latter is damped at a rate that increases more rapidly as $(\bar{u} - c) \to 0$ than is the case for the former.

In their model calculations, HL assumed a Newtonian cooling coefficient that increased linearly from $1/(21$ days$)$ at $z = 0$ (near the 17 km level) to $1/(7$ days$)$ at 30 km and remained constant above 30 km. They also assumed that at $z = 0$ the meridionally averaged EP fluxes for the Kelvin and Rossby-gravity modes were given by

$$F_i(0)/\rho_0 = \mp 4 \times 10^{-3} \, \text{m}^2 \, \text{s}^{-2},$$

respectively, and that the zonal phase speeds were given by $c_1 = -c_2 = 30 \text{ m s}^{-1}$. The zonal wave numbers were chosen to correspond to wave number 1 for the Kelvin wave and wave number 4 for the Rossby-gravity wave, in agreement with observations. The eddy diffusion coefficient in Eq. (8.3.5) was assigned the small constant value $0.3 \text{ m}^2 \text{ s}^{-1}$. For this choice of parameters, Eq. (8.3.5) was integrated numerically, and the resulting time evolution of the mean zonal wind (Fig. 8.10) proved to be very similar to that of the equatorial QBO.

In this integration a semiannual cycle was imposed at the upper boundary at 35 km, but this actually has little influence on the evolution of the QBO. As Fig. 8.10 shows, the HL model generally simulates the major features of the QBO, but it does have defects. In particular, the intensity of the easterly shear zone exceeds that of the westerly shear zone, contrary to observations. This defect appears to be inevitable in the HL model, given the differing dependencies of the Kelvin and Rossby-gravity wave vertical group velocities on the intrinsic phase speed. It is conceivable that a model in which the easterly forcing were partly due to Rossby waves or to higher equatorial modes might produce the proper asymmetry. However, as shown below, a model that includes latitudinal structure can simulate the observed shear zone asymmetries, without appealing to additional easterly forcing sources.

8.3.3 Simulation of the Meridional Structure of the QBO

As illustrated qualitatively in Fig. 8.5, there is a secondary mean meridional circulation associated with the zonal wind and temperature

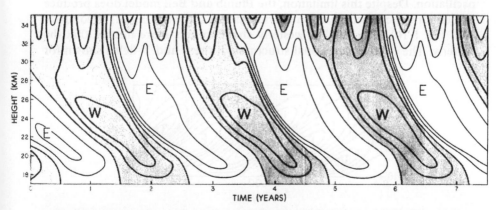

Fig. 8.10. Time-height section of mean zonal wind at the equator in the Holton-Lindzen model of the QBO. Contours are at 10-m s^{-1} intervals, and a semiannual oscillation is imposed at the 35-km level. [From Holton and Lindzen (1972). American Meteorological Society.]

anomalies of the QBO. Although this circulation is not of primary importance to the dynamics of the basic oscillation, it should be considered in a
complete theory. Near the equator the vertical advection due to the secondary circulation should tend to enhance the downward propagation of the
westerly shear zone and retard the easterly shear zone. Away from the
equator the Coriolis torque associated with the residual mean meridional
velocity should enhance the mean wind accelerations below the main shear
zone so that they are greater than the magnitudes due to the equatorial
wave EP fluxes alone. Thus, it is plausible that observed shear zone asymmetries may be at least partly caused by the secondary mean meridional
circulation.

The quantitative influence of the mean meridional circulation was tested
by Plumb and Bell (1982). They used a two-dimensional equatorial β-plane
model for the mean flow in which the meridional as well as the vertical
distributions of $\bar{u}$ were calculated in response to Kelvin and Rossby-gravity
wave forcing. They assumed that the mean zonal wind was symmetric about
the equator and that its evolution was sufficiently slow that the wave fields
were at each instant adjusted to it as steady, damped waves. Thus they were
able to solve for the wave fields at each instant using a numerical boundary-
value technique. The resulting wave forcing was then applied to determine
the EP flux divergence in the mean flow equation. Plumb and Bell formally
posed their model in terms of the small amplitude version of the generalized
Lagrangian mean theory (see Section 3.7). However, this is essentially
equivalent to solving the TEM set [Eqs. (3.5.2)] for the equatorial β-plane.

Unfortunately, due to numerical stability problems, Plumb and Bell were
only able to simulate a QBO of about 40% the amplitude of the observed
oscillation. Despite this limitation, the Plumb and Bell model does produce
a qualitatively correct shear zone asymmetry as shown in Fig. 8.11 (compare
with Fig. 8.10), in which the westerly shear zone is more intense that the
easterly. The overall meridional structures of the $\bar{u}$ and $\bar{T}$ fields shown for

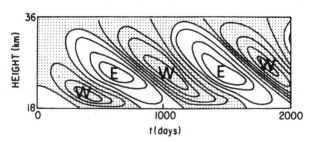

Fig. 8.11. Time–height section of mean zonal wind at the equator in the two-dimensional
model of the QBO of Plumb and Bell. Contours at 2-m s^{-1} intervals, easterlies are shaded.
[After Plumb and Bell (1982).]

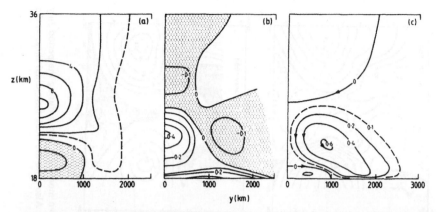

Fig. 8.12. Latitude–height sections showing the mean circulation at the time of the west wind maximum in the Plumb and Bell model of the QBO. (a) Zonal wind in m s^{-1}. (b) Potential temperature deviation in kelvins. (c) Mean meridional mass stream function in m^2 s^{-1}. Compare with the schematic in Fig. 8.5. [From Plumb and Bell (1982), with permission.]

the time of west wind maximum in Fig. 8.12 are also qualitatively in agreement with observations, although the meridional scale is somewhat too narrow (cf. Fig. 8.2). The mean meridional mass circulation presented in Fig. 8.12 shows, as expected, a subsidence in the shear zone near the equator and poleward drift below the shear zone. Thus vertical advection indeed speeds the westerly shear zone descent near the equator, and the Coriolis torque contributes to westerly accelerations below the shear zone away from the equator. Plumb and Bell showed that this meridional circulation quantitatively modifies the momentum budget of the QBO sufficiently to account for the computed shear zone asymmetry in their model. Since the QBO is an eddy-driven circulation, the residual meridional circulation associated with it may be regarded in a general sense as an eddy-driven circulation. But, unlike the *steady* midlatitude case discussed in Section 7.2.1, the residual circulation for the QBO is not directly determined by the EP flux divergence with the diabatic heating rate adjusting in a passive manner. Rather, the temperature change associated with the QBO determines the diabatic heating rate, which in turn helps to force the residual circulation [see Eq. (7.2.14)]. Thus, it is evident that accurate simulation of the detailed structure of the QBO depends at least to some extent on accurate calculation of the diabatic heating rates in the equatorial lower stratosphere.

Although the Plumb and Bell model does not completely reproduce all aspects of the observed QBO, it provides additional confirmation that the basic mechanism of the HL theory is valid even when the meridional

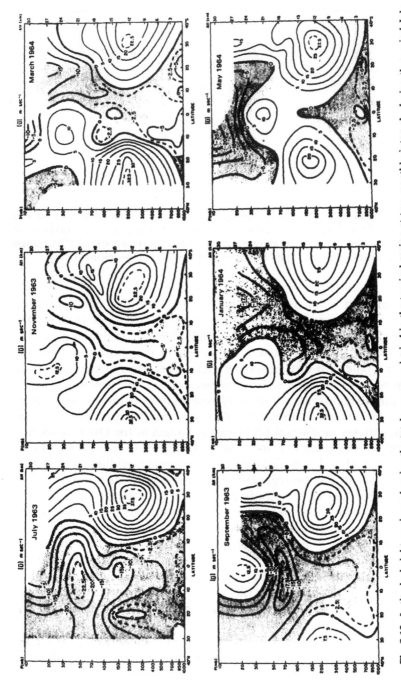

Fig. 8.13. Latitude–height sections showing the total mean zonal wind in the tropical region at two monthly intervals during the period July 1963 to May 1964. Notice the descending QBO westerlies and their relationship to the annual cycle. Also note the weak latitudinal shear at the equator. [From Newell *et al.* (1974), with permission.]

dependence of wave and mean-flow quantities is explicitly computed. There are, however, some important remaining problems associated with the QBO. The most significant of these is perhaps the coupling of the QBO to the annual cycle. The Plumb and Bell model assumed symmetry in the mean zonal wind about the equator. In reality, the Doppler-shifted mean flow "seen" by the Kelvin and Rossby-gravity waves is modulated by the seasonal cycle as shown in Fig. 8.13. This mean wind asymmetry will lead to some asymmetry in the vertically propagating equatorial waves, but perhaps more importantly the westerly winds of the winter hemisphere may provide ducts for the equatorward propagation of Rossby waves. Through "wavebreaking" associated with the zero mean wind critical layer, these waves may influence the QBO, as suggested by Dickinson (1968). Dunkerton (1983a) used a simple numerical model to argue that planetary waves can indeed provide part of the EP flux convergence that forces the easterly phase of the oscillation. The apparent seasonal clustering of the QBO wind reversals shown in Fig. 8.5 suggests that this sort of process may be significant. The tendency for the observed oscillation to have greater meridional width than that of the oscillation in the Plumb and Bell simulation may also be related (among other things) to the neglect of the seasonal cycle in their model.

8.4   Observed Structure of the Equatorial Semiannual Oscillations

A semiannual oscillation (SAO) in the equatorial middle atmosphere was first noted by Reed (1965a) in the course of a study of the amplitude of the QBO above 30 km, as deduced from rocket soundings. Unlike the QBO, whose amplitude is a maximum in the lower stratosphere where radiosonde data are fairly plentiful, the semiannual oscillation studied by Reed is concentrated in the upper stratosphere and lower mesosphere. With the exception of the limited data set from the LIMS experiment on *Nimbus* 7, observation of this semiannual oscillation is limited to meteorological rocket temperature and wind data from relatively few locations. Zonal wind climatologies from the rocket network show that the semiannual harmonic is in fact significant not only in the tropics but at high latitudes as well (Fig. 8.14). However, the high-latitude manifestation of the semiannual harmonic appears to simply reflect a departure from sinusoidal form in the annual march of the winds.

Hirota (1978) showed that the equatorial SAO really consists of two separate oscillations centered at the stratopause and the mesopause, respectively, with an amplitude minimum near 65 km, and an approximate out-of-phase relationship between the mesopause and stratopause maxima as shown in Fig. 8.15.

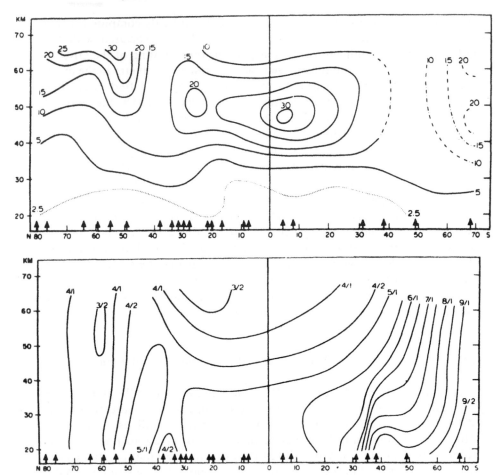

Fig. 8.14. Latitude–height sections showing the amplitude and phase of the SAO in the stratosphere and lower mesosphere. Amplitude (upper) in m s^{-1}; phase (lower) gives date of maximum westerly component in semimonthly periods. [From Belmont *et al.*, (1974), with permission.]

The equatorial SAOs are distinct phenomena that seem to have much in common with the equatorial QBO. In particular, the oscillation centered near the stratopause has its maximum amplitude near the equator and decays away from the equator with a halfwidth of about 25° latitude. As indicated in Fig. 8.15, the oscillation first appears in the mesosphere and propagates downward into the stratosphere. However, the downward propagation is limited mainly to the westerly phase of the semiannual oscillation. The easterly accelerations occur nearly simultaneously over a large depth

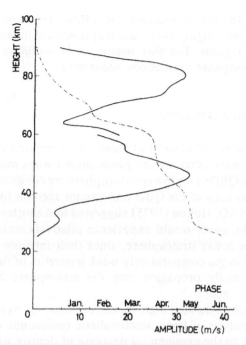

Fig. 8.15. The SAO at Ascension Island (8°S); amplitude (solid lines), phase (dashed line). Phase refers to time of first maximum westerly component in the calendar year. Break near 60 km is caused by separately fitting data above and below that level to sinusoidal curves. [From Hirota (1978). American Meteorological Society.]

centered at the stratopause. Near the stratopause the westerly wind maxima occur at the equinoxes and the easterly wind maxima occur near the solstices. The oscillation centered at the mesopause also propagates downward, but it has its westerly wind maxima near the solstices.

8.5   Dynamics of the Equatorial Semiannual Oscillations

As in the case of the QBO, the semiannual oscillations centered at the stratopause and mesopause both involve generation of westerly winds at the equator and hence the production of angular momentum per unit mass greater than that of the earth. This equatorial "superrotation" cannot be explained by symmetric processes, since the observed zonal mean circulation does not indicate any region that could provide the required angular momentum. It must, therefore, be produced by eddy EP flux divergence. There is now considerable evidence that Kelvin waves provide at least a portion of the forcing for the westerly phase of the stratopause oscillation. They may also be important for the mesopause oscillation. However, the easterly forcing is of less certain origin, and in fact may involve different mechanisms in the two height regions. For this reason, it is useful to consider the stratopause and mesopause oscillations separately.

8.5.1   The Stratopause Oscillation

Observational studies of equatorial waves show beyond doubt that the long-period Kelvin waves that are the predominant wave mode during the easterly phase of the QBO in the lower stratosphere are completely dissipated in the westerly shear zone of the QBO and cannot account for the westerly acceleration of the SAO. Holton (1975) suggested that shorter-period (high-phase-speed) Kelvin waves would experience relatively small damping in passing through the lower stratosphere, since their intrinsic phase speeds would remain large in the comparatively weak westerlies of the QBO. Thus, such waves could easily propagate into the mesosphere and drive the observed SAO.

Unfortunately, the postulated high-speed Kelvin waves were not revealed by observational analysis of lower stratospheric radiosonde data. It might be thought that due to the exponential decrease of density with height, the Kelvin-wave EP flux required to account for the westerly phase of the SAO would be an order of magnitude less than for the QBO, and that the amplitude of the high-speed Kelvin waves would thus be too small to be observable in the lower stratosphere. However, as noted by Dunkerton (1979), the timescale for the SAO is less than one-fourth that of the QBO and the amplitude is somewhat larger, so that the required EP flux at the tropopause to account for the westerly acceleration of the stratopause SAO turns out to be nearly one-third of that required for the QBO.

It is now known from rocket and satellite observations (see Section 4.7) that high-phase-speed Kelvin waves do propagate into the mesosphere. These waves appear to occur primarily in the form of transient wave "packets," an example of which is shown in Fig. 4.27. Such packets would be difficult to detect through time-series analysis of data at a single station. The evidence from the LIMS satellite experiment suggests that Kelvin wave packets do not have sufficient amplitude to completely account for the observed westerly acceleration of the SAO [Hitchman and Leovy (1986)]. The remaining acceleration may be due to gravity-wave modes that cannot be resolved by LIMS.

The easterly phase of the stratopause SAO appears to have a primary driving mechanism that is completely different from that of the QBO. There is no evidence that the Rossby-gravity waves propagate above the lower stratosphere. Indeed, as mentioned in Section 8.4, observations of the easterly acceleration phase of the SAO indicate nearly simultaneous accelerations over a large depth of atmosphere, contrary to the downward-propagating easterly acceleration pattern characteristic of Rossby-gravity wave absorption. The occurrence of simultaneous accelerations in a deep layer centered near the stratopause suggests that the source is meridionally propagating rather than vertically propagating.

On the basis of rocketsonde observations, Hopkins (1975) suggested that the easterly phase was driven by lateral momentum transfer due to the quasi-stationary planetary waves of the winter hemisphere. He argued that absorption of such waves in a critical layer centered at the zero mean zonal wind line in the tropics would lead to easterly accelerations that would be strongly modulated by the seasonal cycle. Thus, the SAO would result from a combination of a quasi-steady background westerly acceleration due to high-speed Kelvin-wave forcing, and a semiannually varying easterly forcing with maxima during both Northern- and Southern-hemisphere winters, and minima at the equinoctial seasons.

Holton and Wehrbein (1980) showed, however, that the residual mean meridional circulation should also be considered as a forcing for the easterly phase of the SAO. As discussed in Chapter 7 (see Fig. 7.2), the residual circulation at the solstices consists of rising motions in the summer hemisphere, a cross-equatorial drift, and sinking in the winter hemisphere; this circulation is primarily driven by eddy forcing in the extratropics. The mean meridional velocity component of this circulation is too weak in the lower stratosphere to have a significant influence. But near the stratopause, the combination of a mean meridional drift of nearly 1 m s^{-1} and a cross-equatorial mean zonal wind shear of order 1 day^{-1} implies an advective mean wind acceleration of the order of $1 \text{ m s}^{-1} \text{ day}^{-1}$ at the equator. Thus, the simple process of advection of the summer easterlies across the equator can itself lead to a significant semiannual oscillation at the equator.

There are, however, two problems with the advective model for the easterly acceleration of the SAO. First, as shown by Holton and Wehrbein (1980), the latitudinal width of the SAO generated by mean wind advection is much less than the observed width. Second, the latitudinal velocity profiles with easterly shear at the equator are strongly inertially unstable (see Section 8.6). When the profiles are adjusted to eliminate the inertial instability, the amplitude of the resulting oscillation is reduced to less than half. Hence, it appears that mean advection alone cannot explain the easterly phase of the stratopause SAO. It is probable that both the mean advection term and planetary-wave forcing associated with vorticity mixing during wave-breaking events (see Sections 5.2.3 and 6.7) contribute to the easterly acceleration. The former may in fact enhance the latter by contributing to the propagation of the zero mean zonal wind line into the winter hemisphere. It is important to note that these processes will both have their maximum influence during the solstices, in agreement with the observed phase of the oscillation (Fig. 8.14).

The overall model of the stratopause SAO presented here is in accord with the simulation reported by Mahlman and Umscheid (1984), who used a high-resolution general circulation model with an annually varying radiative drive. Their zonal mean wind variation at the equator (Fig. 8.16) shows a remarkably realistic stratopause SAO. Diagnosis of their model verified that the westerly accelerations were due to high-speed Kelvin waves (which are strongly excited in their model), while the easterly acceleration was due to a combination of planetary-wave forcing and mean advection.

8.5.2   The Mesopause Oscillation

Observations (Fig. 8.15) indicate a second amplitude maximum in the SAO that is not present in the simulation of Mahlman and Umscheid. Its absence in that model suggests that the oscillation may be driven by processes not simulated in the model. Dunkerton (1982b) suggested that the breaking or "saturation" of vertically propagating gravity waves might provide an explanation for the mesopause oscillation. As discussed in Section 4.6, gravity waves generated in the lower atmosphere that propagate into the mesosphere are expected to become convectively unstable and break in the mesopause region. Such waves should produce mean-flow changes due both to the resulting EP flux divergence and to wave-generated turbulent diffusion.

Dunkerton (1982b) noted that gravity-wave transmissivity through the stratopause region should be strongly modulated by the presence of the SAO, due to the strong dependence of vertical group velocity on the intrinsic frequency. Dunkerton's mechanism is very similar to the HL model for the QBO (or, more precisely, to Plumb's analog of the QBO discussed in Section 8.3.1). If it is assumed that gravity waves of easterly and westerly phase speeds are excited with equal intensity in the troposphere, then the westerly (easterly) waves will be preferentially radiatively damped in passing through the stratopause region during the westerly (easterly) phase of the stratopause SAO. Hence the waves that first break on reaching the mesopause region will be easterly (westerly) during the westerly (easterly) phase of the stratopause SAO. These will tend to drive easterly (westerly) accelerations, and thus there should be an approximate out-of-phase relationship between the stratopause and the mesopause oscillations, as is observed. (A somewhat similar "filtering" process was mentioned in Section 7.3.)

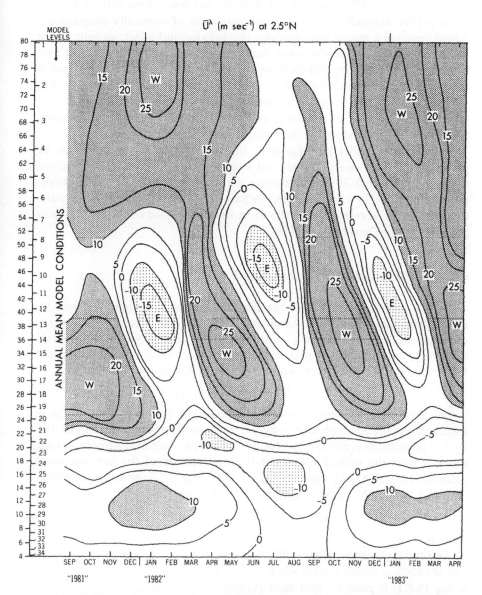

$\bar{U}^{\lambda}$ (m sec^{-1}) at 2.5°N

Fig. 8.16. Time–height cross section of monthly averaged zonal mean winds (m s^{-1}) at 2.5°N in the "SKYHI" general circulation model. [From Mahlman and Umscheid (1984), with permission.]

Dunkerton (1982b) pointed out that the net mean-flow driving by any process that depends on selective transmission of vertically propagating waves requires a model that includes the background winds as well as the oscillating winds, since the wave damping depends on the total intrinsic phase speed—including effects from the background mean wind. If mean wind effects are properly taken into account, then, as shown by Dunkerton (1982b), it is possible to produce a mesopause SAO approximately in accord with the observed in a model that employs Lindzen's (1981) parameterization for gravity-wave breaking as the forcing mechanism.

Dunkerton's model is suggestive, but it depends on specification of gravity-wave parameters that are not easily determined from observations (at least at present). It should also be mentioned that the extremely high-phase-speed Kelvin waves reported by Salby *et al.* (1984) may provide much of the westerly acceleration observed in the mesopause oscillation, as was pointed out by Dunkerton (1982b).

8.6   Inertial Instability in the Equatorial Zone

In this section we consider the possibility that the equatorial middle atmosphere may be unstable to disturbances that are almost independent of longitude. It is known (e.g., Charney, 1973) that a purely zonal basic flow $[\bar{u}(y, z), 0, 0]$ on a β-plane is unstable to zonally symmetric disturbances if $f\bar{P} < 0$ somewhere, where $f(y)$ is the Coriolis parameter and

$$\bar{P} = \rho_0^{-1}[\bar{\theta}_z(f - \bar{u}_y) + \bar{\theta}_y\bar{u}_z]$$

is the Ertel potential vorticity of the basic state [cf. Eq. (3.1.4)]. (It is assumed that the zonal absolute angular momentum of the flow represented by $\bar{u}$ is positive, as is always true for flows in the earth's atmosphere, except possibly near the poles.) Using the β-plane version of the thermal wind equation [Eq. (3.4.1c)], namely, $f\bar{u}_z = -RH^{-1}\bar{\theta}_y e^{-\kappa z/H}$, and the definition [Eq. (3.2.13)] of N^2, this condition for "symmetric instability" becomes

$$\bar{\theta}_z[f(f - \bar{u}_y) - f^2\bar{u}_z^2/N^2] < 0 \qquad \text{somewhere,} \qquad (8.6.1)$$

since $\rho_0 > 0$.

Two types of symmetric instability can be distinguished: *static* or *convective* instability, which occurs when $\bar{\theta}_z < 0$ and the term in square brackets in Eq. (8.6.1) is positive, and *inertial instability*, which occurs when $\bar{\theta}_z > 0$ but

$$f^2(1 - \text{Ri}^{-1}) - f\bar{u}_y < 0 \qquad \text{somewhere,} \qquad (8.6.2)$$

where $\text{Ri} \equiv N^2/\bar{u}_z^2$ is the Richardson number. This condition can be shown to be equivalent to the statement that the absolute angular momentum per unit mass on an isentrope decreases with radial distance from the axis of rotation. Inertial instability arises from an imbalance between the pressure gradient force and the total centrifugal force, and is closely analogous to the more familiar convective instability, which arises from an imbalance between the pressure gradient force and the buoyancy force. For barotropic flow ($\bar{u}_z = 0$ or Ri = ∞), the criterion may be expressed as $f(f - \bar{u}_y) < 0$. Thus (unless $\bar{u}_y = 0$ and $\beta - \bar{u}_{yy} > 0$ at $y = 0$) a zonally symmetric barotropic flow will be inertially unstable somewhere near the equator. For baroclinic flows with Ri < 1, Eq. (8.6.2) shows that instability can occur even without meridional shear. For the zonally averaged circulation of the equatorial middle atmosphere, Ri $\gg$ 1 in general so that baroclinic effects can be neglected. However, although observations (Fig. 8.13) indicate that in the lower stratosphere the meridional shear of the mean zonal wind is very small near the equator, in the upper stratosphere and mesosphere this may not be the case. In this region the cross-equatorial advection by the mean circulation will tend to produce a cross-equatorial shear by advecting the summer-hemisphere easterly flow toward the winter hemisphere. Thus, the mean circulation will tend to produce an inertially unstable zonal-mean flow profile. The study of inertial instability in the equatorial middle atmosphere is, therefore, not merely of academic interest.

An example of inertial instability can be derived from the linear disturbance equations for the equatorial beta-plane discussed in Section 4.7. Following Dunkerton (1981a), we set $\partial/\partial x = 0$, $X' = Y' = Q' = 0$ (so that disturbances are zonally symmetric and conservative), and $\bar{u} = \bar{u}(y)$ in Eqs. (4.7.1); by Eq. (4.7.2), $\bar{\theta}_y = 0$ also. Then it is readily verified that the meridional velocity disturbance satisfies

$$\frac{d^2\hat{v}}{dy^2} + \frac{m^2}{N^2}[\omega^2 - \beta y(\beta y - \bar{u}_y)]\hat{v} = 0. \qquad (8.6.3)$$

This may be reduced to the canonical form of Eq. (4.7.13c) if, following Boyd (1978a), we assume that the mean wind shear is linear: $\bar{u}_y = \gamma$, and shift latitude by letting

$$y_1 = y - \gamma(2\beta)^{-1}$$

and define

$$\omega_1^2 = \omega^2 + \gamma^2/4. \qquad (8.6.4)$$

Then Eq. (8.6.3) becomes equivalent to Eq. (4.7.13c) provided we let

$$\eta = (\beta|m|/N)^{1/2}y_1,$$

and the eigencondition relating ω_1 and m is

$$m^2\omega_1^2/N^2 = (2n + 1)\beta|m|/N \qquad (8.6.5)$$

[cf. Eq. (4.7.15)]. Thus for the lowest mode ($n = 0$) the frequency ω corresponding to a disturbance of wave number m is given from Eqs. (8.6.4) and (8.6.5) as

$$\omega^2 = N\beta/|m| - \gamma^2/4. \tag{8.6.6}$$

For $N\beta/|m| < \gamma^2/4$ the frequency ω is imaginary and approaches an asymptotic value of $\pm i\gamma/2$ as $|m| \to \infty$. Recalling that the perturbation solution was assumed to have a dependence of the form $\exp[i(mz - \omega t)]$, we see that the positive root corresponds to an exponentially growing mode.

For inertial instability, like convection, the fastest-growing mode has infinitesimal scale (i.e., $|m| \to \infty$ in the present case). In the atmosphere it is expected that eddy diffusion will damp the smallest scales so that the maximum growth rate will occur for a finite scale. Dunkerton (1981a) showed that inertially unstable modes have only a weak dependence on the magnitude of the assumed eddy viscosity. (The critical shear for marginal instability depends on the $\frac{1}{5}$ power of the viscosity coefficient, and the critical vertical scale goes as the $\frac{2}{5}$ power.) For a vertical viscosity coefficient of $1 \text{ m}^{-2} \text{s}^{-1}$, marginal instability requires a meridional shear of $\sim 2.8 \text{ day}^{-1}$ and the corresponding lowest mode has vertical wavelength of 2.4 km and meridional scale of 5.25° latitude. Just as convection transfers heat vertically, thereby tending to eliminate static instability, inertially unstable modes transfer momentum meridionally and thus should tend to reduce the lateral shear until the profile is marginally stable. The structure of the resulting lowest-order neutral viscous mode is shown qualitatively in Fig. 8.17. The aspect ratio (vertical/meridional scale) depends weakly on eddy diffusion, so that the ellipses flatten out as viscosity approaches zero.

A structure suggestive of inertial instability has been identified in the LIMS satellite radiance data (Hitchman et al., 1987). This vertically stacked "pancake" pattern in the meridional and zonal velocity fields has also been simulated in the general circulation model of Hunt (1981). The absence of such structure in the model of Holton and Wehrbein (1980) is presumably due to the fact that their model lacked sufficient vertical resolution to resolve the small vertical scales that would characterize inertial instability for the magnitude of damping in the model.

More work is required to determine the true role of inertial instability in the equatorial middle atmosphere. The theory presented here involves *zonally symmetric* disturbances to a *zonally symmetric* basic state. Large-amplitude Kelvin waves are often present in the equatorial easterlies so that the "basic state" zonal flow is *not* zonally symmetric. Furthermore, the disturbances observed by LIMS that have been interpreted as inertial instabilities are also *not* zonally symmetric. There is some theoretical evidence that zonally asymmetric inertial instabilities may occur in the

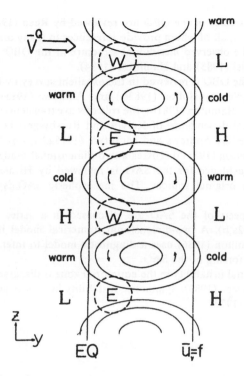

Fig. 8.17. Schematic representation of the lowest order neutral viscous inertial mode on the equatorial β-plane. The residual circulation $\bar{v}^Q$ generates an inertially unstable region between the equator and the latitude where $f = \bar{u}_y$. The elliptical contours indicate the perturbation motion in the meridional plane, and the dashed contours indicate the maxima in the induced perturbation zonal flow. The pressure anomalies are indicated by the symbols L and H for low and high pressures, respectively. [After Dunkerton (1981a). American Meteorological Society.]

presence of a zonally symmetric basic state (Boyd and Christidis, 1982; Dunkerton, 1983b). But the relevance of these theoretical calculations to the atmosphere is uncertain. Nevertheless, present evidence suggests that inertial instability could be an important mechanism for limiting the magnitude of lateral wind shear in the equatorial zone and thereby reducing the mean wind acceleration due to advection by the residual circulation.

References

8.1. Observed seasonal and longer term variations in the lower stratosphere are discussed in detail by Newell *et al.* (1974). Thermal forcing in the equatorial zone is considered by Wallace (1967) and Wallace and Holton (1968).

8.2. Observational studies of the QBO are reviewed by Reed (1965b), Wallace 1973), and Plumb (1984), all of whom provide references to many original sources. Detailed analyses of the observed meridional structure of the QBO are contained in Dunkerton and Delisi (1985) and Hamilton (1984).

8.3. The theory of the QBO is reviewed in the excellent survey by Plumb (1984). Briefer accounts are given by Holton (1983b) and Andrews (1985). The vertical structure is discussed by Hamilton (1981b), the role of wave transience by Dunkerton (1981b). The meridional dependence of Kelvin and Rossby-gravity wave EP flux divergence are discussed by Andrews and McIntyre (1976a,b), Boyd (1978b), and Holton (1979b). Dunkerton (1985) discusses a two-dimensional model of the QBO.

8.4. Observational evidence for the SAO is reviewed by Hirota (1980), who provides references to original sources. The mesospheric SAO is discussed by Hamilton (1982c).

8.5. Theoretical aspects of the SAO are discussed in a series of papers by Dunkerton (1979, 1982a,b). A two-dimensional numerical model is reported by Takahashi (1984). Hamilton (1986) uses a diagnostic model to infer the nature of the distribution of the forcing of the SAO.

8.6. Symmetric inertial instability in the equatorial zone is discussed by Dunkerton (1981a) and by Stevens (1983). Symmetric instability in a more general context is treated by Emanuel (1979).

Chapter 9 | Tracer Transport in the Middle Atmosphere

9.1 Introduction: Types of Tracers

An atmospheric tracer may be defined as any quantity that "labels" fluid parcels. Tracers may be dynamical or chemical, conservative or nonconservative, passive or active. Dynamical tracers consist of derived field variables such as potential temperature and potential vorticity, which are conserved by fluid parcels under suitable conditions. Chemical tracers consist of minor atmospheric constituents that have significant spatial variability. All truly conservative chemical species are well mixed in the lower and middle atmosphere. Only species that have chemical sources and sinks (e.g., ozone) or that undergo phase changes in the atmosphere (e.g., H_2O) maintain significant spatial variability in the presence of the continual tendency for motions on all scales to keep the atmosphere well mixed below the homopause.

The single most important tracer for meteorologists is the potential temperature, defined in Eq. (1.1.9). For motions that are isentropic (i.e., adiabatic), fluid parcels must remain on constant potential temperature surfaces. Since potential temperature varies primarily in the vertical in the middle atmosphere (Fig. 9.1a) and is a monotonic function of height, it may be regarded as a vertical label for fluid parcels; for isentropic motion the problem of following fluid-parcel trajectories is reduced from three spatial dimensions to that of two-dimensional motion on the isentropic surfaces. Thus, the isentropic coordinates introduced in Section 3.8 are particularly useful in problems involving tracer transport.

Further constraints on fluid-parcel trajectories can be imposed by utilizing a second important dynamical tracer, Ertel's potential vorticity, P, defined

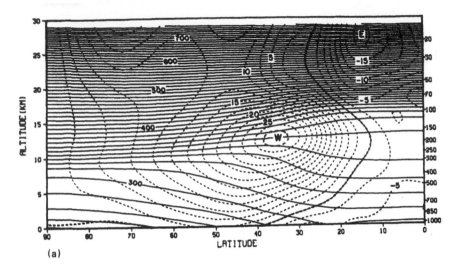

(a)

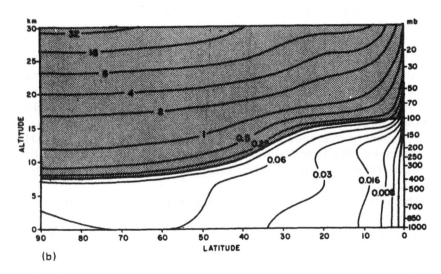

(b)

Fig. 9.1. Northern-Hemisphere zonal–annual mean cross sections for some quasi-conservative tracers. (a) Potential temperature (solid contours, kelvins) and zonal wind component (dashed contours, $m\,s^{-1}$). (b) Ertel potential vorticity in units of $10^{-5}\,K\,m^2\,kg^{-1}\,s^{-1}$. Area above the mean tropopause is shaded. (c) Ozone mixing ratio in parts per million by mass (ppmm). Shading extends from the mean tropopause to the level of maximum mixing ratio. Note that the tropopause [marked by heavy lines in panels (b) and (c)] intersects several of the potential temperature surfaces shown in panel (a). [From Danielsen (1985), with permission.]

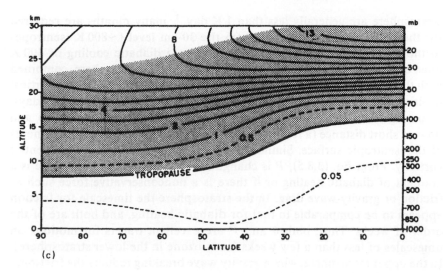

Fig. 9.1 (*continued*)

in Eqs. (3.1.4) and (3.8.4). Potential vorticity tends to vary strongly in both the meridional and vertical directions (Fig. 9.1b) and is conserved for adiabatic, frictionless flow. Thus, if it is assumed that the motion is adiabatic and inviscid, the contours of P on isentropic surfaces are material lines: that is, each contour always comprises the same fluid parcels. These material lines are generally wavy in shape, but tend to be oriented roughly in the east–west direction. The use of such contours in diagnosis of large-scale transport and mixing in the stratosphere was discussed in Sections 5.2.3 and 6.2.4.

Tracers are most valuable when they are conservative, so that their values remain constant following the motion of fluid parcels. In the atmosphere no tracer is truly conservative, since even in the absence of diabatic effects and chemical sources and sinks there will always be some mixing on scales smaller than the "parcel" resolved by a given measurement. It is useful, however, to distinguish those tracers whose rates of change following the motion are very small for the timescales of interest from those that change rapidly following the motion. The former will be referred to below as "quasi-conservative" or "long-lived" tracers. For motions on timescales of several days, the potential temperature and potential vorticity can be considered quasi-conservative tracers, as can ozone in the lower stratosphere (Fig. 9.1c). Thus, for example, since diabatic heating rates in the lower stratosphere are generally less than 1 K day^{-1}, many months are required for the descent of a fluid parcel from the 30-km level ($\sim$800 K isentrope) to the 15-km level ($\sim$400 K isentrope), since a diabatic cooling of 400 K must occur. The rapid change of θ with height in the stratosphere, combined with the slowness of the diabatic heating or cooling, ensures that the vertical position of a fluid parcel can be followed quite accurately for several days, assuming isentropic motion. Of course, such parcels may move up and down a short distance (a few kilometers at most) with vertical displacements of the isentropic surface. Similar considerations hold for Ertel's potential vorticity. From Eq. (3.8.5), P is changed following the motion if there is a gradient of diabatic heating or if there is a nonconservative force such as friction or gravity-wave drag. In the stratosphere the timescale for friction appears to be comparable to that for diabatic heating, and both are of the order of weeks. Thus, both θ and P are excellent tracers for motions on timescales of less than a few weeks (as is ozone in the lower stratosphere). In the upper mesosphere, where gravity wave breaking reduces the frictional timescale to a day or so, P is no longer quasi-conservative.

The dynamical tracers θ and P are both "active" tracers in the sense that they are not simply passively advected by the flow field, but their distributions to a large extent determine the evolution of the flow field. This is especially true for the P field when it has large-scale structure.

An active tracer that is probably more familiar to meteorologists than the Ertel potential vorticity is the "quasi-geostrophic" potential vorticity, q_g. This quantity is conserved following the geostrophic motion on isobaric surfaces and is thus not an approximation to P (see Section 3.8). However, the stream function determining the geostrophic flow is defined in terms of q_g [see Eq. (3.2.15)], so that q_g not only is advected by the geostrophic flow but also determines it. Similar considerations apply to Ertel's potential vorticity, except that the advection is by the three-dimensional flow, and the distribution of P tends to determine the development of the flow in three dimensions (see Hoskins *et al.*, 1985.)

Chemical tracers may also be "active," albeit in a somewhat different sense. Thus, for example, chemical processes may change the ozone distribution, which changes the shortwave diabatic heating distribution, and hence changes the temperature and wind distribution. Other minor species involved in the photochemistry of the ozone layer may also play similar "active" roles, but these are for the most part quite small in the context of circulation changes.

For timescales of several weeks or less, ozone in the lower stratosphere may be considered a passive tracer for most purposes since ozone-induced changes in the circulation are usually second-order effects. For seasonal changes, however, the distribution of ozone is crucial to the determination of the radiative forcing of the circulation. In turn, the circulation has important influences on the distribution of ozone. Passive chemical tracers can be of great value in identifying the history of air parcels. When insufficient data are available for accurate trajectory analyses, multiple chemical tracers can often be used to constrain the origins and trajectories of air parcels. This is especially useful for studies of troposphere-stratosphere exchange and transport within the lower stratosphere (see Section 9.6).

9.2 Long-Lived Chemical Tracers

Much can be learned about the overall mass flow in the middle atmosphere by considering the climatological distribution of quasi-conservative chemical tracers. The distribution of ozone, especially in the Northern Hemisphere, has been studied by balloon and ground-based methods for many years. However, only in the past decade, with the advent of routine satellite measurements, has it been possible to study its global climatology. For other long-lived tracers, such as methane (CH_4), nitrous oxide (N_2O), water vapor, and stratospheric aerosols, limited global measurements have been made by instruments on the *Nimbus* 7 research satellite. For less abundant species (e.g., various halocarbons), some information is available from balloon and aircraft measurements, primarily in midlatitudes.

The role of transport in determining the global distribution of chemical tracers depends on the nature and distribution of the tracer sources and sinks and on the relative magnitude of the timescales for dynamical and chemical processes. In the analysis of global transport it is useful to distinguish between substances whose sources are mainly in the troposphere and those whose sources are mainly in the stratosphere. The former, which include nitrous oxide, methane, and certain halocarbons, are slowly transported into the stratosphere, where they are destroyed by photolysis or oxidation. The latter, which include ozone, cosmogenic radionuclides (e.g., beryllium 7), and stratospheric aerosols, are transported slowly into the troposphere, where they are destroyed by a variety of processes.

The distribution of such substances in the stratosphere depends crucially on competition between dynamics and chemistry. This competition can be qualitatively measured by the ratio of the chemical and dynamical timescales. By "chemical timescale" we mean the characteristic time for replacement or destruction of a species by local sources or sinks. By "dynamical timescale" we are here referring to the time for advective processes (mean motions plus eddies) to transport the tracer through approximately a scale height in the vertical or from equator to pole meridionally. Three cases can be distinguished:

1. The chemical timescale is much less than the dynamical timescale. In this case the species is in local photochemical equilibrium, and transport does not enter directly into the conservation equation for the species. Of course, transport may enter indirectly by affecting the concentration of other species that participate in the photochemical production or loss for the species in question. (See Chapter 10 for further discussion.)

2. The chemical timescale is much greater than the dynamical timescale. In this case the tracer is passively advected and in the absence of localized sources or sinks will eventually become well mixed due to the dispersive effects of transport. The rapid transport and long chemical time constants in the troposphere for species like methane and nitrous oxide are reflected by their uniform tropospheric distributions.

3. Chemical and dynamical time constants are of the same order of magnitude. This is the most interesting case (also the most difficult to handle), since the distribution of such species is determined both by chemistry and by transport. The distributions of methane and nitrous oxide in the meridional plane in the upper stratosphere are examples of species distributions that depend equally on transport and chemistry.

Figure 9.2 shows profiles of photochemical timescales at 30° latitude and equinoctial conditions for the tropospheric source species whose midlatitude

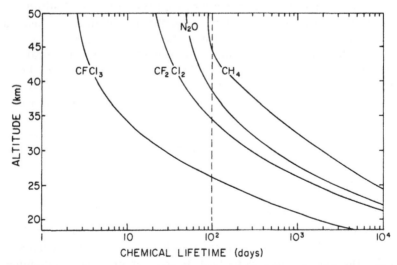

Fig. 9.2. Altitude dependence of lifetimes (i.e., e-folding decay times) for several long-lived trace gases at 30°N and equinoctial conditions calculated with observed ozone and 1983 chemistry. (Courtesy of Dr. J. A. Logan and Dr. M. J. Prather.)

vertical profiles were shown in Fig. 1.8. Methane is primarily destroyed through oxidation by reaction with the hydroxyl radical or through reaction with excited atomic oxygen or chlorine. Nitrous oxide and the chloro-fluoromethanes are destroyed through photolysis by ultraviolet radiation. In all cases the destruction rates increase with height in the middle atmosphere. The mean concentration at any altitude results from a balance between the local destruction and the net convergence of the upward flux. Thus the species with the shortest chemical timescales have concentrations decreasing most rapidly with height.

The influence of transport on the long-lived tropospheric source gases can best be illustrated by considering their distributions in the meridional plane. Examples of monthly means for N_2O and CH_4 from the observations of the SAMS satellite experiment are shown in Fig. 9.3. The mixing-ratio surfaces for both tracers are bulged upward in the tropics and tend to slope downward toward the poles in both hemispheres, although there are important seasonal shifts in the latitude of the bulge and a double bulge occurs during Northern-Hemisphere spring. Recalling from Fig. 9.1 that isentropic surfaces have relatively small slopes in the stratosphere, it is clear that in general the isolines of constant mixing ratio for N_2O and CH_4 slope downward toward the poles more steeply than the isentropes, so that concentration decreases toward the poles on isentropic surfaces.

9.3 Transport in the Meridional Plane

Discussion of the processes through which the general circulation partly determines the observed meridional and vertical distributions of various tracers has often centered on the relative roles of "mean" and "eddy" motions. But, as indicated in Sections 3.5–3.7 and 3.9, the partitioning between mean flow and eddies depends critically on the type of averaging process used. The conventional isobaric Eulerian zonal mean, the transformed Eulerian mean (TEM), the generalized Lagrangian mean (GLM), and the isentropic zonal mean all generally provide different interpretations of "mean flow" and "eddy."

The relative importance of "mean flows" and "eddies" in transport is thus a difficult issue to resolve, even conceptually. One approach would be to compare observed tracer distributions with those obtained in a model that excludes eddy motions altogether (see Section 7.2.2). In such a model the temperature would generally be very close to radiative equilibrium; only the time dependence imposed by the annual cycle in solar heating would drive departures from radiative equilibrium. The flow would then be almost

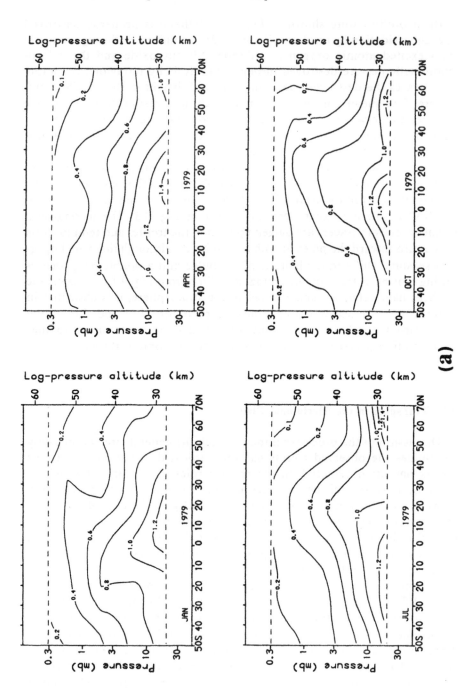

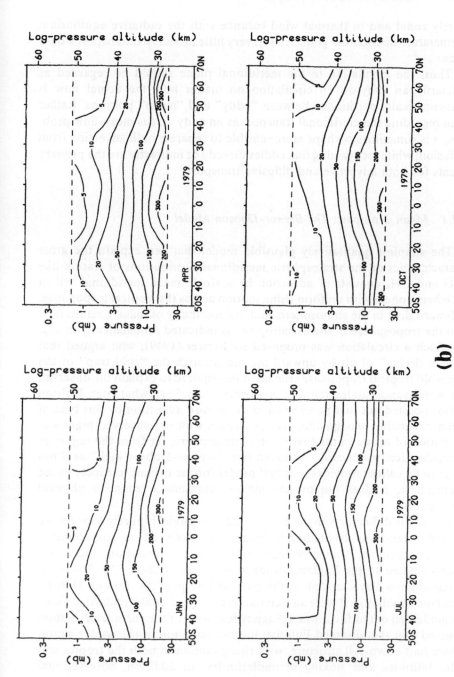

Fig. 9.3. Monthly mean cross sections of (a) CH₄ (ppmv) and (b) N₂O (ppbv) from measurements by the *Nimbus 7* SAMS experiment during 1979. [After Jones and Pyle (1984).]

purely zonal and in thermal wind balance with the radiative equilibrium temperature distribution [Eq. (7.1.1)]; very little meridional transport would occur.

Thus, the circulation in the meridional plane should be regarded as primarily an *eddy*-driven circulation no matter how the actual flow is mathematically partitioned between "eddy" and "mean" portions. Rather than regarding the meridional transport as an eddy versus mean-flow problem, it is sometimes perhaps more sensible to separate bulk advection from diffusion, while recognizing that eddies directly or indirectly are the primary agents for both advective and diffusive transport.

9.3.1 Mean Transport: The Brewer–Dobson Model

The simplest qualitatively plausible model that can explain the gross characteristics of the stratospheric meridional distributions of tracers like N_2O and CH_4 consists of advection by a single mean meridional cell in each hemisphere with uniform rising motion across the tropical tropopause, poleward drift in the stratosphere, and, by continuity of mass, a return flow into the troposphere in the extratropics, as indicated schematically in Fig. 9.4. Such a circulation was proposed by Brewer (1949), who argued that "freeze drying" of air by upward motion through the "cold trap" of the high cold tropical tropopause seemed to be required to explain the observed low water-vapor mixing ratios in the stratosphere. Somewhat later, Dobson (1956) pointed out that poleward and downward advection by this type of mean circulation was qualitatively consistent with the observed high concentration of ozone in the lower polar stratosphere, far from the region of photochemical production. Although this "Brewer–Dobson cell," as it has come to be called, provides a partial model for the overall transport in the stratosphere, it does not by any means represent a complete physical description.

Part of the difficulty with the Brewer–Dobson model is purely conceptual. Since the meridional circulation of this model is derived from consideration of tracer transport, it must be regarded as a *mass* circulation, and hence closely related to the Lagrangian mean flow (see Section 3.7). However, meteorologists have tended in the past to interpret the Brewer–Dobson circulation as though it were an Eulerian mean meridional circulation. Thus, the model fell out of favor when diagnostic studies of the general circulation revealed that the observed Eulerian mean circulation in the winter stratosphere had a two-cell structure, with rising motion in both the tropics and polar latitudes and sinking in midlatitudes. In addition, observational studies of the transport of ozone and radioactive tracers in the lower

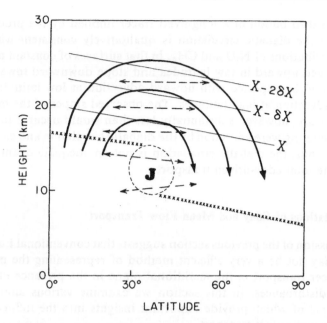

Fig. 9.4. Schematic cross section of transport in the stratosphere. Heavy lines show the mean meridional mass circulation (Brewer–Dobson cell). Dashed lines indicate quasi-isentropic mixing by large-scale eddies. The mean tropopause is indicated by crosses, and J indicates the mean jetstream core. Light lines labeled with mixing-ratio values (X) show mean slope of a long-lived vertically stratified tracer. [After Holton (1986b).]

stratosphere revealed that large-scale eddy motions play an important role in meridional and vertical transport, and that there often tends to be almost complete cancellation between Eulerian eddy and mean-flow transports.

Dunkerton (1978) used an approximate version of the GLM theory reviewed in Section 3.7 to show that the Brewer–Dobson cell should be interpreted as a Lagrangian mean circulation, not an Eulerian mean. He further suggested that the Lagrangian mean could be approximated by the residual circulation of the TEM equations introduced in Section 3.5, which is itself approximately equal to the "diabatic" circulation (see Section 7.2.3). The diabatic circulation is defined simply as the mean meridional circulation for which the adiabatic heating due to the mean vertical motion just balances the diabatic heating associated with the observed zonal mean temperature field: it was illustrated schematically in Fig. 7.2. Because the lower stratosphere is radiatively heated in low latitudes and radiatively cooled at high latitudes, the lower-stratosphere diabatic circulation in the meridional plane should have a structure similar to that of the circulation postulated by Brewer and Dobson (see the lower part of Fig. 7.2).

The mean distribution of a long-lived tracer implied by the process of advection by the diabatic circulation is qualitatively consistent with the observed distributions of N_2O and CH_4, in that surfaces of constant mixing ratio are bulged upward in low latitudes and slope downward toward the poles, as would be expected with upward advection at low latitudes and downward advection at high latitudes. The observed slopes of the mixing-ratio surfaces are, however, substantially less than would occur if the only processes operating were advection by the diabatic circulation and chemical destruction. Thus, the diabatic circulation is not an adequate quantitative model for the total eddy-driven transport.

9.4   Formulations of Eddy and Mean-Flow Transport

The discussion of the previous section suggests that conventional Eulerian averaging may not be a very efficient method of representing the physics of mean tracer transport in the meridional plane in the presence of large scale wave disturbances. In this section we examine various alternative schemes, some of which provide important insights into the influence of large-scale waves on net transport.

9.4.1   The Eulerian Mean Formulation and the Effective Transport Velocity

We start from the continuity equation for tracer mixing ratio:

$$D\chi/Dt = S, \qquad (9.4.1)$$

where S, the net source, designates the sum of all production and loss processes. In the atmosphere the "source" represented by S must include not only chemical production and loss (see Appendix 10A), but also the effects of turbulent diffusion by the unresolved scales of motion. For this reason alone, tracers are never exactly conserved in observational analyses of atmospheric data. However, if the resolved scales include most of the tracer variance, the diffusive source can in many cases be neglected. Note that Eq. (9.4.1) is also the form of the conservation equations [Eqs. (3.1.3e) and (3.1.5)] for θ and P, with S suitably defined.

With the aid of the continuity equation [Eq. (3.1.1d)] we can write Eq. (9.4.1) in flux form as

$$(\rho_0\chi)_t + (\rho_0 u\chi)_x + (\rho_0 v\chi)_y + (\rho_0 w\chi)_z = \rho_0 S, \qquad (9.4.2)$$

where for simplicity we use Cartesian coordinates. Averaging zonally and using Eq. (3.3.2d) yields the conventional Eulerian mean transport equation,

$$\bar{\chi}_t + \bar{v}\bar{\chi}_y + \bar{w}\bar{\chi}_z = \bar{S} - \rho_0^{-1}\nabla\cdot(\rho_0\overline{\mathbf{u}'\chi'}). \qquad (9.4.3)$$

Note that the form of Eq. (9.4.3) is identical to that of the Cartesian coordinate version of the Eulerian mean thermodynamic equation [Eq. (3.3.2e)] when the eddy flux is written in vectorial form.

It was mentioned in Section 9.3 that a hypothetical middle atmosphere containing no eddies would be close to radiative equilbrium, with very weak meridional motions and tracer transport. If $\bar{S}$ were small, then $\bar{\chi}_t$ would also be small, by Eq. (9.4.3). The effect of eddies is to provide not only an eddy flux on the right of Eq. (9.4.3) but also a mean meridional circulation $(\bar{v}, \bar{w})$. These two terms are frequently found to cancel each other, to some extent (see Section 11.4, especially Fig. 11.10). For example, when averaged over a few weeks the tendency term $\bar{\chi}_t$ becomes small; if $\bar{S}$ is small, then the "mean advection" and "eddy" terms in Eq. (9.4.3) must almost balance. The determination of the *net* meridional transport due to the presence of eddies, and thus the calculation of $\bar{\chi}_t$, therefore involves careful consideration. It will be shown in Section 9.4.2 that this net transport, like the eddy-driven mean acceleration (see Section 3.6), depends on departures from steady, linear, conservative wave motion.

The role of eddies in transport of tracers can be elucidated by considering linearized disturbances $[\chi' = O(\alpha)]$ to almost zonal flow $[\bar{v}, \bar{w}, \bar{Q}, \bar{S} = O(\alpha^2)]$, where α is a measure of the eddy amplitude. Equation (9.4.1) shows that χ' satisfies the linear disturbance equation

$$\bar{D}\chi' + \mathbf{u}' \cdot \nabla\bar{\chi} = S' + O(\alpha^2), \tag{9.4.4}$$

where $\bar{D} \equiv \partial/\partial t + \bar{u}\,\partial/\partial x$. Defining parcel displacements (η', ζ') such that

$$\bar{D}(\eta', \zeta') \equiv (v', w') \tag{9.4.5a}$$

[cf. Eq. (3.7.1)] and a source term γ' such that

$$\bar{D}\gamma' \equiv S', \tag{9.4.5b}$$

we obtain from Eq. (9.4.4)

$$\chi' + \boldsymbol{\xi}' \cdot \nabla\bar{\chi} = \gamma', \tag{9.4.6}$$

where $\boldsymbol{\xi}' = (\xi', \eta', \zeta')$ is the parcel displacement vector.

Eliminating χ' in Eq. (9.4.3) with Eq. (9.4.6) we get, after some manipulation,

$$\bar{\chi}_t + \bar{v}^{\dagger}\bar{\chi}_y + \bar{w}^{\dagger}\bar{\chi}_z = \bar{S}^{\dagger} + \rho_0^{-1}\nabla \cdot (\rho_0 \mathbf{K}^{(s)} \cdot \nabla\bar{\chi}) + O(\alpha^3). \tag{9.4.7}$$

In Eq. (9.4.7) $\mathbf{K}^{(s)}$ is the symmetric "diffusion" tensor

$$\mathbf{K}^{(s)} \equiv \begin{bmatrix} K_{yy} & K_{yz} \\ K_{yz} & K_{zz} \end{bmatrix} = \tfrac{1}{2} \begin{bmatrix} \overline{(\eta'^2)}_t & \overline{(\eta'\zeta')}_t \\ \overline{(\eta'\zeta')}_t & \overline{(\zeta'^2)}_t \end{bmatrix}, \tag{9.4.8}$$

a measure of the dispersion of parcels in the meridional plane (subscripts y and z in the components of $\mathbf{K}^{(s)}$ do not represent partial derivatives). Moreover, $(\bar{v}^\dagger, \bar{w}^\dagger)$ is the *effective transport velocity* (Plumb and Mahlman, 1987) defined by

$$\bar{v}^\dagger \equiv \bar{v} + \rho_0^{-1}(\rho_0\Psi)_z, \qquad \bar{w}^\dagger \equiv \bar{w} - \Psi_y, \qquad (9.4.9a)$$

where

$$\Psi \equiv \tfrac{1}{2}(\overline{v'\zeta'} - \overline{w'\eta'}). \qquad (9.4.9b)$$

is a stream function in the meridional plane.

From Eqs. (9.4.9a,b) and the Cartesian equivalent of Eq. (3.3.2d), it follows that

$$(\rho_0\bar{v}^\dagger)_y + (\rho_0\bar{w}^\dagger)_z = 0. \qquad (9.4.9c)$$

Moreover,

$$\bar{S}^\dagger \equiv \bar{S} - \rho_0^{-1}\boldsymbol{\nabla}\cdot(\rho_0\overline{\mathbf{u}'\gamma'}) \qquad (9.4.9d)$$

is a modified Eulerian mean source term. Noting that

$$\overline{v'\gamma'} = (\overline{\eta'\gamma'})_t - \overline{\eta'S'}, \qquad \overline{w'\gamma'} = (\overline{\zeta'\gamma'})_t - \overline{\zeta'S'},$$

we find that if the eddy tracer source is a weak linear relaxation so that $S' = -A\chi'$, and $|A\chi'| \ll |\bar{D}\chi'|$, then

$$S' \approx A\boldsymbol{\xi}'\cdot\boldsymbol{\nabla}\bar{\chi} \qquad (9.4.9e)$$

and Eq. (9.4.9d) yields

$$\bar{S}^\dagger = \bar{S} - \rho_0^{-1}\boldsymbol{\nabla}\cdot(\rho_0\overline{\boldsymbol{\xi}'\gamma'})_t + \rho_0^{-1}\boldsymbol{\nabla}\cdot(\rho_0\mathbf{K}^{(c)}\cdot\boldsymbol{\nabla}\bar{\chi}) + O(A^2\alpha^2) + O(\alpha^3),$$
$$(9.4.10)$$

where

$$\mathbf{K}^{(c)} \equiv A\begin{bmatrix} \overline{\eta'^2} & \overline{\eta'\zeta'} \\ \overline{\eta'\zeta'} & \overline{\zeta'^2} \end{bmatrix}. \qquad (9.4.11)$$

The tensor $\mathbf{K}^{(c)}$ is sometimes referred to as the "chemical diffusion" tensor. Notice that while $\mathbf{K}^{(s)}$ depends on the *rate of increase* of the mean square displacement, $\mathbf{K}^{(c)}$ depends on its amplitude and the tracer relaxation rate.

The complete Eulerian mean tracer transport equation [Eq. (9.4.7)] for linearized disturbances is rather complicated. However, for linear, steady, conservative ($A = 0$) waves, Eq. (9.4.7) simplifies to

$$\bar{\chi}_t + \bar{v}^\dagger\bar{\chi}_y + \bar{w}^\dagger\bar{\chi}_z = \bar{S}; \qquad (9.4.12)$$

thus, in this particular case, the eddy effects are entirely incorporated in the effective transport velocity $(\bar{v}^\dagger, \bar{w}^\dagger)$.

An alternative formulation of the Eulerian mean tracer transport equation can be obtained by using the TEM formalism introduced in Section 3.5. Using the residual mean flow as defined by Eq. (3.5.1) we can rewrite Eq. (9.4.3) in the form

$$\bar{\chi}_t + \bar{v}^* \bar{\chi}_y + \bar{w}^* \bar{\chi}_z = \bar{S} + \rho_0^{-1} \nabla \cdot \mathbf{M}. \qquad (9.4.13)$$

Unlike the effective transport formulation of Eq. (9.4.7), the residual mean form usually does not completely separate advective and diffusive effects of the eddies since $\mathbf{M}$ (defined in Appendix 9A) generally makes a contribution to the total advective transport. The difference between the effective transport circulation and residual circulation [see Eq. (9A.4)] is, however, usually small in the stratosphere. In Appendix 9A it is shown that if the waves are *linear, steady,* and *adiabatic,* then

$$\bar{v}^\dagger = \bar{v}^* \qquad \text{and} \qquad \bar{w}^\dagger = \bar{w}^*.$$

If in addition the tracer eddy is conservative ($\gamma' = S' = 0$), then $\mathbf{M} = (0, 0)$ and the TEM transport equation is identical to the form of Eq. (9.4.12). If furthermore the mean frictional and diabatic terms $\bar{X}$, $\bar{Y}$, $\bar{Q}$ all vanish, and suitable boundary conditions are imposed, then the nonacceleration theorem of Section 3.6 holds. Normally the mean flow then has $\bar{v}^* = \bar{w}^* = 0$ and the TEM transport equation reduces to $\bar{\chi}_t = \bar{S}$. The mean tracer tendency is due only to mean sources and sinks, and no net meridional transport occurs. This is called a *nontransport theorem.* For the same conditions the eddy flux divergence term on the right side in Eq. (9.4.3) will normally not vanish, but will be exactly balanced by mean advection due to the $(\bar{v}, \bar{w})$ field, which is also nonzero in this case. Thus, the TEM formulation of Eq. (9.4.13) or the effective transport formulation of Eq. (9.4.7) can be more efficient than the conventional formulation [Eq. (9.4.3)] for computing the evolution of a chemically active tracer under approximate nonacceleration conditions.

9.4.2   The GLM Formulation

Conceptually, the simplest framework in which to view the actual mass transport in the meridional plane is by use of the generalized Lagrangian mean (GLM) introduced in Section 3.7. In this approach, Eq. (9.4.1) is averaged along a wavy material tube of particles (see Fig. 3.1) and the mean equation has the form

$$\frac{\partial \bar{\chi}^L}{\partial t} + \bar{v}^L \frac{\partial \bar{\chi}^L}{\partial y} + \bar{w}^L \frac{\partial \bar{\chi}^L}{\partial z} = \bar{S}^L. \qquad (9.4.14)$$

In Eq. (9.4.14) there are no flux terms corresponding to resolved eddies, since the averaging is with respect to a tube that follows the parcel motions under the influence of the waves. There is of course still a diffusive effect of unresolved eddy motions, but this must be included in the source term.

This formulation gives a finite-amplitude version of the nontransport theorem mentioned above, which states that for steady, conservative waves $\bar{v}^L = \bar{w}^L = 0$ and hence $\bar{\chi}_t^L = \bar{S}^L$; the net transport again vanishes. A physical interpretation is that under the stated conditions the orbits of resolved fluid parcels, although wavy, are "statistically closed," in the sense that no large-scale dispersion or Lagrangian-mean advection can occur. Hence the parcels can accomplish no systematic meridional transport.

Using the small-amplitude GLM theory introduced in Section 3.7.2, it can be shown that

$$\bar{v}^L = \bar{v}^\dagger + \tfrac{1}{2}\overline{(\eta'^2)}_{ty} + \tfrac{1}{2}\rho_0^{-1}(\rho_0\overline{\eta'\zeta'})_{tz} + O(\alpha^3), \qquad (9.4.15a)$$

$$\bar{w}^L = \bar{w}^\dagger + \tfrac{1}{2}\overline{(\eta'\zeta')}_{ty} + \tfrac{1}{2}\rho_0^{-1}(\rho_0\overline{\zeta'^2})_{tz} + O(\alpha^3). \qquad (9.4.15b)$$

From Eqs. (9.4.15) and (9.4.9c), it is readily verified that

$$(\rho_0\bar{v}^L)_y + (\rho_0\bar{w}^L)_z = \tfrac{1}{2}(\rho_0\overline{\xi_j\xi_k})_{,jkt} = (\rho_0 K_{jk}^{(s)})_{,jk} \qquad (9.4.16)$$

where indices j, k are summed over the values 2, 3 designating the meridional and vertical directions, respectively, and $K^{(s)}$ is the "diffusion" tensor defined in Eq. (9.4.8). Thus, in the GLM system the divergence in the meridional plane does not generally vanish for transient eddies, in which air parcels are dispersing from their mean positions (see Section 3.7.1). If, however, the waves are *linear and steady*, no dispersion occurs, and the divergence vanishes.

In practice the GLM model is not very useful, since when eddy motions are strongly transient the wavy tubes quickly become highly convoluted so that the motion of the "center of mass" defined by the Lagrangian mean may bear little relation to the actual tracer distribution. For adiabatic and frictionless motion the wavy tubes along which the GLM averages are computed coincide with the intersections of θ and P surfaces, and we have already seen in Section 5.2.3 that planetary wave breaking processes lead to rapid distortion, and even "breaking" of such tubes. Only for small-amplitude eddies do these difficulties not arise.

9.4.3   The Isentropic Formulation

The isentropic coordinate approach has some of the conceptual advantages of the GLM formulation without as many of the accompanying technical difficulties. In this system the "vertical velocity" is proportional to the diabatic heating rate, $D\theta/Dt = Q$ [see Eq. (3.1.3e)]. Thus, the tracer continuity equation [Eq. (9.4.1)] is

$$\chi_t + u\chi_x + v\chi_y + Q\chi_\theta = S. \tag{9.4.17a}$$

The Cartesian coordinate version of the continuity equation [Eq. (3.8.1c)] is

$$\sigma_t + (\sigma u)_x + (\sigma v)_y + (\sigma Q)_\theta = 0. \tag{9.4.17b}$$

where the "density" σ is defined in Eq. (3.8.1e). Multiplying Eq. (9.4.17a) by σ and Eq. (9.4.17b) by χ and adding, we obtain the flux form of the tracer continuity equation,

$$(\sigma\chi)_t + (\sigma u\chi)_x + (\sigma v\chi)_y + (\sigma Q\chi)_\theta = (\sigma S). \tag{9.4.18}$$

Averaging zonally (with θ held constant) we get from Eq. (9.4.18)

$$\bar{\sigma}\bar{\chi}_t + \overline{(\sigma v)}\bar{\chi}_y + \overline{(\sigma Q)}\bar{\chi}_\theta = \overline{\sigma S} - \overline{(\sigma'\chi')}_t - [\overline{(\sigma v)'\chi'}]_y - [\overline{(\sigma Q)'\chi'}]_\theta. \tag{9.4.19}$$

As in Eq. (3.9.5), we define a mass-weighted mean for any field A:

$$\bar{A}^* \equiv (\overline{\sigma A})/\bar{\sigma}.$$

Then from Eq. (9.4.17b) we obtain

$$\bar{\sigma}_t + (\bar{\sigma}\bar{v}^*)_y + (\bar{\sigma}\bar{Q}^*)_\theta = 0 \tag{9.4.20}$$

[cf. Eq. (3.9.7c)], while Eq. (9.4.19) can be expressed in the exact form

$$\bar{\chi}_t + \bar{v}^*\bar{\chi}_y + \bar{Q}^*\bar{\chi}_\theta = \bar{S}^* - \bar{\sigma}^{-1}\overline{(\sigma'\chi')}_t - \bar{\sigma}^{-1}\{[\overline{(\sigma v)'\chi'}]_y + [\overline{(\sigma Q)'\chi'}]_\theta\}. \tag{9.4.21}$$

The linearized disturbance equation is

$$\bar{D}\chi' + v'\bar{\chi}_y + Q'\bar{\chi}_\theta = S' + O(\alpha^2), \tag{9.4.22}$$

where we have assumed that $\bar{\chi}_t, \bar{v}, \bar{Q}, \bar{S} = O(\alpha^2)$, and α is as usual a measure of the disturbance amplitude. Here $\bar{D} \equiv \partial/\partial t + \bar{u}\,\partial/\partial x$. It should be noted that since derivatives and averages are taken at constant θ, the definitions of the "disturbance" quantities differ slightly from the z-coordinate versions discussed earlier in this section. We define perturbation quantities η', q', γ' by

$$\bar{D}(\eta', q', \gamma') = (v', Q', S') + O(\alpha^2). \tag{9.4.23}$$

From Eqs. (9.4.22) and (9.4.23) we then obtain

$$\chi' = -\eta'\bar{\chi}_y - q'\bar{\chi}_\theta + \gamma' + O(\alpha^2). \tag{9.4.24}$$

By analogy to our treatment of the conventional Eulerian-mean equation, we define

$$\mathbf{K}^{(s)} \equiv \frac{1}{2}\begin{bmatrix} \overline{(\eta'^2)}_t & \overline{(\eta'q')}_t \\ \overline{(\eta'q')}_t & \overline{(q'^2)}_t \end{bmatrix}, \tag{9.4.25a}$$

$$\Psi \equiv \tfrac{1}{2}\overline{(v'q' - \eta'Q')}, \tag{9.4.25b}$$

$$\bar{v}^\dagger \equiv \bar{v}^* + \bar{\sigma}^{-1}(\bar{\sigma}\Psi)_\theta, \tag{9.4.25c}$$

$$\bar{Q}^\dagger \equiv \bar{Q}^* - \bar{\sigma}^{-1}(\bar{\sigma}\Psi)_y \tag{9.4.25d}$$

and obtain from Eq. (9.4.21) the isentropic coordinate form of the zonal mean tracer transport equation:

$$\bar{\chi}_t + \bar{v}^\dagger\bar{\chi}_y + \bar{Q}^\dagger\bar{\chi}_\theta = \bar{S}^* - \bar{\sigma}^{-1}\overline{(\sigma'\chi')}_t - \bar{\sigma}^{-1}[(\bar{\sigma}\overline{v'\gamma'})_y + (\bar{\sigma}\overline{Q'\gamma'})_\theta]$$
$$+ \bar{\sigma}^{-1}\boldsymbol{\nabla}_\theta \cdot [\bar{\sigma}\mathbf{K}^{(s)} \cdot \boldsymbol{\nabla}_\theta\bar{\chi}] + O(\alpha^3). \tag{9.4.26}$$

If the eddies are *adiabatic* ($Q' = q' = 0$), then the components $K_{y\theta}$, $K_{\theta y}$, and $K_{\theta\theta}$ all vanish, $\Psi = 0$, and $(\bar{v}^\dagger, \bar{Q}^\dagger) = (\bar{v}^*, \bar{Q}^*)$; if in addition we neglect terms in σ', Eq. (9.4.26) simplifies to

$$\bar{\chi}_t + \bar{v}^*\bar{\chi}_y + \bar{Q}^*\bar{\chi}_\theta = \bar{S}^\dagger + \bar{\sigma}^{-1}\frac{\partial}{\partial y}\left[\bar{\sigma}K_{yy}\frac{\partial\bar{\chi}}{\partial y}\right], \tag{9.4.27}$$

where the diffusion coefficient $K_{yy} \equiv \tfrac{1}{2}\overline{(\eta'^2)}_t$ is a measure of dispersion of parcels on isentropic surfaces and

$$\bar{S}^\dagger \equiv \bar{S}^* - \bar{\sigma}^{-1}(\bar{\sigma}\overline{v'\gamma'})_y \tag{9.4.28}$$

is a modified source term [cf. Eq. (9.4.9d)].

When S' can be modeled as a weak linear relaxation, manipulations analogous to those leading to Eq. (9.4.10) yield from Eq. (9.4.28)

$$\bar{S}^\dagger = \bar{S}^* - \bar{\sigma}^{-1}(\bar{\sigma}\overline{\eta'\gamma'})_{ty} + \bar{\sigma}^{-1}\frac{\partial}{\partial y}\left(\bar{\sigma}\hat{K}_{yy}\frac{\partial\bar{\chi}}{\partial y}\right) \tag{9.4.29}$$

where $\hat{K}_{yy} \equiv A\overline{\eta'^2}$. Substitution from Eq. (9.4.29) into Eq. (9.4.27) then gives

$$\bar{\chi}_t + \bar{v}^*\bar{\chi}_y + \bar{Q}^*\bar{\chi}_\theta = \bar{S}^* - \bar{\sigma}^{-1}(\bar{\sigma}\overline{\eta'\gamma'})_{ty} + \bar{\sigma}^{-1}\frac{\partial}{\partial y}\left[\bar{\sigma}K^{(\text{tot})}\frac{\partial\bar{\chi}}{\partial y}\right] \tag{9.4.30}$$

where $K^{(\text{tot})} \equiv K_{yy} + \hat{K}_{yy}$. If the eddies are *steady* as well as being *linear* and *adiabatic*, the second term on the right-hand side in Eq. (9.4.30) vanishes and $K^{(\text{tot})}$ reduces to $\hat{K}_{yy}$.

As mentioned at the end of Section 9.3.1, the observed slopes of mixing ratio surfaces in the meridional plane for long-lived species such as N_2O and CH_4 are substantially less than would occur if the only processes operating were advection by the diabatic circulation and chemical destruction. The isentropic form of the transport equation [Eq. (9.4.30)] clearly indicates that the additional process required to explain the observations is meridional mixing by quasi-isentropic eddies. Thus, as summarized in Fig. 9.4, the gross characteristics of transport in the meridional plane can be modeled in terms of a combination of advection by a mean meridional mass circulation and quasi-isentropic mixing by large-scale eddies.

In particular, the angle between the streamlines of the circulation and the mean mixing ratio surface is related to the eddy diffusion in Eq. (9.4.27) or Eq. (9.4.30). For example, in the steady state, Eq. (9.4.20) implies the existence of a stream function, $\tilde{\Psi}$ say, such that

$$\bar{\sigma}\bar{v}^* = -\tilde{\Psi}_\theta, \qquad \bar{\sigma}\bar{Q}^* = \tilde{\Psi}_y;$$

substitution into Eq. (9.4.27) gives

$$-\tilde{\Psi}_\theta\bar{\chi}_y + \tilde{\Psi}_y\bar{\chi}_\theta = \frac{\partial}{\partial y}\left[\bar{\sigma}K_{yy}\frac{\partial\bar{\chi}}{\partial y}\right]$$

if $\bar{\chi}_t$ and $\bar{S}^\dagger$ are set to zero. The angle between the streamlines and the $\bar{\chi}$ surfaces is thus

$$\sin^{-1}\left[\frac{\dfrac{\partial}{\partial y}\left(\bar{\sigma}K_{yy}\dfrac{\partial\bar{\chi}}{\partial y}\right)}{|\nabla_\theta\tilde{\Psi}|\cdot|\nabla_\theta\bar{\chi}|}\right]$$

in $y\theta$ space: this vanishes for $K_{yy} = 0$, in which case the tracer isolines coincide with streamlines. When $K_{yy} > 0$, the tracer isolines slope less steeply than the streamlines in high latitudes, as shown schematically in Fig. 9.4.

9.5 Dispersive Wave Transport: Irreversible Mixing of Tracers

The derivation of equations like Eqs. (9.4.7) and (9.4.27), exhibiting eddy terms of "diffusive" form, has assumed the eddies to be of small amplitude. Under this assumption, diffusion coefficients like $K_{yy} = \frac{1}{2}(\eta'^2)_t$ represent *reversible* dispersion of fluid parcels: $K_{yy} > 0$ if η'^2 increases with time as parcels disperse, but $K_{yy} < 0$ as η'^2 decreases again if parcels return to their equilibrium latitudes. For finite-amplitude disturbances we can expect more complex behavior, whose representation by equations like Eqs. (9.4.7) and and (9.4.27) may be more difficult to justify rigorously. In Section 5.2.3 we discussed the dynamical implications of so-called planetary wave breaking. We there emphasized that "wave breaking" in this context refers to a rapid irreversible deformation of otherwise wavy material contours. The ability of organized quasi-nondivergent velocity fields to generate such irreversible deformation is illustrated by the classic example of Fig. 9.5, which clearly shows the tendency for flows to string a conservative tracer out into ever longer, thinner laminae, so that ultimately small-scale turbulence can mix the tracer throughout the domain. In strongly nonlinear situations of this kind, parcels certainly do not return to their original latitudes, even in statistically steady flow, and the small-amplitude theory would appear to

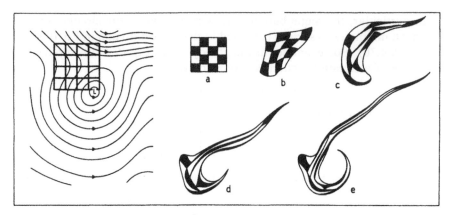

Fig. 9.5. The evolution of the shape of a set of marked fluid parcels initially forming a "checkerboard" pattern under the influence of a two-dimensional deformation field (shown on the left) typical of that occurring in large scale barotropic atmospheric eddies. For all practical purposes the deformation may be regarded as irreversible. [After Welander (1955).]

be invalid. Nevertheless, experiments with a numerical model (Plumb and Mahlman, 1987) suggest that Eqs. (9.4.7) and (9.4.27) may still describe such irreversible mixing quite well, although the components of $\mathbf{K}^{(s)}$ no longer have simple expressions in terms of parcel displacements: see Section 9.7.1.

The planetary wave breaking process, by which isentropic potential vorticity distributions are mixed in the stratospheric surf zone, will thus tend to mix chemical tracers that have meridional gradients on isentropic surfaces. For timescales that are shorter than the chemical and radiative timescales, the behavior of such chemical tracers should be very similar to that of potential vorticity, provided that the mean gradients are similar. The long-lived tracers discussed in Section 9.2 are excellent examples of such tracers, since they all, like potential vorticity, have strong vertical gradients in the lower stratosphere. Global measurements of two of these tracers, N_2O and CH_4, are available from the SAMS instrument on *Nimbus* 7. Despite the limited resolution of the observations, there is clear evidence of material line deformation during planetary wave breaking as shown in the methane maps of Figs. 9.6a,b. (These should be compared with the isentropic potential vorticity charts of Figs. 9.6c,d.)

Similar evidence is available for ozone, as measured by the LIMS experiment. Although ozone has a chemical timescale in the polar night that is much longer than the dissipation timescale for potential vorticity, it does not have a strong vertical gradient in the midstratosphere as does the potential vorticity P. Nevertheless, observations during the strong wave amplification of late January 1979 indicate that ozone and potential

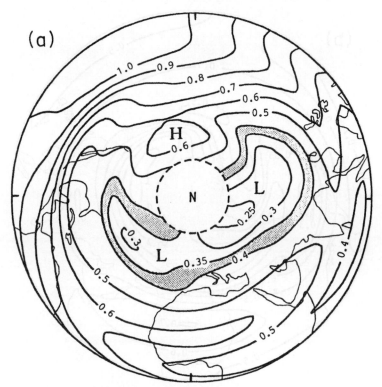

Fig. 9.6. Polar stereographic maps (Greenwich meridian at bottom) of CH_4 (ppmv) at 3 mb on (a) December 1, 1981 and (b) December 6, 1981. (Outer circle, equator; inner circle, $67\frac{1}{2}°N$.) Ertel's potential vorticity P on the 850 K isentropic surface (near 10 mb) in units of $10^{-4}\, K\, m^2\, kg^{-1}\, s^{-1}$ on (c) December 1, 1981 and (d) December 6, 1981. Geostrophic winds are indicated by arrows. [After Jones (1984); (c) and (d) courtesy of A. O'Neill, U.K. Meteorological Office.] *Figure continues.*

vorticity are irreversibly mixed in similar ways (see Section 5.2.3). However, during the major stratospheric warming period of late February 1979, ozone does not follow P quite as well as during the earlier period, apparently due to the fact that the chemical time-scale for ozone in the polar night is much longer than the radiative timescale, so that the potential vorticity distribution tends to recover more rapidly following transport into the polar region than does ozone. A result of the repeated penetration of ozone-enriched tongues of air into the polar night is a gradual buildup of ozone in this region, far from the photochemical source.

In summary, the evidence from SAMS and LIMS is that rapid quasi-isentropic meridional transport is associated with planetary wave breaking events, and that this process tends to reduce the meridional gradients of

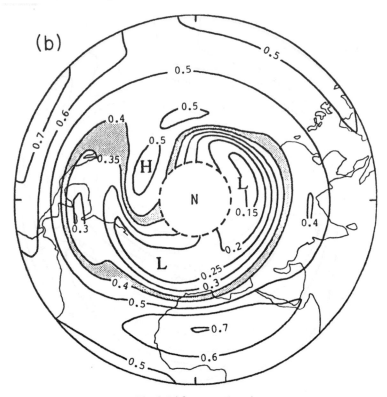

Fig. 9.6 (*figure continues*)

quasi-conserved tracers on isentropic surfaces. In particular, the ozone "hole" observed in the northern polar region during the early winter is gradually filled during the course of the winter by repeated poleward penetrations of low-latitude air parcels advected by breaking waves. Thus, large-amplitude wave motions, through nonlinear processes, produce quasi-isentropic mixing of long-lived stratified tracers, which limits the meridional gradients continually generated by the diabatic circulation.

9.5.1   Vertical Variance of Long-Lived Tracers

In the previous section we emphasized the hemispheric-wide role of planetary wave breaking, and implicitly assumed that the vertical scales of the motions were quite large, since available satellite measurements do not resolve structures with vertical scales less than a few kilometers. However, when the motion field is highly baroclinic, as is often the case in the lower stratosphere in winter, isentropic trajectories for levels separated by only a kilometer or two may differ substantially.

(c)

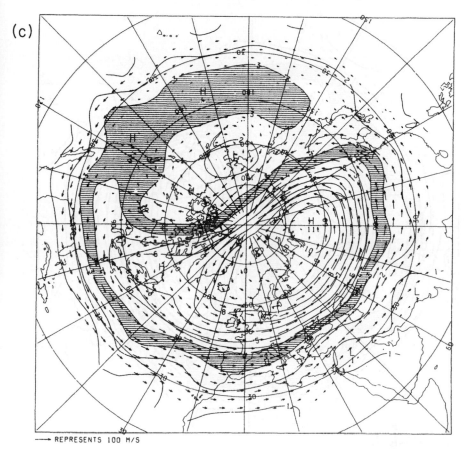

—→ REPRESENTS 100 M/S

Fig. 9.6 (*figure continues*)

A spectacular example of the consequences that this has for local vertical profiles is shown in Fig. 9.7. In this sounding from Laramie, Wyoming, on January 31, 1979, both the ozone and water-vapor traces show distinct minima at the 15-km level. Minima of this type are commonly seen in ozonesonde records and have in the past generally been interpreted as arising from quasi-horizontal transport from the midlatitude upper troposphere (where ozone mixing ratios are very low) into the lower stratosphere. However, the existence of correlated water vapor and ozone minima in the sounding strongly suggests a tropical source, since water-vapor mixing ratios as low as 3 ppmv normally occur only in the lower tropical stratosphere. Furthermore, air originating in the extratropical troposphere would need to be heated diabatically by about 30 K in order to ascend to the 15-km

(d)

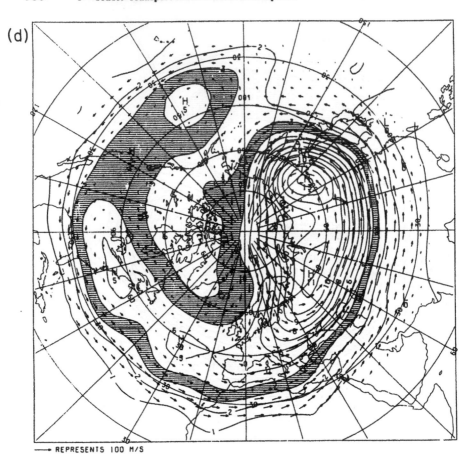

REPRESENTS 100 M/S

Fig. 9.6 (*continued*)

level. Such heating would be very unlikely to occur in the several-day timescale implied by the sharpness of the ozone minimum.

Danielsen and Kley (1987) constructed isentropic trajectories for several levels (marked by the potential temperature values indicated on Fig. 9.7) for a period beginning 5 days prior to the Laramie sounding. The resulting trajectories shown in Fig. 9.8 indicate that at all levels except the 405-K isentrope the air has its recent origins in the extratropical stratosphere. (Note the cross-polar trajectory at 660 K; this is caused by the amplification of the Aleutian anticyclone associated with a minor stratospheric warming.) Only at the level of the observed correlated ozone and water-vapor minima does the air have a tropical origin—in the equatorial Western Pacific. This tropical origin is consistent with the observed positive correlation between

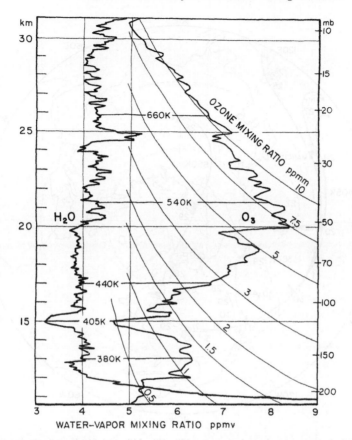

Fig. 9.7. Profiles of water-vapor and ozone mixing ratios in the stratosphere at Laramie, Wyoming, January 31, 1979. To emphasize variability at the lower altitudes the ozone plot is linear in ozone partial pressure; thus, lines of constant ozone mixing ratio are curves sloping upward toward the left. Notice the correlated ozone and water-vapor minima at the 405-K isentropic level. The isentropic levels used for the trajectory analysis in Fig. 9.8 are shown on the diagram. [After Danielsen and Kley (1987).]

the ozone and water-vapor fluctuations, since such air probably was recently "freeze dried" by passage through the cold tropical tropopause.

This example of short-vertical-scale tracer variability suggests one way in which planetary-scale wave breaking may lead to irreversible mixing. The vertical structure in the velocity field allows the fluid deformations to bring tongues of air from vastly different regions into close contact—both horizontally and vertically. Thus very strong gradients are formed, as shown

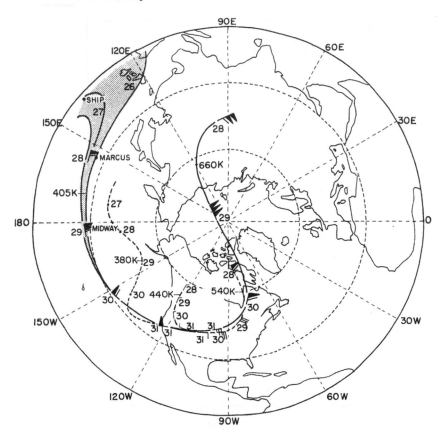

Fig. 9.8. Isentropic trajectories on θ = 380, 405, 540, and 660 K surfaces for the 5-day period ending January 31, 1979. Dated wind barbs show rate of progress of an air parcel along the trajectory. [After Danielsen and Kley (1987).]

in Fig. 9.7. These gradients will inevitably be smoothed by small-scale diffusion so that mid- and low-latitude properties become mixed. Of course, vertical diffusion is generally slow in the stratosphere where the static stability is very large, but even an eddy diffusivity as small as $0.25 \text{ m}^2 \text{ s}^{-1}$ will mix the short-vertical-scale structure of this example in less than 10 days.

9.5.2   Temporal Variance of Stratospheric Trace Species

For the long-lived vertically stratified tracers discussed in Section 9.2, oscillatory vertical motions can produce temporal variance at a given point in physical space merely through vertical displacements of the isentropes.

However, as suggested in Fig. 9.4, such tracers tend to have meridional gradients on isentropic surfaces. Hence, simple meridional sloshing of zonally propagating adiabatic wave motions can produce temporal variance in tracer profiles at a single location; vertical displacements of the isentropes are not required. This phenomenon does not depend on nonlinear wave processes such as planetary wave breaking; nor does it necessarily imply mixing or irreversibility.

Observed temporal variability is conveniently characterized by an "equivalent displacement height" (EDH) (Ehhalt *et al.*, 1983). The EDH is the vertical distance that the time-mean vertical tracer profile would need to be displaced in order to produce the locally observed variance. The use of the EDH to characterize tracer variability should not be taken to imply that it is primarily vertical displacements that act to product the variance. Meridional displacements may actually be more important. The relative roles of meridional and vertical parcel displacements can be assessed approximately by comparing the EDH for potential temperature with that for long-lived chemical tracers, since the meridional tracer slopes are generally greater than the slopes of the isentropes (see Section 9.2).

For conservative waves the linear tracer perturbation equation [Eq. (9.4.6)] for departures from zonal symmetry is

$$\chi' = -\eta'\bar{\chi}_y - \zeta'\bar{\chi}_z. \tag{9.5.1}$$

If it is assumed that the temporal variation at a local point is similar to the spatial variation around a latitude circle, then the EDH can be defined as $\delta_\chi \equiv (\overline{\chi'^2})^{1/2}/\bar{\chi}_z$. Thus using Eq. (9.5.1),

$$\delta_\chi = \overline{[(\eta'[\bar{\chi}_y/\bar{\chi}_z] + \zeta')^2]}^{1/2}, \tag{9.5.2}$$

which shows that the EDH is just the eddy displacement normal to the mean slope of the tracer isopleths. For species with long chemical timescales, the mean tracer slopes in the meridional plane, $[dz/dy]_{\bar{\chi}} = -\bar{\chi}_y/\bar{\chi}_z$ are generally large enough so that the contribution to δ_χ by η' is at least as large as that due to ζ'. For planetary wave breaking, it is indeed quasi-horizontal displacements that play the major role in producing tracer variance.

The altitude dependence of the magnitudes of the EDHs for a number of species measured in Northern Europe in summer are summarized in Fig. 9.9. Measurements of temperature show that the EDH for potential temperature at a given height is an order of magnitude smaller than the EDHs for the trace chemicals. This indicates that the observed variance is not the result of vertical displacement of the isentropic surfaces, as would be produced by internal gravity waves, for example. The variance might,

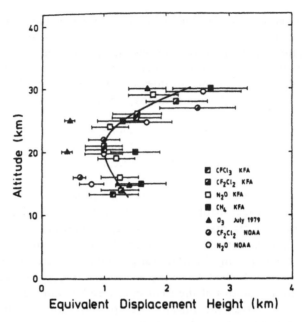

Fig. 9.9. Magnitudes of equivalent displacement heights (EDH) for several long-lived trace gases. See text for definition and significance. [From Ehhalt *et al.* (1983), with permission.]

however, be associated with meridional displacements produced by quasi-barotropic wave motions such as the 5-day wave (see Section 4.4). For observed tracer isopleth slopes, an EDH of 2 km could be produced by meridional displacement of about 1000 km.

It is conceivable, however, that some of the tracer variability observed in the middle stratosphere during the summer is not associated with active wave motions, but represents variance generated by the large meridional and vertical parcel displacements associated with the springtime final warming process that has become "frozen in" and is merely advected around by the symmetric easterly circulation of the summer stratosphere. Such a situation might arise for any tracer whose dissipation timescale is substantially longer than the dynamical timescale. For the final warming, which is marked by breakdown of the winter polar vortex and establishment of the summer polar circulation, the relevant dynamical timescale is the scale on which radiation damps out the wave motions. This scale is of the order of 1–3 weeks in the stratosphere (depending on height) and is substantially shorter than the chemical decay times for the long-lived species considered here. Thus, tongues of air with either enriched or depleted tracer concentrations might possibly persist for a very long time in the summer stratosphere, where diffusion should be quite weak.

9.6 Troposphere–Stratosphere Exchange

Transport of trace substances is important to understanding of tracer behavior in both the troposphere and the stratosphere. Especially important, however, is the vertical flux of trace constituents across the tropopause and through the lowest few kilometers of the stratosphere. It is this region that primarily determines the rate of transport between source and sink regions for both tropospheric and stratospheric source gases. For example, the slowness of transport in this region is responsible for the observed long stratospheric residence times for radioactive tracers, and for the long time needed to establish a steady-state ozone reduction in response to tropospheric release of chlorofluoromethanes (see Section 10.5.3). Thus, the dynamics of exchange across the tropopause merits special consideration. Other aspects of troposphere–stratosphere linkage are considered in Chapter 12.

Conventionally, the tropopause is defined on the basis of the thermal structure of the atmosphere as the level of near discontinuity in the temperature lapse rate that divides the strongly statically stable stratosphere from the weakly statically stable troposphere (Figs. 1.1 and 9.1a). For climatological purposes this definition may be satisfactory. However, lapse rate is not a conservative quantity, and for discussion of exchange of mass between the troposphere and the stratosphere it is preferable to use Reed's (1955) concept of a dynamical tropopause defined in terms of the near discontinuity in Ertel's potential vorticity that separates very large stratospheric values from tropospheric values that are one or two orders of magnitude smaller (see Fig. 9.1b). Thus "exchange" requires that air parcels of stratospheric (tropospheric) origin lose (gain) potential vorticity through mixing or diabatic processes, and the rate of exchange is controlled by the rate at which such processes operate.

Transport across the tropopause may be caused by a number of processes. These include (1) the large-scale mean diabatic circulation (Brewer–Dobson cell), (2) transverse secondary circulations associated with subtropical and polar jetstreams, (3) cumulonimbus clouds penetrating into the stratosphere, (4) tropopause folding and subsequent mixing caused by upper-level cyclogenesis, (5) turbulent mixing associated with gravity-wave breaking or with shear instabilities at the tropopause, and (6) local radiative cooling in the vicinity of high-level cirrus anvil clouds.

Cross tropopause exchange in middle latitudes is dominated by tropopause folding associated with the development of upper-level baroclinic waves in the tropospheric jetstream. The term "folding" is used to describe a process in which the dynamical tropopause (defined in terms of potential vorticity) intrudes deeply into the troposphere along a sloping frontal zone (see Fig. 9.10). Studies of both chemical and dynamical tracers during such folding events have done much to elucidate the nature of this process, which is perhaps the primary mechanism for transport of mass from the stratosphere into the troposphere. The return flow from troposphere to stratosphere appears to take place primarily in the equatorial region, as suggested by the Brewer–Dobson model discussed above. However, the actual mechanism appears to involve cumulus convection rather than gentle large-scale ascent.

9.6.1 Extratropical Exchange: Tropopause Folding

The tropopause folding process has its origins in the deformation field associated with upper-level frontogenesis occurring as a result of baroclinic instability in the westerly jetstreams. According to the accepted theory of frontogenesis (Hoskins, 1982), the geostrophic deformation field associated with the development of a baroclinic wave tends to concentrate preexisting temperature gradients, thus giving rise to a sloping zone of closely spaced isentropes that marks the boundary between cold and warm air masses. This concentration is most intense near the ground and the tropopause. To

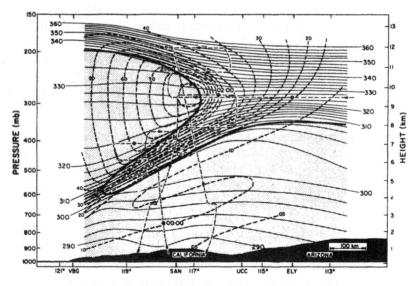

Fig. 9.10. Cross section through tropopause folding event of March 13, 1978: region of tropospheric air stippled; potential temperature (thin solid lines); wind speed (m s^{-1}, dashed lines); research aircraft flight track (thin dashed lines); potential vorticity tropopause (heavy solid line). [From Shapiro (1980). American Meteorological Society.]

maintain the thermal wind balance under such conditions, a secondary transverse ageostrophic circulation arises. Advection by this secondary circulation further concentrates the temperature gradient, which enhances the secondary circulation, and so on. This process leads to a rapid intensification of the temperature gradient and at the same time forces strong parcel subsidence in the frontal zone. This subsidence advects high potential vorticity air parcels of stratospheric origin downward along the sloping isentropic surfaces, so that the dynamical tropopause is folded to form a deep intrusion of stratospheric air into the troposphere.

The development of such an intrusion is shown for a theoretical two-dimensional model in Fig. 9.11. Small arrows in the figure, showing fluid-parcel paths associated with the transverse ageostrophic motion, clearly indicate downward parcel motion along the frontal zone. In tropopause fold cases observed in the atmosphere, the stratospheric intrusions can be much deeper than suggested by the model of Fig. 9.11. The stratospheric air can penetrate deeply into the troposphere in thin laminae of the order of 100 km lateral scale and 1 km vertical scale. The high potential vorticities that exist in the upper portions of such folds are gradually modified by irreversible small-scale mixing with tropospheric air along the boundaries

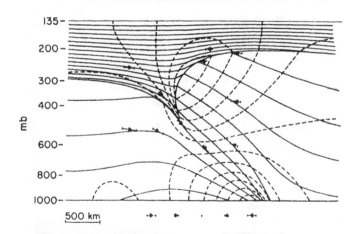

Fig. 9.11. Stratospheric intrusion generated in a model of upper-level frontogenesis caused by an applied deformation field (shown by arrows below lower surface). The initial state had a weak uniform meridional temperature gradient (warm air on right) and constant (but different) static stabilities in the troposphere and the stratosphere. Dashed contours show long front velocity (10.5 m s^{-1} interval), thin solid contours are potential temperature (7.8-K interval), and arrows within fluid show some parcel trajectories. The tropopause marks the discontinuity between widely spaced isentropes in the troposphere and closely spaced isentropes in the stratosphere. [After Hoskins (1972).]

of the fold, so that potential vorticity values and ozone mixing ratios tend to decrease moving downward along the isentropes in the folds.

This point is illustrated by the cross section in Fig. 9.12, which shows ozone mixing-ratio contours for the tropopause folding event depicted in Fig. 9.10. Figure 9.12 indicates that the fold does indeed contain air with stratospheric concentrations of ozone, but that there is substantial mixing with tropospheric air since the ozone mixing ratio decreases downward in the fold. Although Ertel's potential vorticity is not plotted, it can be deduced from the large static stability $(\partial\theta/\partial z)$ and large shear vorticity present in the fold region, as depicted in Fig. 9.10, that the potential vorticity must be large in the fold.

It should be noted that in addition to the intrusion process that brings stratospheric air into the troposphere, the transverse mass circulation associated with jetstreams also transports some tropospheric air into the lower stratosphere, particularly on the anticyclonic shear (equatorward) side of the jet. Since such air must have rather high water-vapor mixing ratios, and the observed mixing ratios are very low in the stratosphere, it is believed that such return flow only transfers tropospheric air into the lowest stratospheric layers, where it may be quickly returned to the troposphere through

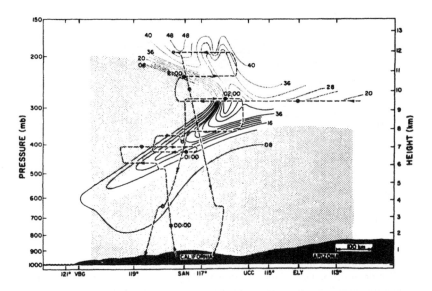

Fig. 9.12. Ozone concentration in parts per hundred million by volume for tropopause folding event shown in Fig. 9.10. Upper-flight track analysis shown in dotted lines, lower-flight track in solid lines. Stippled region indicates the troposphere. Note: warm air is on the left side, opposite to the plotting convention used in Fig. 9.11. [From Shapiro (1980). American Meteorological Society.]

the tropopause folding process. Although precise estimates are difficult, it is thought that nearly 10% of the stratospheric mass is injected into the troposphere by the folding process each year, but that most of this mass originates in the lowest part of the stratosphere.

9.6.2 Tropical Exchange: The Role of Cumulus Convection

The Brewer–Dobson model suggests that upward transfer from the troposphere into the stratosphere occurs only in the tropics, and that the rate of mass transfer is related to the diabatic heating rate. There can be little doubt that freeze drying due to upward passage of air parcels through the tropical cold trap is a qualitatively reasonable explanation for the observed extreme aridity of the stratosphere. It is certainly in general accord with the zonal mean water vapor distribution as deduced from the LIMS experiment. The LIMS water-vapor cross section shown in Fig. 9.13 clearly

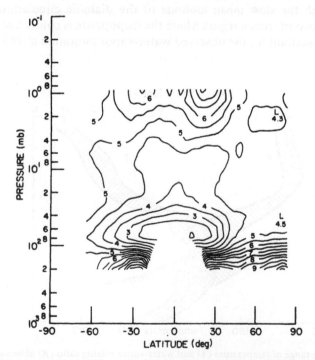

Fig. 9.13. Zonal mean water-vapor mixing ratio cross section (ppmv) for December 1978, from the *Nimbus* 7 LIMS experiment. [From Remsberg *et al.* (1984b). American Meteorological Society.]

indicates a minimum in the lower tropical stratosphere, with larger values in the upper stratosphere and at high latitudes in the lower stratosphere. This is consistent with a model in which the diabatic circulation advects a "plume" of dry air upward and poleward in both hemispheres, thus maintaining a low mixing ratio despite the presence of an upper-stratospheric source of water vapor due to methane oxidation.

Evidence from LIMS and *in situ* sampling indicates that the mean mixing ratio for water vapor in the lower stratosphere is about 3–4 ppmv. Climatological data suggest, however, that only in rather restricted regions of the tropics are the temperatures at the tropical tropopause sufficiently cold to allow freeze drying of upward-moving air to mixing ratios of less than 4 ppmv. Furthermore, observations in the Panama Canal Zone (Fig. 9.14) indicate that in that location the water-vapor minimum occurs not at the tropopause but several kilometers higher, and is also significantly smaller than the saturation mixing ratio at the tropopause level. Thus, it appears that the water-vapor distribution of the stratosphere cannot be accounted for by a model in which tropospheric air is transferred into the stratosphere locally through the slow mean motions of the diabatic circulation. Only horizontal transport from a region where the tropopause is colder and higher can plausibly account for the observed water-vapor minimum at 19 km over Panama.

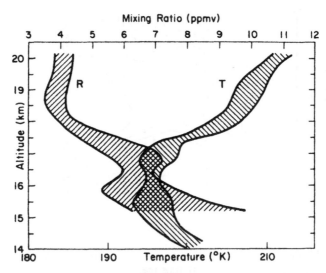

Fig. 9.14. The range of temperature (T) and water-vapor mixing ratio (R) observations for eight flights during the NASA Panama Experiment in September 1980. Areas enclosed by the diagonal lines show plus and minus 1 standard deviation from the mean. [After Kley *et al.* (1982).]

Newell and Gould-Stewart (1981) suggested that observations are consistent with a model, which they called the "stratospheric fountain," in which most of the flux of mass from the troposphere into the stratosphere is concentrated in relatively limited regions of the tropics—primarily the so-called "maritime continent" of Indonesia during the Northern-Hemisphere winter, and the Indian monsoon region during Northern-Hemisphere summer. The tropical tropopause is on the average highest and coldest at those locations and times.

It is known that much of the upward mass flux in the tropical troposphere is concentrated in the updrafts of individual cumulonimbus convective clouds. These so-called "hot towers" tend to overshoot their levels of neutral buoyancy and penetrate into the lower stratosphere, where they may mix with stratospheric air. Thus, it is likely that exchange in the tropics is associated with the convective motion scale rather than a slow uniform ascent as envisioned in the usual model for the Brewer–Dobson cell. However, the precise mechanisms of tropical exchange remain to be elucidated.

9.7 Transport Modeling

The major motivation for the development of tracer transport models in recent years has been the problem of the photochemistry of the ozone layer. Detailed analyses of this problem often involve over 100 reactions, including several dozens of chemical species. Since nearly all such species are directly or indirectly influenced by transport in the meridional and vertical directions, proper representation of transport processes is a crucial aspect of ozone-layer modeling.

Enormous computational resources would be required to compute the time-dependent ozone distribution in three dimensions. Although some progress has been made in modeling certain aspects of long-lived tracer transport in three dimensions using stratospheric general circulation models (see Chapter 11), most photochemical modeling of the ozone layer to date has been limited to either two-dimensional (latitude–height) models that average all variables in longitude or one-dimensional models that implicitly average all variables in both longitude and latitude.

The two-dimensional modeling approach is attractive for photochemical modeling because the vertical and meridional variations in species concentrations and photochemical reaction rates generally far exceed longitudinal variations. As emphasized in Section 9.3, the distribution of species in the meridional plane is not in itself determined solely by mean advection.

In two-dimensional models it is necessary to parameterize the chemical and dynamical effects of eddy motions in terms of the mean fields.

In the past, most photochemical models have been one-dimensional models in which only the global average tracer concentration can be predicted at each height. The "transport" in such models is represented entirely by vertical eddy diffusion. A rational basis for determining such one-dimensional diffusion coefficients is given in Section 9.7.3. It is first advantageous, however, to analyze the formulation of two-dimensional models.

9.7.1 Two-Dimensional Models

In Section 9.4 we discussed the formulation of two-dimensional models for tracer transport. Such models can, for example, be based on either the traditional Eulerian mean formulation, the TEM formulation, or an isentropic coordinate formulation. In each case there are certain eddy flux terms that must be parameterized and the mean flow must be specified or computed.

A few of the two-dimensional models in current use include the full set of dynamics equations in zonally averaged form. In these models the mean meridional circulation field (either the traditional Eulerian or the TEM version) is calculated from the dynamical equations for use in the tracer continuity equation. In such cases it is necessary to parameterize the eddy heat and momentum (or potential vorticity) fluxes in addition to the eddy tracer flux. Only if these fluxes are accurately represented can the mean zonal flow and mean meridional wind fields be expected to approximate reality. A simpler approach, which has been used in most two-dimensional photochemical models, is simply to externally specify the mean meridional circulation.

The earliest two-dimensional stratospheric transport models were based on the Reed and German (1965) eddy-flux parameterization, which essentially followed the Prandtl mixing-length hypothesis used in small-scale turbulence modeling. In this model it is assumed that

$$-\overline{\mathbf{u}'\chi'} = \mathbf{K}^{(r)} \cdot \nabla \bar{\chi},$$

where

$$\mathbf{K}^{(r)} \equiv \begin{bmatrix} K_{yy}^{(r)} & K_{yz}^{(r)} \\ K_{yz}^{(r)} & K_{zz}^{(r)} \end{bmatrix}.$$

Thus the entire effect of the eddies is incorporated in the symmetric diffusion tensor $\mathbf{K}^{(r)}$. From Eq. (9.4.3), this leads to an Eulerian mean transport equation of the form

$$\bar{\chi}_t + \bar{v}\bar{\chi}_y + \bar{w}\bar{\chi}_z = \bar{S} + \rho_0^{-1}\nabla \cdot (\rho_0 \mathbf{K}^{(r)} \cdot \nabla \bar{\chi}), \qquad (9.7.1)$$

and the elements of the symmetric diffusion tensor $\mathbf{K}^{(r)}$ are empirically determined to fit observed tracer distributions.

The symmetric tensor $\mathbf{K}^{(r)}$ can be diagonalized by a coordinate transformation to principal axes orientated at an angle ε to the yz axes given by

$$\varepsilon = \frac{1}{2}\tan^{-1}\left[\frac{K_{yz}^{(r)}}{\frac{1}{2}(K_{yy}^{(r)} - K_{zz}^{(r)})}\right] \simeq \tan^{-1}\left[\frac{K_{yz}^{(r)}}{K_{yy}^{(r)}}\right], \qquad (9.7.2)$$

where in the last expression we have assumed that $K_{zz}^{(r)} \ll K_{yy}^{(r)}$ and $K_{yz}^{(r)} \ll K_{yy}^{(r)}$. The symmetric tensor thus represents diffusion primarily along a sloping "mixing surface" at an angle ε to the y axis.

The Reed and German model of Eq. (9.7.1) neglects the eddy-flux contribution to the advection. This contribution, represented by the stream function Ψ defined in Eq. (9.4.9b), may be thought of as the antisymmetric component of an eddy "diffusion" tensor,

$$\mathbf{K} \equiv \mathbf{K}^{(s)} + \begin{bmatrix} 0 & \Psi \\ -\Psi & 0 \end{bmatrix}. \qquad (9.7.3)$$

In terms of $\mathbf{K}$ the "effective transport velocity" formulation of Eq. (9.4.7) can be rewritten as

$$\bar{\chi}_t + \bar{v}\bar{\chi}_y + \bar{w}\bar{\chi}_z = \bar{S}^{\dagger} + \rho_0^{-1}\mathbf{\nabla} \cdot (\rho_0\mathbf{K} \cdot \mathbf{\nabla}\bar{\chi}); \qquad (9.7.4)$$

comparing Eq. (9.7.1) with Eq. (9.7.4), we see that a conventional two-dimensional Eulerian model should have an *asymmetric* diffusion tensor and an eddy-modified source if the mean meridional velocity is taken to be the observed Eulerian velocity ($\bar{v}, \bar{w}$). In models based on the Reed and German (1965) theory, only the symmetric components of the diffusion tensor are included; it is then necessary to empirically specify the distribution of K_{yz} to be large and negative (in the Northern Hemisphere) so that the angle ε in Eqs. (9.7.2) slopes downward steeply toward the pole. This is not surprising, since in such a model the "diffusion" must represent both the dispersive and advective effects of the wave fluxes.

In principle it would be possible to use the "correct" form of the effective transport formulation [Eq. (9.4.7)], which includes the advective effects of the eddies and the portion of the diffusion that depends on the photochemical damping rate. However, $\mathbf{K}^{(s)}$ and Ψ, which involve parcel displacements, are difficult to estimate from observed data, although as mentioned in Section 9.5, Plumb and Mahlman (1987) calculated them for the lower and middle stratosphere from a three-dimensional model. Instead, most recent Eulerian-mean models have tended to use the TEM formulation [Eq. (9.4.13)] with the residual mean ($\bar{v}^*, \bar{w}^*$) approximated by the diabatic circulation, and a diffusion tensor on the right-hand side that is again determined empirically. With this type of formulation it is not necessary to specify a large K_{yz} component of the diffusion tensor, since the poleward-downward transport is provided by the advection due to the diabatic circulation.

9.7.2 Isentropic Two-Dimensional Models

The isentropic coordinate approach discussed in Section 9.4.7 has both conceptual and practical advantages. Since the large-scale motions that are the dominant eddies in the stratosphere are quasi-adiabatic, reasonable accuracy in modeling stratospheric tracer distributions is possible using the simplified version of the zonally averaged isentropic transport equation, Eq. (9.4.30). The main assumption is that the "eddy diffusion" formalism in Eq. (9.4.30) can still be used to represent the irreversible mixing of tracers by large-amplitude eddies. It was mentioned in Section 9.5 that three-dimensional model experiments have given some justification for this. In such a case the relations $K_{yy} = \frac{1}{2}(\eta'^2)_t$ and $\hat{K}_{yy} = A\eta'^2$ can no longer be expected to hold, and $K^{(tot)}$ has to be determined empirically or with a numerical model. Its value depends on the chemical relaxation rate, and hence separate $K^{(tot)}$ distributions should be used for each tracer diffused. However, including this dependence and the time-differentiated source-term modification in Eq. (9.4.30) does not seem to be essential to modeling the overall global distribution of long-lived tracers.

The other main component of the model is the advection by the isentropic-coordinate residual circulation $(\bar{v}^*, \bar{Q}^*)$. For adiabatic eddies, $Q' = 0$ and $\bar{Q}^*$ reduces to $\bar{Q}$ [see Eq. (3.9.5)], which can be calculated from *observed* temperatures. The resulting residual circulation is thus essentially the observed diabatic circulation (see Sections 7.2 and 9.3.1). It should be noted that since the diabatic or residual circulation is primarily eddy-driven (see Chapter 7), it should not, in principle, be specified independently of the eddy diffusion coefficient K_{yy}.

A simple model for the stream function of the diabatic circulation for both solstice and equinox conditions is shown in Fig. 9.15. The diabatic circulation is dominated below 25 km by a two-cell pattern, with rising in the equatorial region, poleward motion in both hemispheres and sinking at extratropical latitudes. Above 30 km at the solstices the diabatic circulation is dominated by a single cell, with rising in the summer hemisphere, a cross-equatorial drift, and sinking in the winter hemisphere (cf. the schematic diagram of Fig. 7.2). Clearly, in the annual mean the overall pattern should resemble the equinoctial distribution of Fig. 9.15, and the mean tracer isolines for a tracer in equilibrium with the time-mean diabatic

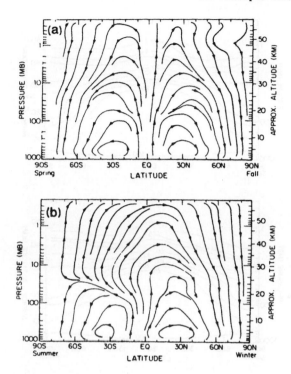

Fig. 9.15. Streamlines of a theoretical estimate of the diabatic circulation for (a) equinox and (b) solstice. [After Ko *et al.* (1985).]

circulation should thus slope downward toward the poles much more steeply than is observed for long-lived tracers such as N_2O and CH_4 (see Fig. 9.3). This suggests, as has been stressed earlier, that quasi-isentropic transport by eddies is essential to the modeling of observed tracer distributions.

The isentropic form of the two-dimensional transport equation [Eq. (9.4.27)] clearly exhibits the balance among the three key processes that determine the distribution of tracers relative to the isentropes in the meridional plane: advection by the diabatic circulation, isentropic transport by large-scale eddies, and chemical source and sinks.

Once the diabatic circulation is specified, it is still necessary, of course, to determine suitable values for K_{yy}. As noted above, the diffusion rate should in principle be related to the diabatic circulation; in particular it should depend on latitude, height, and season. However, Ko *et al.* (1985) have found that surprisingly good simulations of the N_2O distribution can be obtained by employing a constant value of $K_{yy} = 1 \times 10^5 \, \text{m}^2 \, \text{s}^{-1}$. The equinox and solstice distributions shown in Fig. 9.16 are in general accord with the observations shown in Fig. 9.3b. Such a model, of course, does

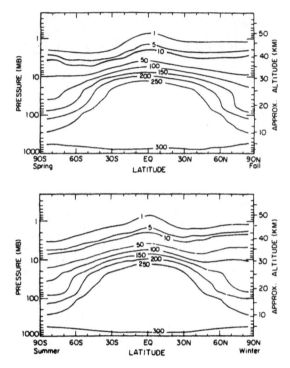

Fig. 9.16. Latitude–height distribution of N_2O mixing ratio in ppbv computed with a two-dimensional model using the diabatic circulation of Fig. 9.15 and a horizontal diffusion coefficient of $K_{yy} = 10^5 \text{ m}^2 \text{ s}^{-1}$. [After Ko *et al.* (1985).]

not reproduce all of the seasonal structure shown in the observations, and clearly should not be expected to do so since the strong mixing of the "surf zone" is not represented.

A minor disadvantage of isentropic coordinates is that the isentropes move up and down in physical space as temperatures fluctuate (e.g., with the annual cycle). Thus, the distribution of a trace species with respect to latitude and physical height may vary in time even if it remains steady in isentropic space.

9.7.3   One-Dimensional Transport Modeling

One-dimensional models that treat a vertical column of the atmosphere have been the most popular type of model for simulating photochemical cycles in the middle atmosphere. Such models by necessity can only represent transport processes in a very crude form. Formally, a one-dimensional model can be obtained by taking a global average in the horizontal coordinates:

$$\langle\cdots\rangle = \frac{1}{4\pi}\int_0^{2\pi}\int_{-\pi/2}^{\pi/2}(\cdots)\cos\phi\,d\phi\,d\lambda.$$

When this global averaging operation is applied to the spherical coordinate version of Eq. (9.4.18), so that the averaging is done on an isentropic surface, the resulting equation has the form

$$\frac{\partial}{\partial t}\langle\sigma\chi\rangle + \frac{\partial}{\partial\theta}\langle\sigma Q\chi\rangle = \langle\sigma S\rangle. \tag{9.7.5}$$

If the diabatic heating rate is then approximated by the zonal mean, $\bar{Q}^*$, we obtain

$$\frac{\partial}{\partial t}\langle\chi\rangle + \frac{1}{\sigma_0}\frac{\partial}{\partial\theta}\langle\sigma_0\bar{Q}^*\bar{\chi}\rangle = \langle S\rangle, \tag{9.7.6}$$

where we have also let $\bar{\sigma} = \sigma_0(\theta)$ (i.e., we define a global standard "density" that depends only on the vertical coordinate). We assume that there is an approximate balance among the last three terms in Eq. (9.4.27), and that the diffusion and chemical sinks can be parameterized in terms of linear damping at rates of τ_d^{-1} and τ_c^{-1}, respectively. Then we have

$$\bar{Q}^*\,\partial\langle\chi\rangle/\partial\theta \simeq -\bar{\chi}(\tau_d^{-1} + \tau_c^{-1}), \tag{9.7.7}$$

where we have also assumed that $\bar{\chi} \approx \langle\chi\rangle$ in the vertical advection term. Solving Eq. (9.7.7) for $\bar{\chi}$, and substituting into Eq. (9.7.6) to eliminate $\bar{\chi}$ in the vertical flux term, yields

$$\frac{\partial}{\partial t}\langle\chi\rangle - \frac{1}{\sigma_0}\frac{\partial}{\partial\theta}\left[\sigma_0\hat{K}_{zz}\frac{\partial\langle\chi\rangle}{\partial\theta}\right] = \langle S\rangle, \tag{9.7.8}$$

where

$$\hat{K}_{zz} \equiv \langle(\bar{Q}^*)^2(\tau_d^{-1} + \tau_c^{-1})^{-1}\rangle. \tag{9.7.9}$$

Usually $\hat{K}_{zz}$ has been empirically determined to fit observed tracer profiles. But Eq. (9.7.9) shows that if $\bar{Q}^*$ is known and τ_d and τ_c can be determined, then the vertical transport coefficient is known. Unlike most proposed one-dimensional eddy-diffusion parameterizations, the form derived here explicitly incorporates the dependence of transport on the chemical time-scale for the individual tracer. Tracers with short stratospheric lifetimes

have smaller $\hat{K}$ values than do those with longer lifetimes. Some specific
examples are given in Fig. 9.17.

9.7.4 Mechanistic Three-Dimensional Transport Modeling

In discussions of three dimensional models it is useful to distinguish
between complete general circulation models (GCMs) and so-called
"mechanistic" models. The former attempt to simulate the complete large-
scale circulation, and they feature rather detailed representations of the
various physical processes that drive the flow. The latter attempt to focus
on certain aspects of the circulation by deliberately simplifying the dynamical or physical formulation. For example, some mechanistic models of the
middle atmosphere do not attempt to include self-consistent calculations
of the tropospheric circulation, but merely impose a specified geopotential
height variation at a level near the tropopause. Other mechanistic models
simplify the radiative heating and cooling by using a Newtonian cooling
formulation. Dynamical simplifications may include use of the quasi-geostrophic formulation, or severely truncating the zonal wave-number spec-

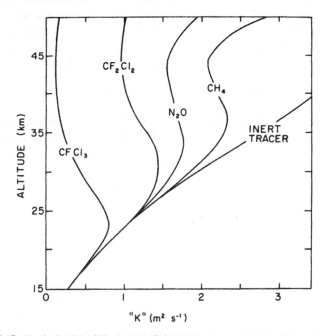

Fig. 9.17. Vertical eddy-diffusion coefficients for the trace gases whose stratospheric lifetimes are given in Fig. 9.2, based on the one-dimensional formulation of Eq. (9.7.9) with a
diabatic circulation similar to the equinoctial circulation of Fig. 9.15. [After Holton (1986a).]

trum to include only one or two zonal harmonic waves interacting with the zonal flow. Examples of such models as applied to purely dynamical problems appeared in some earlier chapters. In this section we consider mechanistic models applicable to various aspects of the three-dimensional transport problem; general circulation models will be discussed in Chapter 11.

One of the most enlightening mechanistic models of transport is the simplified general circulation model of Kida (1983). This model is based on the primitive equations. It includes only a single hemisphere with a wall at the equator, and has a grid resolution of 3° longitude by 2.5° latitude. There are 12 layers in the vertical, extending from the surface to the 1-mb level. Although in terms of resolution this model is similar to a typical GCM, it is here regarded as a mechanistic model because the representation of physical processes in the troposphere and the stratosphere is greatly simplified. For example, diabatic heating is given by the sum of a specified term plus a Newtonian relaxation to a specified temperature profile. There is no topography, so planetary waves are excited only by land–sea thermal contrasts.

Despite these simplifications, Kida's model does a reasonably good job of representing the Northern-Hemisphere winter circulation, including the development of upper-tropospheric baroclinic eddies. Kida used this model to analyze the long-term evolution of the position of air parcels initially located along a narrow band confined between latitudes 2.5 and 7.5°N at the 100-mb level (just above the tropical tropopause). Thus, in his experiment a three-dimensional Eulerian model is used to compute Lagrangian parcel trajectories. The long-term advection and dispersion of the initial "ribbon" of parcels is shown in the form of the parcel positions projected on the meridional plane in Fig. 9.18. The figure confirms the conceptual model of transport in the meridional plane that was introduced in Section 9.3. The parcels tend to rise in the tropics and sink at extratropical latitudes due to the diabatic circulation. But superposed on the slow diabatic drift there is a quasi-horizontal meridional dispersion that is very rapid just above the tropopause. Also indicated clearly is the preferred region of return flow from stratosphere to troposphere associated with the tropospheric subtropical jetstream. Although the model cannot resolve the actual scale of upper-level fronts and tropopause folds, it nevertheless shows that such processes are at least crudely reflected in the parcel trajectories. The overall meridional dispersion rate calculated in this model is probably an underestimate of the actual rate in the "surf zone" of the winter hemisphere because the planetary wave activity is less than observed. However, the model does provide vivid confirmation of our overall view of transport in the meridional plane.

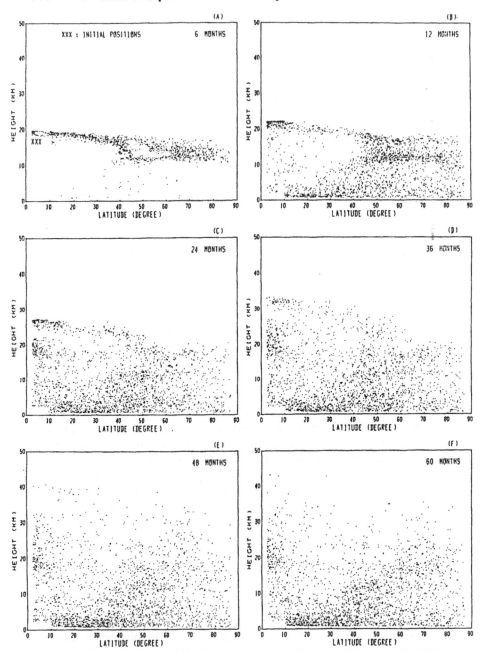

Fig. 9.18. The long-term evolution of the positions as projected on a meridional plane for air parcels originally located along a zonal band as marked by the XXX in frame (A). [After Kida (1983).]

Rapid meridional transport by planetary waves has been studied with the aid of mechanistic models that are much cruder than Kida's model. Typically such models resolve only one or two zonal harmonic waves in longitude, but retain fairly high resolution in the latitudinal and vertical direction. These models can to some extent simulate the meridional advection and zonal stretching of "tongues" of tracer associated with planetary wave breaking, of the sort pictured in Figs. 5.6 and 9.6 for the examples of potential vorticity and methane mixing ratio. However, the nonlinear cascade of tracer variance towards higher zonal wave numbers obviously cannot be simulated by such models. Despite this important defect, the Lagrangian parcel motions in such models during the evolution of simulated sudden stratospheric warmings are quite instructive.

Figure 9.19 shows an example for a simulated wave-number 1 sudden warming from the model of Hsu (1980) (see Section 6.3.2). As the anticyclonic disturbance amplifies, the air parcels that are initially evenly distributed around the 30°N latitude circle are drawn poleward and eastward around the northern flank of the high, and equatorward and westward around the southern flank. As the warming develops the parcels are advected in a narrow tongue across the polar region and are gradually "wrapped up" around the anticyclone. When viewed in the meridional plane (Fig. 9.20), the parcel evolution indicates a rapid meridional dispersion and a pronounced downward advection in the polar region. The latter is not due to the diabatic circulation (radiative effects are weak in this model), but rather represents the adiabatic downward displacement of isentropic surfaces in the polar region as a result of the sudden warming. The large vertical range of parcels in high latitudes reflects the strong longitudinal variation in vertical velocity (and temperature) associated with the wave disturbance. (Figures 9.19 and 9.20 should be compared with Figs. 6.10 and 6.11, which are for a wave-number 2 warming. The latter also show polar descent and meridional dispersion, but focus more on the splitting of the vortex and the high-latitude behavior of air parcels.)

This model has also been used to simulate ozone transport during a sudden stratospheric warming (Jou, 1985). For this purpose a very simple linear photochemistry was employed and the model-derived velocity fields were used to solve for the evolution of the ozone mixing ratio using the three-dimensional ozone continuity equation, but retaining the same zonal truncation—the zonal mean plus a single wave-number 1 disturbance. The model, despite its severe zonal truncation, was able to simulate the poleward and eastward advection of ozone-rich tongues of air, as was observed during the 1979 warming (see Section 9.5). These results suggest that sudden warmings may contribute a rather large fraction of the total poleward tracer transport during the winter season, and hence may introduce subseasonal

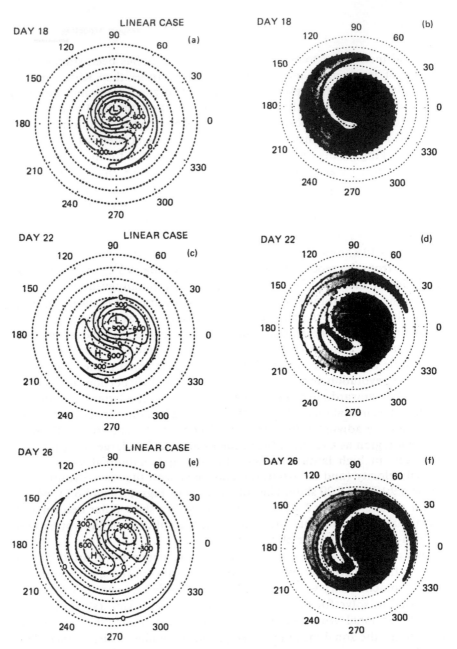

Fig. 9.19. Parcel motions during a simulated sudden stratospheric warming in a quasi-linear model in which a forced wave-number 1 disturbance interacts with the zonal flow. The left-hand frames show the evolution of the geopotential field at the log-pressure altitude $z = 30$ km. The right side shows the projection on the $z = 30$ km surface of the positions of a set of marked parcels that were initially spaced uniformly along the 30°N latitude circle. Air that was originally poleward of 30°N is shown by shading. [After Hsu (1981). American Meteorological Society.]

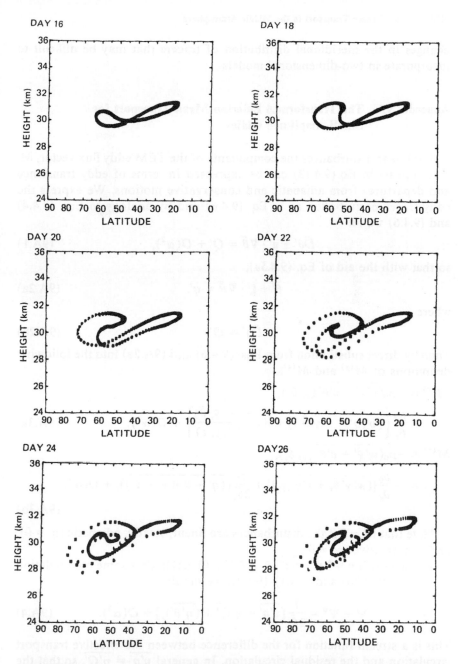

Fig. 9.20. As in the right-hand panels of Fig. 9.19, but parcel projections onto the meridional plane. Note the poleward and downward drift associated with the downward displacement of the isentropes during the sudden warming. [After Hsu (1981). American Meteorological Society.]

changes in the meridional distribution of tracers that may be difficult to incorporate in two-dimensional models.

Appendix 9A   The Transformed Eulerian-Mean Transport for Small-Amplitude Eddies

For linear disturbances the components of the TEM eddy flux vector, $\mathbf{M}$, that appears in Eq.(9.4.13) can be expressed in terms of eddy transience and departures from adiabatic and conservative motions. We express the eddy tracer perturbation as in Eq. (9.4.6) and by analogy to Eqs. (9.4.4) and (9.4.6) we let

$$\bar{D}\theta' + \mathbf{u}' \cdot \nabla \bar{\theta} = Q' + O(\alpha^2), \qquad (9A.1)$$

so that with the aid of Eq. (9.4.5a),

$$\theta' + \boldsymbol{\xi}' \cdot \nabla \bar{\theta} = q'. \qquad (9A.2a)$$

where

$$\bar{D}q' \equiv Q'. \qquad (9A.2b)$$

Then by direct substitution from Eqs. (9.4.6) and (9A.2a) into the following definitions of $M^{(y)}$ and $M^{(z)}$:

$$M^{(y)} \equiv -\rho_0(\overline{v'\chi'} - \overline{v'\theta'}\,\bar{\chi}_z/\bar{\theta}_z),$$

$$= \frac{\rho_0}{\bar{\theta}_z}\left\{\overline{[v'(q'\bar{\chi}_z - \gamma'\bar{\theta}_z)]} + \tfrac{1}{2}\overline{(\eta'^2)}_t \frac{\partial(\bar{\chi}, \bar{\theta})}{\partial(y, z)}\right\} + O(\alpha^3); \qquad (9A.3a)$$

$$M^{(z)} \equiv -\rho_0(\overline{w'\chi'} + \overline{v'\theta'}\,\bar{\chi}_y/\bar{\theta}_z),$$

$$= -\frac{\rho_0}{\bar{\theta}_z}\{(\overline{w'\gamma'}\,\bar{\theta}_z + \overline{v'q'}\,\bar{\chi}_y)\} + \frac{\rho_0}{2\bar{\theta}_z}\overline{[(q' - \theta')(\gamma' - \chi')]}_t + O(\alpha^3). \qquad (9A.3b)$$

Note that $\mathbf{M} = 0$ if the disturbances are linear, steady, adiabatic ($q' = 0$), and conservative ($\gamma' = 0$).

We now define $\Psi^* \equiv -\overline{v'\theta'}/\bar{\theta}_z$ [cf. Eq. (3.5.1)]; after some manipulation using Eqs. (9.4.9b) and (9A.1)–(9A.2), we obtain

$$\Psi - \Psi^* = \frac{1}{2\bar{\theta}_z}[\overline{v'q'} - \overline{\eta'Q'} + \overline{(\eta'\theta')}_t] + O(\alpha^3). \qquad (9A.4)$$

This is a stream function for the difference between the effective transport circulation and the residual circulation. In general $\overline{v'q'} \neq \overline{\eta'Q'}$, so that the residual and the effective transport circulations are equal only when the waves are steady, linear, and adiabatic. The equality of the circulations does not, however, require that the tracer be conservative ($\gamma' \neq 0$, $S' \neq 0$).

References

9.1. The history of transport ideas is reviewed by Mahlman *et al.* (1984), who give extensive references to original literature. The use of isentropic potential vorticity maps is discussed thoroughly by Hoskins *et al.* (1985).

9.2. The chemistry of trace gases in the middle atmosphere is lucidly discussed by Brasseur and Solomon (1984), who provide much useful information on chemical timescales.

9.3. An interpretation of the Brewer–Dobson model in terms of the GLM theory is given by Dunkerton (1978).

9.4. The generalized Lagrangian mean (GLM) theory is discussed in the context of tracer transport by McIntyre (1980b). A lucid account of the theoretical basis for two-dimensional transport modeling is given by Tung (1982).

9.5. McIntyre and Palmer (1984) discuss the concept of "planetary wave breaking" and its role in irreversible transport. These ideas are applied to observational data by Leovy *et al.* (1985).

9.6. Stratosphere-troposphere exchange is reviewed by Reiter (1975). The problem of tropical exchange is reviewed by Holton (1984b). Implications of the *Nimbus* 7 LIMS water-vapor measurements for exchange are discussed by Remsberg *et al.* (1984 b).

9.7. One of the earliest two-dimensional transport models for stratospheric tracers was that of Reed and German (1965). Two-dimensional models that include the complete dynamical equations are discussed by Harwood and Pyle (1975) and Garcia and Solomon (1983). Contributions to the formulation of the eddy diffusion tensor include Plumb (1979), Matsuno (1980), Danielsen (1981), and Holton (1981). For more detailed treatment of the one-dimensional modeling approach described in Section 9.7.3, see Holton (1986a).

Chapter 10 | The Ozone Layer

10.1 Introduction

It has been known for more than 50 years that ozone in the atmosphere is largely confined to the stratosphere, with the maximum molecular concentration occurring near 22 km and the maximum mixing ratio near 35 km (see Figs. 1.6 and 1.7). Ozone, by its absorption of solar radiation of wavelengths less than 300 nm, provides the heat source that is responsible for the global mean increase of temperature between the tropopause and the stratopause; in the absence of ozone there would be no stratosphere! The ozone layer, through its absorption of harmful ultraviolet radiation, is also essential for the health of plant and animal life. Thus, study of the formation, maintenance, and stability of the ozone layer is a critical aspect of middle atmosphere science.

Chapman (1930) formulated the first plausible model for the existence and vertical structure of the ozone layer. The Chapman mechanism begins with the photolysis of O_2 by solar ultraviolet radiation of wavelengths less than 250 nm (see Section 2.3.1). The atomic oxygen thus produced combines with ground-state molecular oxygen to form ozone. The amount of short-wave radiation available for photodissociation of O_2 decreases downward from the top of the atmosphere, with the most rapid rate of decrease occurring at high latitudes where the optical path is long [see Eq. (2.2.16)]. Since the amount of O_2 available for reaction with atomic oxygen decreases exponentially away from the earth's surface, the production of ozone must be a maximum in a layer (sometimes called the Chapman layer; see Section 2.7) centered in low latitudes at some intermediate altitude. Chapman proposed that photochemical production of ozone was balanced by destruction through the reaction of ozone with atomic oxygen. His scheme is now known to be a very incomplete model of ozone layer photochemistry (see Section 10.4). Nevertheless, it provides a useful conceptual model.

Although the existence of ozone depends on photochemical reactions, its distribution in the atmosphere can only be explained by taking account of the role of atmospheric motions in transporting ozone from its source region in the tropical upper stratosphere to the high-latitude lower stratosphere where it behaves as a quasi-conservative tracer. The motions responsible for this transport are strongly influenced by the heating distribution associated with solar absorption in the ozone layer. Thus, it is necessary to consider the three-way interactions among chemistry, radiation, and dynamics in order to understand the distribution and temporal variations of ozone.

10.2   The Climatology of Ozone

There is a long history of ozone observations from ground-based instruments, balloons, rockets, and satellites. Ground-based measurements make use of the fact that the absorption of solar radiation by ozone is highly wavelength-dependent. Thus, comparison of the atmospheric absorption in several wavelength regions makes it possible to deduce the total column of ozone. Similar measurements can be made from satellites by observing the ultraviolet radiation backscattered from the surface and clouds. Ozone also absorbs in the infrared, so that total ozone can be deduced from space by measuring the total extinction of upwelling thermal infrared radiation in appropriate wavelength bands.

10.2.1   Total Ozone

The total column abundance of ozone determines the amount of radiation that can reach the surface of the earth in the near-ultraviolet region (290–310 nm), and hence the potential for damage to the biosphere. Total ozone is usually expressed in terms of the equivalent thickness of the ozone layer at standard temperature and pressure (STP; 0°C, 1013.25 mb). This thickness is about 3 mm (STP) for a column containing the global mean amount of ozone (8×10^{22} molecules m^{-2}). Maps of the distribution of total ozone often have the column abundance expressed in terms of Dobson units (DU), where 1 DU $= 10^{-5}$ m (STP). Thus, the global mean total ozone is about 300 DU.

A 4-year average total ozone map based on data from the TOMS satellite backscatter ultraviolet instrument on *Nimbus* 7 is shown in Fig. 10.1. There is a general increase of total ozone from equator to high latitudes in both hemispheres, but with larger amounts in the Northern Hemisphere. Although longitudinal variations are fairly small in this long-term average, there are on average distinct maxima in midlatitudes over the east coasts of Asia and North America and over Eastern Europe during winter, corresponding to the mean positions of the troughs in the stationary planetary-wave pattern of the upper troposphere.

The seasonal variation of the zonal-mean total ozone field derived from the TOMS data is shown in Fig. 10.2. There is a large-amplitude annual cycle in the extratropical regions of both hemispheres, but with a substantially larger range in the Northern Hemisphere. Maxima occur just after the spring equinox near 90°N and 60°S; a minimum occurs near 90°S (the Antarctic ozone hole).

Although the time mean and seasonal variability of the total ozone field primarily depend on latitude, longitudinal variations in total ozone can be quite large on individual days. Since most of the ozone molecules in a column are confined to the lower stratosphere, the synoptic pattern of total ozone reflects the variations in tropopause height associated with tropospheric disturbances. In particular, total ozone tends to be highest on the cyclonic side of upper level jetstreams (see Section 9.6.1) and in the region of isolated cyclonic vortices (cutoff lows). Thus, to some extent total ozone is a tracer for large-scale meteorological processes in the upper troposphere.

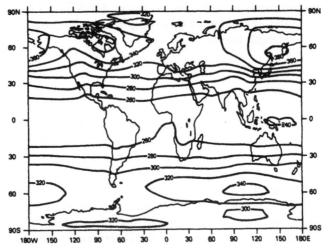

Fig. 10.1. Global distribution of total ozone (Dobson units) based on 4 years of observation with the TOMS instrument on the *Nimbus* 7 satellite. [After Bowman and Krueger (1985).]

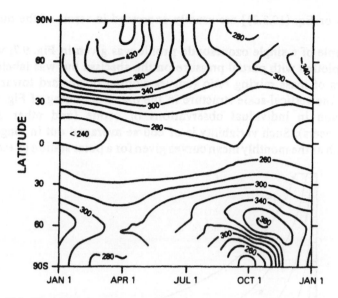

Fig. 10.2. Time–latitude section showing the seasonal variation of total ozone (Dobson units) based on TOMS data. Note the springtime maxima near 90°N and 60°S and the minimum near 90°S. [After Bowman and Krueger (1985).]

10.2.2   The Vertical Distribution

The vertical distribution of ozone in the troposphere and stratosphere up to about the 10-mb level has been observed *in situ* by balloon-borne ozonesondes. Above the 10-mb level, limited *in situ* measurements are available from rockets. These balloon and rocket data have been used to establish the standard profiles shown in Figs. 1.6 and 1.7. Recently satellite observations on *Nimbus* 7 and the Solar Mesosphere Explorer (SME) satellites have provided global measurements of the ozone distribution in both the stratosphere and the mesosphere.

Before discussing the observed distribution, it is useful to consider the various conventions used in presenting ozone data. Ozonesonde profile data are usually plotted in terms of the partial pressure of ozone in nanobars (nb). The mass mixing ratio and the volume mixing ratio given by Eq. (1.3.3) are also in common use. For ozone these are expressed in parts per million by mass (ppmm) and parts per million by volume (ppmv), respectively. Satellite data are usually expressed in ppmv. Recall from Eq. (1.3.3) that the volume mixing ratio is equal to the partial pressure of ozone divided by the air pressure. Also, note that the volume mixing ratio for ozone is about 0.6 times the mass mixing ratio. By contrast, chemical equations [e.g.,

Eq. (10.3.2) or Eq. (10.3.3)] are usually expressed in terms of the number density.

An example of a single ozonesonde profile was given in Fig. 9.7, where ozone was plotted with partial pressure on the abscissa but was labeled by the contours of mass mixing ratio (curves sloping downward toward the right). The fine vertical-scale structure in the ozone sounding of Fig. 9.7 is quite common in individual observations of ozone (and other quasi-conserved tracers). Such variability is of course averaged out in long-term averages such as the monthly mean curves given for a polar station (Resolute,

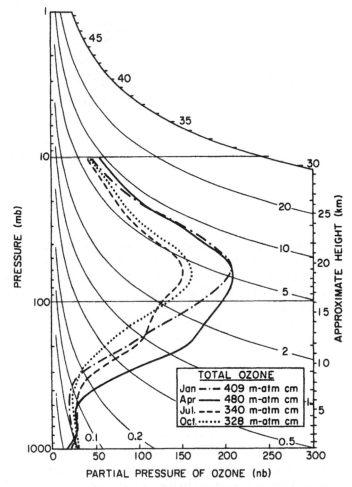

Fig. 10.3. Vertical profiles of ozone for four seasons at a polar station (Resolute, 75°N). Lines curving downward toward the right are isolines of constant mixing ratio (ppm). [Courtesy of Professor J. London. Adapted from London (1985).]

75°N) in Fig. 10.3. As this figure shows, the spring maximum in total ozone that was discussed earlier is largely accounted for by the seasonal variation of the ozone concentration in the 50- to 100-mb height range (where the maximum concentration lies); above 50 mb the mixing ratio actually decreases slightly from January to April, while below 300 mb there is a small decrease from October to January.

Satellite observations over the past 20 years have provided a global climatology of ozone mixing ratio in the stratosphere and mesosphere. In most cases the satellite data cannot provide accurate values below the ozone peak (30 mb), so that information from ozonesondes and total ozone measurements must be used to provide complete vertical distributions.

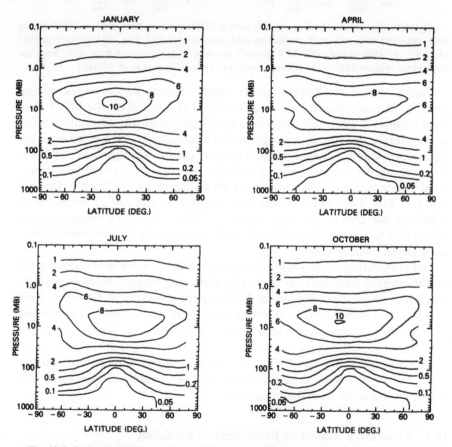

Fig. 10.4. Latitude–height sections of the ozone mixing ratio (ppmv) for January, April, July, and October 1979, as observed by the *Nimbus* 7 SBUV experiment. [After McPeters *et al.* (1984).]

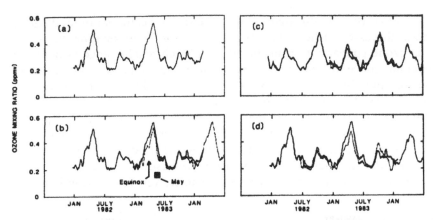

Fig. 10.5. Variation of the ozone mixing ratio at 0.01 mb during 1982–1983: (a) 45°N during 1982–1983; (b) 45°N with 1982 overlaid on 1983 (dotted); (c) 45°S with 1982 overlaid on 1983 (dotted); (d) 45°N (solid) with 45°S (dotted) overlaid with a 6-month time shift. Note the remarkable repeatability in the seasonal variation. [After Thomas *et al.* (1984).]

Figure 10.4 shows zonal-monthly mean sections of the ozone mixing ratio for January, April, July, and October 1979, based on the SBUV data from *Nimbus* 7. These indicate that in the photochemically controlled region of the middle and upper stratosphere the mixing ratio is a maximum in the equatorial region, but that in the transport-controlled region of the lower stratosphere the constant-mixing-ratio surfaces have poleward and downward slopes qualitatively consistent with transport by the Brewer–Dobson cell (cf. Fig. 9.4). These slopes are steepest in the spring hemisphere, reflecting the gradual buildup of ozone in the lower stratosphere during winter due to the rapid meridional transport and long photochemical timescale for the winter season.

In the extratropics the seasonal variation in the stratosphere is dominated by the annual harmonic; at most latitudes and heights there is a single maximum occurring in the spring. In the upper mesosphere, however, there is a dramatic semiannual cycle with a large peak in the spring, a secondary peak in the autumn, and minima at the solstices (Fig. 10.5). The ozone concentration at the spring equinox is more than twice that at the solstices. Such an extraordinary variability cannot be accounted for by photochemistry alone; it must reflect seasonal variability in dynamical processes that influence transport near the mesopause.

10.3   Elementary Aspects of Photochemical Modeling

The ozone climatology documented in the previous section cannot be understood without considering photochemical production and loss processes, as well as the redistribution of ozone (and other trace gases involved in ozone chemistry) by atmospheric motions on all scales. The relative importance of photochemical and dynamical processes is highly dependent on altitude, latitude, and season. In this section we first indicate how the concept of chemical lifetime introduced in Section 9.2 can be employed to simplify the analysis of the behavior of ozone and other chemically active trace species. We then discuss the Chapman mechanism for formation of the ozone layer, which provides a simple illustration of the use of lifetime information, and hence provides a background for the modern theory of ozone photochemistry, which is briefly discussed in Section 10.4.

10.3.1 Some Scaling Considerations

The flux form of the tracer continuity equation for a chemically active tracer of volume mixing ratio χ can be written (see Appendix 10A) in the form

$$\chi_t + \rho_0^{-1} \nabla \cdot (\rho_0 \mathbf{u} \chi) = D + \rho_0^{-1}(T/T_s) N_a^{-1} M(P - L). \quad (10.3.1)$$

Here D designates diffusion by unresolved scales of motion, M is the molecular weight of air, N_a is Avogadro's number, T is temperature, T_s is a reference temperature (see Section 1.1.1), and P and L are the photochemical production and loss rates, respectively.

The importance of photochemistry in determining the distribution of any trace species can be assessed by comparing the magnitude of the production and loss terms with that of the flux divergence due to transport by motions of all scales. The simplest example is a tracer that is produced at the ground and photodissociated in the atmosphere (e.g., $CFCl_3$). For such a tracer, P vanishes and $L = J_A A$, where J_A designates the photodissociation rate coefficient (s^{-1}), and A designates the tracer concentration (molecules m^{-3}). Thus,

$$\left(\frac{\partial A}{\partial t}\right)_{\text{chemical}} \equiv P - L = -J_A A. \quad (10.3.2)$$

and the e-folding time scale for chemical decay of species A is simply $\tau_A = J_A^{-1}$. The relative roles of chemistry and transport in this situation can be assessed by comparing τ_A with the relevant dynamical time scale τ_d. The latter is given by the smaller of L_A/U or H_A/W, where L_A and H_A designate the horizontal and vertical scales of variation for the tracer mixing ratio and U and W are characteristic horizontal and vertical velocities. The dynamical timescale defined in this manner can vary enormously depending on the spatial scale of the tracer mixing ratio variation. For example, in the lower stratosphere ozone has substantial variation on the synoptic scale ($L_A \approx 10^3$ km, $U \approx 10$ m s^{-1}), and the dynamical timescale for such variations is on the order of 1 day. On the other hand, for hemispheric-scale variations ($L_A \approx 10^4$ km) in the summer stratosphere (where eddy motions are very weak), the relevant velocity scale is the residual mean meridional velocity ($U \approx 10$ cm s^{-1}), and the dynamical timescale is in excess of 100 days. Thus, the relative importance of the dynamical and chemical terms in the tracer continuity equation depends crucially on the scale of variation considered.

In the case of hemispheric-scale transport, Fig. 9.2 indicates that for $CFCl_3$ $\tau_d \ll \tau_{CFCl_3}$ below 20 km, while $\tau_d \gg \tau_{CFCl_3}$ above about 30 km. Thus, since there is no chemical source for this trace gas in the stratosphere, the concentration decreases rapidly above 30 km. In the region 20 km $< z <$ 30 km, the distribution is determined by both transport and photochemical effects. Below 20 km, $CFCl_3$ is transported as a passive tracer.

An e-folding time scale for chemical decay can also be defined for molecules that are destroyed through bimolecular or termolecular reactions. Let A, B, and C stand for the concentrations of three different trace gases (molecules m^{-3}). If A and B react with reaction rate k_1 (m^3 molecule^{-1} s^{-1}) while A and C react in association with a third body of concentration M with reaction rate k_2 (m^6 molecule^{-2} s^{-1}), then neglecting production processes,

$$\left(\frac{\partial A}{\partial t}\right)_{\text{chemical}} = -k_1 AB - k_2 ACM \equiv -A\tau_A^{-1}, \qquad (10.3.3)$$

where $\tau_A \equiv (k_1 B + k_2 CM)^{-1}$, and it is here assumed that the concentrations B and C remain fixed on the timescale τ_A.

When the chemical time scale τ_A, defined on the basis of the e-folding rate for loss processes as in Eqs. (10.3.2) or (10.3.3), is short compared to the relevant dynamical timescale τ_d, then the loss must be approximately offset by photochemical production if the concentration A does not become negligible, in order that there be a balance in the continuity equation [Eq. (10.3.1)]. When the balance between production and loss is exact, the species is said to be in *photochemical equilibrium*. Such an equilibrium condition does not, of course, mean that transport is of no importance, since the concentrations of various species involved in the balance among production and loss processes may be strongly influenced by transport. As we shall see in the next subsection, at 30 km atomic oxygen, O, is in approximate photochemical equilibrium with O_3, but the concentration of total *odd oxygen*, defined as the sum of the concentrations of O and O_3 and designated by $O_x (O_x = O + O_3)$, is strongly influenced by transport processes.

10.3.2   The Chapman Mechanism

The Chapman reactions can be written in the following symbolic form:

$$(J_2) \qquad O_2 + h\nu \to 2O, \tag{10.3.4a}$$

$$(k_2) \qquad O + O_2 + M \to O_3 + M, \tag{10.3.4b}$$

$$(J_3) \qquad O_3 + h\nu \to O + O_2, \tag{10.3.4c}$$

$$(k_3) \qquad O + O_3 \to 2O_2. \tag{10.3.4d}$$

Here J_2 and J_3 are the rate coefficients for O_2 and O_3 photolysis, respectively; k_2 and k_3 are the reaction-rate coefficients for Eqs. (10.3.4b) and (10.3.4d). In the stratosphere, Eqs. (10.3.4b,c) are "fast" reactions, while Eqs. (10.3.4a,d) are 'slow." The net result of Eqs. (10.3.4a,b) is simply

$$3O_2 \to 2O_3.$$

Thus, these two reactions govern the production of O_3.

The reaction (10.3.4c) does not actually destroy O_3, since the atomic oxygen thus formed immediately reacts with O_2 to form ozone again, by Eq. (10.3.4b). This pair of fast reactions conserves O_x but interconverts O and O_3 and hence influences the partitioning of total odd oxygen between O and O_3: see Eq. (10.3.8).

In the original Chapman scheme, Eq. (10.3.4d) is the only reaction that serves to destroy ozone. It is now known that this reaction is much too slow to provide the rate of ozone destruction required to establish the observed ozone concentration in the stratosphere. A number of catalytic cycles involving "free radicals" (molecular fragments with unpaired electrons) of the nitrogen, chlorine, and hydrogen families are now believed to furnish the additional required sinks of O_3. These cycles are discussed in Section 10.4.

The net chemical tendencies for the O_3 and O concentrations as given by the Chapman reactions [Eqs. (10.3.4)] are

$$\left(\frac{\partial[O_3]}{\partial t}\right)_{\text{chemical}} = -J_3[O_3] - k_3[O][O_3] + k_2[O][O_2][M],$$

$$\tag{10.3.5a}$$

and

$$\left(\frac{\partial[O]}{\partial t}\right)_{\text{chemical}} = -k_3[O][O_3] - k_2[O][O_2][M] + J_3[O_3] + 2J_2[O_2],$$

$$\tag{10.3.5b}$$

where the concentrations are denoted by square brackets. At the 30-km level the timescales for chemical loss given by these reactions are (Brasseur and Solomon, 1984, pp. 205–206)

$$\tau_{O_3} = (J_3 + k_3[O])^{-1} = 2000 \quad s$$

and

$$\tau_O = (k_2[O_2][M] + k_3[O_3])^{-1} = 0.04 \quad s.$$

These are much shorter than the dynamical timescale, and can only be balanced by chemical production. Thus the net chemical sources, which determine the changes in mixing ratio following the motion, are given by very small differences between large terms of opposite sign. This is somewhat similar to the problem that arises in integration of the primitive equations in weather-prediction models. In that case the rate of change of momentum must be estimated from small differences between the pressure gradient force and the Coriolis force; small inaccuracies in either can produce huge errors in the computed tendency. As we saw in Chapter 3, this problem can be avoided in some situations by approximating the primitive equations by a prediction equation for a quasi-conserved quantity—quasi-geostrophic potential vorticity. Similarly, prediction of the evolution of the ozone field can be simplified by observing that the largest terms on the right of Eqs. (10.3.5) are those associated with the fast reactions [Eqs. (10.3.4b,c)], which represent cycling between O and O_3 but do not affect the total odd oxygen concentration. Thus, O_x should be more conservative than either O or O_3 alone.

Summing the two equations of Eqs. (10.3.5) we obtain

$$\left(\frac{\partial[O_x]}{\partial t}\right)_{chemical} = -2k_3[O][O_3] + 2J_2[O_2]. \tag{10.3.6}$$

In the stratosphere $[O] \ll [O_3]$, so that $[O_x] = [O] + [O_3] \approx [O_3]$ to a very good approximation. Thus, Eq. (10.3.6) provides a close approximation to the net source or sink for O_3 as given by the Chapman mechanism. The chemical loss timescale in Eq. (10.3.6) is

$$\tau_{O_x} = (2k_3[O])^{-1} \approx \text{weeks at 30 km,}$$

which confirms that O_x is indeed far more conservative than O or O_3.

Noting that atomic oxygen has an extremely short chemical time constant, we can assume that [O] is in photochemical equilibrium with $[O_3]$ so that setting the left side to zero in Eq. (10.3.5b) gives

$$[O] = \frac{J_3[O_3] + 2J_2[O_2]}{k_3[O_3] + k_2[O_2][M]}. \tag{10.3.7}$$

In the stratosphere, $J_3[O_3] \gg J_2[O_2]$ and $k_2[O_2][M] \gg k_3[O_3]$, so that from Eq. (10.3.7),

$$[O]/[O_3] \approx J_3/(k_2[O_2][M]). \tag{10.3.8}$$

Thus, the $[O]/[O_3]$ ratio is closely determined by the fast reactions of Eqs. (10.3.4b,c).

In summary, for the Chapman chemistry, in the lower and middle stratosphere the ozone mixing ratio can be predicted from Eq. (10.3.1) with the chemical production and loss terms approximated by the right-hand side of Eq. (10.3.6), while the corresponding concentration of O can be computed at any instant from the equilibrium expression Eq. (10.3.7). In the upper stratosphere and mesosphere different balances hold (Brasseur and Solomon, 1984).

10.4 Photochemistry of Ozone: Catalytic Cycles

By substitution from Eq. (10.3.8), the loss rate for odd oxygen in Eq. (10.3.6) can be expressed in terms of the ozone concentration as

$$L = -2k_3[O_3]^2 J_3/(k_2[O_2][M]).$$

For observed ozone concentrations this rate is less than one-fourth of the ozone production rate in the stratosphere. The difference cannot be accounted for by transport processes; comparison of total global production and destruction rates indicates that ozone is formed 5 times faster than it is destroyed by the Chapman process.

The major loss of ozone is now known to occur through a number of catalytic reaction cycles, the simplest of which can be represented symbolically as follows:

$$X + O_3 \rightarrow XO + O_2 \tag{10.4.1a}$$

$$XO + O \rightarrow X + O_2 \tag{10.4.1b}$$

$$\text{Net:} \quad O + O_3 \rightarrow 2O_2$$

where X is a catalytic molecule that remains unchanged at the end of the cycle. The net effect of the catalytic cycles represented by Eqs. (10.4.1) is thus the same as that of Eq. (10.3.4d).

The efficacy of the cycle of Eqs. (10.4.1) is dependent on the speed of the slower of the two reactions. This is referred to as the *rate-limiting step*. The major free radicals in the stratosphere that can provide fast rates of reaction in *both* steps of Eqs. (10.4.1) are members of the odd hydrogen, odd nitrogen, and odd chlorine families (conventionally designated by HO_x,

NO_x, and Cl_x). Replacing the X in Eqs. (10.4.1) by OH, NO, or Cl yields a specific catalytic reaction cycle that is important in the stratosphere. In each of these cycles the rate coefficient for the rate-limiting step is typically several orders of magnitude larger than that for Eq. (10.3.4d).

The effectiveness of any such cycle is of course limited by the concentration of the catalyst as well as the reaction rate. In Fig. 10.6 the percentages of the odd oxygen production that are balanced by the HO_x, NO_x, and Cl_x cycles are shown as functions of altitude together with that of the Chapman mechanism. Because of the relatively large concentration of nitric oxide (NO), the NO_x cycle dominates throughout much of the stratosphere.

The concentrations of various catalytic free radicals depend on the processes that lead to their formation and removal. The source molecule for stratospheric NO_x is nitrous oxide (N_2O). Nitrous oxide, which is of biological origin, is well mixed in the troposphere with a volume mixing ratio of about 300 ppbv and is destroyed in the stratosphere primarily by ultraviolet photolysis to yield $N_2 + O$. A small fraction (1–2%) of the N_2O reacts with excited oxygen atoms in the stratosphere to produce NO. The main removal process for NO_x involves the reaction of nitrogen dioxide with the hydroxyl radical:

$$HO + NO_2 + M \rightarrow HNO_3 + M. \tag{10.4.2}$$

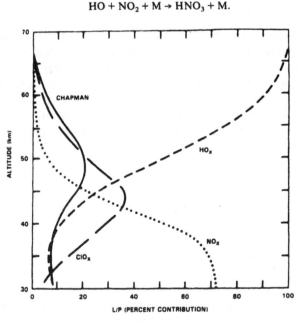

Fig. 10.6. Ratio (in percent) of the odd oxygen loss rate (L) due to the Chapman HO_x, NO_x, and Cl_x mechanisms to the odd oxygen production rate (P) based on midlatitude diurnally averaged conditions. [From Watson *et al.* (1986), with permission.]

The nitric acid (HNO_3) formed in this reaction has a fairly long chemical lifetime and is removed by transport into the troposphere, where it dissolves in water droplets and is rained out.

The hydroxyl radical (OH) participates in many important chemical reactions in the troposphere and stratosphere. OH is formed primarily through the reactions of $O(^1D)$ with water vapor and methane. An important path for removal is by coupling to the NO_x cycle through the reaction of Eq. (10.4.2).

The source for chlorine in the natural stratosphere is the upward transport of methyl chloride (CH_3Cl), whose main source is the world ocean. Methyl chloride reacts with the OH radical in the atmosphere to provide free chlorine for participation in the Cl_x cycles. In addition, methyl chloride is photolyzed above 30 km; this process also forms Cl_x.

Because it is chemically attacked by the OH radical, which is present in the troposphere as well as the stratosphere, methyl chloride has a relatively short tropospheric lifetime; most of it is destroyed before it can reach the stratosphere. The same is not, however, true of the chlorofluoromethanes ($CFCl_3$, and CF_2Cl_2). These synthetic species are almost completely inert in the troposphere but are photolyzed in the stratosphere, where they now provide the predominant source of chlorine. The atmospheric concentration of the chlorofluoromethanes is increasing at a rate of about 5% per year. Thus, they potentially could cause a substantial perturbation to the natural balance of the ozone layer.

The principal sink for stratospheric chlorine is the reaction of chlorine with methane to form hydrochloric acid:

$$Cl + CH_4 \rightarrow HCl + CH_3.$$

HCl is relatively long-lived in the stratosphere and, like HNO_3, is transported to the troposphere and rained out.

Chemical models of the ozone layer must account for the sources and sinks of the catalytic molecules, as well as the partitioning that occurs due to the many reactions among the various species. The rates of ozone destruction by the HO_x, NO_x, and Cl_x families of reactions cannot simply be added together to give the total catalytic destruction rate. Because of the coupling among the catalytic cycles, the rate of destruction for one set of reactions may depend on the concentration of another catalytic species. In addition to the cycles considered here, which involve atomic oxygen and thus are primarily important only in the middle and upper stratosphere where atomic oxygen is present in significant amounts, there are cycles in which ozone replaces atomic oxygen as a reactant. There are also null cycles that involve interconversion among the reacting species, and reactions that sequester free radicals in temporary reservoir species so that they are unavailable for participation in ozone-destroying cycles. ($ClONO_2$ is a notable example of such a temporary reservoir.) There are even some chemical cycles that actually produce ozone in the lower stratosphere and troposphere. Detailed models of the ozone layer may involve more than 100 chemical and photochemical equations. Only about a dozen species or families of species are directly affected by transport processes. But others are indirectly affected, since their photochemical equilibrium states depend on the concentrations of species that are themselves partly transport controlled.

10.5   Models of the Natural and Perturbed Ozone Layer

Realization that human activities might lead to depletion of the ozone layer has in recent years provided motivation for extensive attempts to model the natural and perturbed ozone distribution. As indicated in the last section, when the catalytic ozone destruction cycles involving the HO_x, NO_x, and Cl_x families and the coupling among these are included, the computation of photochemical production and loss of ozone becomes quite complicated. Partly because of the resulting enormous computational demands of three-dimensional modeling, most simulations of the ozone layer have been based on either one-dimensional or two-dimensional models. The former can only provide information on the horizontally and annually averaged vertical profile, while the latter can include latitudinal and seasonal dependence.

One-dimensional models can only account for transport in a very crude fashion by parameterizing all transport effects in terms of vertical diffusion. Such models can, however, include rather complete representation of photochemical processes and can be integrated for long periods of simulated time with rather modest computational resources. Two-dimensional models can in principle simulate the effects of meridional and vertical transport as well as meridional variations in photochemistry. The fidelity of such simulations depends, of course, on the accuracy with which the zonally averaged eddy transports can be represented.

10.5.1   Two-Dimensional Modeling of the Natural Ozone Layer

A number of two-dimensional models of the ozone layer have been formulated over the past two decades. The earliest of these were based on the conventional Eulerian-mean formulation. In most cases such models employed the tracer continuity equation [Eq. (9.7.1)] with a specified Eulerian-mean meridional circulation ($\bar{v}, \bar{w}$) estimated with the aid of observational data, and an empirically determined symmetric diffusion tensor.

In this formulation there should be a large degree of cancellation between two poorly known properties, the "eddy" and the "mean" transport. For this reason the traditional Eulerian-mean approach has been abandoned by most modelers. Currently, an approximate TEM formulation [Eq. (9.4.7) with the effective transport velocity approximated by the residual circulation] or an isentropic formulation [see Eq. (9.4.27)] is favored. Either of these tends to reduce the cancellation between "eddy" and "mean" transports that occurs in the conventional Eulerian mean formulation.

A further important distinction among the two-dimensional models is between those in which the meridional circulation is specified and those in which it is computed self-consistently from the complete zonally averaged equations [i.e., Eqs. (3.5.2) in the case of the TEM formulation]. In the former case it is usual to assume that the residual circulation is approximated by the diabatic circulation and to use theoretically derived estimates of the diabatic heating rate, using observed temperatures, as a function of latitude, height, and season to obtain the $(\bar{v}^*, \bar{w}^*)$ field.

Models in which the zonal mean dynamical equations are solved must, of course, employ parameterizations of the eddy forcing in the zonal mean momentum and thermodynamic equations. It is also necessary to solve for the zonal mean diabatic heating corresponding to the seasonally varying temperature and ozone distributions computed in the model. An example of this type is the model of Garcia and Solomon (1983), in which the parameterized planetary waves are assumed to be steady, linear, and adiabatic so that their EP flux divergence vanishes. Thus, the model does not include parameterized meridional mixing by transient planetary waves. It does include weak meridional and vertical eddy diffusion terms in the thermodynamic equation and tracer continuity equation to parameterize the effects of mixing by small-scale disturbances. The momentum budget is approximately accounted for by a balance between the Coriolis torque of the residual meridional motion and a strongly altitude-dependent Rayleigh friction that is intended to model gravity-wave drag. The mean meridional mass stream function for the residual circulation given by this model is shown in Fig. 10.7. The corresponding Northern-Hemisphere winter solstice ozone distribution is shown in Fig. 10.8; this may be compared with the observed distribution of Fig. 10.4. Clearly this type of model, in which transport is dominated by the advection due to the residual circulation, does a surprisingly good job of modeling the distribution of ozone in the meridional plane.

The TEM version of the tracer continuity equation used by Garcia and Solomon (1983) does not completely neglect meridional transport by

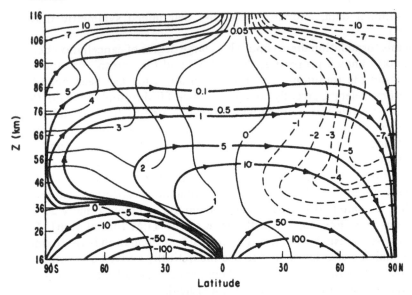

Fig. 10.7. Mass stream function for the residual meridional circulation (kg m^{-1} s^{-1}) shown in heavy solid lines and the diabatic heating distribution (light solid and dashed lines; K day^{-1}) for Northern-Hemisphere solstice conditions as computed in a two-dimensional model. [After Garcia and Solomon (1983).]

planetary waves. Even though transient eddy effects are ignored, the model does include a treatment of the flux divergence due to eddy chemical dissipation by steady planetary waves. This is modeled in a manner similar to the weak linear relaxation treatment of Section 9.4.1. Garcia and Solomon also include effects of temperature-dependent reaction rates in the photochemistry of ozone. Thus, they let

$$S' = -a\chi' - b\theta'$$

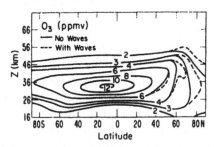

Fig. 10.8. Ozone mixing ratio computed in the two-dimensional model whose residual meridional circulation was indicated in Fig. 10.7. The dashed contours show the effects of transport due to chemical dissipation in steady planetary waves (see text for details). [After Garcia and Solomon (1983).]

where $b = \bar{\chi}\,\partial a/\partial\bar{\theta}$ (units of $K^{-1}\,s^{-1}$). In that case it may be verified by the methods of Section 9.4.1 that the modified Eulerian source of Eq. (9.4.10) must have an additional term of the form

$$\rho_0^{-1}\,\boldsymbol{\nabla}\cdot(\rho_0\mathbf{K}^{(T)}\cdot\boldsymbol{\nabla}\bar{\theta}) \qquad\qquad (10.5.1)$$

where $\mathbf{K}^{(T)} \equiv ba^{-1}\mathbf{K}^{(c)}$. Thus, for ozone the source modification (sometimes referred to as the "chemical eddy" effect) involves two flux divergence terms: the first has the flux proportional to the zonal-mean tracer mixing ratio gradient; the second has the flux proportional to the zonal-mean potential temperature gradient. The modification to the Northern-Hemisphere winter solstice ozone distribution due to the chemical eddy effect is shown in Fig. 10.8. Only at relatively high latitudes, where the stationary planetary wave amplitudes are large, is the planetary-wave chemical eddy effect significant.

10.5.2 Modeling Mesospheric Ozone

In Section 10.2 we pointed out that observations indicate that near the mesopause the seasonal variation of ozone is characterized by a strong maximum at the spring equinox, a secondary maximum at the autumn equinox, and minimum at the solstices. If ozone at this level were under photochemical control, the maximum would occur at the summer solstice, since the temperature-dependent reactions that destroy ozone would be slowest in the cold summer mesosphere and the production rate would be a maximum. Thus, the observed distribution must be dynamically controlled.

As discussed in Section 4.6.2, gravity wave breaking is thought to generate a strong zonal drag force and large vertical diffusion in the mesopause region. Although little is known about the climatology of gravity-wave sources, it is clear that gravity-wave transmissivity must depend on the mean wind distribution in the stratosphere and mesosphere. During the winter season, when stationary gravity waves can propagate through the bulk of the middle atmosphere, strong gravity wave breaking is expected to occur in the upper mesosphere. During the summer season, when a zero wind line in the lower stratosphere separates tropospheric westerlies from stratospheric easterlies, the transmission of low-phase-speed wave modes will be inhibited; only waves with westerly phase speeds great enough to escape the troposphere will be able to penetrate to the mesopause. Since such waves are expected to have lower amplitudes than the low-phase-speed waves, the wave-breaking region will be higher than in the winter. During the equinoctial seasons the zonal winds tend to be weak westerly in the stratosphere and weak easterly in the mesosphere. Under these conditions only the high-phase-speed "tail" of the gravity-wave spectrum can propagate to the mesopause region; wave breaking and turbulence should then be much reduced.

Thomas *et al.* (1984) argued that the seasonal variability in the vertical diffusion at the mesopause generated by breaking gravity waves could, by altering the rate of vertical tracer transport, account for the observed equinoctial peaks in the ozone distribution near the mesopause. Reduced vertical transport at the equinox should have two important effects. First, the downward flux of atomic oxygen below the 80-km level would be reduced. This fact, combined with the more or less steady downward diffusion from the thermosphere, would lead to a buildup of atomic oxygen near the mesopause. Second, there would be a reduced upward transport of water vapor from the lower mesosphere. The former effect would tend to increase the rate of ozone formation; the latter would tend to reduce the amount of OH available for catalytic destruction of ozone.

In Fig. 10.9 the equinox to solstice ozone ratio observed by the SME satellite is compared with results from a two-dimensional model that includes the simple parameterization of gravity wave breaking and diffusion discussed in Section 4.6.2. The extratropical equinoctial ozone maximum is well simulated, suggesting that seasonal variation of turbulent diffusion by breaking gravity waves is the correct explanation for the equinoctial ozone maximum.

10.5.3   One-Dimensional Models of Ozone Perturbations

The one-dimensional model remains a popular tool for studying the response of the ozone layer to various postulated secular changes in trace gases. In most one-dimensional models the tracer concentrations are carried as independent variables. The concentration (number density) of the ith species, n_i, is then predicted from an equation of the form

$$\frac{\partial n_i}{\partial t} = \frac{\partial}{\partial z^*}\left[\hat{K}_{zz}\rho\frac{\partial}{\partial z^*}(n_i/\rho) \right] + P_i - L_i \tag{10.5.2}$$

where P_i and L_i are the photochemical production and loss rates for the ith species, $\hat{K}_{zz}$ is the vertical diffusion coefficient, z^* designates the geometric height, and ρ is the air density.

In most one-dimensional formulations, $\hat{K}_{zz}$ is determined empirically so that Eq. (10.5.2) provides the steady-state observed distribution of CH_4 at 30°N latitude. As discussed in Section 9.7.3, the diffusion coefficient in a one-dimensional model should not be the same for every species, but should be smaller for those species with shorter lifetimes (see Fig. 9.17). Most formulations to date have, however, employed the same value of $\hat{K}_{zz}$ for all species.

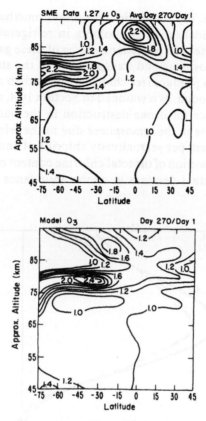

Fig. 10.9. The ratio of O_3 concentration at the Southern-Hemisphere spring equinox to that at the southern summer solstice as measured by the SME satellite (upper) and simulated in a two-dimensional model that includes parameterized gravity wave diffusion (lower; see text for details). [After Garcia and Solomon (1985).]

The concentrations governed by Eq. (10.5.2) should be interpreted as global averages at constant height. In most cases global data are not available for chemical tracers, and it is not possible to compute global means of the production and loss terms without knowledge of the meridional distribution of the tracers. Thus, one-dimensional models have often been regarded as providing vertical profiles at 30°N, a location for which there is a considerable data base, and production and loss terms are computed for annual mean conditions at 30° latitude.

The major use of one-dimensional photochemical models is to provide information on possible perturbations of the ozone layer due to trends in various trace gases that participate in chemical reactions that may affect ozone. The gases that are currently deemed to pose the greatest threat to the ozone layer are the synthetic chlorofluoromethanes, $CFCl_3$ and CF_2Cl_2, which are widely used as coolants in refrigeration, in industrial foam blowing, and as aerosol propellants. Most of these gases are eventually released into the troposphere and transported into the stratosphere, where they are photolyzed to produce free chlorine, which then reacts catalytically to destroy ozone. Although, as mentioned in Section 10.4, chlorine chemistry is not the major source for ozone destruction in the natural stratosphere, the increase of chlorine in the atmosphere due to the release of $CFCl_3$ and CF_2Cl_2 is several percent per year; already chlorine of anthropogenic origin constitutes the major portion of the total chlorine content of the stratosphere. Estimates of the eventual decrease in ozone abundance that would result

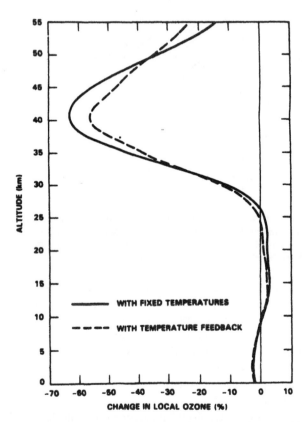

Fig. 10.10. One-dimensional model predictions (Lawrence Livermore National Laboratory model) of the percentage change in ozone concentration at steady state as a function of height for continuous release of chlorofluoromethanes at 1980 rates, relative to the atmosphere with no chlorofluoromethanes. [From Watson *et al.* (1986).]

from release of chlorine at a constant rate indefinitely into the future have varied markedly as information on reaction rates has been refined. Current estimates suggest that there should be only a few percent depletion in the ozone column at steady state for release at the 1980 rate. This steady state would not be reached, however, until late in the next century.

The predicted small net change in the ozone column results from a "self-healing" effect in which large reductions in the upper and middle stratosphere induced by the catalytic chlorine destruction reactions are to a large extent cancelled by increases at the lower levels (Fig. 10.10). The predicted increase in the lower stratosphere is due both to increased production associated with the greater penetration of solar ultraviolet radiation and to formation of the "reservoir" species chlorine nitrate, which reduces the amount of Cl and NO available to participate in the catalytic ozone destruction cycles. For current release rates of chlorofluoromethanes, this process is very effective in limiting the total change in column abundance of ozone. But if the rate of chlorofluoromethane release were to continue to increase, eventually all of the NO_x in the lower stratosphere would be transferred to the chlorine nitrate reservoir, and further increases of chlorine would then be expected to cause rapid decreases in ozone.

Although most perturbation studies have been limited to examining the effects of changes in a single radiatively or chemically important species, the effects of individual perturbing gases should not be considered in isolation, since significant coupling effects might occur. Some of these are briefly discussed in Chapter 12.

Appendix 10A The Continuity Equation for Chemical Species

In atmospheric chemistry the tracer continuity equation is usually expressed in terms of the number density n (molecules m^{-3}) of the tracer. It takes the form [e.g., Brasseur and Solomon (1984), Eq. (3.36)]

$$\frac{\partial n}{\partial t} + \nabla^* \cdot (n\mathbf{u}^*) = P - L, \tag{10A.1}$$

when P and L are the "production" and "loss" terms (molecules $m^{-3} s^{-1}$) and asterisks indicate the use of geometric-height (z^*) coordinates, rather than the log-pressure (z) coordinates that are mostly adopted in this book. Now by Eqs. (1.3.1) and (1.3.3a),

$$n = (N_a M^{-1})\rho\chi, \tag{10A.2}$$

where ρ is the air density, χ is the volume mixing ratio, M is the molecular weight of air, and N_a is Avogadro's number. On substituting Eq. (10A.2) into Eq. (10A.1) and using the mass continuity equation in z^* coordinates,

$$\frac{\partial \rho}{\partial t} + \nabla^* \cdot (\rho \mathbf{u}^*) = 0, \tag{10A.3}$$

we obtain

$$\frac{D\chi}{Dt} \equiv \frac{\partial \chi}{\partial t} + \mathbf{u}^* \cdot \nabla^* \chi = \rho^{-1} N_a^{-1} M(P - L). \tag{10A.4}$$

Note that derivatives with respect to x, y, and t are here taken at constant z^*, not at constant z. However, D/Dt is a coordinate-independent operator and can alternatively be expressed in its z-coordinate form (see Section 3.1.1). Using the ideal gas law [Eq. (1.1.1)] and the expression $\rho_0(z) = p/RT_s$ (Section 3.1.1), we can thus write Eq. (10A.4) in the z-coordinate form

$$\frac{D\chi}{Dt} \equiv \chi_t + \mathbf{u} \cdot \nabla \chi = \rho_0^{-1}(T/T_s) N_a^{-1} M(P - L) \equiv S, \tag{10A.5}$$

where x, y, t derivatives are now taken at constant z [cf. Eq. (9.4.1)]. An alternative z-coordinate form, derived from Eqs. (10A.5) and (3.1.3d), is

$$(\rho_0 \chi)_t + \nabla \cdot (\rho_0 \chi \mathbf{u}) = (T/T_s) N_a^{-1} M(P - L) = \rho_0 S \tag{10A.6}$$

[cf. Eq. (9.4.2)]. In practice, a further term representing small-scale diffusion should also be added to the right-hand side of Eq. (10A.6): see Eq. (10.3.1).

References

10.2. London (1985) presents an excellent review of observational studies of the climatological mean and variability of atmospheric ozone, with an emphasis on ground-based observations. His paper contains an extensive bibliography.

10.3. The concept of chemical time scales is developed thoroughly in Brasseur and Solomon (1984). They also provide a general discussion of the chemistry of the middle atmosphere.

10.4. The fundamentals of ozone photochemistry are presented clearly in Wayne (1985). Other treatments of the subject include Thrush (1980) and Whitten and Prasad (1985).

10.5. Watson et al. (1986) present a summary of model calculations of the perturbed ozone layer and also review many aspects of the radiative, dynamical, and chemical processes that are of importance in the stratosphere. Their work is essentially a summary of the detailed presentation given in WMO (1986).

Chapter 11 | General Circulation Modeling

The *general circulation* of the atmosphere can be defined as the totality of motions that characterize the global-scale atmospheric flow. Since many of the eddy motions in the middle atmosphere have their origins in the troposphere, it is not possible to isolate the general circulation of the middle atmosphere from that of the atmosphere as a whole. The problem of modeling the general circulation is one of numerically simulating the global circulation of the atmosphere using the primitive equations, with only a minimum of assumptions and parameterizations. Ideally, only the external conditions (solar radiation, sea surface temperature, and permanent ice cover) should be specified and the model should self-consistently compute the temperature, wind, cloud, and trace species distributions as functions of three-dimensional space and time. In practice, at the present stage of model development, many compromises are necessary both for scientific reasons and for reasons of computational efficiency.

General circulation models (GCMs) can provide stringent tests of our understanding of the atmospheric system. If all relevant physical processes are correctly represented, then a GCM should provide a faithful simulation of the three-dimensional circulation. Thus, one of the primary roles of the GCM in middle atmosphere research is as an "experimental" tool to test the adequacy of our understanding of various physical processes. Externally specified parameters (e.g., orography) can be altered to provide "controlled" experiments that cannot be done on the real atmosphere. Deficiencies in model simulations can provide indications that important processes have been omitted or inadequately treated (see the discussion of the cold polar bias in Section 11.2.1).

415

A long-term objective of middle atmosphere GCM research is to develop simulations of the atmosphere that are suitable for use in predicting possible anthropogenic perturbations of the ozone layer. Although completely inter-active radiative–chemical–dynamical models suitable for such predictions are not yet available, general circulation models have in the past several years provided many fundamental dynamical insights as well as information on the interaction of radiation and dynamics. To date comparatively little general circulation model research has been devoted to the subtle problems of chemical–dynamical interactions, although wind "data" from GCM simulations have been used in studies of trace constituent transport. There is little doubt, however, that GCMs are the most powerful technique avail-able for developing an understanding of the full range of radiative, dynami-cal, and chemical interactions in the middle atmosphere. These models also offer perhaps the best methodology for elucidating the dynamical inter-actions between the troposphere and stratosphere, both for assessing the influence of the middle atmosphere on the climate of the troposphere and for studying the driving of the middle atmosphere by the circulation of the troposphere.

The overwhelming majority of general circulation model studies have been devoted to the tropospheric climate problem. Most general circulation models do, however, include at least the lower stratosphere in order to avoid problems that would result from imposing a rigid-lid upper boundary condition at the tropopause. Such models have been used to investigate various aspects of the stratospheric circulation, including sudden strato-spheric warmings. In fact, the ability of a given model to generate realistic midwinter warmings has often been taken as a key test of its suitability for stratospheric studies.

Efforts to model the bulk of the middle atmosphere are in a much less mature state than is tropospheric general circulation modeling. The very large demands on computer resources required for multilevel models extending over many scale heights in altitude have forced middle atmosphere modelers to make a number of simplifications. For example, some models have been based on quasi-geostrophic dynamics, some have included only a single hemisphere, some have used greatly simplified radiative heating algorithms, and some have featured highly simplified tropospheres.

The primary focus in this chapter will be on those models that can be regarded as true general circulation models. By this, we mean primitive equation models that attempt to simulate the entire global atmosphere with horizontal resolution adequate to resolve synoptic-scale eddies, and with radiative processes calculated from accurate radiative transfer codes. How-ever, we shall also mention some models that, while not meeting all these criteria, have provided important insights into the dynamics of the middle atmosphere.

11.1 Models of the Lower Stratosphere

The first primitive-equation general circulation model that attempted to resolve the lower stratosphere was that of Smagorinsky *et al.* (1965), developed at the Geophysical Fluid Dynamics Laboratory (GFDL). This model had nine prediction levels in the vertical, with the top three levels at standard heights of 12, 18, and 31.6 km. The horizontal domain was limited to the Northern Hemisphere and represented by a rectangular grid on a polar stereographic map. In this model, as in some subsequent GFDL models, the solar radiation was set at its annual mean value. The hydrological cycle was also omitted and the lower boundary was taken to be a uniform land surface with zero heat capacity. The radiatively active constituents (ozone, water vapor, carbon dioxide, and clouds) were all assigned zonally and annually averaged climatological mean values.

A model without topography could hardly be expected to generate vertically propagating planetary waves of sufficient strength to properly account for the eddy forcing required to satisfy the zonal mean climatology of the winter stratosphere (cf. Chapter 7). Nevertheless, the Smagorinsky model did at least qualitatively simulate the zonal mean temperature distribution in the lower stratosphere, the high cold tropical tropopause, and the low polar tropopause. However, there was also a notable deficiency: the polar stratospheric temperatures were significantly lower than the observed annual mean values, and (in agreement with the thermal wind balance) the mean zonal wind was significantly stronger than in the observed annual mean.

Manabe and Hunt (1968) used a model similar in structure to that of Smagorinsky *et al.* (1965), but with 18 levels in the vertical extending from the surface to 37.5 km altitude. The computed zonal mean temperatures and winds for this model were in better agreement with observations than were those for the Smagorinsky model. However, the polar cold bias remained. In fact, the computed polar temperature distribution agreed better with the observed winter mean than with the observed annual mean, even though annual mean radiative forcing was used.

The Manabe and Hunt experiments showed that increased vertical resolution in the stratosphere can improve the stratospheric simulation. Nevertheless, the incorporation of annual mean radiative forcing, the absence of topography, the restriction of the model domain to a single hemisphere, and other model limitations prevented the model from fully simulating important aspects of the circulation of the stratosphere.

In recent years several groups have formulated global general circulation models that have incorporated realistic topography and have at least crudely resolved the lower stratosphere. In most cases, these models have utilized the "sigma" coordinate system, in which the vertical coordinate is defined as $\sigma = p/p^*$, where $p^*(x, y, t)$ is the surface pressure. This system has the advantage that the coordinate surface $\sigma = 1$ coincides with the earth's surface everywhere, so that the lower boundary condition becomes simply $D\sigma/Dt = 0$. The discretization of the equations in the horizontal has commonly been in the form of finite differencing on a spherical grid with typically about 5° to 10° latitude and longitude grid spacing. Spectral transform models have become popular. A well-known example of a spectral model is the Community Climate Model (CCM) of the National Center for Atmospheric Research (NCAR), which is based on expansion in spherical harmonics with truncation typically at wave number 15.

The details of the physical parameterizations used in the various models differ, but they tend to share certain common features:

1. Sea surface temperatures are externally specified using observed climatological values. In some models these are specified to vary according to the annual cycle, but in many cases either annual means or fixed January monthly means are used. Surface temperatures over land are computed from a surface energy-balance equation.

2. Boundary-layer fluxes of momentum, heat, and moisture are parameterized by the *bulk aerodynamic method*. In this method the flux for each of these fields is specified to be proportional to the magnitude of the horizontal wind at the lowest model level times the difference between the value of the field at the boundary and the lowest model level.

3. Sub-grid-scale horizontal mixing is usually parameterized either by nonlinear diffusion or by second- or fourth-order linear diffusion; occasionally mathematical filtering is employed.

4. The hydrological cycle is included, with predicted water vapor and precipitation explicitly determined for forced synoptic-scale uplift, and parameterized for convective overturning.

5. Radiative heating and cooling by both solar and longwave radiation are computed using parameterized versions of the radiative transfer equations. Specified zonally averaged distributions of the radiatively active constituents are often used. Some models (e.g., the CCM) do, however, compute cloud-radiation effects interactively using the zonally varying cloudiness predicted by the model.

Despite the many improvements that have been made in general circulation model simulations in the past several years, the cold polar winter stratosphere bias has remained as an almost universal model deficiency.

Although it has sometimes been suggested that the cold bias is due simply to poor vertical resolution or to problems associated with the upper boundary condition, the problem remains even in the GFDL "SKYHI" model, which has excellent vertical resolution and an upper boundary near the mesopause.

There are basically only two plausible sources for the cold bias—radiative and dynamical. Radiative sources of the bias could involve either errors in the computed radiative cooling rates for given departures from the radiatively determined temperature, or errors in the radiatively determined temperature itself (i.e., in the temperature that would be computed in the model in the absence of dynamical heating; see Section 7.1). Dynamical sources of the cold bias could include problems in representation of eddy sources and sinks for either large- or small-scale eddies so that the net dynamical heating is underestimated and, as a consequence, the simulated winter stratospheres are too close to radiative equilibrium.

It is certainly true that radiative cooling computations for the winter stratosphere must be done with great care. Seemingly small differences in parameterizations of the radiative effects can produce rather large differences in the model climatology, as has been pointed out by Ramanathan *et al.* (1983). They showed that the NCAR CCM was able to produce a greatly improved winter lower stratosphere simulation when certain seemingly minor changes were made in the treatment of both solar and longwave radiation. These included the following:

1. Careful treatment of the upper boundary condition for solar heating by ozone.
2. Inclusion of correct temperature dependence for the 15-μm CO_2 band.
3. A reduction in the assumed H_2O mixing ratio in the polar stratosphere.
4. Allowance for variable cirrus-cloud emissivity dependent on the cloud water content.

Simulations using this improved radiation package for January mean conditions were compared with simulations using the same dynamics, but the earlier radiation code. As shown in Fig. 11.1, the new radiation code resulted in a remarkably improved simulation of the climate. The tropospheric subtropical and stratospheric polar night jets were well separated, in agreement with observations (Figs. 1.4 and 5.2), and the polar temperatures were in good agreement with those observed (Figs. 1.3 and 5.1).

Although these results might tempt one to conclude that better modeling of the radiative forcing is the key to improved simulations of the middle atmosphere, the actual situation is not so simple. The large differences between the two simulations shown in Fig. 11.1 cannot be attributed to radiation alone. In the polar stratosphere the temperature difference between

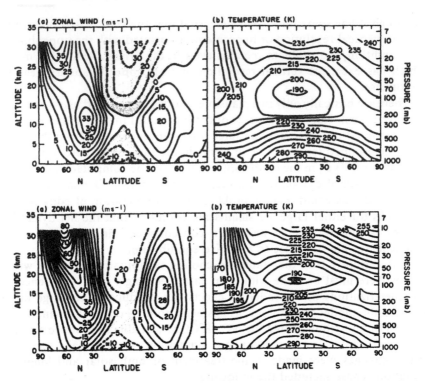

Fig. 11.1. Mean zonal winds (m s^{-1}) and zonally averaged temperatures in the NCAR CCM with improved (top) and degraded (bottom) radiation algorithms. [After Ramanathan *et al.* (1983). American Meteorological Society.]

the two cases shown in Fig. 11.1 is about three times the amount that can be attributed to direct effects of changes in the radiation code (i.e., to changes in the radiative equilibrium temperature distribution, and in cooling rates for a given departure from radiative equilibrium). The remaining difference must be due to an increase in the eddy forcing, which results in a much greater dynamical heating in the simulation with the improved radiation scheme. Thus, not only is the radiative equilibrium temperature somewhat warmer in the model with the improved radiative code, but the eddies have a different structure so that the departure from radiative equilibrium is maintained at a considerably higher value than in the original version.

We must conclude that a satisfactory simulation of the winter polar stratosphere is only possible if the model produces sufficient eddy forcing to maintain the required departure from radiative equilibrium. The distribution of eddy forcing required to maintain a given radiative heating rate can be estimated (within quasi-geostrophic theory) by taking the time average of Eq. (3.5.7) or Eq. (7.2.4c):

$$(\nabla \cdot \mathbf{F} + \rho_0 \bar{X})_y = (\rho_0 f_0 \bar{Q} / \bar{\theta}_{0z})_z. \qquad (11.1.1)$$

The heating rate $\bar{Q}$ is approximately proportional to the deviation of $\bar{\theta}$ from its radiative equilibrium value $\bar{\theta}_r$. Thus, from Eq. (11.1.1) the magnitude of the departure from radiative equilibrium is approximately linearly proportional to the magnitude of the eddy forcing. The precise form of the eddy forcing is not important for the present argument. Either the Eliassen–Palm (EP) flux divergence due to the resolved eddies or the zonal mean drag due to parameterized small-scale eddies can provide the required forcing. It is only necessary that the *total* forcing be sufficient in amplitude.

That the *amount* rather than the *form* of the eddy forcing is the key to simulating a realistic climatology in the winter stratosphere was shown nicely by Boville (1985) in a series of simulations using a 14-level version of the NCAR CCM with improved resolution in the lower stratosphere. Now, Eq. (3.6.5) indicates that in a time mean the EP flux divergence is proportional to the wave damping if nonlinear effects can be neglected. In the CCM both radiative damping and mechanical (diffusive) damping are present. Boville showed that the magnitude of the EP flux divergence was approximately proportional to the product of the square of the wave amplitude and the inverse time scale for radiative plus mechanical damping. In this model, as in most GCMs, the amplitude of the stationary planetary waves simulated in the winter stratosphere was less than observed in the atmosphere. Thus, a larger damping rate was required to produce the same amount of EP flux divergence as in the atmosphere. Boville showed by controlled experiments that the parameterized momentum diffusion in the CCM (which provides wave damping at about the same rate as the computed radiative damping, and also directly damps the zonal mean flow) was essential for simulating a realistic climatology. Omission of the mechanical damping resulted in a smaller total EP flux divergence, and hence the temperatures became closer to radiative equilibrium unless the radiative damping of the eddies was artifically enhanced. Increasing the damping rate in the stratosphere does, however, reduce the propagation of wave activity into the mesosphere; thus alone it cannot improve the simulation throughout the entire middle atmosphere.

If the model were to produce larger-amplitude planetary waves, then it is conceivable that radiative wave damping could provide a sufficient amplitude of EP flux divergence to maintain the observed zonal mean temperature distribution. Present indications are, however, that some mechanical dissipation by small scale motions must be present in the stratosphere, and for the mesosphere there is little doubt that gravity wave breaking must be a dominant factor in the momentum balance.

11.2 The GFDL SKYHI model

The general circulation model that currently provides the most complete representation of the middle atmosphere is the SKYHI model, developed at GFDL (Fels *et al.*, 1980). This model has 40 prediction levels extending from the surface of the earth to about the 80 km level. The vertical coordinate system, shown schematically in Fig. 11.2, is a hybrid system in which the sigma surfaces are defined as

$$\sigma = \frac{p - p_c}{p^* - p_c}, \quad p > p_c; \quad \sigma = \frac{p - p_c}{1013.25 - p_c}, \quad p < p_c$$

where $p^*(x, y, t)$ is the surface pressure and $p_c = 353.55$ mb. Thus, the coordinates are equivalent to the usual terrain-following σ coordinates below about 350 mb and become isobaric coordinates above that level. The vertical spacing of prediction levels increases gradually from the surface to the top boundary; it is about 1 km standard height in the midtroposphere and about 3 km at the stratopause. The horizontal grid mesh is based on spherical coordinates with Fourier filtering (i.e., removal of the high zonal wave-number components) near the poles. In early experiments a rather coarse resolution of 9° latitude by 10° longitude was employed. More recent experiments have employed versions with 5° by 6°, 3° by 3.6°, and even 1° by 1.2° resolution.

The treatment of physical processes in the troposphere is more extensive than in the early GFDL models. The model includes topography, realistic distributions of continents and oceans, and a hydrological cycle. Predicted fields include the wind, temperature, water vapor, ground temperature, soil moisture, snow cover, and surface pressure. However, cloudiness, pack ice, and ocean surface temperatures are all externally specified using climatological mean data.

The radiative heating and cooling is calculated once every 12 hr using diurnally averaged solar insolation. The solar heating in the middle atmosphere is based on the parameterization of Lacis and Hansen (1974), which includes heating by ozone and molecular oxygen. The infrared atmospheric radiation is computed using the scheme of Fels and Schwarzkopf (1981). A zonally averaged ozone distribution is specified from the surface to 34 km; above that level, temperature-dependent photochemistry is included by allowing the ozone to deviate from a specified climatological mean in response to temperature fluctuations, using a temperature-dependent parameterization of the photochemical equilibrium.

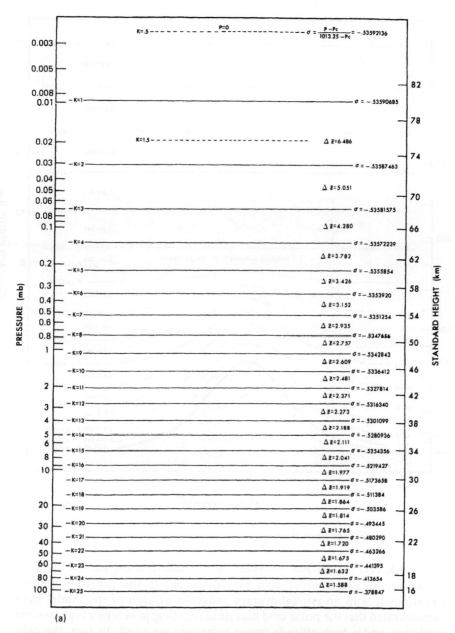

Fig. 11.2. (a) Upper and (b) lower portions of the SKYHI model's vertical coordinate system. The standard height labeling is based on a midlatitude standard atmosphere temperature distribution. [After Fels *et al.* (1980). American Meteorological Society.] *Figure continues.*

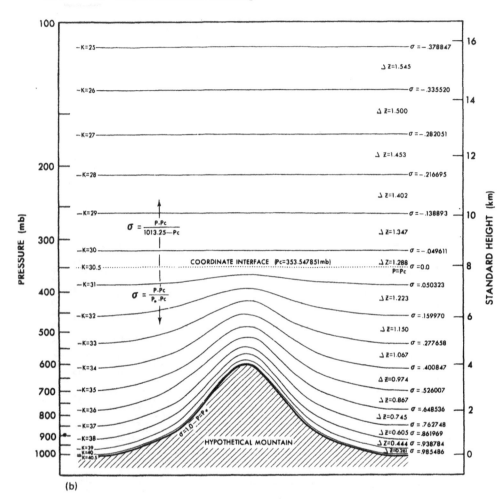

(b)

Fig. 11.2 (*continued*)

11.2.1 The Cold Winter Pole

The SKYHI model has provided useful information on the possible causes of the winter polar cold bias discussed in the previous section. Experiments with an annual average radiation version of the model have demonstrated that the polar cold bias does not disappear when high vertical resolution and a high-altitude upper boundary are used. In fact, the cold bias turns out to be a bigger problem in this model than in models with upper boundaries in the stratosphere. This should not be surprising, since, as discussed in Chapter 7, it is believed that gravity-wave drag and diffusion play primary roles in the momentum balance of the mesosphere. There is no parameterization of gravity-wave drag in the SKYHI model, and the experiments with horizontal resolutions of about 9°, 5°, and 3° are all too coarse to resolve the several-hundred-kilometer horizontal wavelength internal gravity waves that appear to be the primary breaking waves in the extratropics. Early results from a 1° resolution model, which can resolve the longer gravity waves, indicate that the extra dynamical heating provided by such waves does improve the simulation in the polar winter stratosphere.

Varying the horizontal resolution in the model has provided important insights into the momentum balance of the middle atmosphere. In the early 9°-by-10° horizontal resolution version, the mean zonal wind increases monotonically with height in the middle atmosphere so that the maximum speed is at the top level in the model. In the 5°-by-6° version, on the other hand, the maximum wind speed occurs in jet cores at about 64 km altitude with easterly shear above that level. Thus, the mesospheric polar temperatures deviate substantially from radiative equilibrium. The dependence of the polar temperature profile on horizontal resolution is illustrated in Fig. 11.3. Similarly dramatic changes occur in the zonal wind profile. In

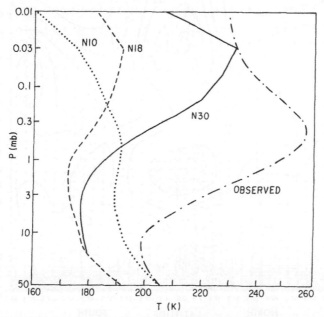

Fig. 11.3. Vertical profiles of the December mean temperature at the north pole simulated by 10°, 5°, and 3° resolution (labeled N10, N18, and N30, respectively) versions of the GFDL SKYHI general circulation model. Chain line shows observed December mean temperature at 80°N from Barnett and Corney (1985a). [Adapted from Fels (1985).]

evaluating the effects of model resolution, it should be noted that the model has a resolution-dependent parameterization of sub-grid-scale diffusion. Thus, in the coarse grid model the planetary waves (which are weaker than observed even in the troposphere) are damped out rapidly in the lower stratosphere; there is little EP flux divergence in the mesosphere, and the mesospheric temperatures are extremely cold. As resolution is improved, the planetary waves penetrate to increasing heights and are dissipated in the mesosphere. The mesospheric temperatures are then greatly improved, but the temperature in the polar stratosphere cools toward radiative equilibrium due to the reduced wave damping there. That planetary wave drag in the mesosphere is important for the 5°-by-6° model is illustrated by the EP flux cross section (Fig. 11.4), which shows a wave drag force of $15 \text{ m s}^{-1} \text{ day}^{-1}$ near the 80-km level. Although this is perhaps only 20% of

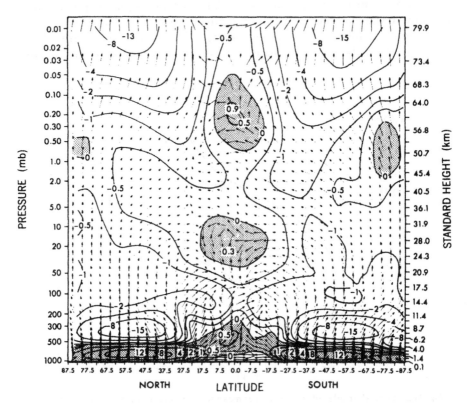

Fig. 11.4. EP flux vector direction (arrows) and contours of EP flux divergence normalized as zonal force per unit mass (10^{-5} m s^{-2}) for a 30-day mean in the SKYHI model with annual mean forcing. [After Andrews *et al.* (1983). American Meteorological Society.]

that required to produce the observed deviation from radiative equilibrium, it is sufficient to produce a substantial deceleration of the winds.

11.2.2 Equatorial Waves

As was indicated in Section 8.5, in its seasonal cycle version the SKYHI model simulates the semiannual oscillation of the zonal winds in the equatorial stratosphere quite realistically. There is little question that the westerly phase of the oscillation in the model, and the atmosphere, is driven by an EP flux divergence associated with upward-propagating equatorial Kelvin waves. A detailed analysis of the equatorial waves simulated in an annual mean 5°-by-6° horizontal resolution version of the SKYHI model is contained in Hayashi *et al.* (1984). In the annual mean simulation, the mean wind is westerly throughout most of the equatorial middle atmosphere and there is, of course, no semiannual oscillation.

A space–time power spectral density plot for the zonal wave-number 1 temperature disturbance in the 5° latitude band centered at the equator is shown in Fig. 11.5. Analysis of the meridional structure confirms that the spectral peaks corresponding to *eastward*-moving waves are indeed Kelvin waves. It is interesting to note that the characteristic period of the waves decreases with increasing height just as found in observations. Thus, the lower stratosphere has waves of period approximately 15 days, in agreement with the observations of Wallace and Kousky (1968). These are damped out below 10 mb. The upper-stratospheric Kelvin waves have periods of 5–7 days (phase speed 60–90 m s^{-1}), while the mesospheric waves have periods of 3–4 days (phase speeds 115–150 m s^{-1}). The corresponding vertical wavelengths are about 10, 20, and 40 km, respectively. These characteristics are in remarkable agreement with the observed waves reported by Salby *et al.* (1984). It is thus not surprising that the model simulates the semiannual oscillation in the upper stratosphere (see Fig. 8.16). However, despite the fact that the model simulates the observed long-period Kelvin waves in the lower stratosphere, there is no evidence of a simulated equatorial quasi-biennial oscillation. This failure is probably due to lack of sufficient vertical resolution. (Recall that the vertical scale of the shear zone in the QBO is only about 2 km.)

In addition to the clear Kelvin wave signal at periods of greater than 2 days, the model also simulates a broad spectrum of short-period equatorial gravity waves that propagate both eastward and westward. These have zonal wave numbers 1–30 and periods of about $\frac{1}{2}$ to 2 days. The contribution of such waves to the momentum balance in the model's middle atmosphere turns out to be at least as large as that of the longer-period Kelvin waves.

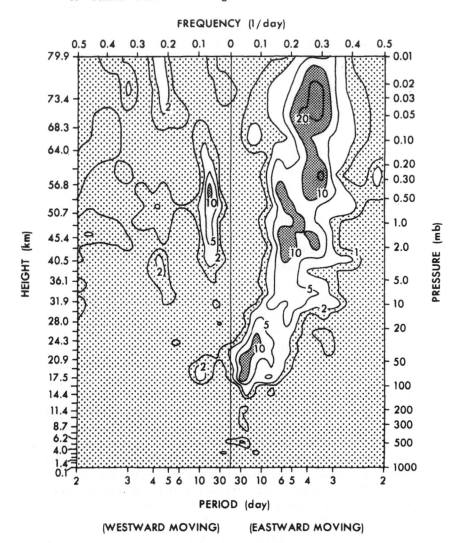

Fig. 11.5. Frequency-height distribution (2.5°S–2.5°N) of space-time power spectral density (K^2 day) of temperature for zonal wave number 1. [After Hayashi *et al.* (1984). American Meteorological Society.]

11.3 Forecasting of Sudden Stratospheric Warmings

There are two basic modes for utilizing a general circulation model: the climate mode, and the forecast mode. In the former mode the model is used to simulate climatological distributions of the atmospheric fields, and the simulation is validated by comparison with statistics of the observed fields such as time means and variances. In the second mode the model is initialized with real data, and an attempt is made to forecast the evolution of the actual flow over the course of several days.

A number of studies of the dynamics of stratospheric sudden warmings have been carried out using numerical models of varying complexity. Most such investigations have been process-oriented, employing mechanistic models rather than full GCMs; some examples were given in Chapter 6. There have also been studies that have reported sudden warmings generated spontaneously in general circulation models (O'Neill, 1980; Grose and Haggard, 1981; Mahlman and Umscheid, 1984): in all of these examples the models have been run in the climate mode.

The earliest attempt to utilize a GCM to forecast a sudden warming was due to Miyakoda et al. (1970). They used a nine-level version of the GFDL model with horizontal grid spacing of about 250 km and the top prediction level at about 9 mb. They attempted to forecast the wave-number 2 final warming of spring 1965 by initializing the model with data from 5 days prior to the observed splitting of the polar vortex. Their forecast did manage to predict the vortex splitting, but the subsequent evolution was not correctly predicted. In the forecast the polar vortex reformed and no vortex break-down or polar warming occurred. The reasons for the failure of this attempt were not clear at the time. However, it is now known (Simmons and Strüfing, 1983) that the radiation scheme used in this early GFDL model caused much too rapid cooling near the winter pole in the lower stratosphere, and thus tended to rapidly reform the polar vortex. Recent experiments also suggest that a good forecast of the troposphere is a necessary precondition to a successful stratospheric forecast, and the tropospheric forecast in this early experiment was quite poor after a few days.

In recent years the quality of extended-range forecasts has improved dramatically, due both to improved models and to improved initial data. Sudden warmings occurred in January and February 1979, during the Global Weather Experiment (a period of enhanced global observations). Forecasting experiments for this period were carried out by Simmons and Strüfing (1983) and Mechoso et al. (1985). Simmons and Strüfing used a modified version of the operational forecast model of the European Centre for Medium Range Weather Forecasts (ECMWF). This model had 18 levels with the top level at 10 mb, and had a horizontal grid spacing of 1.875° in latitude and longitude. A number of 10-day forecasts were carried out for both the January wave-number 1 minor warming and the late-February wave-number 2 major warming. In Fig. 11.6 we show results for 4- and 10-day forecasts initialized on February 13, 1979 (about 5 days prior to vortex splitting). The breakdown of the vortex is forecast quite accurately

Analysis   Forecast

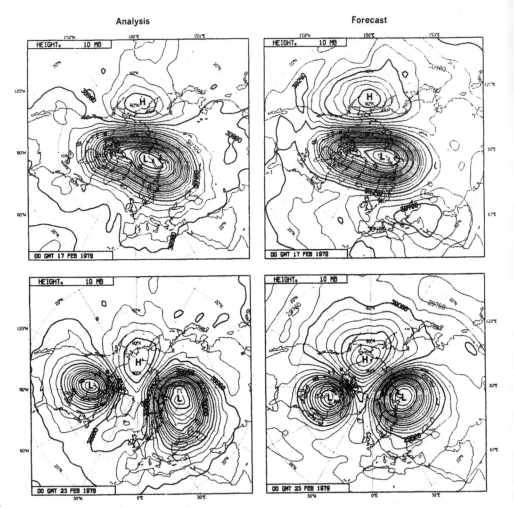

Fig. 11.6. Maps of the 10-mb height for February 17 and 23, 1979 compared with 4- and 10-day forecasts initialized on February 13, 1979. [After Simmons and Strüfing (1983).]

out to 10 days, although the strength of the Aleutian high is somewhat overestimated. Forecasts initialized for other days in this period were also successful, as indicated by the meridional cross sections for several 8-day forecasts shown in Fig. 11.7.

In forecasts of the same sudden warming carried out with a 15-level version of the University of California at Los Angeles (UCLA) general circulation model with an upper boundary at 1 mb, Mechoso *et al.* (1985) found that the influence of horizontal grid resolution was quite important.

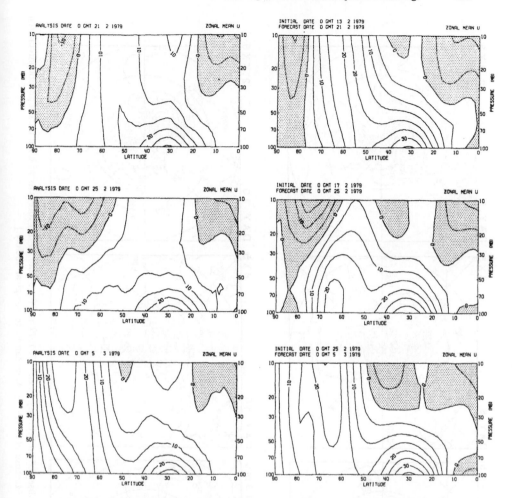

Fig. 11.7. Latitude–height sections of mean zonal winds for February 21, February 25, and March 5, 1979 (left-hand side) and 8-day forecasts verifying for the same dates (right-hand side). [After Simmons and Strüfing (1983).]

They compared forecasts based on a 4° latitude by 5° longitude model with those in a 2.4°-by-3° model. A 10-day forecast with the high-resolution model initialized on February 17, 1979 correctly predicted the vortex break-down, as shown in Fig. 11.8. The coarse-resolution model predicted substantially less amplification of the wavenumber two pattern. Mechoso *et al.* attributed this difference to the poorer tropospheric forecast in the latter model. In particular, they found that the refractive index squared for wave number 2 (cf. Section 4.5.4) had a negative region in the high-latitude upper

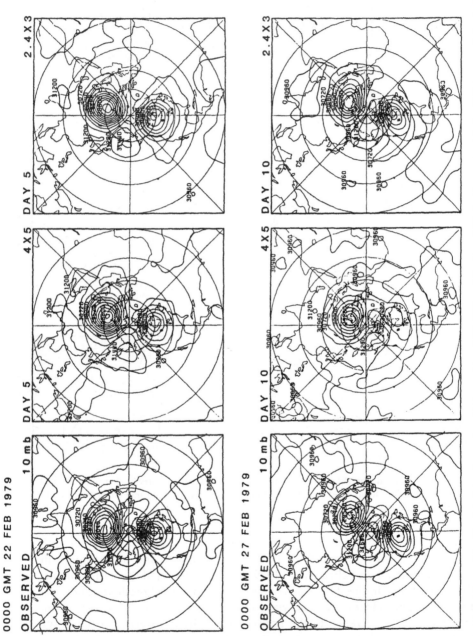

Fig. 11.8. Maps of the 10-mb height observed on February 22 and 27, 1979 and for 5- and 10-day forecasts with the UCLA general circulation model initialized on February 17. [After Mechoso *et al.* (1985). American Meteorological Society.]

troposphere in the low-resolution model that was not present in the high-resolution case. Hence, vertical propagation into the polar stratosphere was underpredicted in the low-resolution model. In conclusion, it appears that, despite the fact that the stratosphere is dominated by planetary scales of motion, a very accurate tropospheric forecast is a minimum requirement for the forecast of a sudden stratospheric warming. Such accuracy depends on many factors, including horizontal resolution.

11.4 Transport Modeling

Simulation of the climatological distribution of ozone and its temporal and spatial variability has long been a goal of middle atmosphere general circulation models. Indeed, the early hemispheric model of Hunt and Manabe (1968) was used by Hunt (1969) to study the influence of transport on ozone in a simulation that included a very simple parameterization of ozone photochemistry. A somewhat more sophisticated treatment of photochemistry was used in a global model by Cunnold et al. (1975), but their model was based on the quasi-geostrophic equations and was not a true general circulation model. Schlesinger and Mintz (1979) used the Cunnold et al. (1975) photochemistry in a global primitive-equation model with 12 prediction levels, including seven levels in the stratosphere between 100 and 1 mb. The ozone distribution computed in the model was used in the solar radiative heating algorithm. Thus, the model involved feedback among the dynamical, photochemical, and radiative processes. Unfortunately, however, they were unable to integrate the model for a sufficiently long simulated time to establish an equilibrium ozone distribution.

In all of the above studies the tracer (ozone) was carried as an additional dependent scalar variable in the general circulation model. This has the advantage that the calculations for the tracer evolution can be made simply by solving an additional continuity equation to obtain the tracer mixing-ratio distribution at each timestep. If feedback of the tracer onto the dynamics is to be included, this is the only possible strategy. However, for study of the evolution of a *passive* tracer it is rather inefficient to carry the tracer as a variable of the GCM, since each time that a different initial tracer distribution or different photochemical loss rate is tested an entirely new GCM simulation is required. Rather, it is far more efficient to save the winds from the GCM as a "data" set and to solve the tracer continuity equation independently of the GCM using the GCM derived winds as specified advecting velocities. It is this strategy that has been adopted in a number of recent studies of stratospheric tracers carried out at GFDL.

The GFDL tracer model is based on winds obtained from the seasonally varying GFDL global model described by Manabe and Mahlman (1976).

The model has horizontal grid spacing of about 250 km and 11 levels in the vertical arranged to provide good resolution in the boundary layer and in the lower stratosphere. The top level is at 10 mb, so the model is limited to tropospheric and lower-stratospheric tracer studies. However, for many trace species (e.g., ozone) it is the lower stratosphere where transport processes are dominant, while photochemistry controls the distributions at higher altitudes (see Chapter 10). Thus, the model can be used for studying stratospheric transport processes despite the low altitude of the top boundary. The results must be interpreted with caution, however, in view of the cold polar bias in the model.

The details of the tracer model are given by Mahlman and Moxim (1978). Briefly, the model solves a continuity equation for mixing ratio similar to Eq. (9.4.2), but using the terrain-following sigma-coordinate system. In addition to tracer flux divergences due to the explicitly specified winds, the model contains sub-grid-scale horizontal and vertical diffusion, provisions for source and sink terms appropriate to particular tracers, and a "filling" term to eliminate local negative values of mixing ratio that may appear due to finite-differencing truncation errors. The input fields for the tracer model are obtained from an annual cycle run of the GCM in which the wind and surface pressure fields are averaged over 6-hr periods and saved once every 6 hr. The tracer continuity equation is integrated over the annual cycle using a 24-min time step. Additional years of integration can be done by recycling the annually varying wind fields.

Among the studies carried out with this tracer model is a "stratified tracer" experiment (Mahlman et al., 1980) designed to elucidate the role of transport in determining the lower-stratospheric distribution of an ozone-like tracer. For this experiment the tracer mixing ratio was held fixed at 7.5 ppm at the top model level (10 mb) and a simple tracer removal was specified in the troposphere. The tracer was initialized with a horizontally uniform distribution that increased almost linearly with height from the surface to the 10 mb level. The model was integrated over 4 simulated years, which was sufficient to establish an annually varying equilibrium distribution. Results at 3-month intervals for the fourth year are shown in Fig. 11.9. The zonal mean distribution for all seasons shows an upward bulge in the mixing-ratio surfaces in the tropics, and downward slope toward the poles in both hemispheres, consistent with the diabatic circulation of the lower stratosphere discussed in Chapter 9. The poleward–downward slope is greater in the Northern Hemisphere at all seasons, consistent with the model's stronger eddy activity and stronger diabatic circulation in the Northern Hemisphere (Manabe and Mahlman, 1976). A comparison of the terms in the zonal mean tracer budget equations for the northern and southern winters (Fig. 11.10) shows that indeed the eddy tracer fluxes in

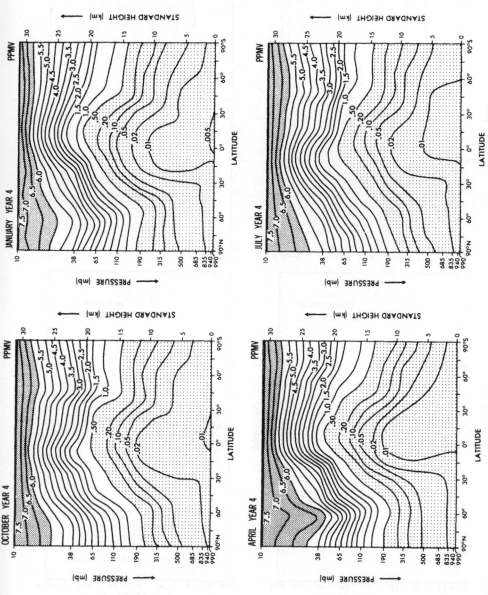

Fig. 11.9. Zonal-mean mixing ratio (ppmv) at 3-month intervals for the "stratified tracer" experiment described in the text. [After Mahlman et al. (1980). American Meteorological Society.]

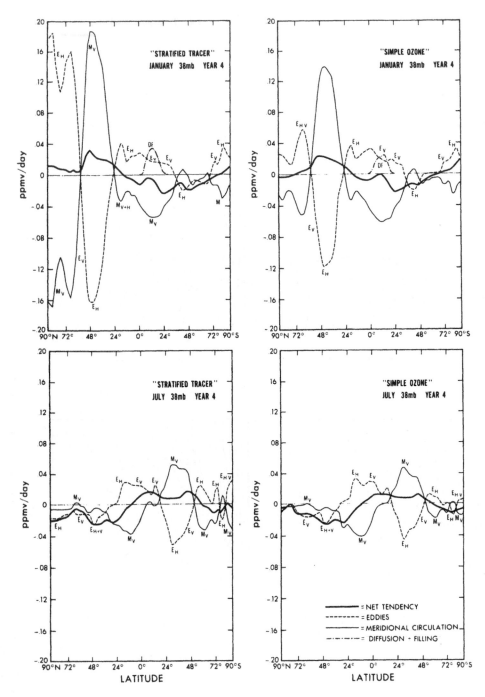

Fig. 11.10. Eulerian-mean balances in the tracer continuity equation at the 38-mb level for two experiments with the GFDL tracer model. [After Mahlman *et al.* (1980). American Meteorological Society.]

the extratropics are much greater in the Northern Hemisphere in January than in the Southern Hemisphere in July.

The Eulerian mean tracer budget of Fig. 11.10 also illustrates the strong compensation between the eddy flux divergence and advection by the mean meridional circulation, which was discussed extensively in Chapter 9. Note especially that near 45°N in January the net tracer tendency is due to a very small excess of mean meridional advection over eddy flux divergence. Thus, conditions in the model at this time and latitude are quite close to those of the nontransport theorem.

In order to better elucidate the net transport, Mahlman *et al.* (1980) used a simple Lagrangian analysis in which they followed short-term isobaric trajectories of a number of parcels advected by the January monthly mean winds from the tracer model. These trajectories show the parcel motions relative to the time mean flow (stationary waves) and do not correspond to the total parcel motions, which depart from isobaric surfaces and are significantly influenced by wave transience (cf. Section 9.4.2). Nevertheless, the schematic parcel drift relative to the mean flow shown in Fig. 11.11

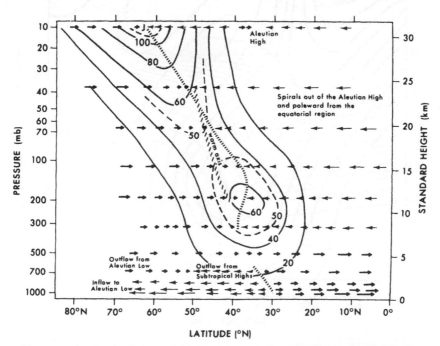

Fig. 11.11. Schematic view of parcel drift relative to the January-mean jetstream simulated in a GCM. Note strong convergence of parcels at and to the poleward side of the time-mean jetstream axis (dashed lines). Diagonal dashed line shows axis of warmest air. [After Mahlman *et al.* (1980). American Meteorological Society.]

should be qualitatively correct. The figure, which shows a cross section at 150°E longitude through the core of the West Pacific jetstream, reveals that the parcel drifts are toward the jetstream core from both the equatorial and polar sides. The effect of this parcel drift is to intensify the gradient of the tracer across the jet (Fig. 11.12) by poleward advection of low mixing ratios and equatorward advection of high mixing ratios. Comparison of Fig. 11.12 with Fig. 9.12 indicates that the parcel drifts in the model are attempting to produce a tracer distribution similar to that of the observed ozone distribution associated with a folded tropopause. The limited horizontal and vertical resolution in the model prevents formation of an actual fold. Nevertheless, it is clear that the model is able to simulate the large-scale environment required for formation of tropopause folds.

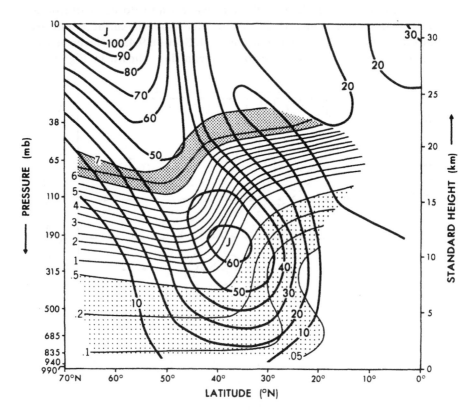

Fig. 11.12. January-mean mixing ratio calculated relative to the simulated jetstream at 150°E for the stratified tracer experiment described in the text. Note the downward displacement of high values on the poleward side of the jet. [After Mahlman *et al.* (1980). American Meteorological Society.]

The stratified tracer experiment represents a study of a stratospheric source gas (i.e., a tracer that is produced in the stratosphere and is transported to the troposphere, where it is destroyed). Another important class of tracers are the long-lived tropospheric source gases, which are produced in the troposphere and destroyed in the stratosphere; some typical vertical profiles of such tracers were shown in Fig. 1.8. The GFDL tracer model has been used by Mahlman *et al.* (1986) to study the transport of a representative tropospheric source gas (N_2O). They examined the dependence of the mean tracer distribution on the chemical timescale, by running simulations with three differing assumed rates of stratospheric destruction. The "slow-sink" case, whose January monthly mean meridional cross section is shown in Fig. 11.13, represents the best estimate of the actual atmospheric destruction rates, with a vertical profile of chemical lifetime similar to that of Fig. 9.2. The general pattern in Fig. 11.13 is in agreement with observations (cf. Fig. 9.3); the model simulates the familiar poleward–downward slopes of the constant-mixing-ratio surfaces, with slopes very similar to those of the stratified tracer experiment shown in Fig. 11.9. However, in the model, unlike the atmosphere, the equatorial bulge remains centered south of the equator for almost the entire year.

The major differences apparent between the slow sink and a "fast-sink" simulation (with chemical lifetimes about half those in the slow sink case) are in the meridional and vertical gradients, which are much steeper in the fast sink case. The meridional tracer slope,

$$\left(\frac{\partial z}{\partial y}\right)_{\bar{\chi} = \text{constant}} = -\bar{\chi}_y / \bar{\chi}_z, \tag{11.4.1}$$

is, however, nearly the same for the two cases. This similarity of equilibrium slopes is typical of all tracers for which the chemical timescale is long relative to the transport timescale. The similarity of slopes also extends to the longitudinal direction. Thus, the topography of time-mean mixing ratio surfaces is similar for all long-lived tracers. Comparison of the model with observations of N_2O shows that the meridional slopes in the model are about 30% smaller than the observed slopes. Mahlman *et al.* argue that this discrepancy is due to the fact that the eddies in the model stratosphere are not as active as in the real atmosphere, and hence the diabatic circulation is underestimated. This defect is, of course, directly related to the polar cold bias of the GCM discussed in Section 11.2.

Global data from satellite observations are available for only a very limited number of tracers. Thus, comparisons of model simulated global distributions with observed distributions are not always possible. Some comparisons can still be made, however, by using the time series of "data"

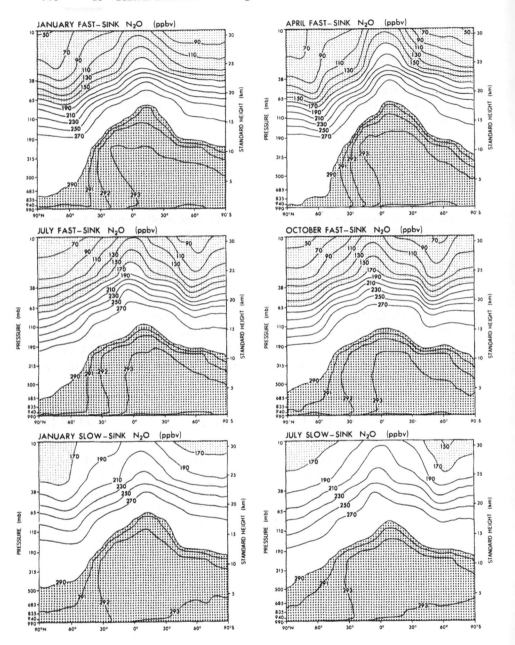

Fig. 11.13. Height–latitude distribution of simulated N_2O in the GFDL tracer model. [After Mahlman *et al.* (1986).]

generated at a particular grid location in the model to compare the temporal variability of the simulated tracer with that found in long series of observations. Mahlman *et al.* (1986) evaluated the "equivalent displacement height" defined in Eq. (9.5.2) for their N_2O simulation using the grid point nearest Laramie, Wyoming, a location for which there are a number of balloon-borne tracer measurements. The vertical profiles from the model and from the observations (shown in Fig. 11.14) are remarkably similar, indicating that the model does produce about the right amount of temporal variance. Synoptic maps of the N_2O distribution simulated in the model reveal that large pulses of temporal variability can be accounted for by rapid meridional transport associated with planetary wavebreaking, just as in the cases of observed ozone, methane, and potential vorticity discussed in Chapter 9.

The GFDL tracer model has demonstrated that three-dimensional general circulation models are powerful tools for the study of tracer transport. As these models become increasingly sophisticated, it should be possible to use them to simulate the annually varying ozone distribution with accurate photochemical and radiative heating algorithms and thus to achieve a primary goal for middle atmosphere science: a completely coupled radiative, dynamical, and photochemical model.

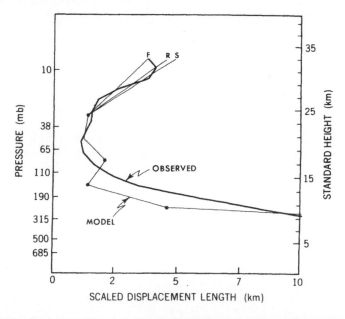

Fig. 11.14. Vertical profiles of equivalent displacement heights for N_2O simulated in the GFDL tracer model compared to observed values for Laramie, Wyoming. [After Mahlman *et al.* (1986).]

References

11.1. A review of models of the middle atmospheric circulation is given by Geller (1984). Haltiner and Williams (1980) provide a nice survey of the various numerical techniques and physical parameterizations used in GCMs. General circulation modeling by spectral methods is reviewed by Bourke *et al.* (1977). A stratospheric GCM developed at the United Kingdom Meteorological Office is described by O'Neill *et al.* (1982).

11.2. The basic structure of the GFDL models is discussed in Holloway and Manabe (1971). Results from these models are given in a number of papers, including Manabe and Terpstra (1974) Manabe *et al.* (1974), and Manabe and Mahlman (1976). For the SKYHI model, see Fels *et al.* (1980), Mahlman and Umscheid (1984), and Miyahara *et al.* (1986).

11.3. Initialization of global models for real data forecasts is reviewed by Daley (1981). The ECMWF forecast model is described by Burridge and Haseler (1977).

11.4. The interpretation of tracer transport experiments in isobaric and isentropic coordinates is discussed by Mahlman (1985). A detailed analysis of water-vapor transport in a GCM is given in Allam and Tuck (1984).

Chapter 12 | Interaction between the Middle Atmosphere and the Lower Atmosphere

12.1 Introduction

The middle and lower atmospheres are linked dynamically, radiatively, and chemically: the general circulation of the middle atmosphere is to a large degree controlled dynamically by motions that originate in the troposphere; the thermal structure of the middle atmosphere is influenced by upwelling thermal radiation originating in the troposphere, and by solar radiation backscattered from the surface and clouds; and the chemical composition of the middle atmosphere is influenced by tropospheric source gases that are transported into the stratosphere and by the loss of photochemically active gases through transport into the troposphere. Thus, the troposphere strongly influences the physics and chemistry of the middle atmosphere in a variety of ways. The influence of the middle atmosphere on the troposphere, on the other hand, is far more subtle. There is, however, growing evidence from both theory and observation that the middle atmosphere does play at least a minor role in the general circulation and radiative balance of the troposphere and a crucial role in the chemistry of the troposphere.

The most obvious link between the middle and lower atmospheres is, perhaps, the direct exchange of radiation between the stratosphere and troposphere (or ground) due to emission and absorption of radiation by various trace gases. Changes in the distributions of any of these may influence the temperature profile in both the troposphere and stratosphere, and hence may possibly influence the general circulation in both regions.

The radiative linkage is complicated by the fact that many radiatively active trace gases also participate in photochemical processes that may

443

influence the budget of ozone. For example, the synthetic chlorofluoro-methanes have direct radiative effects in the troposphere due to their strong absorption in the infrared. They also may indirectly affect the temperature in the troposphere through their photochemical influence on the ozone layer. The actual role of such processes depends not only on radiation and photochemistry, but also on the manner in which transport influences the distributions of photochemically and radiatively significant tracers within both the troposphere and the stratosphere.

Tracer transport is, of course, only one manner by which the motion field provides linkages between the lower and upper atmospheres. Meteorologically more important are the links provided by the vertical propagation of wave activity as indicated, for example, by the EP flux. It is through such vertical wave propagation that the troposphere provides a first-order control on the circulation of the middle atmosphere (see Chapter 7).

In this final chapter we examine some of the radiative and dynamical processes that link the lower and upper atmospheres and discuss possible climatological consequences of such linkages.

12.2   Radiative Links: Deductions from Simple Models

The major solar and atmospheric radiative processes, which together establish the vertical temperature structure of the middle atmosphere, were discussed in Chapter 2. To a first approximation the radiative equilibrium temperature distribution is established by the balance among absorption of solar ultraviolet radiation by O_3 and the absorption and emission of thermal radiation by CO_2, O_3, and H_2O. There are, however, a number of minor trace gases that have strong absorption features in the 7- to 13-μm region where the atmosphere is otherwise rather transparent; together these may have a significant impact on the radiative balance. The concentrations of several of the radiatively active gases (including tropospheric ozone) are increasing in time due to various human impacts; the concentration of stratospheric O_3, on the other hand, is predicted to decrease (see Chapter 10). Stratospheric aerosols, which also influence the radiative balance, vary dramatically in time, mainly owing to volcanic emissions.

The possible impacts of perturbations in the concentrations of radiatively active species on atmospheric and surface temperatures have been estimated primarily with the aid of one-dimensional radiative convective equilibrium models. These solve for the vertical temperature profile in an atmospheric column by considering only the vertical transfers of heat due to radiation and convection. Normally, the temperature profiles computed are intended to represent global mean profiles. (Recall, however, that such a model was used to compute the latitudinally dependent "radiatively determined" temperature distribution shown in Fig. 1.2.) Typically, these models are solved iteratively to determine, for a given distribution of trace species, the temperature profile that satisfies the following conditions: (1) a balance between the upward infrared radiation and downward solar radiation at the top of the atmosphere, (2) radiative equilibrium at each level in the stratosphere, (3) a fixed lapse rate (usually 6.5 K km^{-1}) in the troposphere, and (4) a surface energy balance. The last may include explicit calculation of the surface–atmosphere exchange of latent heat (Ramanathan, 1981). Such models greatly simplify the actual situation by neglecting horizontal variability and many of the feedbacks among radiation, dynamics, and chemistry. Nevertheless, they are valuable tools for exploring the sensitivity of the atmosphere to changes in radiatively active species.

CO_2 is the best known example of a radiatively active gas whose concentration is changing in time. Current projections indicate that the present atmospheric abundance of CO_2 will increase by 50% during the next 50 years, due to burning of fossil fuels. Such an increase is expected to cause an increase in global mean surface temperature of about 1 K, while temperatures in the stratosphere will *decrease* by several degrees due to the increased thermal emission to space caused by the increased CO_2 concentrations in the stratosphere. The increases currently being observed in several of the minor traces gases—N_2O, CH_4, and the chlorofluoromethanes— together may produce a surface temperature increase nearly as large as that predicted for CO_2 alone (Ramanathan *et al.*, 1985). Such optically thin minor species have little direct radiative effect on the temperature of the stratosphere, but they may have an indirect influence through their influence on the photochemical budget of the ozone layer (see Chapter 10). Radiative convective models predict that increases in CO_2 should produce temperature decreases in the stratosphere that are several times the tropospheric temperature increase. An example is given in Fig. 12.1, which shows predicted vertical profiles of the atmospheric temperature change from 1980 to 2030 given by a one-dimensional radiative convective equilibrium model using a current "best estimate" scenario for changes in the concentrations of various trace gases during that period.

The large temperature decrease in the stratosphere given for the "all trace gases" case of Fig. 12.1 is primarily due to the decrease in stratospheric ozone due to chlorine chemistry. The predicted ozone change is strongly height-dependent, with maximum decreases expected near the 40-km level. Such ozone depletion should cause a substantial decrease in the solar heating due to absorption by O_3 and hence in the temperature of the upper stratosphere. Although increased ultraviolet radiation at the ground, rather than

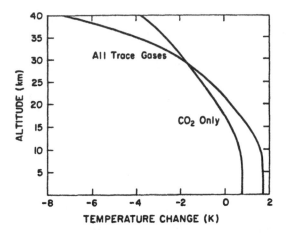

Fig. 12.1. Predicted change from 1980 to 2030 in the vertical distribution of temperature due to increase in CO_2 alone and in CO_2 along with all other trace gases that are thought to be increasing or decreasing in concentration. [After Ramanathan *et al.* (1985).]

surface-temperature change, is the primary environmental concern related to ozone depletion, radiative convective models do suggest that a decrease in the stratospheric ozone not only will decrease the temperature in the stratosphere due to the reduction of the absorption of solar radiation by ozone, but also may tend to cause a small increase in temperature at the ground in low and middle latitudes due to the increased penetration of downward-directed solar ultraviolet radiation.

At higher latitudes, surface warming due to increased solar flux at the ground is more than compensated by surface cooling due to a reduced downward flux of infrared radiation. The latter is caused by the temperature decrease in the stratosphere, which leads to smaller infrared emission for both ozone and carbon dioxide. This, then, is one example (although rather minor) of a situation in which human impacts on the stratosphere may cause changes in the troposphere.

12.3   Radiative Links: Deductions from GCMs

The prediction based on one-dimensional radiative convective models that stratospheric temperatures should be strongly affected by perturbations in the concentrations of ozone or carbon dioxide is supported by a series of three-dimensional general circulation model experiments reported by Fels *et al.* (1980). They used the GFDL SKYHI model described in Section 11.2 to examine the annual mean temperature changes that would result from a doubling of CO_2 or a halving of O_3. For these experiments the low-resolution 9°-by-10° version of the model was used. This version underestimates the eddy activity in the stratosphere and hence produces a climatology that is too close to radiative equilibrium, but it should still provide qualitatively valid information.

For both the ozone and CO_2 perturbation experiments, the zonally averaged temperature perturbations predicted in the GCM runs were compared to predictions not only from a radiative convective model applied at each latitudinal gridpoint in the GCM, but also from a two-dimensional model, called the fixed dynamical heating (FDH) model. In the FDH model a heat balance is assumed to hold at each height and latitude such that the sum of solar heating and long-wave cooling is balanced by dynamical heating or cooling:

$$\bar{Q}_{SW} + \bar{Q}_{LW}(\bar{T}) + \bar{Q}_{DYN} = 0, \qquad \bar{Q}_{DYN} \equiv -(\overline{D\theta/Dt}). \quad (12.3.1)$$

The FDH model assumes that for any perturbation in the trace gases that alters $\bar{Q}_{SW}$, to $\bar{Q}_{SW}^{(1)}$, say, the temperature profile is altered to $\bar{T}^{(1)}$ such that a new balance arises in the form

$$\bar{Q}_{SW}^{(1)} + \bar{Q}_{LW}(\bar{T}^{(1)}) + \bar{Q}_{DYN} = 0. \qquad (12.3.2)$$

Thus, in the FDH model the zonal-mean dynamical heating at each altitude and latitude is assumed to remain fixed; the perturbed temperature is found by subtracting Eq. (12.3.1) from Eq. (12.3.2) and performing a purely radiative calculation.

The arguments presented in Chapter 7 concerning the role of atmospheric eddies in establishing the temperature structure of the middle atmosphere suggest that the fixed dynamical heating assumption is a reasonable first approximation. There it was shown that the departure of the temperature distribution from radiative equilibrium is dependent on the intensity of the eddy forcing. Thus the net radiative heating [the sum of the first two terms in Eq. (12.3.1)] should depend on dynamics and will be altered only by processes that alter the dynamics. Specifically, a change of the EP flux divergence or small-scale heating or momentum forcing terms is required to support a change in the departure of temperature from radiative equilibrium [see Eq. (7.2.4c)]. Now, most of the eddy activity in the stratosphere is believed to result from vertical propagation of eddies generated in the troposphere. Thus, perturbations in radiatively active trace gases that do not influence the climate of the troposphere should not change the intensity of the eddy sources for the stratosphere. As the eddies propagate into the stratosphere, their propagation characteristics depend on the zonal mean

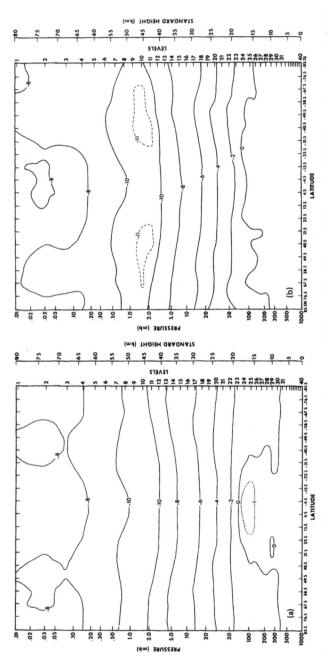

Fig. 12.2. Comparison of temperature changes simulated by (a) the FDH model and (b) a GCM for doubled CO_2 and annual mean forcing. [After Fels et al. (1980). American Meteorological Society.]

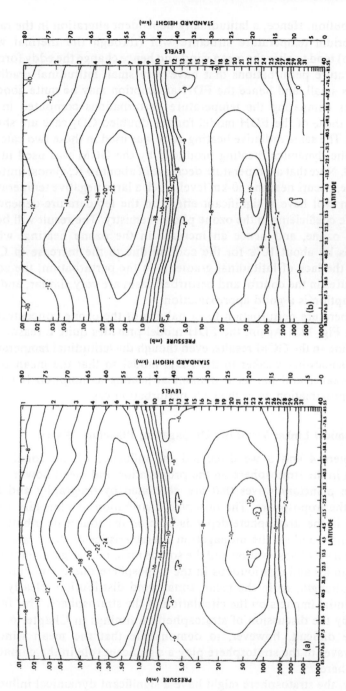

Fig. 12.3. Same as Fig. 12.2 but for 50% ozone reduction. [After Fels *et al.* (1980). American Meteorological Society.]

wind distribution. Hence, a latitudinally dependent alteration in the radiative equilibrium temperature structure will (through the thermal wind relationship) tend to alter the mean wind and hence change the eddy forcing. For temperature perturbations that have very small latitudinal gradients this effect is small, and hence the FDH assumption may be quite good.

Examples comparing the temperature perturbations computed in the GCM with those of the FDH model for the doubled CO_2 case are shown in Fig. 12.2. The same radiative heating code was used in these two calculations, and the dynamical heating produced by the GCM was used in the FDH model. Note that a temperature decrease of about 10 K, almost uniform with latitude, occurs near the 50-km level. Such a large negative temperature perturbation will have a significant effect on the temperature-dependent reaction-rate coefficients in the ozone photochemistry. The result will be an increase of ozone, and hence an increase in the ozone heating, which compensates by about 10% for the cooling due to the increase in CO_2. Because of the lack of latitudinal gradients in the perturbation, the zonal wind structures in the control and perturbed cases are very similar, and the FDH assumption is a good approximation.

The temperature perturbation for a halving of the ozone concentration is shown in Fig. 12.3. Again, the FDH model provides quite a reasonable approximation to the GCM results, even though the latitudinal temperature gradient is somewhat weaker in the FDH model so that the mean zonal winds are weaker and the fixed dynamical heating assumption is not as good an approximation as in the CO_2 perturbation case.

12.4 Dynamical Links: Vertically Propagating Planetary Waves

In Chapters 4 and 5 we discussed the influence of the zonal wind distribution in the stratosphere on the propagation and mean-flow interactions of the planetary waves that are generated by orography and heat sources in the troposphere. The observed distribution of planetary waves at any level in the stratosphere depends both on the source distribution in the troposphere and on the propagation characteristics of the atmosphere between the level of observation and the source region. Changes in planetary wave amplitudes and distributions in the troposphere due to anomalies in the source strength, or in the mean zonal wind distribution, clearly may have substantial impacts on the circulation of the stratosphere. This fact is illustrated by the discussion of stratospheric warmings in Chapter 6. It is much more difficult, however, to demonstrate that the mean wind or planetary waves in the stratosphere play a significant role in the climate of the troposphere.

In theory, the stratosphere might have a significant dynamical influence on the troposphere even if planetary wave sources were limited entirely to the troposphere, provided that some reflection of vertically propagating wave activity were to occur. An ozone reduction that was primarily limited to high latitudes, as predicted in some two-dimensional models, would alter the latitudinal gradient of the radiative equilibrium temperature, and hence should perturb the mean zonal wind distribution. This might in turn change the amount or distribution of wave reflection. However, the quasi-geostrophic model of Schoeberl and Strobel (1978) showed only a small planetary-wave response for even a large high-latitude ozone reduction.

Only if wave reflection were to create a resonant amplification of tropospheric waves does it seem possible that mean wind changes in the stratosphere could influence the climate of the troposphere. Tung and Lindzen (1979) argued that such a resonant amplification could occur if for some reason the stratospheric polar night jet were to weaken and at the same time descend to a lower than normal elevation, as observed during the initial phase of some stratospheric sudden warmings. They proposed that this linear resonance could account for planetary-scale "blocking" events in the troposphere. Plumb (1981) (see Section 6.3.4) suggested a nonlinear feedback mechanism in which amplifying planetary waves alter the mean wind distribution in the stratosphere so that it becomes closer to a resonant state, causing waves to grow more rapidly.

Resonance is a mechanism through which in principle the stratosphere might strongly influence the planetary wave distribution throughout the depth of the atmosphere, including the troposphere, and hence play a significant role in the production of climate anomalies. Thus, resonance might seem to provide a qualitative explanation for the apparent relationship between the occurrence of sudden stratospheric warmings and tropospheric blocking. As yet, however, there has been no quantitative demonstration that such stratosphere–troposphere resonances can occur in the real atmosphere. Indeed, detailed calculations of the linear stationary planetary wave response to realistic topography and heating distributions argue against the resonance theory. Using a high-resolution global primitive-equation model, Jacqmin and Lindzen (1985) demonstrated that the wave response in the troposphere is quite insensitive to changes in the basic state in the stratosphere, provided that the zonal-mean wind variation in the troposphere is limited to the range of values observed. Even large alterations in the distribution of the polar night jet in the stratosphere caused only very small changes in the stationary circulation of the troposphere in their model, although large changes occurred in the planetary-wave distribution in the stratosphere. Little evidence of planetary wave reflection or resonance was found in the several cases that they studied.

Boville (1984), on the other hand, showed on the basis of experiments with a general circulation model that unrealistically large changes in the

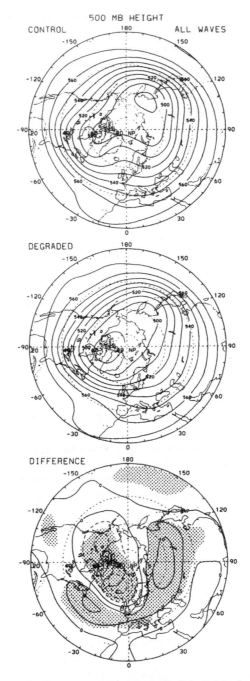

Fig. 12.4. Average 500-mb geopotential height fields (labelled in decameters) for (top) a standard January simulation with the NCAR spectral GCM, (middle) a simulation with a degraded climatology in the lower stratosphere, and (bottom) the degraded minus control difference. [After Boville (1984). American Meteorological Society.]

mean zonal wind distribution of the upper troposphere and lower stratosphere could produce dramatic alterations in the midtropospheric wave fields. These changes affected not only the stationary planetary wave structure, as shown in Fig. 12.4, but the transient cyclone-scale waves as well. The cyclone waves were apparently altered because the "storm tracks" along which they develop and evolve are strongly influenced by the stationary planetary-wave fields. Boville's results indicate that even if in reality the stratospheric circulation does not significantly affect the troposphere, it is very important to properly model the lower stratosphere in simulations of the general circulation, and in particular to avoid the common model defect of the cold winter pole bias (see Section 11.2.1) in order to avoid adverse impacts on the simulated tropospheric circulation.

12.5 Interannual Variability in the Stratosphere

It has been known for many years that the long-period variation in the circulation of the equatorial lower stratosphere is dominated not by the annual cycle but by the quasi-biennial oscillation (see Chapter 8). In recent years, with the advent of satellite temperature sounding of the stratosphere it has become clear that the extratropical winter stratosphere also contains considerable variation on interannual time scales. This interannual variability is clearly apparent in the monthly averaged Northern-Hemisphere mean zonal wind sections shown in Fig. 12.5. The figure shows December, January, and February conditions for four separate years. In each year the zonal mean circulation for December is stronger than that for January or February, but there is substantial year-to-year variation in the circulation for each month and in the month-to-month changes.

At present it is not known whether the source of this interannual variability is in the troposphere, the stratosphere, or both. Some idealized models (e.g., Holton and Mass, 1976) have indicated that even in the presence of constant tropospheric forcing, the circulation in the winter stratosphere might oscillate irregularly between two quasi-equilibrium states. In reality, however, the general circulation of the troposphere itself has considerable interannual variability, and the observed interannual variability in the extratropical stratosphere may primarily reflect variability in the tropospheric forcing rather than variability generated in the stratosphere.

12.5.1 The Extratropical QBO

Although the interannual variability in the extratropical stratosphere is much less regular than is the QBO in the equatorial lower stratosphere, a

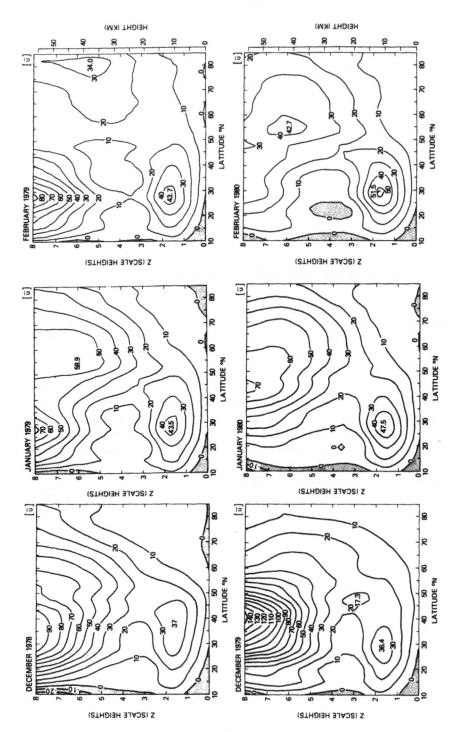

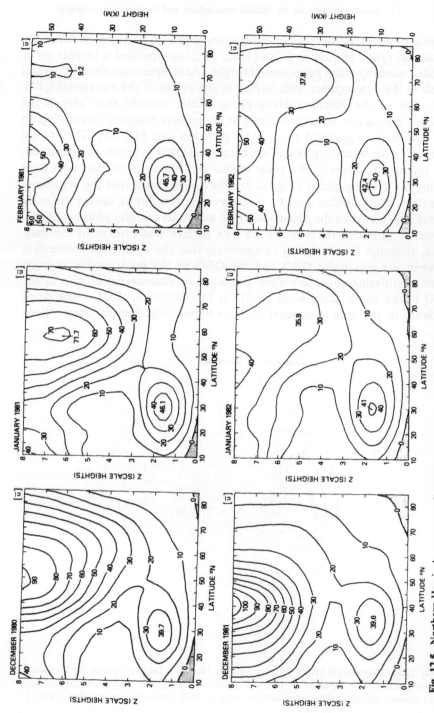

Fig. 12.5. Northern-Hemisphere average December, January, and February mean zonal winds (m s^{-1}) for the four winters of 1978–1979 through 1981–1982. [After Geller *et al.* (1983). American Meteorological Society.]

portion of the extratropical variability does appear to be correlated with the equatorial QBO. Holton and Tan (1980, 1982) composited a 16-year set of gridded monthly mean geopotential height and temperature data for several levels in the stratosphere with respect to the phase of the equatorial QBO. For each of the Northern-Hemisphere winter months they placed the monthly means into either a westerly or an easterly category, depending on the sign of the equatorial zonal wind at 50 mb (see Fig. 8.1). They found that during the westerly phase of the equatorial QBO at 50 mb, the geopotential heights are lower in the polar region and higher in midlatitudes than during the easterly phase (Fig. 12.6). They also found that the stationary planetary wave number 1 had a larger average amplitude during the early winter in years when the equatorial QBO was in its easterly phase at 50 mb. However, little difference in the EP flux pattern was evident in the data. Thus, although it is tempting to speculate that the dynamical connection between the equatorial QBO and the QBO in the polar vortex strength involves differing planetary-wave propagation characteristics due to the QBO in the mean zonal wind profile at low latitudes, there is not much evidence in the data to support such an hypothesis. Rather, observations

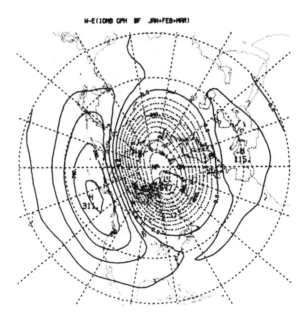

Fig. 12.6. Geopotential height difference (m) at 10 mb for January–March for 16 years of data composited with respect to the phase of the equatorial QBO at 50 mb (westerly category minus easterly category). The outer latitude circle is 20°N. [After Holton and Tan (1982).]

indicate that the extratropical QBO is primarily a zonally symmetric "seesaw" oscillation in which the oscillations in the polar height and thickness fields are out of phase with those in midlatitudes; there is no clear evidence for eddy driving of this seesaw oscillation.

A similar sort of QBO has been observed in the total ozone field. Quasi-biennial oscillations in total ozone have been reported from individual observing stations for many years. Recently the availability of multiyear satellite observations has made it possible to deduce the global pattern of the QBO in total ozone. In the equatorial region, positive ozone deviations occur a few months before the maximum westerlies at 50 mb, consistent with downward advection by the residual circulation associated with the zonal wind QBO (see Fig. 8.5). In midlatitudes the QBO in total ozone appears to have substantial asymmetry in phase between the Northern and Southern Hemispheres, as shown in Fig. 12.7. It should be noted, however,

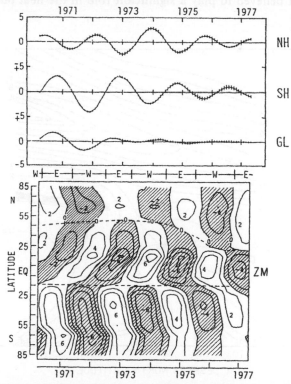

Fig. 12.7. QBO in total ozone (Dobson units) for the Northern Hemisphere (NH), the Southern Hemisphere (SH), global mean (GL), and latitudinally dependent zonal mean (ZM). Easterly (E) and westerly (W) phases of the zonal wind QBO at 50 mb are indicated above ZM. [After Hasebe (1983).]

that much longer data records will be needed to confirm that this phase behavior is an intrinsic aspect of the QBO in total ozone.

12.5.2   The Influence of Volcanic Emissions

An important, but highly irregular, source of interannual variability in the stratosphere is the aerosol mass loading that can result from volcanic eruptions. Normally the stratosphere has a thin aerosol layer centered several kilometers above the tropopause, as illustrated in Fig. 12.8. This layer is believed to consist primarily of sulfuric acid particles that result from transport of gaseous carbonyl sulfide (COS) across the tropopause, followed by photochemical conversion to sulfuric acid, nucleation, and particle growth by condensation and coalescence. This background stratospheric aerosol is not believed to play a significant role in the heat budget of the atmosphere.

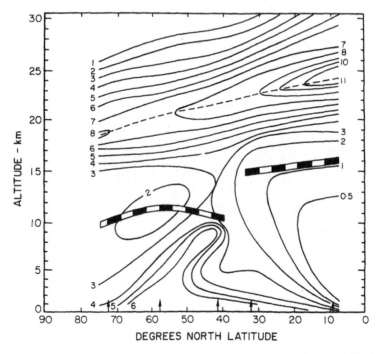

Fig. 12.8. Latitude–height cross section of the background aerosol mixing ratio (particles per milligram of air), June 1973. Heavy broken line marks the conventional tropopause, and dashed line marks the center of the stratospheric aerosol layer. [After Rosen *et al.* (1975). American Meteorological Society.]

Occasionally the stratospheric aerosol density is enhanced by an order of magnitude or more due to massive injections from volcanic eruptions of sulfur gases, which are rapidly converted to sulfuric acid aerosols. The eruptions of Mt. Agung in 1963 and El Chichon in 1982 both caused enormous enhancements in the aerosol loading of the stratosphere. For nearly 6 months following the April eruption of El Chichon the enhanced aerosol was concentrated in a zonal ring at about the 24-km level and in a latitudinal range of about 0° to 30°N. A vertical profile of the lidar scattering ratio (roughly proportional to aerosol mixing ratio) more than 6 months following the eruption is shown in Fig. 12.9. Shibata *et al.* (1984) showed that the very slow vertical spread of the aerosol layer is consistent with a vertical diffusion coefficient of $K_{zz} \approx 10^{-2} \, \mathrm{m^2 \, s^{-1}}$, which implies a very low magnitude of turbulent diffusion in the lower-stratospheric easterlies. The lack of significant transport across 30°N is consistent with the weak eddy activity and near radiative equilibrium temperature distribution expected for the extratropical summer stratosphere (see Section 7.2). The absence of cross-equatorial transport may in part be due to the fact that the QBO was in its easterly phase in the lower equatorial stratosphere during this period, so that Kelvin waves should be the dominant equatorial wave mode present. These waves, as discussed in Sections 4.7.1 and 8.3.2, are symmetric about the equator and produce negligible meridional parcel displacements.

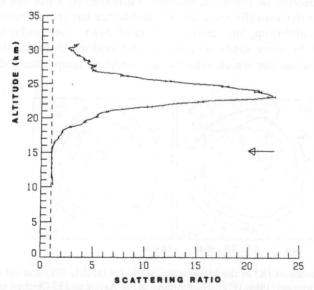

Fig. 12.9. Lidar scattering-ratio profile for October 22, 1982, at 17°N and 82°W. Arrow marks tropopause. [After McCormick and Osborn (1985).]

Such a persistent enhancement in the aerosol layer should influence the radiative equilibrium of the lower stratosphere by scattering solar radiation and by absorption of both solar and terrestrial radiation. For the eruption of El Chichon, radiative–convective equilibrium calculations predict that absorption should raise the radiative equilibrium temperature in the aerosol layer in the lower stratosphere by 3.5 K, while the net effect of the reduced solar radiation and enhanced downward thermal radiation reaching the surface should lead to insignificant temperature changes at the ground, except in the interior of continents (Pollack and Ackerman, 1983). For both the Mt. Agung and El Chichon eruptions, observations (Fig. 12.10) indicate warmings at 30 mb in the tropical belt affected by the aerosol of about 4–6 K. According to Labitzke and Naujokat (1983), these perturbations are more than 3 standard deviations from the 18-year average (1964–1982) at 10°N and 30 mb. Thus, they are clearly associated with the volcanic aerosol layers and do not simply represent normal fluctuations in the strength of the equatorial QBO.

12.5.3 Temperature Trends in the Stratosphere

Because of the coupling between temperature and photochemistry, the existence (or otherwise) of long-term trends in temperature in the stratosphere is a question of some significance. Furthermore, since the summer stratosphere is dynamically undisturbed, and hence has temperatures close to radiative equilibrium, any temperature trend due to the secular increase in CO_2 might be most easily detected in this region. Unfortunately, the rather short period for which reliable stratospheric temperature data are

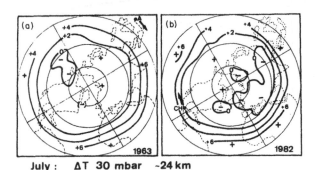

July :   ΔT 30 mbar   ~24 km

Fig. 12.10. Deviations (K) of the 30-mb temperatures for (a) July 1963 and (b) July 1982 from the 10-year average (1964–1973). The positions of Mt. Agung and El Chichon are shown, together with the mean wind directions for the summers of 1963 and 1982, respectively. [After Labitzke and Naujokat (1983).]

Departures ($\tfrac{1}{10}$K) of the 30–mbar July Temperatures, averaged over
2-years, from the 18-year mean 1964–1981

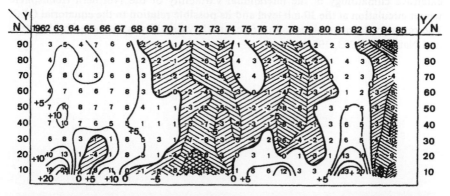

Fig. 12.11. Time–latitude plot of the departures (0.1 K) of the 30-mb July temperatures (averaged over 2 years to remove the QBO) from the 18-year mean of 1964–1981. Negative departures are shaded. [Updated from Labitzke and Naujokat (1983), courtesy of K. Labitzke.]

available, combined with the "red noise" character of natural interannual variability, makes the detection of any such secular trends difficult. The 23-year time series of July mean temperatures at the 30 mb level shown in Fig. 12.11 indicates substantial variation on a decadal timescale. It is not clear, however, whether the overall cooling observed during this period indicates a systematic decrease (as should eventually occur due to the secular increase of CO_2) or merely represents part of a longer-period oscillation. Because of the limited coverage provided by the meteorological radiosonde network, well-calibrated long-term monitoring by satellite is required for definitive detection of global trends in the temperature of the middle atmosphere.

References

12.1. The influence of perturbations in the middle atmosphere on climate is reviewed in Chapter 7 of Brasseur and Solomon (1984).

12.2. Radiatively induced climate perturbations associated with trace gas perturbations are discussed by Ramanathan *et al.* (1985) and by Wang *et al.* (1986). Both papers have extensive bibliographies.

12.3. CO_2 climate change computed in a spectral GCM is discussed by Washington and Meehl (1983).

12.4. Possible stratosphere–troposphere links due to modulations of planetary waves by solar-induced changes in the circulation of the middle atmosphere are analyzed by Schoeberl and Strobel (1978) and Geller and Alpert (1980).

12.5. Interannual variability in the winter stratosphere is discussed by Smith (1983), Geller *et al.* (1984), and Hamilton (1982b). Labitzke (1982) presents an extensive climatology of the interannual variability of the Northern-Hemisphere winter circulation at the 30-mb level and its possible relation to the equatorial QBO. The role of volcanic emissions in stratospheric variability is discussed by Labitzke and Naujokat (1983, 1984).

Bibliography

Abel, P. G., P. J. Ellis, J. T. Houghton, G. Peckham, C. D. Rodgers, S. D. Smith, and E. J. Williamson (1970). Remote sounding of atmospheric temperatures from satellites. II. The selective chopper radiometer for Nimbus D. *Proc. R. Soc. London, Ser. A* **320**, 35–55.

Akmaev, R. A., and G. M. Shved (1982). Parameterization of the radiative flux divergence in the 15 μm CO_2 band in the 30 to 75 km layer. *J. Atmos. Terr. Phys.* **44**, 993–1004.

Al-Ajmi, D. N., R. S. Harwood, and T. Miles (1985). A sudden warming in the middle atmosphere of the southern hemisphere. *Q. J. R. Meteorol. Soc.* **111**, 359–389.

Allam, R. J., and A. F. Tuck (1984). Transport of water vapour in a stratosphere–troposphere general circulation model. I: Fluxes, II: Trajectories. *Q. J. R. Meteorol. Soc.* **110**, 321–392.

Allen, M., and J. E. Frederick (1982). Effective photo-dissociation cross sections for molecular oxygen and nitric oxide in the Schumann–Runge bands. *J. Atmos. Sci.* **39**, 2066–2075.

Alpert, J. C., M. A. Geller, and S. K. Avery (1983). The response of stationary planetary waves to tropospheric forcing. *J. Atmos. Sci.* **40**, 2467–2483.

Andrews, D. G. (1983). A finite-amplitude Eliassen–Palm theorem in isentropic coordinates. *J. Atmos. Sci.* **40**, 1877–1883.

Andrews, D. G. (1985). Wave, mean-flow interaction in the middle atmosphere. *Adv. Geophys.* **28A**, 249–275.

Andrews, D. G., and M. E. McIntyre (1976a). Planetary waves in horizontal and vertical shear: the generalized Eliassen–Palm relation and the mean zonal acceleration. *J. Atmos. Sci.* **33**, 2031–2048.

Andrews, D. G., and M. E. McIntyre (1976b). Planetary waves in horizontal and vertical shear: asymptotic theory for equatorial waves in weak shear. *J. Atmos. Sci.* **33**, 2049–2053.

Andrews, D. G., and M. E. McIntyre (1978a). Generalized Eliassen–Palm and Charney–Drazin theorems for waves on axisymmetric mean flows in compressible atmospheres. *J. Atmos. Sci.* **35**, 175–185.

Andrews, D. G., and M. E. McIntyre (1978b). An exact theory for nonlinear waves on a Lagrangian-mean flow. *J. Fluid Mech.* **89**, 609–646.

Andrews, D. G., and M. E. McIntyre (1978c). On wave-action and its relatives. *J. Fluid Mech.* **89**, 647–664.

Andrews, D. G., J. D. Mahlman, and R. W. Sinclair (1983). Eliassen–Palm diagnostics of wave-mean flow interaction in the GFDL "SKYHI" general circulation model. *J. Atmos. Sci.* **40**, 2768–2784.

Apruzese, J. P., D. F. Strobel, and M. R. Schoeberl (1984). Parameterization of IR cooling in a middle atmosphere dynamics model 2. Non-LTE radiative transfer and the globally averaged temperature of the mesosphere and lower thermosphere. *J. Geophys. Res.* **89**, 4917-4926.

Bailey, P. L., and J. C. Gille (1986). Inversion of limb radiance measurements: an operational algorithm. *J. Geophys. Res.* **91**, 2757-2774.

Balsley, B. B., W. L. Ecklund, and D. C. Fritts (1983). VHF echoes from the high-latitude mesosphere and lower thermosphere: observations and interpretations. *J. Atmos. Sci.* **40**, 2451-2466.

Barnett, J. J. (1974). The mean meridional temperature behaviour of the stratosphere from November 1970 to November 1971 derived from measurements by the selective chopper radiometer on Nimbus IV. *Q. J. R. Meteorol. Soc.* **100**, 505-530.

Barnett, J. J., and M. Corney (1985a). Middle atmosphere reference model derived from satellite data. *Handbook MAP* **16**, 47-85.

Barnett, J. J., and M. Corney (1985b). Planetary waves. *Handbook MAP* **16**, 86-137.

Batchelor, G. K. (1967). "An Introduction to Fluid Dynamics," Cambridge Univ. Press, London and New York.

Belmont, A. D., D. G. Dartt, and G. D. Nastrom (1974). Periodic variations in stratospheric zonal wind from 20 to 65 km, at 80°N to 70°S. *Q. J. R. Meteorol. Soc.* **100**, 203-211.

Bevilacqua, R. M., J. J. Olivero, P. R. Schwartz, C. J. Gibbins, J. M. Bologna, and D. J. Thacker (1983). An observational study of water vapor in the mid-latitude mesosphere using ground-based microwave techniques. *J. Geophys. Res.* **88**, 8523-8534.

Bhartia, P. K., K. F. Klenk, C. K. Wong, D. Gordon, and A. J. Fleig (1984). Intercomparison of the Nimbus 7 SBUV/TOMS total ozone data sets with Dobson and M83 results. *J. Geophys. Res.* **89**, 5239-5247.

Blumen, W. (1978). A note on horizontal boundary conditions and stability of quasi-geostrophic flow. *J. Atmos. Sci.* **35**, 1314-1318.

Bourke, W., B. McAvaney, K. Puri, and R. Thurling (1977). Global modeling of atmospheric flow by spectral methods. *Methods Comput. Phys.* **17**, 267-324.

Boville, B. A. (1984). The influence of the polar night jet on the tropospheric circulation in a GCM. *J. Atmos. Sci.* **41**, 1132-1142.

Boville, B. A. (1985). The influence of wave damping on the winter lower stratosphere. *J. Atmos. Sci.* **42**, 904-916.

Bowman, K. P., and A. J. Krueger (1985). A global climatology of total ozone from the Nimbus 7 total ozone mapping spectrometer. *J. Geophys. Res.* **90**, 7967-7976.

Boyd, J. P. (1976). The noninteraction of waves with the zonally averaged flow on a spherical earth and the interrelationships of eddy fluxes of energy, heat and momentum. *J. Atmos. Sci.* **33**, 2285-2291.

Boyd, J. P. (1978a). The effects of latitudinal shear on equatorial waves. Part I: Theory and methods. *J. Atmos. Sci.* **35**, 2236-2258.

Boyd, J. P. (1978b). The effects of latitudinal shear on equatorial waves. Part II: Application to the atmosphere. *J. Atmos. Sci.* **35**, 2259-2267.

Boyd, J. P. (1982a). The influence of meridional shear on planetary waves. Part 1: Nonsingular wind profiles. *J. Atmos. Sci.* **39**, 756-769.

Boyd, J. P. (1982b). The influence of meridional shear on planetary waves. Part 2: Critical latitudes. *J. Atmos. Sci.* **39**, 770-790.

Boyd, J. P. (1985). Barotropic equatorial waves: the nonuniformity of the equatorial beta-plane. *J. Atmos. Sci.* **42**, 1965-1967.

Boyd, J. P., and Z. D. Christidis (1982). Low wavenumber instability on the equatorial beta-plane. *Geophys. Res. Lett.* **9**, 769-772.

Brasseur, G., and S. Solomon (1984). "Aeronomy of the Middle Atmosphere." Reidel, Dordrecht, Netherlands.

Bretherton, F. P. (1966a). Critical layer instability in baroclinic flows. *Q. J. R. Meteorol. Soc.* **92**, 325–334.

Bretherton, F. P. (1966b). The propagation of groups of internal gravity waves in a shear flow. *Q. J. R. Meteorol. Soc.* **92**, 466–480.

Bretherton, F. P. (1969). Momentum transport by gravity waves. *Q. J. R. Meteorol. Soc.* **95**, 213–243.

Bretherton, F. P., and C. J. R. Garrett (1968). Wavetrains in inhomogeneous moving media. *Proc. R. Soc. London Ser. A* **302**, 529–554.

Brewer, A. W. (1949). Evidence for a world circulation provided by the measurements of helium and water vapor distribution in the stratosphere. *Q. J. R. Meteorol. Soc.* **75**, 351–363.

Brownscombe, J. L., J. Nash, G. Vaughan, and C. F. Rogers (1985). Solar tides in the middle atmosphere. I: Description of satellite observations and comparison with theoretical calculations at equinox. *Q. J. R. Meteorol. Soc.* **111**, 677–689.

Burch, D. E., D. Gryvnak, and D. Williams (1960). "Infrared Absorption by Carbon Dioxide," Rep. No. 778, Ohio State Univ. Res. Found., Columbus.

Burridge, D. M., and J. Haseler (1977). A model for medium range weather forecasting—Adiabatic formulation. ECMWF Tech. Rep. No. 4.

Butchart, N., and E. E. Remsberg (1986). The area of the stratospheric polar vortex as a diagnostic for tracer transport on an isentropic surface. *J. Atmos. Sci.* **43**, 1319–1339.

Butchart, N., S. A. Clough, T. N. Palmer, and P. J. Trevelyan (1982). Simulations of an observed stratospheric warming with quasigeostrophic refractive index as a model diagnostic. *Q. J. R. Meteorol. Soc.* **108**, 475–502.

Chandrasekhar, S. (1950). "Radiative Transfer." Oxford Univ. Press, London and New York. (Reprinted by Dover.)

Chang, C.-P. (1976). Forcing of stratospheric Kelvin waves by tropospheric heat sources. *J. Atmos. Sci.* **33**, 740–744.

Chapman, S. (1930). A theory of upper atmospheric ozone. *R. Meteorol. Soc. Mem.* **3**, 103–125.

Chapman, S. (1931). The absorption and dissociative or ionizing effect of monochromatic radiations in an atmosphere on a rotating Earth. *Proc. Phys. Soc., London* **43**, 26–45.

Chapman, S., and R. S. Lindzen (1970). "Atmospheric Tides." Reidel, Dordrecht, Netherlands.

Charney, J. G. (1973). Planetary fluid dynamics. *In* "Dynamic Meteorology" (P. Morel, ed.), pp. 97–352. Reidel, Dordrecht, Netherlands.

Charney, J. G., and J. G. DeVore (1979). Multiple flow equilibria in the atmosphere and blocking. *J. Atmos. Sci.* **36**, 1205–1216.

Charney, J. G., and P. G. Drazin (1961). Propagation of planetary-scale disturbances from the lower into the upper atmosphere. *J. Geophys. Res.* **66**, 83–109.

Charney, J. G., and A. Eliassen (1949). A numerical method for predicting the perturbations of the middle latitude westerlies. *Tellus* **1**(2), 38–54.

Charney, J. G., and M. E. Stern (1962). On the stability of internal baroclinic jets in a rotating atmosphere. *J. Atmos. Sci.* **19**, 159–172.

Clough, S. A., N. S. Grahame, and A. O'Neill (1985). Potential vorticity in the stratosphere derived using data from satellites. *Q. J. R. Meteorol. Soc.* **111**, 335–358.

Cogley, A. C., and W. J. Borucki (1976). Exponential approximation for daily average solar heating or photolysis. *J. Atmos. Sci.* **33**, 1347–1356.

Coy, L., and M. H. Hitchman (1984). Kelvin wave packets and flow acceleration: a comparison of modeling and observations. *J. Atmos. Sci.* **41**, 1875–1880.

Cunnold, D., F. Alyea, N. Phillips, and R. Prinn (1975). A three-dimensional dynamical chemical model of atmospheric ozone. *J. Atmos. Sci.* **32**, 170–194.

Curtis, A. R. (1952). Discussion of "A statistical model for water-vapour absorption" by R. M. Goody. *Q. J. R. Meteorol. Soc.* **78**, 638–640.

Curtis, A. R. (1956). The computation of radiative heating rates in the atmosphere. *Proc. R. Soc. London, Ser. A* **236**, 156–159.

Curtis, A. R., and R. M. Goody (1956). Thermal radiation in the upper atmosphere. *Proc. R. Soc. London, Ser. A* **236**, 193–206.

Curtis, P. D., J. T. Houghton, G. D. Peskett, and C. D. Rodgers (1974). Remote sounding of atmospheric temperature from satellites. V. The pressure modulator radiometer for Nimbus F. *Proc. R. Soc. London, Ser. A* **337**, 135–150.

Daley, R. (1981). Normal mode initialization. *Revs. Geophys. Space Phys.* **19**, 450–468.

Danielsen, E. F. (1981). An objective method for determining the generalized-transport tensor for two dimensional Eulerian models. *J. Atmos. Sci.* **38**, 1319–1339.

Danielsen, E. F. (1985). Ozone transport. *In* "Ozone in the Free Atmosphere" (R. C. Whitten and S. S. Prasad, eds.), pp. 161–194. Van Nostrand-Reinhold, Princeton, New Jersey.

Danielsen, E. F., and D. Kley (1987). A tropical cumulonimbus source for correlated water vapor and ozone minima in extratropical stratosphere. *Q. J. R. Meteorol. Soc.* (submitted).

Davies, H. C. (1981). An interpretation of sudden warmings in terms of potential vorticity. *J. Atmos. Sci.* **38**, 427–445.

Dickinson, R. E. (1968). Planetary Rossby waves propagating vertically through weak westerly wind wave guides. *J. Atmos. Sci.* **25**, 984–1002.

Dickinson, R. E. (1969). Theory of planetary wave-zonal flow interaction. *J. Atmos. Sci.* **26**, 73–81.

Dickinson, R. E. (1970). Development of a Rossby wave critical level. *J. Atmos. Sci.* **27**, 627–633.

Dickinson, R. E. (1972). Infrared radiative heating and cooling in the Venusian mesosphere. I: Global mean radiative equilibrium. *J. Atmos. Sci.* **29**, 1531–1556.

Dickinson, R. E. (1973). Method of parameterization for infrared cooling between altitudes of 30 and 70 km. *J. Geophys. Res.* **78**, 4451–4457.

Dickinson, R. E. (1984). Infrared radiative cooling in the mesosphere and lower thermosphere. *J. Atmos. Terr. Phys.* **46**, 995–1008.

Dobson, G. M. B. (1956). Origin and distribution of the polyatomic molecules in the atmosphere. *Proc. R. Soc. London, Ser. A* **236**, 187–193.

Dopplick, T. G. (1972). Radiative heating of the global atmosphere. *J. Atmos. Sci.* **29**, 1278–1294.

Dopplick, T. G. (1979). Radiative heating of the global atmosphere: corrigendum. *J. Atmos. Sci.* **36**, 1812–1817.

Drayson, S. R. (1967). Calculation of long-wave radiative transfer in planetary atmospheres. Ph.D. Thesis, Coll. Eng., Univ. of Michigan, Ann Arbor.

Drayson, S. R., P. L. Bailey, H. Fischer, J. C. Gille, A. Girard, L. L. Gordley, J. E. Harries, W. G. Planet, E. E. Remsberg, and J. M. Russell, III (1984). Spectroscopy and transmittances for the LIMS experiment. *J. Geophys. Res.* **89** 5141–5146.

Dritschel, D. G. (1986). The nonlinear evolution of rotating configurations of uniform vorticity. *J. Fluid Mech.* **172**, 157–182.

Dunkerton, T. J. (1978). On the mean meridional mass motions of the stratosphere and mesosphere. *J. Atmos. Sci.* **35**, 2325–2333.

Dunkerton, T. J. (1979). On the role of the Kelvin wave in the westerly phase of the semiannual zonal wind oscillation. *J. Atmos. Sci.* **36**, 32–41.

Dunkerton, T. J. (1981a). On the inertial stability of the equatorial middle atmosphere. *J. Atmos. Sci.* **38**, 2354–2364.

Dunkerton, T. J. (1981b). Wave transience in a compressible atmosphere. Part II: Transient equatorial waves in the quasi-biennial oscillation. *J. Atmos. Sci.* **38**, 298–307.

Dunkerton, T. J. (1982a). Wave transience in a compressible atmosphere, Part III: The saturation of internal gravity waves in the mesosphere. *J. Atmos. Sci.* **39**, 1042-1051.

Dunkerton, T. J. (1982b). Theory of the mesopause semiannual oscillation. *J. Atmos. Sci.* **39**, 2681-2690.

Dunkerton, T. J. (1983a). Laterally-propagating Rossby waves in the easterly acceleration phase of the quasi-biennial oscillation. *Atmosphere-Ocean* **21**, 55-68.

Dunkerton, T. J. (1983b). A nonsymmetric equatorial inertial instability. *J. Atmos. Sci.* **40**, 807-813.

Dunkerton, T. J. (1984). Inertia-gravity waves in the stratosphere. *J. Atmos. Sci.* **41**, 3396-3404.

Dunkerton, T. J. (1985). A two-dimensional model of the quasi-biennial oscillation. *J. Atmos. Sci.* **42**, 1151-1160.

Dunkerton, T. J., and D. P. Delisi (1985). Climatology of the equatorial lower stratosphere. *J. Atmos. Sci.* **42**, 376-396.

Dunkerton, T. J., and D. P. Delisi (1986). Evolution of potential vorticity in the winter stratosphere of January-Frebruary 1979. *J. Geophys. Res.* **91**, 1199-1208.

Dunkerton, T. J., C.-P. F. Hsu, and M. E. McIntyre (1981). Some Eulerian and Lagrangian diagnostics for a model stratospheric warming. *J. Atmos. Sci.* **38**, 819-843.

Dutton, J. A. (1976). "The Ceaseless Wind." McGraw-Hill, New York.

Edmon, H. J., Jr., B. J. Hoskins, and M. E. McIntyre (1980). Eliassen-Palm cross sections for the troposphere. *J. Atmos. Sci.* **37**, 2600-2616; also corrigendum, *J. Atmos. Sci.* **38**, 1115 (1981).

Ehhalt, D. G., E. P. Röth, and U. Schmidt (1983). On the temporal variance of stratospheric gas concentrations. *J. Atmos. Chem.* **1**, 27-51.

Eliassen, A., and E. Palm (1961). On the transfer of energy in stationary mountain waves. *Geofys. Publ.* **22**(3), 1-23.

Elson, L. S. (1986). Ageostrophic motions in the stratosphere from satellite observations. *J. Atmos. Sci.* **43**, 409-418.

Emanuel, K. A. (1979). Inertial instability and mesoscale convective systems. Part I: Linear theory of inertial instability in rotating viscous fluids. *J. Atmos. Sci.* **36**, 2425-2449.

Ertel, H. (1942). Ein neuer hydrodynamischer Wirbelsatz. *Meteorol. Z.* **59**, 271-281.

Fels, S. B. (1979). Simple strategies for inclusion of Voigt effects in infrared cooling rate calculations. *Appl. Opt.* **18**, 2634-2637.

Fels, S. B. (1982). A parameterization of scale-dependent radiative damping rates in the middle atmosphere. *J. Atmos. Sci.* **39**, 1141-1152.

Fels, S. B. (1984). The radiative damping of short vertical scale waves in the mesosphere. *J. Atmos. Sci.* **41**, 1755-1764.

Fels, S. B. (1985). Radiative-dynamical interactions in the middle atmosphere. *Adv. Geophys.* **28A**, 277-300.

Fels, S. B., and M. D. Schwarzkopf (1981). An efficient, accurate algorithm for calculating CO_2 15 μm band cooling rates. *J. Geophys. Res.* **86**, 1205-1232.

Fels, S. B., J. D. Mahlman, M. D. Schwarzkopf, and R. W. Sinclair (1980). Stratospheric sensitivity to perturbations in ozone and carbon dioxide: radiative and dynamical response. *J. Atmos. Sci.* **37**, 2265-2297.

Fjørtoft, R. (1950). Application of integral theorems in deriving criteria of stability for laminar flows and for the baroclinic circular vortex. *Geofys. Publ.* **17**(6), 1-52.

Fomichev, V. I., and G. M. Shved (1985). Parameterization of the radiative flux divergence in the 9.6 μm O_3 band. *J. Atmos. Terr. Phys.* **47**, 1037-1049.

Fomichev, V. I., G. M. Shved, and A. A. Kutepov (1986). Radiative cooling of the 30-110 km atmospheric layer. *J. Atmos. Terr. Phys.* **48**, 529-544.

Forbes, J. M. (1984). Middle atmosphere tides. *J. Atmos. Terr. Phys.* **46**, 1049-1067.

Forbes, J. M., and H. B. Garrett (1979). Theoretical studies of atmospheric tides. *Rev. Geophys. Space Phys.* **17**, 1951-1981.

Frederiksen, J. S. (1982). Instability of the three-dimensional distorted stratospheric polar vortex at the onset of the sudden warming. *J. Atmos. Sci.* **39**, 2313-2329.

Fritts, D. C. (1984). Gravity wave saturation in the middle atmosphere: a review of theory and observations. *Rev. Geophys. Space Phys.* **22**, 275-308.

Fritz, S., and S. D. Soules (1972). Planetary variations of stratospheric temperatures. *Mon. Weather Rev.* **100**, 582-589.

Gage, K. S., and B. B. Balsley (1984). MST radar studies of wind and turbulence in the middle atmosphere. *J. Atmos. Terr. Phys.* **46**, 739-753.

Garcia, R. R., and S. Solomon (1983). A numerical model of zonally averaged dynamical and chemical structure of the middle atmosphere. *J. Geophys. Res.* **88**, 1379-1400.

Garcia, R. R., and S. Solomon (1985). The effect of breaking waves on the dynamics and chemical composition of the mesosphere and lower thermosphere. *J. Geophys. Res.* **90**, 3850-3868.

Geisler, J. E., and R. E. Dickinson (1976). The five-day wave on a sphere with realistic zonal winds. *J. Atmos. Sci.* **33**, 632-641.

Geller, M. A. (1984). Modeling the middle atmosphere circulation. In "Dynamics of the Middle Atmosphere" (J. R. Holton and T. Matsuno, ed.), pp. 467-500. Terrapub, Tokyo.

Geller, M. A., and J. C. Alpert (1980). Planetary wave coupling between the troposphere and the middle atmosphere as a possible sun-weather mechanism. *J. Atmos. Sci.* **37**, 1197-1215.

Geller, M. A., M.-F. Wu, and M. E. Gelman (1983). Troposphere-stratosphere (surface-55 km) monthly winter general circulation statistics for the northern hemisphere—four year averages. *J. Atmos. Sci.* **40**, 1334-1352.

Geller, M. A., M.-F. Wu, and M. E. Gellman (1984). Troposphere-stratosphere (surface-55 km) monthly winter general circulation statistics for the Northern Hemisphere—interannual variations. *J. Atmos. Sci.* **41**, 1726-1744.

Ghazi, A., P.-H. Huang, and M. P. McCormick (1985). A study on radiative damping of planetary waves utilizing stratospheric observations. *J. Atmos. Sci.* **42**, 2032-2042.

Gill, A. E. (1982). "Atmosphere-Ocean Dynamics." Academic Press, New York.

Gille, J. C., and F. B. House (1971). On the inversion of limb radiance measurements, I, Temperature and thickness. *J. Atmos. Sci.* **28**, 1427-1442.

Gille, J. C., and L. V. Lyjak (1984). An overview of wave-mean flow interactions during the winter of 1978-79 derived from LIMS observations. *In* "Dynamics of the Middle Atmosphere" (J. R. Holton and T. Matsuno, eds.), pp. 289-306. Terrapub, Tokyo.

Gille, J. C. and L. V. Lyjak (1986). Radiative heating and cooling rates in the middle atmosphere. *J. Atmos. Sci.* **43**, 2215-2229.

Gille, J. C., J. M. Russell, III, P. L. Bailey, L. L. Gordley, E. E. Remsberg, J. H. Lienesch, W. G. Planet, F. B. House, L. V. Lyjak, and S. A. Beck (1984). Validation of temperature retrievals obtained by the limb infrared monitor of the stratosphere (LIMS) experiment on Nimbus 7. *J. Geophys. Res.* **89**, 5147-5160.

Gilmore, F. R. (1964). "Potential Energy Curves for N_2, NO, O_2, and Corresponding Ions," Memo. R-4034-PR. RAND Corp., Santa Monica, California.

Godson, W. L. (1953). The evaluation of infra-red radiative fluxes due to atmospheric water vapour. *Q. J. R. Meteorol. Soc.* **79**, 367-379.

Goldman, A. (1968). On simple approximations to the equivalent width of a Lorentz line. *J. Quant. Spectrosc. Radiat. Transfer* **8**, 829-831.

Goody, R. M. (1952). A statistical model for water vapor absorption. *Q. J. R. Meteorol. Soc.* **78**, 165-169.

Goody, R. M. (1964). "Atmospheric Radiation I. Theoretical Basis." Oxford Univ. Press (Clarendon), London and New York.

Green, J. S. A. (1965). Atmospheric tidal oscillations: an analysis of the mechanics. *Proc. R. Soc. London, Ser. A* **288**, 564–574.

Grose, W. L., and K. V. Haggard (1981). Numerical simulation of a sudden stratospheric warming with a three-dimensional spectral, quasi-geostrophic model. *J. Atmos. Sci.* **38**, 1480–1497.

Groves, G. V. (1976). Rocket studies of atmospheric tides. *Proc. R. Soc. London, Ser. A* **351**, 437–469.

Haltiner, G. J., and R. T. Williams (1980). "Numerical Prediction and Dynamic Meteorology," 2nd Ed. Wiley, New York.

Hamilton, K. (1981a). Latent heat release as a possible forcing mechanism for atmospheric tides. *Mon. Weather Rev.* **109**, 3–17.

Hamilton, K. (1981b). The vertical structure of the quasi-biennial oscillation: Observations and theory. *Atmosphere-Ocean* **19**, 236–250.

Hamilton, K. (1982a). Stratospheric circulation statistics. NCAR Tech. Note NCAR/TN-191 + STR.

Hamilton, K. (1982b). Some features of the climatology of the northern hemisphere stratosphere revealed by NMC upper atmosphere analyses. *J. Atmos. Sci.* **39**, 2737–2749.

Hamilton, K. (1982c). Rocketsonde observations of the mesospheric semiannual oscillation at Kwajalein. *Atmosphere-Ocean* **20**, 281–286.

Hamilton, K. (1983). Diagnostic study of the momentum balance in the northern hemisphere winter stratosphere. *Mon. Weather Rev.* **111**, 1434–1441.

Hamilton, K. (1984). Mean wind evolution through the quasi-biennial cycle in the tropical lower stratosphere. *J. Atmos. Sci.* **41**, 2113–2125.

Hamilton, K. (1985). A possible relationship between tropical ocean temperatures and the observed amplitude of the atmospheric (1,1) Rossby normal mode. *J. Geophys. Res.* **90**, 8071–8074.

Hamilton, K. (1986). Dynamics of the stratospheric semi-annual oscillation. *J. Meteorol. Soc. Jpn.* **64**, 227–244.

Hanel, R. A., B. Schlachman, D. Rogers, and D. Vanous (1971). Nimbus 4 Michelson interferometer. *Appl. Opt.* **10**, 1376–1382.

Hartmann, D. L. (1979). Baroclinic instability of realistic zonal-mean states to planetary waves. *J. Atmos. Sci.* **36**, 2336–2349.

Hartmann, D. L. (1983). Barotropic instability of the polar night jet stream. *J. Atmos. Sci.* **40**, 817–835.

Hartmann, D. L. (1985). Some aspects of stratospheric dynamics. *Adv. Geophys.* **28A**, 219–247.

Hartmann, D. L., C. R. Mechoso, and K. Yamazaki (1984). Observations of wave–mean flow interaction in the southern hemisphere. *J. Atmos. Sci.* **41**, 351–362.

Harwood, R. S., and J. A. Pyle (1975). A two-dimensional mean circulation model for the atmosphere below 80 km. *Q. J. R. Meteorol. Soc.* **101**, 723–747.

Hasebe, F. (1983). Interannual variations of global total ozone revealed from Nimbus 4 BUV and ground-based observations. *J. Geophys. Res.* **88**, 6819–6834.

Hayashi, Y., and D. G. Golder (1978). The generation of equatorial transient planetary waves: control experiments with a GFDL general circulation model. *J. Atmos. Sci.* **35**, 2068–2082.

Hayashi, Y., D. G. Golder, and J. D. Mahlman (1984). Stratospheric and mesospheric Kelvin waves simulated by the GFDL "SKYHI" general circulation model. *J. Atmos. Sci.* **41**, 1971–1984.

Haynes, P. H. (1985). Nonlinear instability of a Rossby-wave critical layer. *J. Fluid Mech.* **161**, 493–511.

Haynes, P. H., and M. E. McIntyre (1987). On the evolution of vorticity and potential vorticity in the presence of diabatic heating and frictional or other forces. *J. Atmos. Sci.* **44**, 828–841.

Held, I. M. (1983). Stationary and quasi-stationary eddies in the extratropical troposphere: theory. *In* "Large-Scale Dynamical Process in the Atmosphere" (B. J. Hoskins and R. P. Pearce, eds.), pp. 127–168. Academic Press, London.

Held, I. M., and B. J. Hoskins (1985). Large-scale eddies and the general circulation of the troposphere. *Adv. Geophys.* **28A**, 3–31.

Held, I. M., and A. Y. Hou (1980). Nonlinear axially symmetric circulations in a nearly inviscid atmosphere. *J. Atmos. Sci.* **37**, 515–533.

Herman, J. R., and J. E. Mentall (1982). O_2 absorption cross sections (187–225 nm) from stratospheric solar flux measurements. *J. Geophys. Res.* **87**, 8967–8975.

Hess, P. G., and J. R. Holton (1985). The origin of temporal variance in long-lived trace constituents in the summer stratosphere. *J. Atmos. Sci.* **42**, 1455–1463.

Hirooka, T., and I. Hirota (1985). Normal mode Rossby waves observed in the upper stratosphere. Part II: second antisymmetric and symmetric modes of zonal wavenumbers 1 and 2. *J. Atmos. Sci.* **42**, 536–548.

Hirota, I. (1978). Equatorial waves in the upper stratosphere and mesosphere in relation to the semiannual oscillation of the zonal wind. *J. Atmos. Sci.* **35**, 714–722.

Hirota, I. (1980). Observational evidence of the semiannual oscillation in the tropical middle atmosphere—a review. *Pure Appl. Geophys.* **118**, 217–238.

Hirota, I., and J. J. Barnett (1977). Planetary waves in the winter mesosphere—preliminary analysis of Nimbus 6 PMR results. *Q. J. R. Meteorol. Soc.* **103**, 487–498.

Hirota, I., and T. Hirooka (1984). Normal mode Rossby waves observed in the upper stratosphere. Part I: First symmetric modes of zonal wavenumbers 1 and 2. *J. Atmos. Sci.* **41**, 1253–1267.

Hirota, I., and T. Niki (1985). A statistical study of inertia-gravity waves in the middle atmosphere. *J. Meteorol. Soc. Jpn*, **63**, 1055–1066.

Hitchman, M. H., and C. B. Leovy (1986). Evolution of the zonal mean state in the equatorial middle atmosphere during October 1978–May 1979. *J. Atmos. Sci.* **43**, 3159–3176.

Hitchman, M. H., C. B. Leovy, J. C. Gille, and P. L. Bailey (1987). Quasi-stationary zonally asymmetric circulations in the equatorial lower mesosphere. *J. Atmos. Sci.* (in press).

Holl, P. (1970). Die Vollständigkeit des Orthogonalsystems der Houghfunktionen. *Nachr. Akad. Wiss. Goettingen, Math.-Phys. Kl.*, 2, 7, 159–168. (Engl. transl., The completeness of the orthogonal system of the Hough functions. Avail. from Dep. Atmos. Sci., Colorado State Univ., Fort Collins.)

Holloway, J. L., and S. Manabe (1971). Simulation of climate by a global general circulation model: I. Hydrologic cycle and heat balance. *Mon. Weather Rev.* **99**, 335–370.

Holopainen, E. O. (1983). Transient eddies in mid-latitudes: observations and interpretation. *In* "Large-Scale Dynamical Processes in the Atmosphere" (B. J. Hoskins and R. P. Pearce, eds.), pp. 201–233. Academic Press, London.

Holton, J. R. (1972). Waves in the equatorial stratosphere generated by tropospheric heat sources. *J. Atmos. Sci.* **29**, 368–375.

Holton, J. R. (1973). On the frequency distribution of atmospheric Kelvin waves. *J. Atmos. Sci.* **30**, 499–501.

Holton, J. R. (1974). Forcing of mean flows by stationary waves. *J. Atmos. Sci.* **31**, 942–945.

Holton J. R. (1975). "The Dynamic Meteorology of the Stratosphere and Mesosphere," Meteorol. Monogr. No. 37. Am. Meteorol. Soc., Boston, Massachusetts.

Holton J. R. (1976). A semi-spectral numerical model for wave-mean flow interactions in the stratosphere: application to sudden warmings. *J. Atmos. Sci.* **33**, 1639–1649.

Holton, J. R. (1979a). "An Introduction to Dynamic Meteorology." Academic Press, New York.

Holton, J. R. (1979b). Equatorial wave-mean flow interaction: A numerical study of the role of latitudinal shear. *J. Atmos. Sci.* **36**, 1030-1040.

Holton, J. R. (1980). The dynamics of sudden stratospheric warmings. *Annu. Rev. Earth Planet. Sci.* **8**, 169-190.

Holton, J. R. (1981). An advective model for two-dimensional transport of stratospheric trace species. *J. Geophys. Res.* **86**, 11,989-11,994.

Holton, J. R. (1982). The role of gravity wave induced drag and diffusion in the momentum budget of the mesosphere. *J. Atmos. Sci.* **39**, 791-799.

Holton, J. R. (1983a). The influence of gravity wave breaking on the general circulation of the middle atmosphere. *J. Atmos. Sci.* **40**, 2497-2507.

Holton, J. R. (1983b). The stratosphere and its links to the troposphere. *In* "Large-Scale Dynamical Processes in the Atmosphere" (B. J. Hoskins and R. P. Pearce, eds.), pp. 277-303. Academic Press, London.

Holton, J. R. (1984a). The generation of mesospheric planetary waves by zonally asymmetric gravity wave breaking. *J. Atmos. Sci.* **41**, 3427-3430.

Holton, J. R. (1984b). Troposphere-stratosphere exchange of trace constituents: the water vapor puzzle. *In* "Dynamics of the Middle Atmosphere"(J. R. Holton and T. Matsuno, eds.), pp. 369-385. Terrapub, Tokyo.

Holton, J. R. (1986a). A dynamically based transport parameterization for one-dimensional photochemical models of the stratosphere. *J. Geophys. Res.* **91**, 2681-2686.

Holton, J. R. (1986b). Meridional distribution of stratospheric trace constituents. *J. Atmos. Sci.* **43**, 1238-1242.

Holton, J. R., and R. S. Lindzen (1972). An updated theory for the quasi-biennial cycle of the tropical stratosphere. *J. Atmos. Sci.* **29**, 1076-1080.

Holton, J. R., and C. Mass (1976). Stratospheric vacillation cycles. *J. Atmos. Sci.* **33**, 2218-2225.

Holton, J. R., and H.-C. Tan (1980). The influence of the equatorial quasi-biennial oscillation on the global circulation at 50 mb. *J. Atmos. Sci.* **37**, 2200-2208.

Holton, J. R., and H.-C. Tan (1982). The quasi-biennial oscillation in the Northern Hemisphere lower stratosphere. *J. Meteorol. Soc. Jpn.* **60**, 140-148.

Holton, J. R., and W. M. Wehrbein (1980). A numerical model of the zonal mean circulation of the middle atmosphere. *Pageoph* **118**, 284-306.

Holton, J. R., and X. Zhu (1984). A further study of gravity wave induced drag and diffusion in the mesosphere. *J. Atmos. Sci.* **41**, 2653-2662.

Hopkins, R. H. (1975). Evidence of polar-tropical coupling in upper stratospheric zonal wind anomalies. *J. Atmos. Sci.* **32**, 712-719.

Hoskins, B. J. (1972). Non-Boussinesq effects and further development in a model of upper tropospheric frontogenesis. *Q. J. R. Meteorol. Soc.* **98**, 532-541.

Hoskins, B. J. (1982). The mathematical theory of frontogenesis. *Annu. Rev. Fluid Mech.* **14**, 131-151.

Hoskins, B. J. (1983). Modelling of the transient eddies and their feedback on the mean flow. *In* "Large-Scale Dynamical Processes in the Atmosphere" (B. J. Hoskins and R. P. Pearce, eds.), pp. 169-199. Academic Press, London.

Hoskins, B. J., I. Dragici, and H. C. Davies (1978). A new look at the omega-equation. *Q. J. R. Meteorol. Soc.* **104**, 31-38.

Hoskins, B. J., M. E. McIntyre, and A. W. Robertson (1985). On the use and significance of isentropic potential vorticity maps. *Q. J. R. Meteorol. Soc.* **111**, 877-946.

Houghton, J. T. (1969). Absorption and emission by carbon dioxide in the mesosphere. *Q. J. R. Meteorol. Soc.* **95**, 1-20.

Houghton, J. T. (1978). The stratosphere and mesosphere. *Q. J. R. Meteorol. Soc.* **104**, 1-29.

Houghton, J. T., and S. D. Smith (1966). "Infra-Red Physics." Oxford Univ. Press, London and New York.

Houghton, J. T., and S. D. Smith (1970). Remote sounding of atmospheric temperature from satellites. I. Introduction. *Proc. R. Soc. London, Ser. A* **320**, 23–33.

Houghton, J. T., F. W. Taylor, and C. D. Rodgers (1984). "Remote Sounding of Atmospheres." Cambridge Univ. Press, London and New York.

Houze, R. A., Jr. (1982). Cloud clusters and large-scale vertical motions in the tropics. *J. Meteorol. Soc. Jpn.* **60**, 396–410.

Hsu, C.-P. F. (1980). Air parcel motions during a numerically simulated sudden stratospheric warming. *J. Atmos. Sci.* **37**, 2768–2792.

Hsu, C.-P. F. (1981). A numerical study of the role of wave–wave interactions during sudden stratospheric warmings. *J. Atmos. Sci.* **38**, 189–214.

Hunt, B. G. (1969). Experiments with a stratospheric general circulation model, III. Large scale diffusion of ozone including photochemistry. *Mon. Weather Rev.* **97**, 287–306.

Hunt, B. G. (1981). The maintenance of the zonal mean state of the upper atmosphere as represented in a three-dimensional general circulation model extending to 100 km. *J. Atmos. Sci.* **38**, 2172–2186.

Hunt, B. G., and S. Manabe (1968). Experiments with a stratospheric general circulation model, II. Large-scale diffusion of tracers in the stratosphere. *Mon. Weather Rev.* **96**, 503–539.

Jacqmin, D., and R. S. Lindzen (1985). The causation and sensitivity of northern winter planetary waves. *J. Atmos. Sci.* **42**, 724–745.

Jones, R. L. (1984). Satellite measurements of atmospheric composition: three years' observations of CH_4 and N_2O. *Adv. Space Res.* **4**(4), 121–130.

Jones, R. L., and J. A. Pyle (1984). Observations of CH_4 and N_2O by the Nimbus 7 SAMS: a comparison with *in situ* data and two-dimensional numerical model calculations. *J. Geophys. Res.* **89**, 5263–5279.

Jou, J.-D. (1985). A numerical study of the distribution and transport of ozone in the stratosphere. Ph.D. Thesis, Univ. of Washington, Seattle.

Kanzawa, H. (1982). Eliassen–Palm flux diagnostics and the effect of the mean wind on planetary wave propagation for an observed sudden stratospheric warming. *J. Meteorol. Soc. Jpn.* **60**, 1063–1073.

Kanzawa, H. (1984). Four observed sudden warmings diagnosed by the Eliassen–Palm flux and refractive index. *In* "Dynamics of the Middle Atmosphere" (J. R. Holton and T. Matsuno, eds.), pp. 307–331. Terrapub, Tokyo.

Karoly, D. J., and B. J. Hoskins (1982). Three dimensional propagation of planetary waves. *J. Meteorol. Soc. Jpn.* **60**, 109–123.

Kida, H. (1983). General circulation of air parcels and transport characteristics derived from a hemispheric GCM. Part 2. Very long-term motions of air parcels in the troposphere and stratosphere. *J. Meteorol. Soc. Jpn.* **61**, 510–523.

Kiehl, J. T., and V. Ramanathan (1983). CO_2 radiative parameterization used in climate models: comparison with narrow band models and with laboratory data. *J. Geophys. Res.* **88**, 5191–5202.

Killworth, P. D., and M. E. McIntyre (1985). Do Rossby-wave critical layers absorb, reflect or over-reflect? *J. Fluid Mech.* **161**, 449–492.

Kley, D., A. L. Schmeltekopf, K. Kelly, R. H. Winkler, T. L. Thompson, and M. McFarland (1982). Transport of water vapor through the tropical tropopause. *Geophys. Res. Lett.* **9**, 617–620.

Ko, M. K. W., K. K. Tung, D. K. Weisenstein, and N. D. Sze (1985). A zonal-mean model of stratospheric tracer transport in isentropic coordinates: Numerical simulations for nitrous oxide and nitric acid. *J. Geophys. Res.* **90**, 2313–2329.

Kockarts, G. (1971). Penetration of solar radiation in the Schumann-Runge bands of molecular oxygen. *In* "Mesospheric Models and Related Experiments" (G. Fiocco, ed.), pp. 160-176. Reidel, Dordrecht, Netherlands.

Kockarts, G. (1976). Absorption and photodissociation in the Schumann-Runge bands of molecular oxygen in the terrestrial atmosphere. *Planet. Space Sci.* **24**, 589-604.

Kondratyev, K. Ya. (1969). "Radiation in the Atmosphere." Academic Press, New York.

Kuhn, W. R., and J. London (1969): Infrared radiative cooling in the middle atmosphere (30-110 km). *J. Atmos. Sci.* **26**, 189-204.

Kurzeja, R. J. (1981). The transport of trace chemicals by planetary waves in the stratosphere. Part 1: Steady waves. *J. Atmos. Sci.* **38**, 2779-2788.

Kutepov, A. A., and G. M. Shved (1978). Radiative transfer in the 15 μ CO_2 band with breakdown of local thermodynamic equilibrium in the Earth's atmosphere. *Atmos. Ocean Phys.* **14**, 18-30.

Labitzke, K. (1981a). The amplification of height wave 1 in January 1979: a characteristic precondition for the major warming in February. *Mon. Weather Rev.* **109**, 983-989.

Labitzke, K. (1981b). Stratospheric-mesospheric midwinter disturbances: a summary of observed characteristics. *J. Geophys. Res.* **86**, 9665-9678.

Labitzke, K. (1982). On the interannual variability of the middle stratosphere during the northern winters. *J. Meteorol. Soc. Jpn.* **60**, 124-139.

Labitzke, K., and B. Naujokat (1983). On the variability and on trends of the temperature in the middle stratosphere. *Contrib. Atmos. Phys.* **56**, 495-507.

Labitzke, K., and B. Naujokat (1984). On the effect of the volcanic eruptions of Mount Agung and El Chichon on the temperature of the stratosphere. *Geofys. Int.* **23**(2), 223-232.

Labitzke, K., J. J. Barnett, and B. Edwards (1985). Atmospheric structure and its variation in the Region 20 to 120 km. Draft of a new reference middle atmosphere. *Handbook for MAP* (Middle Atmosphere Program), Vol. 16. SCOSTEP, Univ. of Illinois, Urbana-Champaign.

Lacis, A. A., and J. E. Hansen (1974). A parameterization for the absorption of solar radiation in the earth's atmosphere. *J. Atmos. Sci.* **31**, 118-133.

Leovy, C. B. (1964a). Radiative equilibrium of the mesosphere. *J. Atmos. Sci.* **21**, 238-248.

Leovy, C. B. (1964b). Simple models of thermally driven mesospheric circulation. *J. Atmos. Sci.* **21**, 327-341.

Leovy, C. B. (1984). Infrared radiative exchange in the middle atmosphere in the 15 micron band of carbon dioxide. *In* "Dynamics of the Middle Atmosphere" (J. R. Holton and T. Matsuno, eds.), pp. 355-366. Terrapub, Tokyo.

Leovy, C. B., C.-R. Sun, M. H. Hitchman, E. E. Remsberg, J. M. Russell, III, L. L. Gordley, J. C. Gille, and L. V. Lyjak (1985). Transport of ozone in the middle stratosphere: evidence for planetary wave breaking. *J. Atmos. Sci.* **42**, 230-244.

Levine, J. N. (1974). "Molecular Spectroscopy." Wiley, New York.

Lighthill, J. (1978). "Waves in Fluids." Cambridge Univ. Press, London and New York.

Lin, B.-D. (1982). The behavior of winter stationary planetary waves forced by topography and diabatic heating. *J. Atmos. Sci.* **39**, 1206-1226.

Lindzen, R. S. (1967). Thermally driven diurnal tide in the atmosphere. *Q. J. R. Meteorol. Soc.* **93**, 18-42.

Lindzen, R. S. (1968). The application of classical atmospheric tidal theory. *Proc. R. Soc. London Ser. A* **303**, 299-316.

Lindzen, R. S. (1971a). Atmospheric tides. *Lect. Appl. Math.* **14**, 293-362.

Lindzen, R. S. (1971b). Equatorial planetary waves in shear: Part I. *J. Atmos. Sci.* **28**, 609-622.

Lindzen, R. S. (1972). Equatorial planetary waves in shear: Part II. *J. Atmos. Sci.* **29**, 1452-1463.

Lindzen, R. S. (1979). Atmospheric tides. *Annu. Rev. Earth Planet. Sci.* **7**, 199-225.

Lindzen, R. S. (1981). Turbulence and stress owing to gravity wave and tidal breakdown. *J. Geophys. Res.* **86**, 9707-9714.

Lindzen, R. S., and J. R. Holton (1968). A theory of the quasi-biennial oscillation. *J. Atmos. Sci.* **25**, 1095-1107.

Lindzen, R. S., and D. I. Will (1973). An analytic formula for heating due to ozone absorption. *J. Atmos. Sci.* **30**, 513-515.

Liou, K.-N. (1980). "An Introduction to Atmospheric Radiation." Academic Press, New York.

London, J. (1980). Radiative energy sources and sinks in the stratosphere and mesosphere. *Atmos. Ozone, Proc. NATO Adv. Study Inst.* (A. C. Aiken, ed.), pp. 703-721. Fed. Aviat. Admin., U.S. Dep. Transp. Washington, D.C.

London, J. (1985). The observed distribution of atmospheric ozone and its variations. *In* "Ozone in the Free Atmosphere" (R. C. Whitten and S. S. Prasad, eds.), pp. 11-80. Van Nostrand-Reinhold, Princeton, New Jersey.

Longuet-Higgins, M. S. (1968). The eigenfunctions of Laplace's tidal equations over a sphere. *Philos. Trans. R. Soc. London. Ser. A* **262**, 511-607.

McCormick, M. P. (1983). Aerosol measurements from Earth orbiting spacecraft. *Adv. Space Res.* **2**(5), 73-86.

McCormick, M. P., and M. T. Osborn (1985). Airborne lidar measurements of El Chichon stratospheric aerosols. NASA Ref. Publ. No. 1136.

McCormick, M. P., P. Hamill, T. J. Pepin, W. P. Chu, T. J. Swissler, and L. R. McMaster (1979). Satellite studies of the stratospheric aerosol. *Bull. Am. Meteorol. Soc.* **60**, 1038-1046.

McCormick, M. P., T. J. Swissler, E. Hilsenrath, A. J. Krueger, and M. T. Osborn (1984). Satellite and correlative measurements of stratospheric ozone: comparison of measurements made by SAGE, ECC balloons, chemiluminescent and optical rocketsondes. *J. Geophys. Res.* **89**, 5315-5320.

McInturff, R. M., ed. (1978). Stratospheric warmings: synoptic, dynamic and general-circulation aspects. NASA Ref. Publ. No. 1017.

McIntyre, M. E. (1980a). An introduction to the generalized Lagrangian-mean description of wave, mean-flow interaction. *Pure Appl. Geophys.* **118**, 152-176.

McIntyre, M. E. (1980b). Towards a Lagrangian-mean description of stratospheric circulations and chemical transports. *Philos. Trans. R. Soc. London, Ser. A* **296**, 129-148.

McIntyre, M. E. (1981). On the "wave momentum" myth. *J. Fluid Mech.* **106**, 331-347.

McIntyre, M. E. (1982). How well do we understand the dynamics of stratospheric warmings? *J. Meteorol. Soc. Jpn*, **60**, 37-65.

McIntyre, M. E., and T. N. Palmer (1983). Breaking planetary waves in the stratosphere. *Nature (London)* **305**, 593-600.

McIntyre, M. E., and T. N. Palmer (1984). The "surf zone" in the stratosphere. *J. Atmos. Terr. Phys.* **46**, 825-849.

McIntyre, M. E., and M. A. Weissman (1978). On radiating instabilities and resonant over-reflection. *J. Atmos. Sci.* **35**, 1190-1196.

McPeters, R. D., D. F. Heath, and P. K. Bhartia (1984). Average ozone profiles for 1979 from the NIMBUS 7 SBUV instrument. *J. Geophys. Res.* **89**, 5199-5214.

Madden, R. A. (1979). Observations of large-scale traveling Rossby waves. *Rev. Geophys. Space Phys.* **17**, 1935-1949.

Madden, R. A., and K. Labitzke (1981). A free Rossby wave in the troposphere and stratosphere during January 1979. *J. Geophys. Res.* **86**, 1247-1254.

Madden, R. F. (1957). Study of CO_2 absorption spectra between 15 and 18 microns. Unpubl. rep., Johns Hopkins Univ., Baltimore, Maryland.

Mahlman, J. D. (1969). Heat balance and mean meridional circulations in the polar stratosphere during the sudden warming of January 1958. *Mon. Weather Rev.* **97**, 534-540.

Mahlman, J. D. (1985). Mechanistic interpretation of stratospheric tracer transport. *Adv. Geophys.* **28A**, 301-323.

Mahlman, J. D., and W. J. Moxim (1978). Tracer simulation using a global general circulation model: Results from a midlatitude instantaneous source experiment. *J. Atmos. Sci.* **35**, 1340-1374.

Mahlman, J. D., and L. J. Umscheid (1984). Dynamics of the middle atmosphere: successes and problems of the GFDL "SKYHI" general circulation model. *In* "Dynamics of the Middle Atmosphere." (J. R. Holton and T. Matsuno, eds.), pp. 501-525. Terrapub, Tokyo.

Mahlman, J. D., H. Levy, II. and W. J. Moxim (1980). Three-dimensional tracer structure and behavior as simulated in two ozone precursor experiments. *J. Atmos. Sci.* **37**, 655-685.

Mahlman, J. D., D. G. Andrews, D. L. Hartmann, T. Matsuno, and R. J. Murgatroyd (1984). Transport of trace constituents in the stratosphere. *In* "Dynamics of the Middle Atmosphere" (J. R. Holton and T. Matsuno, eds.), pp. 387-416. Terrapub, Tokyo.

Mahlman, J. D., H. Levy, II, and W. J. Moxim (1986). Three-dimensional simulations of stratospheric N_2O: predictions for other trace constituents. *J. Geophys. Res.* **91**, 2687-2707 (also corrigendum **91**, 9921).

Malkmus, W. (1967). Random Lorentz band model with exponential-tailed line intensity distribution. *J. Opt. Soc. Am.* **57**, 323-329.

Manabe, S., and B. G. Hunt (1968). Experiments with a stratospheric general circulation model. I. Radiative and dynamic aspects. *Mon. Weather Rev.* **96**, 477-502.

Manabe, S., and J. D. Mahlman (1976). Simulation of seasonal and interhemispheric variations in the stratospheric circulation. *J. Atmos. Sci.* **33**, 2185-2217.

Manabe, S., and T. B. Terpstra (1974). The effects of mountains on the general circulation of the atmosphere as identified by numerical experiments. *J. Atmos. Sci.* **31**, 3-42.

Manabe, S., D. G. Hahn, and J. L. Holloway, Jr. (1974). The seasonal variation of the tropical circulation as simulated by a global model of the atmosphere. *J. Atmos. Sci.* **31**, 43-83.

Manson, A. H., and C. E. Meek (1986). Dynamics of the middle atmosphere at Saskatoon (52°N, 107°W): a spectral study during 1981, 1982. *J. Atmos. Terr. Phys.* **48**, 1039-1055.

Manson, A. H., C. E. Meek, M. J. Smith, and G. J. Fraser (1985). Direct comparisons of prevailing winds and tidal wind fields (24h-12h) in the upper middle atmosphere (60-105 km) during 1978-1980 at Saskatoon (52°N, 107°W) and Christchurch (44°S, 173°E). *J. Atmos. Terr. Phys.* **47**, 463-476.

Matsuno, T. (1966). Quasi-geostrophic motions in the equatorial area. *J. Meteorol. Soc. Jpn.* **44**, 25-43.

Matsuno, T. (1970). Vertical propagation of stationary planetary waves in the winter northern hemisphere. *J. Atmos. Sci.* **27**, 871-883.

Matsuno, T. (1971). A dynamical model of the stratospheric sudden warming. *J. Atmos. Sci.* **28**, 1479-1494.

Matsuno, T. (1980). Lagrangian motion of air parcels in the stratosphere in the presence of planetary waves. *Pure Appl. Geophys.* **118**, 189-216.

Matsuno, T. (1982). A quasi one-dimensional model of the middle atmosphere circulation interacting with internal gravity waves. *J. Meteorol. Soc. Jpn.* **60**, 215-226.

Matsuno, T. (1984). Dynamics of minor stratospheric warmings and "preconditioning." *In* "Dynamics of the Middle Atmosphere" (J. R. Holton and T. Matsuno, eds.), pp. 333-351. Terrapub, Tokyo.

Mechoso, C. R., and D. L. Hartmann (1982). An observational study of traveling planetary waves in the southern hemisphere. *J. Atmos. Sci.* **39**, 1921-1935.

Mechoso, C. R., K. Yamazaki, A. Kitoh, and A. Arakawa (1985). Numerical forecasts of stratospheric warming events during the winter of 1979. *Mon. Weather Rev.* **113**, 1015-1029.

Meier, R. R., D. E. Anderson, Jr., and M. Nicolet (1982). Radiation field in the troposphere and stratosphere from 240 to 1000 nm. I. General analysis. *Planet. Space Sci.* **30**, 923-933.

Miyahara, S. (1984). A numerical simulation of the zonal mean circulation of the middle atmosphere including effects of solar diurnal tidal waves and internal gravity waves; solstice condition. *In* "Dynamics of the Middle Atmosphere" (J. R. Holton and T. Matsuno, eds.), pp. 271-287. Terrapub, Tokyo.

Miyahara, S. (1985). Suppression of stationary planetary waves by internal gravity waves in the mesosphere. *J. Atmos. Sci.* **42**, 100-107.

Miyahara, S., Y. Hayashi, and J. D. Mahlman (1986). Interactions between gravity waves and planetary scale flow simulated by the GFDL "SKYHI" general circulation model. *J. Atmos. Sci.* **43**, 1844-1861.

Miyakoda, K., R. F. Strickler and G. D. Hembree (1970). Numerical simulation of the breakdown of a polar-night vortex in the stratosphere. *J. Atmos. Sci.* **27**, 139-154.

Murgatroyd, R. J., and R. M. Goody (1957). Sources and sinks of radiative energy from 30 to 90 km. *Q. J. R. Meteorol. Soc.* **84**, 225-234.

Murgatroyd, R. J., and F. Singleton (1961). Possible meridional circulations in the stratosphere and mesosphere. *Q. J. R. Meteorol. Soc.* **87**, 125-135.

Nastrom, G. D., B. B. Balsley, and D. A. Carter (1982). Mean meridional winds in the mid- and high-latitude summer mesosphere. *Geophys. Res. Lett.* **9**, 139-142.

Naujokat, B. (1986). An update of the observed quasi-biennial oscillation of the stratospheric winds over the tropics. *J. Atmos. Sci.* **43**, 1873-1877.

Newell, R. E., and S. Gould-Stewart (1981). A stratospheric fountain? *J. Atmos. Sci.* **38**, 2789-2796.

Newell, R. E., J. W. Kidson, D. G. Vincent, and G. J. Boer (1974). "The General Circulation of the Tropical Atmosphere and Interactions with Extratropical Latitudes." MIT Press, Cambridge, Massachusetts.

Newman, P. A., and J. L. Stanford (1985). Short meridional scale anomalies in the lower stratosphere and upper troposphere. *J. Atmos. Sci.* **42**, 2081-2092.

Nicolet, M., R. R. Meier, and D. E. Anderson (1982). Radiation field in the troposphere and stratosphere. II. Numerical analysis. *Planet. Space Sci.* **30**, 935-983.

Nielsen, H. H. (1941). The near infra-red spectrum of water vapor. Part I. The perpendicular bands ν_2 and $2\nu_2$. *Phys. Res.* **59**, 565-575.

O'Neill, A. (1980). The dynamics of stratospheric warmings generated by a general circulation model of the troposphere and stratosphere. *Q. J. R. Meteorol. Soc.* **106**, 659-690.

O'Neill, A., and C. E. Youngblut (1982). Stratospheric warmings diagnosed using the transformed Eulerian-mean equations and the effect of the mean state on wave propagation. *J. Atmos. Sci.* **39**, 1370-1386.

O'Neill, A., R. L. Newson, and R. J. Murgatroyd (1982). An analysis of the large scale features of the upper troposphere and stratosphere in a global, three dimensional, general circulation model. *Q. J. R. Meteorol. Soc.* **108**, 25-53.

Oort, A. H. (1983). Global atmospheric circulation statistics, 1958-1973. NOAA Prof. Pap. No. 14.

Palmer, T. N. (1981a). Diagnostic study of a wavenumber-2 stratospheric sudden warming in a transformed Eulerian-mean formalism. *J. Atmos. Sci.* **38**, 844-855.

Palmer, T. N. (1981b). Aspects of stratospheric sudden warmings studied from a transformed Eulerian-mean viewpoint. *J. Geophys. Res.* **86**, 9679-9687.

Palmer, T. N. (1982). Properties of the Eliassen-Palm flux for planetary scale motions. *J. Atmos. Sci.* **39**, 992-997.

Palmer, T. N., and C.-P. F. Hsu (1983). Stratospheric sudden coolings and the role of nonlinear wave interactions in preconditioning the circumpolar flow. *J. Atmos. Sci.* **40**, 909-928.

Paltridge, G. W., and C. M. R. Platt (1976). "Radiative Processes in Meteorology and Climatology." Elsevier, Amsterdam.

Pedlosky, J. (1979). "Geophysical Fluid Dynamics." Springer-Verlag, Berlin and New York.

Phillips, N. A. (1973). Principles of large scale numerical weather prediction. *In* "Dynamic Meteorology" (P. Morel, ed.), pp. 1–96. Reidel, Dordrecht, Netherlands.

Plass, G. N. (1956a). The influence of the 9.6 micron ozone band on the atmospheric infrared cooling rate. *Q. J. R. Meteorol. Soc.* **82**, 30–44.

Plass, G. N. (1956b). The influence of the 15 micron carbon dioxide band on the atmospheric infrared cooling rate. *Q. J. R. Meteorol. Soc.* **82**, 310–324.

Plumb, R. A. (1977). The interaction of two internal waves with the mean flow: implications for the theory of the quasi-biennial oscillation. *J. Atmos. Sci.* **34**, 1847–1858.

Plumb, R. A. (1979). Eddy fluxes of conserved quantities by small–amplitude waves. *J. Atmos. Sci.* **36**, 1699–1704.

Plumb, R. A. (1981). Instability of the distorted polar night vortex: a theory of stratospheric warmings. *J. Atmos. Sci.* **38**, 2514–2531.

Plumb, R. A. (1982). The circulation of the middle atmosphere. *Aust. Meteorol. Mag.* **30**, 107–121.

Plumb, R. A. (1983). Baroclinic instability of the summer mesosphere: a mechanism for the quasi-two-day wave? *J. Atmos. Sci.* **40**, 262–270.

Plumb, R. A. (1984). The quasi-biennial oscillation. *In* "Dynamics of the Middle Atmosphere" (J. R. Holton and T. Matsuno, eds.), pp. 217–251. Terrapub, Tokyo.

Plumb, R. A., and R. C. Bell (1982). A model of the quasi-biennial oscillation on an equatorial beta-plane. *Q. J. R. Meteorol. Soc.* **108**, 335–352.

Plumb, R. A., and A. D. McEwan (1978). The instability of a forced standing wave in a viscous stratified fluid: a laboratory analogue of the quasi-biennial oscillation. *J. Atmos. Sci.* **35**, 1827–1839.

Plumb, R. A., and J. D. Mahlman (1987). The zonally-averaged transport characteristics of the GFDL general circulation/transport model. *J. Atmos. Sci.* **44**, 298–327.

Pollack, J. B., and T. C. Ackerman (1983). Possible effects of the El Chichon volcanic cloud on the radiation budget of the northern tropics. *Geophys. Res. Lett.* **10**, 1057–1060.

Pollack, J. B., and C. P. McKay (1985). The impact of polar stratospheric clouds on the heating rates of the winter polar stratosphere. *J. Atmos. Sci.* **42**, 245–262.

Prata, A. J. (1984). The 4-day wave. *J. Atmos. Sci.* **41**, 150–155.

Quiroz, R. S. (1975). The stratospheric evolution of sudden warmings in 1969–74 determined from measured infrared radiation fields. *J. Atmos. Sci.* **32**, 211–224.

Quiroz, R. S. (1986). The association of stratospheric warmings with tropospheric blocking. *J. Geophys. Res.* **91**, 5277–5285.

Quiroz, R. S., A. J. Miller, and R. M. Nagatani (1975). A comparison of observed and simulated properties of sudden stratospheric warmings. *J. Atmos. Sci.* **32**, 1723–1736.

Ramanathan, V. (1976). Radiative transfer within the Earth's troposphere and stratosphere: a simplified radiative–convective model. *J. Atmos. Sci.* **33**, 1330–1346.

Ramanathan, V. (1981). The role of ocean–atmospheric interactions in the CO_2 climate problem. *J. Atmos. Sci.* **38**, 918–930.

Ramanathan, V., E. J. Pitcher, R. C. Malone, and M. L. Blackmon (1983). The response of a spectral general circulation model to refinements in radiative processes. *J. Atmos. Sci.* **40**, 605–630.

Ramanathan, V., R. J. Cicerone, H. B. Singh, and J. T. Kiehl (1985). Trace gas trends and their potential role in climate change. *J. Geophys. Res.* **90**, 5547–5566.

Reed, R. J. (1955). A study of a characteristic type of upper-level frontogenesis. *J. Meteorol.* **12**, 226–237.

Reed, R. J. (1965a). The quasi-biennial oscillation of the atmosphere between 30 and 50 km over Ascension Island. *J. Atmos. Sci.* **22**, 331–333.

Reed, R. J. (1965b). The present status of the 26-month oscillation. *Bull. Am. Meteorol. Soc.* **46**, 374–387.

Reed, R. J. (1972). Further analysis of semidiurnal tidal motions between 30 and 60 km. *Mon. Weather Rev.* **100**, 579–581.

Reed, R. J., and K. E. German (1965). A contribution to the problem of stratospheric diffusion by large-scale mixing. *Mon. Weather Rev.* **93**, 313–321.

Reed, R. J., W. J. Campbell, L. A. Rasmussen, and D. G. Rogers (1961). Evidence of downward-propagating annual wind reversal in the equatorial stratosphere. *J. Geophys. Res.* **66**, 813–818.

Reed, R. J., M. J. Oard, and M. Sieminski (1969). A comparison of observed and theoretical diurnal motions between 30 and 60 km. *Mon. Weather Rev.* **97**, 456–459.

Reiter, E. R. (1975). Stratospheric–tropospheric exchange processes. *Rev. Geophys. Space Phys.* **13**, 459–474.

Remsberg, E. E., J. M. Russell, III. J. C. Gille, L. L. Gordley, P. L. Bailey, W. G. Planet, and J. E. Harries (1984a). The validation of Nimbus 7 LIMS measurements of ozone. *J. Geophys. Res.* **89**, 5161–5178.

Remsberg, E. E., J. M. Russell, III, L. L. Gordley, J. C. Gille, and P. L. Bailey (1984b). Implications of the stratospheric water vapor distribution as determined from the NIMBUS 7 LIMS experiment. *J. Atmos. Sci.* **41**, 2934–2945.

Rodgers, C. D. (1976a). Evidence for the five-day wave in the upper stratosphere. *J. Atmos. Sci.* **33**, 710–711.

Rodgers, C. D. (1976b). Retrieval of atmospheric temperature and composition from remote measurements of thermal radiation. *Rev. Geophys. Space Phys.* **14**, 609–624.

Rodgers, C. D., and A. J. Prata (1981). Evidence for a travelling two-day wave in the middle atmosphere. *J. Geophys. Res.* **86**, 9661–9664.

Rodgers, C. D., and C. D. Walshaw (1966). The computation of infrared cooling rate in planetary atmospheres. *Q. J. R. Meteorol. Soc.* **92**, 67–92.

Rodgers, C. D., and A. P. Williams (1974). Integrated absorption of a spectral line with the Voigt profile. *J. Quant. Spectrosc. Radiat. Transfer* **14**, 319–323.

Rosen, J. M., D. J. Hofmann, and J. Laby (1975). Stratospheric aerosol measurements II: The worldwide distribution. *J. Atmos. Sci.* **32**, 1457–1462.

Rosenfield, J. E., M. R. Schoeberl, and M. A. Geller (1986). A computation of the stratospheric diabatic residual circulation using an accurate radiative transfer model. *J. Atmos. Sci.* **44**, 857–876.

Rossby, C.-G. (1939). Relation between variations in the intensity of the zonal circulation of the atmosphere and the displacements of the semipermanent centers of action. *J. Mar. Res.* **2**, 38–55.

Rossby, C.-G. (1940). Planetary flow patterns in the atmosphere. *Q. J. R. Meteorol. Soc.* **66**, Suppl., 68–87.

Rothman, L. S., and L. D. G. Young, 1981. Infrared energy levels and intensities of carbon dioxide—II. *J. Quant. Spectrosc. Radiat. Transfer* **25**, 505–524.

Rothman, L. S., R. R. Gamaché, A. Barbe, A. Goldman, J. R. Gillis, L. R. Brown, R. A. Toth, J.-M. Flaud, and C. Camy-Peyret (1983). AFGL atmospheric absorption line parameters compilation: 1982 edition. *Appl. Opt.* **22**, 2247–2256.

Russell, J. M., III, E. E. Remsburg, L. L. Gordley, J. C. Gille, and P. L. Bailey (1984). Implications of the stratospheric water vapor distribution as determined from the NIMBUS-7 LIMS experiment. *J. Atmos. Sci.* **41**, 2934–2945.

Salby, M. L. (1979). On the solution of the homogeneous vertical structure problem for long-period oscillations. *J. Atmos. Sci.* **36**, 2350–2359.

Salby, M. L. (1980). The influence of realistic dissipation on planetary normal structures. *J. Atmos. Sci.* **37**, 2186–2199.

Salby, M. L. (1981a). Rossby normal modes in nonuniform background configurations. Part I: Simple fields. *J. Atmos. Sci.* **38**, 1803–1826.

Salby, M. L. (1981b). Rossby normal modes in nonuniform background configurations. Part II: Equinox and solstice conditions. *J. Atmos. Sci.* **38**. 1827–1840.

Salby, M. L. (1984). Survey of planetary-scale traveling waves: the state of theory and observations. *Rev. Geophys. Space Phys.* **22**, 209–236.

Salby, M. L., D. L. Hartmann, P. L. Bailey, and J. C. Gille (1984). Evidence for equatorial Kelvin modes in Nimbus-7 LIMS. *J. Atmos. Sci.* **41**, 220–235.

Sasamori, T., and J. London (1966). The decay of small temperature perturbations by thermal radiation in the atmosphere. *J. Atmos. Sci.* **23**, 543–554.

Schlesinger, M. E., and Y. Mintz (1979). Numerical simulation of ozone production, transport and distribution with a global atmospheric general circulation model. *J. Atmos. Sci.* **36**, 1325–1361.

Schmidt, U., G. Kulessa, A. Khedim, D. Knapska, and J. Rudolph (1984). Sampling of long lived trace gases in the middle and upper stratosphere. *ESA Symp. Eur. Rocket Balloon Programmes,* 6*th* ESA SP-183, pp. 141–145.

Schoeberl, M. R. (1978). Stratospheric warmings: observations and theory. *Rev. Geophys. Space Phys.* **16**, 521–538.

Schoeberl, M. R., and M. A. Geller (1977). A calculation of the structure of stationary planetary waves in winter. *J. Atmos. Sci.* **34**, 1235–1255.

Schoeberl, M. R., and D. F. Strobel (1978). The response of the zonally averaged circulation to stratospheric ozone reductions. *J. Atmos. Sci.* **35**, 1751–1757.

Schoeberl, M. R., and D. F. Strobel (1984). Nonzonal gravity wave breaking in the winter mesosphere. *In* "Dynamics of the Middle Atmosphere" (J. R. Holton and T. Matsuno, eds.), pp. 45–64. Terrapub, Tokyo.

Schoeberl, M. R., D. F. Strobel, and J. P. Apruzese (1983). A numerical model of gravity wave breaking and stress in the mesosphere. *J. Geophys. Res.* **88**, 5249–5259.

Shapiro, M. A. (1980). Turbulent mixing within tropopause folds as a mechanism for the exchange of chemical constituents between the stratosphere and troposphere. *J. Atmos. Sci.* **37**, 994–1004.

Shibata, T., M. Fujiwara, and M. Hirono (1984). El Chichon volcanic cloud in the stratosphere: lidar observation at Fukuoka and numerical simulation. *J. Atmos. Terr. Phys.* **46**, 1121–1146.

Shiotani, M., and I. Hirota (1985). Planetary wave–mean flow interaction in the stratosphere: a comparison between northern and southern hemispheres. *Q. J. R. Meteorol. Soc.* **111**, 309–334.

Simmons, A. J. (1974). Planetary-scale disturbances in the polar winter stratosphere. *Q. J. R. Meteorol. Soc.* **100**, 76–108.

Simmons, A. J., and R. Strüfing (1983). Numerical forecasts of stratospheric warming events using a model with a hybrid vertical coordinate. *Q. J. R. Meteorol. Soc.* **109**, 81–111.

Smagorinsky, J. (1953). The dynamical influence of large-scale heat sources and sinks on the quasi-stationary mean motions of the atmosphere. *Q. J. R. Meteorol. Soc.* **79**, 342–366.

Smagorinsky, J., S. Manabe, and J. L. Holloway (1965). Numerical results from a nine-level general circulation model of the atmosphere. *Mon. Weather Rev.* **93**, 727–768.

Smith, A. K. (1983). Stationary waves in the winter stratosphere: seasonal and·interannual variability. *J. Atmos. Sci.* **40**, 245–261.

Smith, A. K., and L. V. Lyjak (1985). An observational estimate of gravity wave drag from the momentum balance in the middle atmosphere. *J. Geophys. Res.* **90**, 2233–2241.

Smith, E. V. P., and D. M. Gottlieb (1974). Solar flux and its variations. *Space Sci. Rev.* **16**, 771–802.

Solomon, S., J. T. Kiehl, B. J. Kerridge, E. E. Remsberg, and J. M. Russell, III (1986). Evidence for non-local thermodynamic equilibrium in the ν_3 mode of mesospheric ozone. *J. Geophys. Res.* **91**, 9865–9876.

Spiegel, E. A. (1957). The smoothing of temperature fluctuations by radiative transfer. *Astrophys. J.* **126**, 202–207.

Stevens, D. E. (1983). On symmetric stability and instability of zonal mean flows near the equator. *J. Atmos. Sci.* **40**, 882–893.

Stewartson, K. (1978). The evolution of the critical layer of a Rossby wave. *Geophys. Astrophys. Fluid Dyn.* **9**, 185–200.

Straus, D. M. (1981). Long-wave baroclinic instability in the troposphere and stratosphere with spherical geometry. *J. Atmos. Sci.* **38**, 409–426.

Strobel, D. F. (1978). Parameterization of the atmospheric heating rate from 15 to 120 km due to O_2 and O_3 absorption of solar radiation. *J. Geophys. Res.* **83**, 6225–6230.

Swider, W., and M. E. Gardner (1967). "On the Accuracy of Certain Approximations for the Chapman Function," Environ. Res. Pap. No. 272. Air Force Cambridge Res. Lab., Bedford, Massachusetts.

Takahashi, M. (1984). A 2-dimensional numerical model of the semi-annual zonal wind oscillation. *In* "Dynamics of the Middle Atmosphere" (J. R. Holton and T. Matsuno, eds.), pp. 253–269. Terrapub, Tokyo.

Taylor, F. W., J. T. Houghton, G. D. Peskett, C. D. Rodgers, and E. J. Williamson (1972). Radiometer for remote sounding of the upper atmosphere. *Appl. Opt.* **11**, 135–141.

Taylor, G. I. (1915). Eddy motion in the atmosphere. *Philos. Trans. R. Soc. London, Ser. A* **215**, 1–26.

Thomas, R. J., C. A. Barth, and S. Solomon (1984). Seasonal variations of ozone in the upper mesosphere and gravity waves. *Geophys. Res. Lett.* **11**, 673–676.

Thrush, B. A. (1980). The chemistry of the stratosphere. *Philos. Trans. R. Soc. London, Ser. A* **296**, 149–160.

Tung, K. K. (1982). On the two-dimensional transport of stratospheric trace gases in isentropic coordinates. *J. Atmos. Sci.* **39**, 2230–2355.

Tung, K. K. (1986). Nongeostrophic theory of zonally averaged circulation. Part I: Formulation. *J. Atmos. Sci.* **43**, 2600–2618.

Tung, K. K., and R. S. Lindzen (1979). A theory of stationary long waves. Part II: Resonant Rossby waves in the presence of realistic vertical shears. *Mon. Weather Rev.* **107**, 735–750.

"U.S. Standard Atmosphere" (1976), U.S. Gov. Print. Off., Washington, D.C.

Valley, S. L., ed. (1965). "Handbook of Geophysics and Space Environments," Air Force Cambridge Res. Lab., Bedford, Massachusetts.

Veryard, R. G., and R. A. Ebdon (1961). Fluctuations in tropical stratospheric winds. *Meteorol. Mag.* **90**, 125–143.

Vincent, R. A. (1984). MF/HF radar measurements of the dynamics of the mesopause region—a review. *J. Atmos. Terr. Phys.* **46**, 961–974.

Vincent, R. A., and I. M. Reid (1983). HF Doppler measurements of mesospheric gravity wave momentum fluxes. *J. Atmos. Sci.* **40**, 1321–1333.

Wallace, J. M. (1967). A note on the role of radiation in the biennial oscillation. *J. Atmos. Sci.* **24**, 598–599.

Wallace, J. M. (1973). General circulation of the tropical lower stratosphere. *Rev. Geophys. Space Phys.* **11**, 191–222.

Wallace, J. M. (1983). The climatological mean stationary waves: observational evidence. *In* "Large-Scale Dynamical Processes in the Atmosphere" (B. J. Hoskins and R. P. Pearce, eds.), pp. 27–53. Academic Press, London.

Wallace, J. M., and P. V. Hobbs (1977). "Atmospheric Science: An Introductory Survey." Academic Press, New York.

Wallace, J. M., and J. R. Holton (1968). A diagnostic numerical model of the quasi-biennial oscillation. *J. Atmos. Sci.* **25**, 280–292.

Wallace, J. M., and V. E. Kousky (1968). Observational evidence of Kelvin waves in the tropical stratosphere. *J. Atmos. Sci.* **25**, 900–907.

Wallace, J. M., and R. F. Tadd (1974). Some further results concerning the vertical structure of atmospheric tidal motions within the lowest 30 kilometers. *Mon. Weather Res.* **102**, 795–803.

Walterscheid, R. L. (1980). Traveling planetary waves in the stratosphere. *Pure Appl. Geophys.* **118**, 239–265.

Wang, W.-C., D. J. Wuebbles, W. M. Washington, R. G. Isaacs, and G. Molnar (1986). Trace gases and other potential perturbations to global climate. *Rev. Geophys. Space Phys.* **24**, 110–140.

Warn, T., and H. Warn (1976). On the development of a Rossby wave critical level. *J. Atmos. Sci.* **33**, 2021–2024.

Warn, T., and H. Warn (1978). The evolution of a nonlinear critical level. *Stud. Appl. Math.* **59**, 37–71.

Washington, W. M., and G. A. Meehl (1983). General circulation model experiments on the climatic effects due to a doubling and a quadrupling of CO_2 concentration. *J. Geophys. Res.* **88**, 6600–6610.

Waters, J. W., J. C. Hardy, R. F. Jarnot, and H. M. Pickett (1981). Chlorine monoxide radical, ozone, and hydrogen peroxide: stratospheric measurements by microwave limb sounding. *Science* **214**, 61–64.

Watson, R. T., M. A. Geller, R. S. Stolarski, and R. F. Hampson (1986). Present state of knowledge of the upper atmosphere: an assessment report. NASA Ref. Publ. No. 1162.

Wayne, R. P. (1985). "Chemistry of Atmospheres." Oxford Univ. Press, London and New York.

Wehrbein, W. M., and C. Leovy (1982). An accurate radiative heating and cooling algorithm for use in a dynamical model of the middle atmosphere. *J. Atmos. Sci.* **39**, 1532–1544.

Welander, P. (1955). Studies on the general development of motion in a two-dimensional ideal fluid. *Tellus* **7**, 141–156.

White, A. A. (1982). Zonal translation properties of two quasi-geostrophic systems of equations. *J. Atmos. Sci.* **39**, 2107–2118.

Whitten, R. C., and I. G. Poppoff (1971). "Fundamentals of Aeronomy." Wiley, New York.

Whitten, R. C., and S. S. Prasad (1985). Ozone photochemistry in the stratosphere. *In* "Ozone in the Free Atmosphere" (R. C. Whitten and S. S. Prasad, eds.) pp. 81–121. Van Nostrand Reinhold, Princeton, New Jersey.

Williams, A. P. (1971). Relaxation of the 2.7 and 4.3 micron bands of carbon dioxide. *In* "Mesospheric Models and Related Experiments" (G. Fiocco, ed.), pp. 177–187, Reidel, Dordrecht, Netherlands.

WMO (1986). "Ozone Assessment Report, 1985." World Meteorol. Organ., Geneva.

Wu, M.-F., M. A. Geller, J. G. Olson, and M. E. Gelman (1984). Troposphere–stratosphere (surface–55 km) monthly general circulation statistics for the northern hemisphere—four year averages. *NASA Tech. Memo.* NASA TM-X-86182.

Yanai, M., and T. Maruyama (1966). Stratospheric wave disturbances propagating over the equatorial Pacific. *J. Meteorol. Soc. Jpn.* **44**, 291–294.

Yarger, D. N., and C. L. Mateer (1976). Inversion of backscattered solar ultraviolet radiation measurements to infer vertical profiles of atmospheric ozone. *Proc. Int. Radiat. Comm. Meet., Ft. Collins, Colo.*

Wallace, J. M., and F. V. Hobbs (1977) "Atmospheric Science: An Introductory Survey." Academic Press, New York.

Wallace, J. M., and J. K. Bolton (1984) A diagnostic numerical model of the quasi-biennial oscillation. J. Atmos. Sci. 41, 760-292.

Waltscor, J. M., and V. E. Kousky (1967) Observational evidence of Kelvin waves in the tropical stratosphere. J. Atmos. Sci. 25, 900-907.

Webster, F. J., and J. R. Holton (1982) Some theoretical aspects... interaction of symmetric disturbances with the mean flow. J. Atmos. Sci. 39, 723-737.

Webster, F. J., and J. L. Keller ... interaction and the general circulation. Quart. J. Roy. Meteor. Soc.

Webster, F. J. (1981) Mechanisms of monsoon low frequency variability: surface hydrological effects and the intrannual period of the tropical atmosphere. J. Atmos. Sci.

Wehr, S., and H. Wehr (1974) The evolution of a nonlinear critical level. Stud. Appl. Math. 59, 1-71.

Wergurson, S. M., and G. A. Meehl (1983) Tropical circulation model experiments on the ... a model describing a nondividing ... concentration. J. Geophys. Res. 88, 6641-6659.

Weinstein, A. J., S. B. Fels, and J. D. Mahlman (1961) Comparison rate-distant vector and the wave radiative ... transport of ... experimental and related fields in a deep atmosphere. Science 214, 83-85.

Woodman, R. F., T. E. Van Zandt, P. K. Rastogi, and J. L. Green (1981) Vertical scale of turbulence in the free atmosphere ... radar observation report. NOAA.

Wurtele, M. (1961) Treatment of atmospheric ... Cambridge Univ. Press, London and New York.

Way, Solsinan, W. H., and C. Liang (1982) An acoustic radiation heating and cooling algorithm for use in a dynamic circulation model of the upper atmosphere. J. Atmos. Sci. 39, 1577-1591.

Welander, P. (1955) Studies on the general development of motion in a two-dimensional ideal fluid. Tellus 7, 141-156.

White, A. A. (1982) Zonal translation properties of a quasi-geostrophic system of equations. J. Atmos. Sci. 39, 2107-2118.

Whitten, R. C., and I. G. Poppoff (1971) "Fundamentals of Aeronomy." Wiley, New York.

Whitten, R. C., and S. Prasad (1985) Ozone photochemistry in the stratosphere. In "Ozone in the Free Atmosphere" (R. C. Whitten and S. S. Prasad, eds.) pp. 81-121. Van Nostrand Reinhold, New York.

Williams, G. P. (1971) Baroclinic annulus waves. J. Fluid Mech. ...

Williams, G. P. and ...

Wirth, D. G., ... Middle atmosphere dynamics ...

Wulff, C. H. ... "Modern..."

Wunsch, C. (1975) ... Deep-sea exchange between the troposphere and stratosphere. J. Geophys. Res. ...

Young, R. E., and H. Houben (1986) ... baroclinic waves in ...

Yulaeva, E., and J. M. Wallace ...

Zhu, Xun, ... and J. R. Holton (1986) Mean fields of tropospheric structure ...

Voigt, D. F., and E. L. Benjey (1976) Properties of the deep positions in the stratosphere relating measurements of the chemical profiles of atmospheric flame. NCAR, Nat. Cent. Atmos. Research, Boulder, Colo.

Index